T4-AVD-304

Advances in

ECOLOGICAL RESEARCH

VOLUME 42

Advances in Ecological Research

Series Editor: GUY WOODWARD
School of Biological and Chemical Sciences
Queen Mary University of London
London, UK

Advances in

ECOLOGICAL RESEARCH

VOLUME 42

Ecological Networks

Edited by

GUY WOODWARD

School of Biological and Chemical Sciences,
Queen Mary University of London,
London E1 4NS, UK

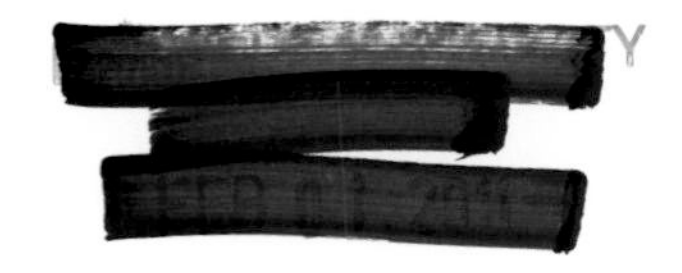

ELSEVIER

AMSTERDAM • BOSTON • HEIDELBERG • LONDON
NEW YORK • OXFORD • PARIS • SAN DIEGO
SAN FRANCISCO • SINGAPORE • SYDNEY • TOKYO
Academic Press is an imprint of Elsevier

Academic Press is an imprint of Elsevier

32 Jamestown Road, London NW1 7BY, UK
Linacre House, Jordan Hill, Oxford OX2 8DP, UK
Radarweg 29, PO Box 211, 1000 AE Amsterdam, The Netherlands
30 Corporate Drive, Suite 400, Burlington, MA 01803, USA
525 B Street, Suite 1900, San Diego, CA 92101-4495, USA

First edition 2010

ISBN: 978-0-12-381363-3
ISSN: 0065-2504

For information on all Academic Press publications
visit our website at elsevierdirect.com

Printed and bound in UK

10 11 12 10 9 8 7 6 5 4 3 2 1

Contents

Scaling of Food-Web Properties with Diversity and Complexity Across Ecosystems

JENS O. RIEDE, BJÖRN C. RALL, CAROLIN BANASEK-RICHTER, SERGIO A. NAVARRETE, EVIE A. WIETERS, MARK C. EMMERSON, UTE JACOB, AND ULRICH BROSE

Temporal Variability in Predator–Prey Relationships of a Forest Floor Food Web

ÓRLA B. McLAUGHLIN, TOMAS JONSSON, AND MARK C. EMMERSON

Food Web Structure and Stability in 20 Streams Across a Wide pH Gradient

KATRIN LAYER, JENS O. RIEDE, ALAN G. HILDREW, AND GUY WOODWARD

Manipulating Interaction Strengths and the Consequences for Trivariate Patterns in a Marine Food Web

EOIN J. O'GORMAN AND MARK C. EMMERSON

Contributors to Volume 42

CAROLIN BANASEK-RICHTER, *Environmental Research Institute, University College Cork, Lee Road, Cork, Ireland and School of Biology, Earth and Environmental Sciences, University College Cork, Distillery Fields, North Mall, Cork, Ireland.*

JONATHAN P. BENSTEAD, *Department of Biological Sciences, University of Alabama, Tuscaloosa, AL 35487, USA.*

OLIVER S. BEVERIDGE, *Department of Animal and Plant Sciences, University of Sheffield, Western Bank, Sheffield, S10 2TN, United Kingdom. and School of Biological and Biomedical Sciences, Durham University, South Road, Durham, DH1 3LE, United Kingdom.*

JULIA BLANCHARD, *Centre for Environment, Fisheries and Aquaculture Science, Lowestoft Laboratory, Lowestoft NR33 OHT, United Kingdom.*

THOMAS BREY, *Alfred Wegener Institute for Polar and Marine Research, PO 120161, 27515, Bremerhaven, Germany.*

ULRICH BROSE, *Systemic Conservation Biology Group, J.F. Blumenbach Institute of Zoology and Anthropology, Georg-August-University Goettingen, Germany.*

LEE E. BROWN, *School of Geography, University of Leeds, Woodhouse Lane, Leeds, LS2 9JT, United Kingdom.*

WYATT F. CROSS, *Department of Ecology, Montana State University, Bozeman, MT 59717, USA.*

YOKO L. DUPONT, *Department of Biological Sciences, Aarhus University, Ny Munkegade 114, DK-8000 Aarhus C, Denmark.*

MARK C. EMMERSON, *Environmental Research Institute, University College Cork, Lee Road, Cork, Ireland; School of Biology, Earth and Environmental Sciences, University College Cork, Distillery Fields, North Mall, Cork, Ireland; Department of Zoology, Ecology and Plant Sciences, University College Cork, Distillery Fields, North Mall, Cork, Ireland and Queens University Belfast, School of Biological Sciences, Medical Biology Centre, 97 Lisburn Road, Belfast, BT9 7BL, Northern Ireland.*

NIKOLAI FRIBERG, *National Environmental Research Institute, Department of Freshwater Ecology, Aarhus University, Vejlsøvej 25, DK-8600 Silkeborg, Denmark.*

ALAN G. HILDREW, *School of Biological & Chemical Sciences, QueenMary University of London, London, E1 4NS, United Kingdom.*

THOMAS C. INGS, *School of Biological & Chemical Sciences, Queen Mary University of London, London, E1 4NS, United Kingdom.*

UTE JACOB, *Institute for Hydrobiology and Fisheries Science, University of Hamburg, Grosse Elbstrasse 133, D-22767 Hamburg, Germany.*

SIMON JENNINGS, *Centre for Environment, Fisheries and Aquaculture Science, Lowestoft Laboratory, Lowestoft NR33 OHT, United Kingdom.*

TOMAS JONSSON, *Ecological Modelling Group, Research Centre for Systems Biology, University of Skövde, P.O. Box 408, SE-541 28 Skövde, Sweden.*

KATRIN LAYER, *School of Biological & Chemical Sciences, Queen Mary University of London, London, E1 4NS, United Kingdom.*

MARK E. LEDGER, *School of Geography, Earth and Environmental Sciences, University of Birmingham, Edgbaston, Birmingham, B15 2TT, United Kingdom.*

ÓRLA B. MCLAUGHLIN, *Environmental Research Institute, University College Cork, Lee Road, Cork, Ireland and School of Biology, Earth and Environmental Sciences, University College Cork, Distillery Fields, North Mall, Cork, Ireland.*

CARLOS J. MELIÁN, *National Center for Ecological Analysis and Synthesis, 735 State Street, Suite 300, Santa Barbara, CA 93101.*

ALEXANDER M. MILNER, *School of Geography, Earth and Environmental Sciences, University of Birmingham, Edgbaston, Birmingham, B15 2TT, United Kingdom.*

JOSE M. MONTOYA, *I Institute of Marine Sciences, Consejo Superior de Investigaciones Científicas, Passeig Maritim Barceloneta 37–49, 08003, Barcelona, Spain.*

SERGIO A. NAVARRETE, *Estación Costera de Investigaciones Marinas & Center for Advanced Studies in Ecology & Biodiversity, Depto. de Ecología Pontificia Universidad Católica de Chile, Casilla 114-D, Santiago, Chile.*

EOIN O'GORMAN, *School of Biology and Environmental Science, Science Centre West, University College Dublin, Belfield, Dublin 4, Ireland.*

JENS M. OLESEN, *Department of Biological Sciences, Aarhus University, Ny Munkegade 114, DK-8000 Aarhus C, Denmark.*

OWEN L. PETCHEY, *Department of Animal and Plant Sciences, University of Sheffield, Western Bank, Sheffield, S10 2TN, United Kingdom.*

DORIS E. PICHLER, *School of Biological & Chemical Sciences, Queen Mary University of London, London, E1 4NS, United Kingdom.*

BJÖRN C. RALL, *Systemic Conservation Biology Group, J.F. Blumenbach Institute of Zoology and Anthropology, Georg-August-University Goettingen, Germany.*

DANIEL C. REUMAN, *Imperial College London, Silwood Park Campus, Buckhurst Road, Ascot, Berkshire, SL5 7PY, United Kingdom.*

JENS O. RIEDE, *Systemic Conservation Biology Group, J.F. Blumenbach Institute of Zoology and Anthropology, Georg-August-University Goettingen, Germany.*

CLAUS RASMUSSEN, *Department of Biological Sciences, Aarhus University, Ny Munkegade 114, DK-8000 Aarhus C, Denmark.*

MURRAY S.A. THOMPSON, *School of Biological & Chemical Sciences, QueenMary University of London, London, E1 4NS, United Kingdom and Natural History Museum, Cromwell Road, London SW7 5BD, United Kingdom.*

KRISTIAN TRØJELSGAARD, *Department of Biological Sciences, Aarhus University, Ny Munkegade 114, DK-8000 Aarhus C, Denmark.*

FRANK J. F. VAN VEEN, *School of Bioscience, The University of Exeter, Exeter, Devon, EX4 4SBUnited Kingdom.*

EVIE A. WIETERS, *Estación Costera de Investigaciones Marinas & Center for Advanced Studies in Ecology & Biodiversity, Depto. de Ecología Pontificia Universidad Católica de Chile, Casilla 114-D, Santiago, Chile.*

GUY WOODWARD, *School of Biological & Chemical Sciences, Queen Mary University of London, London, E1 4NS, United Kingdom.*

GABRIEL YVON-DUROCHER, *School of Biological & Chemical Sciences, Queen Mary University of London, London,E1 4NS, United Kingdom.*

Preface

Ecological Networks: Simple Rules for Complex Systems in a Changing World?

This Thematic Volume of *Advances in Ecological Research* is dedicated to the study of ecological networks—the webs of antagonistic or mutualistic interactions that occur between individuals and, ultimately, also at the higher levels of biological organisation (e.g. among species populations or size-classes). Darwin first employed the metaphor of the '*entangled bank*' to describe a network of interacting species in the *Origin of Species* in 1859, and the first recognisable ball-and-stick diagram of a food web was described by Camerano (1880) a couple of decades later. Many of the eminent pioneering ecologists of their day also turned their attention to these systems, including Elton (1927), Lindeman (1942), MacArthur (1955), and Hutchinson (1959), who were among the first to suggest that complexity might confer stability on natural systems and that body size plays a key role in structuring communities and in determining the strength of interactions. These early ideas linking complexity to stability were subsequently challenged by the seminal modelling work of May (1972, 1973) and many others, who demonstrated mathematically that complexity should decrease stability. This raised the long-standing question as to what might be the '*devious strategies*' that allow the complex food webs we see in nature to persist. A new generation of food webs that were constructed with far greater sampling effort and better taxonomic resolution appeared from the 1990s onwards, in conjunction with new mathematical models that demonstrated how complex systems could be stable—for instance, if most links were weak (e.g. McCann, Hastings and Huxel, 1998). Since then, numerous studies have demonstrated, both empirically and theoretically, that the body-size of consumers and their resources is a key determinant of interaction strength, network structure and stability (e.g. Berlow *et al.*, 2009; Emmerson and Raffaelli, 2004).

In the last two decades, the study of ecological networks has undergone a dramatic renaissance, which has been manifested by the opening of new research vistas and an exponential rise in the number of publications (Ings *et al.*, 2009). The catalogue of high-quality data has now grown sufficiently for ecologists to be able to undertake meaningful meta-analysis and to search for macroecological patterns in network structure and dynamics across a wide range of systems, two areas that had, until recently, been hampered by limited or inconsistent data collection, as highlighted by the paper by Riede and co-authors in this Volume. In addition, the study of 'traditional' food webs and host–parasitoid networks has been complemented by the recent blossoming of research into mutualistic networks (e.g. plant–pollinator webs) and the attendant cross-fertilisation of ideas emerging from the study of networks in the social sciences, which are covered in Olesen *et al.*'s comprehensive review and synthesis paper. We are now edging closer to being able to make some (fairly) firm generalisations about network structure and dynamics, and to start to build predictive frameworks to anticipate how they might respond to natural and anthropogenic perturbations. One recurrent theme that has (re)emerged in recent years and that runs through much of this Volume and echoes the ideas of the early ecologists, particularly those of Elton and Hutchinson, is that body size plays a key role (e.g. Cohen *et al.*, 2003) and that its correlates (e.g. basal metabolic rate) have effects that can ramify through the food web.

The first paper in the Volume (Olesen *et al.*, 2010) provides an overview of the current and emerging perspectives in eight key areas of network research, including an in-depth consideration of the structuring of networks, and how this varies with both space and time. In the second paper, by Woodward *et al.* (2010), a new theoretical framework, based on first-principles related primarily to metabolic constraints and ecological stoichiometry, is suggested as a possible means of developing a more mechanistic understanding of network-level responses to climate change. The subsequent papers are based on detailed empirical data from a range of different systems, including marine, freshwater and terrestrial food webs. The first of these, by Riede *et al.* (2010), is a meta-analysis of network properties using a new, high-quality dataset, which reveals strong evidence of power-law scaling between network size and a range of food web parameters. The paper by McLaughlin *et al.* (2010) explores network properties and size-structuring within a highly resolved terrestrial above- and below-ground food web. The penultimate paper, by Layer *et al.* (2010) is a macroecological and modelling study of 20 stream food webs across a wide pH gradient, and is one of the first attempts to understand how structure and stability of networks is shaped by an environmental stressor. The final paper, by O'Gorman and Emmerson (2010), considers the role of body size within a benthic marine food web, and represents one of the first studies to characterise highly resolved networks within a replicated field experiment. In summary, the six papers that comprise this Volume cover a wide range of ecosystems (marine, freshwater and terrestrial), network types (food webs, mutualistic and host–parasitoid networks) and also combine theoretical and empirical approaches. Although the role of body size is addressed in each contribution, this does not in any way imply that it is the sole variable of interest or importance, but rather it represents a useful principal component that can capture much of the biologically relevant variation within a network. Examining the 'residuals about the line' can therefore provide invaluable additional information about a wide range of other important ecological properties and phenomena (e.g. idiosyncratic species traits, adaptive behaviour, and phylogenetic constraints) that would otherwise be masked. In many cases, and especially in aquatic systems, body size provides a relatively simple set of rules that can be used to help understand and (ultimately) predict the structure and dynamics of complex ecological networks and how they might respond to future change. This is a rather daunting, but also exciting, prospect for future ecological research, and the compilation of papers within this Volume represent a step towards this goal.

Guy Woodward

REFERENCES

Berlow, E.A., Dunne, J.A., Martinez, N.D., Starke, P.B., Williams, R.J., and Brose, U. (2009). Simple prediction of interaction strengths in complex food webs. *Proc. Natl. Acad. Sci. USA* **106**, 187–191.

Camerano, L. (1880). Dell' equilibrio dei viventi mercé la reciproca distruzione. *Atti della Reale Accademia delle Scienze di Torino* **15**, 393–414.

Cohen, J.E., Jonsson, T., and Carpenter, S.R. (2003). Ecological community description using the food web, species abundance, and body size. *Proc. Natl. Acad. Sci. USA* **100**, 1781–1786.

Darwin, C. (1859). On the Origin of Species by Means of Natural Selection, or the Preservation of Favoured Races in the Struggle for Life.

Elton, C.S. (1927). *Animal Ecology*. Sedgewick and Jackson, London.

Emmerson, M., and Raffaelli, D. (2004). Predator–prey body size, interaction strength and the stability of a real food web. *J. Anim. Ecol.* **73**, 399–409.

Hutchinson, G.E. (1959). Homage to Santa Rosalia or why are there so many kinds of animals? *Am. Nat.* **93**, 145–159.

Ings, T.C., Montoya, J.M., Bascompte, J., Bluthgen, N., Brown, L., Dormann, C.F., Edwards, F., Figueroa, D., Jacob, U., Jones, J.I., Lauridsen, R.B., , Ledger, M.E. *et al.* (2009). Ecological networks—Beyond food webs. *J. Anim. Ecol.* **78**, 253–269.

Lindeman, R.L. (1942). The trophic–dynamic aspect of ecology. *Ecology* **23**, 399–418.

Layer, K., Riede, J.O., Hildrew, A.G., and Woodward, G. (2010). Food web structure and stability in 20 streams across a wide pH gradient. *Adv. Ecol. Res.* **42**, 267–301.

MacArthur, R. (1955). Fluctuations of animal populations and a measure of community stability. *Ecology* **36**, 533–536.

May, R.M. (1972). Will a large complex system be stable? *Nature* **238**, 413–414.

May, R.M. (1973). Stability and Complexity in Model Ecosystems. Princeton University Press, Princeton.

McCann, K., Hastings, A., and Huxel, G.R. (1998). Weak trophic interactions and the balance of nature. *Nature* **395**, 794–798.

McLaughlin, O.B., Jonsson, T., and Emmerson, M.C. (2010). Temporal variability in predator–prey relationships of a forest floor food web. *Adv. Ecol. Res.* **42**, 173–266.

O'Gorman, E.J., and Emmerson, M.C. (2010). Manipulating interaction strengths and the consequences for trivariate patterns in a marine food web. *Adv. Ecol. Res.* **42**, 303–421.

Olesen, J.M., Dupont, Y.L., O'Gorman, E.J., Ings, T.C., Layer, K., , Melián, C.J., *et al.* (2010). From Broadstone to Zackenberg: Space, time and hierarchies in ecological networks. *Adv. Ecol. Res.* **42**, 1–70.

Riede, J.O., Rall, B.C., Banasek-Richter, C., Navarrete, S.A., Wieters, E.A., and Brose, U. (2010). Scaling of food-web properties with diversity and complexity across ecosystems. *Adv. Ecol. Res.* **42**, 141–172.

Woodward, G., Benstead, J.P., Beveridge, O.S., Blanchard, J., Brey, T., , Brown, L.E., *et al.* (2010). Ecological networks in a changing climate. *Adv. Ecol. Res.* **42**, 71–139.

From Broadstone to Zackenberg: Space, Time and Hierarchies in Ecological Networks

JENS M. OLESEN, YOKO L. DUPONT, EOIN O'GORMAN, THOMAS C. INGS, KATRIN LAYER, CARLOS J. MELIÁN, KRISTIAN TRØJELSGAARD, DORIS E. PICHLER, CLAUS RASMUSSEN AND GUY WOODWARD

SUMMARY

Ecological networks are typically complex constructions of species and their interactions. During the last decade, the study of networks has moved from static to dynamic analyses, and has attained a deeper insight into their internal structure, heterogeneity, and temporal and spatial resolution. Here, we review, discuss and suggest research lines in the study of the spatio-temporal heterogeneity of networks and their hierarchical nature. We use case study data from two well-characterized model systems (the food web in Broadstone Stream in England and the pollination network at Zackenberg in Greenland), which are complemented with additional information from other studies. We focus upon eight topics: temporal dynamic space-for-time substitutions linkage constraints habitat borders network modularity individual-based networks invasions of networks and super networks that integrate different network types. Few studies have explicitly examined temporal change in networks, and we present examples that span

ADVANCES IN ECOLOGICAL RESEARCH VOL. 42 0065-2504/10 $35.00

DOI: 10.1016/S0065-2504(10)42001-2

from daily to decadal change: a common pattern that we see is a stable core surrounded by a group of dynamic, peripheral species, which, in pollinator networks enter the web via preferential linkage to the most generalist species. To some extent, temporal and spatial scales are interchangeable (i.e. networks exhibit 'ergodicity') and we explore how space-for-time substitutions can be used in the study of networks. Network structure is commonly constrained by phenological uncoupling (a temporal phenomenon), abundance, body size and population structure. Some potential links are never observed, that is they are 'forbidden' (fully constrained) or 'missing' (a sampling effect), and their absence can be just as ecologically significant as their presence. Spatial habitat borders can add heterogeneity to network structure, but their importance has rarely been studied: we explore how habitat generalization can be related to other resource dimensions. Many networks are hierarchically structured, with modules forming the basic building blocks, which can result in self-similarity. Scaling down from networks of species reveals another, finer-grained level of individual-based organization, the ecological consequences of which have yet to be fully explored. The few studies of individual-based ecological networks that are available suggest the potential for large intraspecific variance and, in the case of food webs, strong size-structuring. However, such data are still scarce and more studies are required to link individual-level and species-level networks. Invasions by alien species can be tracked by following the topological 'career' of the invader as it establishes itself within a network, with potentially important implications for conservation biology. Finally, by scaling up to a higher level of organization, it is possible to combine different network types (e.g. food webs and mutualistic networks) to form super networks, and this new approach has yet to be integrated into mainstream ecological research. We conclude by listing a set of research topics that we see as emerging candidates for ecological network studies in the near future.

I. INTRODUCTION

At first glance, ecological networks often appear paradoxical: they are exceedingly complex in their overall structure and behaviour, and yet their only building blocks are nodes (usually species) and links or interactions. In mathematics, theoretical statistics and social sciences, the formalized study of networks started over 50 years ago and proceeded at a relatively slow and steady pace until about one decade ago, when a series of seminal papers on large, empirical networks were published. This caused a paradigm shift in the study of complex networks, with the papers of Watts and Strogatz (1998), Barabási and Albert (1999) and Albert *et al.* (2000) being especially influential. With more than 10,000 citations (ISI: April 2010), these three publications have pushed empirical and theoretical network analysis forward at a rapid rate and into many new scientific fields.

In ecology, these new approaches to the analysis of complex networks were initiated by Strogatz (2001), Solé and Montoya (2001), Montoya and Solé (2002), Williams *et al.* (2002), and Dunne *et al.* (2002a,b), although many earlier ecological papers were stepping stones to this development (e.g. Jordano, 1987; Memmott, 1999) more than 100 years after the publication of the first recognizable ball-and-stick diagrammatic representation of a food web (Camerano, 1880). Food-web ecology, which has underpinned much of community and ecosystem ecology for decades, developed along rather different lines from those of network analysis as defined in the aforementioned papers, with much of the focus on the interrelationships between interaction strength, complexity and dynamic stability (e.g. May, 1972, 1973; McCann *et al.*, 1998) or simple structural measures (e.g. connectance, food chain length). It has become increasingly apparent, however, that it was not simply complexity that was important in food webs, but the positioning and patterning of the links that was a key determinant of the stability of these and other types of ecological network. More sophisticated analyses started to emerge in the last decade, with a movement away from connectance to the study of, for instance, path length, clustering coefficients and other 'small-world' properties (Ings *et al.*, 2009). Many of the mathematical expressions that have emerged in the last decade or so have long been intuitively familiar to many ecologists, but often only as a verbal construct, such as Darwin's 'entangled bank' (Darwin, 1859). One of the first ecologists, Forbes (1887), was also struck as to how plants and animals were connected to each other in a complex manner and noted: '. . .whatever affects any species belonging to it [the "organic complex"], must have its influence of some sort upon the whole assemblage'. It is impossible to study isolated species, he concluded: one has to focus upon the entire community. In the last few years, interest in ecological networks, especially food webs, host–parasitoid webs, and mutualistic networks, has increased dramatically, as reflected by the exponential rise in the number of publications (e.g. Bascompte and Jordano, 2007; Ings *et al.*, 2009; Montoya *et al.*, 2006; Pascual and Dunne, 2006; Waser and Ollerton, 2006; Woodward *et al.*, 2005a).

During the past decade, pollination ecology at the network level has also attracted considerable attention (e.g. Bascompte and Jordano, 2007) over which time it has developed apace with its older and more familiar cousin: food-web ecology. Despite this recent rise to prominence, however, pollination-community studies have a long history that goes back at least as far as Clements and Long (1923). A pollination network consists of all species of plants and their flower-visiting animals within a researcher-defined space and time. If a few plant and animal species are unconnected from the main network ('the giant component'), they are most often excluded from the network and its analysis. Jordano (1987) carried out a pioneering meta-analysis of a set of pollination and seed-dispersal community studies and concluded that facultative interactions of high generality were the general rule. That is, most species establish several links and the strength of any one link is 'weak'. By the strength of a link between a mutualistic animal species and a plant species, Jordano (1987) meant visitation rate, that is no. of flower visits (or no. of flower–visiting animals) per flower per time, which differs somewhat from the definitions of interaction strength typically used in food-web ecology (Ings *et al.*, 2009). In fact, Jordano's work was a comparative or macro-ecological network study, but it was made more than a decade before anybody really discussed 'networks' *per se* in ecology, at least in the sense used by Watts and Strogatz (1998), Barabási and Albert (1999) and Albert *et al.* (2000). An obvious consequence of Jordano's conclusions was that research in interaction biology has to take place at the network level. More recently, general network theory has provided a new framework to describe and quantify interactions in species-rich pollination networks (Bascompte *et al.*, 2003, 2006; Jordano *et al.*, 2003, 2006; Memmott and Waser, 2002; Olesen and Jordano, 2002; Olesen *et al.*, 2006; Ollerton and Cranmer, 2002; Rezende *et al.*, 2009; Thompson, 2005; Vázquez and Aizen, 2004, and many more, Figure 1). These and other studies point towards both variant and invariant patterns across network size, habitats and biogeographical regions.

Here, we focus upon communities of interacting species, including both mutualists (e.g. plant–pollinator networks) and antagonists (e.g. predator–prey food webs). Although we apply this to sets of different species, similar arguments may be made for assemblages of conspecific, interacting individuals (see SectionVII). We have selected some of our own studies to highlight promising avenues that lie ahead for future network research. We see these in research areas about spatial and temporal dynamics and their scale and extent, and in the studies of nature's hierarchical organization. Importantly, from a network perspective, the latter may be seen from several angles, including hierarchies of aggregations of natural units (individuals, guilds, functional groups, etc.), the classic textbook hierarchy of biological organization (cell, organ, individual, population, etc.) and the trophic-level organization in food webs. Our choices of data and examples are certainly biased, and most come from our own published and unpublished work, from the small but intensively studied Broadstone Stream in England to the world's largest national park at Zackenberg in NE Greenland. However, we have used these data sets as model systems because they provide specific examples of the general points discussed in this chapter, and many of the phenomena we highlight are widely manifested in other systems beyond our case studies.

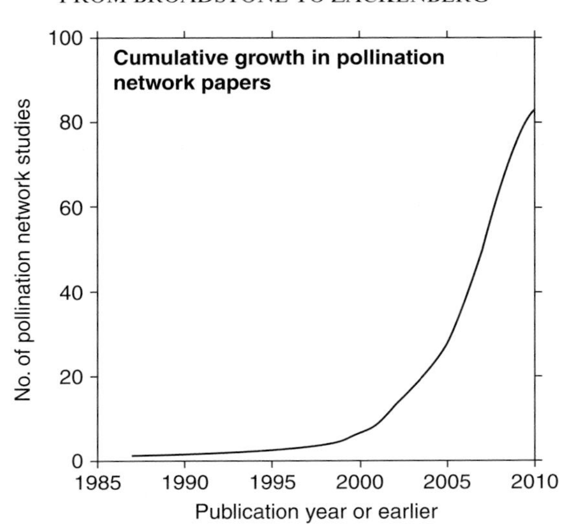

Figure 1 Cumulative growth in pollination-network papers since Jordano (1987); included are analyses of 1-more networks and studies of 1-more network parameters across a larger sample of networks. The list of papers can be downloaded from http://mit.biology.au.dk/~biojmo.

II. TEMPORAL DYNAMICS

Ecological networks are typically studied as slices or snapshots in space and time (Polis *et al.*, 1997; Rooney *et al.*, 2008; Winemiller, 1990, 1996; Winemiller and Jepsen, 1998; Winemiller and Layman, 2005). Despite their inferred dynamical attributes, they are most often depicted and analysed statically based on data accumulated over one 'thick' spatio-temporal slice or from summary data aggregated over multiple sampling occasions (e.g. Woodward and Hildrew, 2001; but see McLaughlin *et al.*, 2010). As a consequence, their spatio-temporal extent and resolution become limited. This pooling of data hampers our understanding of temporal development and spatial structure of networks, and, consequently, our ability to predict and mitigate the impacts of disturbances. Thus, to understand the dynamics of ecological networks, we need spatio-temporally well-resolved data (Hildrew *et al.*, 2004; McLaughlin *et al.*, 2010; O'Gorman and Emmerson, 2010; Speirs *et al.*, 2000) and, ideally, individual-level data, collected on both larger and finer scales (Ings *et al.*, 2009; Vacher *et al.*, 2008; Woodward and Warren, 2007), for example studies spanning decades, seasonal studies resolved daily and studies covering several habitats, that is spatially coupled networks. Only by doing so may we ultimately reach an understanding of the mechanisms underlying the observed structure of individual- and species-based networks in space and time. Such studies could also provide the necessary background to understand robustness and resilience of natural systems, for example in response to climate change (Woodward *et al.*, 2010b), the degree of coupling between individual- and species-level networks (Bolnick *et al.*, 2002; Roughgarden, 1972), and they also offer a means to address the perpetual question in ecology: what 'devious strategies' (May, 1972) confer stability upon complex ecosystems (Ings *et al.*, 2009; May, 1973)?

In this section, we focus upon time, 'a close relative of space' (*cit.* McCann *et al.*, 2005). To some extent, temporal data accumulation in network studies is necessary in order to obtain robust results, especially because several environmental and demographic factors may be driving individual-level interactions: for instance, many predators have empty guts at any given time, so the chances of observing a link between a rare predator and a rare prey soon become vanishingly small, even though such an interaction could be dynamically important for the prey species (Woodward *et al.*, 2005b). At the same time, increasing the temporal scale also blurs the picture, because in the accumulation process species that do not co-occur in time can get pooled together: in extreme cases, this could potentially create artefactual groups of species and their links within the network that don't exist in reality (at least not simultaneously). Thus, data accumulation may increase the incidence of temporal uncoupling between species within the longer term network as a whole (e.g. Hegland *et al.*, 2009), and in pollination networks, this can vary from 1 week (Mosquin and Martin, 1967) to 4 years (Petanidou *et al.*, 2008), whilst in some food webs, data may have been aggregated over several decades (Hall and Raffaelli, 1993). However, a few data sets have been partitioned into successive time slices making it possible to study or infer the seasonal or annual dynamics in more detail (e.g. Benke *et al.*, 2001; Closs and Lake, 1994; Lundgren and Olesen, 2005; McLaughlin *et al.*, 2010; O'Gorman *et al.*, 2010; Olesen *et al.*, 2008; Petanidou *et al.*, 2008; Tavares-Cromar and Williams, 1996; Thompson and Townsend, 1999; Warren, 1989; Woodward *et al.*, 2005a).

Here, we draw on our research from the area around the Zackenberg Research Station (74°30′ N, 20°30′ W) in the northeast Greenland National Park, which is the largest park in the world, with an area of 972,000 km^2, almost twice the size of Madagascar (Meltofte *et al.*, 2008). In a study at Zackenberg from 1996, covering the entire season, Olesen *et al.* (unpublished data a,b) focused upon the modules in the entire network and their seasonal development. A module is defined as a small group of plants and pollinators that are more tightly linked to each other than they are to species in other modules in the network. Thus, a module is a kind of mesoscopic property of the network, placed somewhere between the levels of the entire network and the single species. The Zackenberg network was nested, that is it consisted of a core of generalists and two tails of specialists linked to the core (here defined by their linkage level L, which is the number of links from a species to other species; $L > 2$ for generalists and $L \leq 2$ for specialists, Figure 2). Nestedness is essentially a measure of the 'order' of the network (Ulrich *et al.*, 2009). A link was noted if an insect visited a flower of a plant species (visitation data) or if pollen of a plant species were found on the body surface of a sampled insect (pollen-load data). Olesen *et al.* (unpublished data a) described the seasonal assembly and disassembly of the modules in the network and suggested 'rules' along which this took place. They used both cumulative data, encompassing the total season, and time-slice data, whereby each network only included links observed during one day. The length of season over which the network was constructed was 43 days, and it contained 61 pollinator and 31 plant species organized into five modules. The 'life-span' range of individual modules was 31–42 days, that is the modules were present during most of the life of the wider network. Despite this, modules could be sorted into a seasonal series, peaking sequentially in numbers of species and links. Thus, as the season progressed, new modules developed and old ones slowly vanished. Olesen *et al.* (unpublished data a) identified core and tail species in both the network and in each module. A module-core species is defined as one that has > 2 links to other species within its module, and a tail species has ≤ 2 links. The cumulative network had twice as many tail species per core species as the cumulative modules and, in addition, module tail species had a phenophase five times shorter than core species. The phenophase of a species is the period in which it is a member of the network or the periodical extent of an event in its life cycle, for example the flowering of a plant species. Thus, at any given time, tail species were an important part of the structure of the entire network, but much less so for modules.

Seasonal change is also expected to be prominent in host–parasitoid networks, which have high species turnover within a year (although to a lesser extent than plant–pollinator systems) (Lewis *et al.*, 2002). We might expect to see somewhat less extreme seasonal change in benthic riverine food webs, such as Broadstone Stream, because outside the oviposition period the insects that dominate these systems tend to be present as larvae, and few of these enter resting phases over the winter, so they are active components of the network for most of the year. Freshwater food webs can and do display considerable seasonal variation, especially in their quantitative rather than binary form: that is although most species might be present throughout the year, their abundance, biomass and interaction strengths will shift markedly over the year (Tavares-Cromar and Williams, 1996; Warren, 1989; Woodward *et al.*, 2005b). Such seasonal patterns are likely to be far less pronounced at lower latitudes, but highly resolved, species-level food-web data from these systems are still scarce, making it currently difficult to test such generalizations on a more global scale.

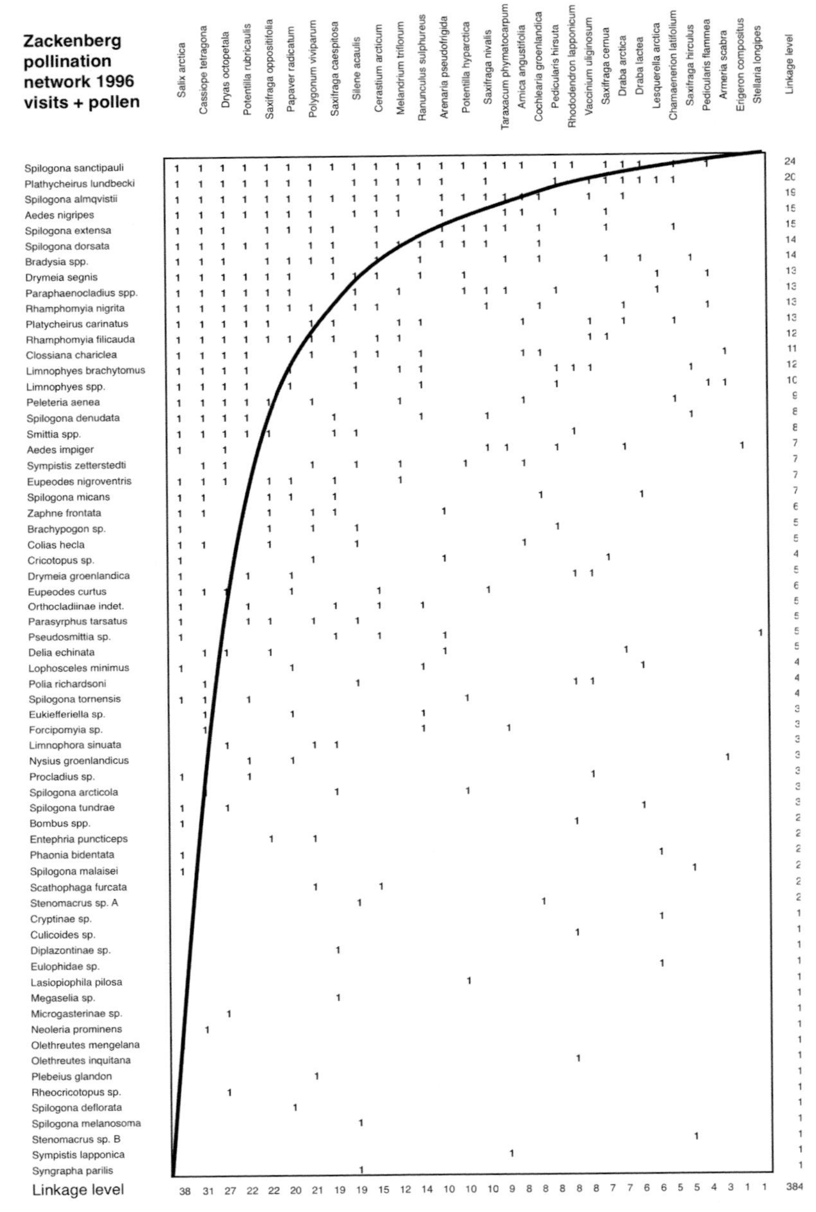

Figure 2 Pollination matrix from Zackenberg, northeast Greenland, based on data on both flower visit and pollen load on insect-body surface sampled over 1 year. Pollinators in left column and plants in uppermost row. Linkage level L is given for each species. Species are sorted in a nested way, that is according to descending L from upper left corner of matrix. The curved line represents the isocline of perfect nestedness. If a matrix shows perfect nestedness, all links are to the left of the line. The position of the line is obtained using the software Nestedness Temperature Calculator (e.g. Bascompte *et al.*, 2003).

Broadstone Stream in Ashford Forest in southeast England is a long-studied and well-characterized model system, which we will refer to in most of our topical sections. Its food web is most complex and its interactions are most intense in the summer months (Hildrew, 2009; Woodward and Hildrew, 2002b; Woodward *et al.*, 2005a). This is during the larval recruitment period, when abundance of most species is highest and generations overlap to give the maximum gradient within the community size spectrum and hence the greatest opportunity for both upper and lower size-refugia to be overcome: for example small individuals of 'large' predator species coexist with large individuals of 'small' predator species, increasing the opportunities for intra-guild predation and mutual feeding loops to be manifested (Figure 3; Woodward and Warren, 2007). The number of omnivorous, cannibalistic and mutual predator loops (e.g. *a* eats *b* eats *a*) all peak during summer, and the network then disassembles progressively over the winter into a much simpler form (Woodward and Hildrew, 2002b; Woodward *et al.*, 2005b). As is the case in many temperate freshwater food webs, seasonal differences in relative abundances within years are often far more pronounced than those within a given month between years, even when separated by several decades (Woodward *et al.*, 2002a). However, it is among the tail of rare and/

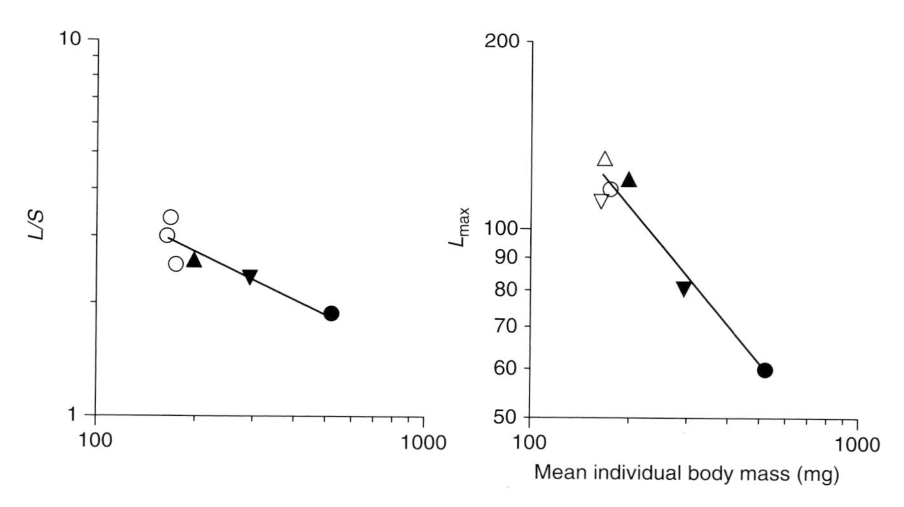

Figure 3 Seasonal shifts in the structure of the Broadstone Stream food web as a function of mean body mass of predators and prey. Left panel depicts the number of links per species on the *y*-axis ($r^2 = 0.75$; $P < 0.001$); the right panel depicts the predicted asymptotic number of feeding links for the predator assemblage ($r^2 = 0.89$; $P < 0.01$). The summer months (June, August, October) are shown as open symbols and the winter months (December, February, April) as closed symbols. The web is least connected in the winter and spring, when the larger prey species are able to attain size refugia by outgrowing their potential predators (Woodward *et al.*, 2005a).

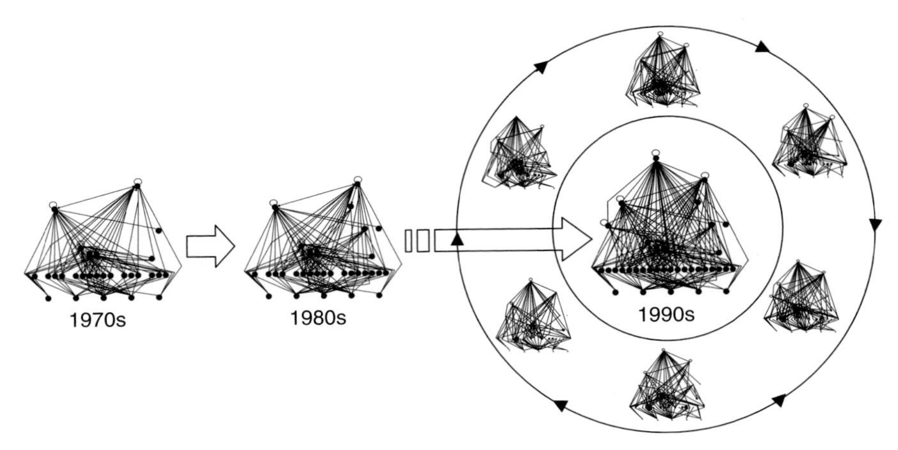

Figure 4 Decadal and seasonal change in the Broadstone Stream food web. The smaller webs encircling the 1990s summary web depict snapshots constructed for six alternate months over 1 year (August is at the top of the circle). Nodes for the seasonal webs are quantified: the area is proportional to numerical abundance per metre square on each date, redrawn after Woodward *et al.* (2005a).

or transient species that most of the turnover in the 2-mode network structure is manifested at these longer timescales ('2-mode' means the network consists of two interacting communities). In Figure 4, the summary webs become increasingly more complex from the 1970s because of the occasional colonization by these taxa in the 1980s and 1990s as the acidity of the stream has ameliorated in response to long-term reductions in acidifying emissions across Europe (Hildrew, 2009; Woodward *et al.*, 2002).

The seasonal dynamics scenario for the modular pollination network from Zackenberg follows two rules (Olesen *et al.*, unpublished data a): (1) the seasonal dynamics of the entire network consists of a seasonal series of modules peaking successively in species and link number (see also Dupont and Olesen, submitted for publication). Modules may also exhibit fission in the beginning of the season and fusion at the end. (2) The first species to appear in a module are also those that have the longest phenophase (flowering or foraging period), reach the highest abundance, become generalists and thus act as core species both in the network and in their module and maybe also, through their many links, connect different modules and make the entire network more cohesive (Olesen *et al.*, 2007). These connecting species blur modularity in a similar way to the so-called higher order generalist consumers in food webs, such as migratory vertebrate predators (McCann *et al.*, 2005). The level of modularity tells how distinct the different modules of the network are, that is more links within than between modules translates into higher modularity of the network. The first-arriving core species become the structural backbone of the modules onto which the tails of specialists are connected. Each tail species or specialist only stays in its module for a few days, but it is soon succeeded by other, topologically similar (and maybe also functionally similar species): that is there is a high turnover of specialists. The two characteristic tails of specialists observed in a traditional portraiture of cumulative networks (Figure 2) disappear when we switch to the time-slice view because tail species are short lived and only co-occur to a limited extent. Or, to phrase it differently, the two tails of specialist species seen in a matrix organized in a nested way lengthen with increasing time-slice thickness. Consequently, when we talk about network and module core species surrounded by swarms of specialists it is, at least to some extent, an artefact of data accumulation (e.g. Olesen *et al.*, 2007). There is also compelling evidence for the presence and importance of core species in other network types: for instance, there is a core of eight well-connected species that have been recorded on all sampling occasions in the Broadstone Stream over three decades, onto which is attached a tail of progressively rarer and less connected species (Woodward *et al.*, 2002), and analogous patterns are also evident over two decades of sampling the Lochnagar food web (Layer *et al.*, 2010a).

Spatio-temporal variation in the structure of a network can have a critical impact upon its dynamics (e.g. McCann *et al.*, 2005; Melián *et al.*, 2005; Rezende *et al.*, 2009). Climate change, especially in the Arctic, causes profound alterations in the phenology of species and this may affect the dynamics of entire ecosystems: in Zackenberg, Hoye *et al.* (2007) found that many species during the last decade exhibited an earlier phenophase, by up to as much as 3 weeks. The consequences of these fluctuations in species richness are unknown: What is the relationship between onset and length of phenophase? What is the phenological response to climate change in different species? And what are the network repercussions of all these to connectance, nestedness, modularity and stability (Bascompte *et al.*, 2006; Fortuna *et al.*, 2010; May *et al.*, 2008; K. Trøjelsgaard and J. M. Olesen, unpublished data; Olesen *et al.*, 2007; Woodward *et al.*, 2010b)?

A few studies have explored in detail the temporal dynamics of pollination networks from one short time slice to the next, for example from week to week (e.g. Petanidou, 1991 and later analyses of her data set, in Alarcón *et al.*, 2008; Basilio *et al.*, 2006; Kaiser-Bunbury *et al.*, 2010; Lundgren and Olesen, 2005; Medan *et al.*, 2002, 2006; Nielsen and Bascompte, 2007; Olesen *et al.*, 2008). Olesen *et al.* (2008) analysed the Zackenberg pollination network on a daily basis over two seasons to characterize how new species (i.e. species that initiated flowering or began active foraging) forged links to old species already established in the network. To understand the formation or assembly of networks, theoreticians have developed models of network assembly and one of these is based on the concept of preferential linkage or attachment (Barabási *et al.*, 1999). According to this model, a new species in a network is most likely to link to species in the network that already has many links. Some of the ecological drivers behind preferential linkage, which further leads to scale-free networks, are species abundance and phenology, that is the more individuals a species has and the longer they are active, the more links the species will ultimately accumulate, as would be expected under neutral assumptions. In a scale-free network, all species are close to each other, because most are linked to a few central core or hub species. A small world network (a concept related to the classic 'six degrees of separation', Barabási, 2002) has a cumulative linkage-level (L) frequency distribution, which follows a power law, and this originates from (1) the continuous growth in the number of species and links, and (2) the preferential linkage of new species to already well-connected species. Preferential linkage also takes place in social, genetic and technological networks (Jeong *et al.*, 2003; Newman, 2001; Redner, 1998), so it appears to be a phenomenon ubiquitous to all developing networks, not just restricted to ecological systems.

Ecological networks are often not perfect small worlds, because their L-frequency distribution follows a power law, which becomes truncated for higher values of L (e.g. Dunne *et al.*, 2002a; Jordano *et al.*, 2003; Montoya and Solé, 2002). The small-world property is also expressed by the slow, or logarithmic, increase of the mean shortest path between species pairs in the network, with increasing number of species (Nooy *et al.*, 2005). The reasons for this truncation are various linkage constraints (see Section IV, and Olesen *et al.*, 2008).

In the Zackenberg network, several properties calculated for an entire season were stable between years, for example species number. Thus, while at a 'global', macroscopic or network scale, structure seemed stable, the identity and composition of links and species were quite dynamic (the local or microscopic scale), echoing some of the patterns seen in the Broadstone Stream and Gearagh food webs (McLaughlin *et al.*, 2010; Reuman and Cohen, 2004; Woodward *et al.*, 2002). A tension between global stability and local instability is reported from various kinds of network, including those that display fractal growth, that is the repeated occurrence of similar patterns on different organizational levels (Burgos *et al.*, 2007; Song *et al.*, 2006). At Zackenberg, one-fifth of the pollinator species and two-thirds of all links were only observed in any one of the two years of study. These species were specialized and often rare and their links made up the tails of the link pattern in the nested matrix version. Again, matrix dynamics were mainly confined to the nestedness tails, as has also been suggested in other pollination studies (Basilio *et al.*, 2006; Lundgren and Olesen, 2005; Medan *et al.*, 2006; Petanidou and Potts, 2006). The total numbers of species and links in the Zackenberg network accumulated linearly throughout most of the season, but at the end of the year numbers stabilized. Species-linkage probability was not perfectly linear, rather 'sub-linear', that is intermediate between preferential linkage and complete independence of L. This dynamical pattern turns out to be fully compatible with the observed truncated power-law linkage-level distribution, because the most generalized species do not receive the high number of links they would have received if the season has been longer, as the onset of frost and new snow truncates the linkage-level distribution.

When constructing binary networks, the probability of detecting a link according to abundance alone might account for some of the widely reported patterns in the food-web literature, such as link-scaling coefficients of <2 (Ings *et al.*, 2009). If linkage of one individual of a new species to an individual of an old species i in the network is random, only abundance of i determines the linkage probability of i. Thus, linkage among individuals is neutral or unconstrained, whereas linkage at the species level becomes constrained by abundance: abundance frequency distributions have a few common species (McGill *et al.*, 2007) that will attain a high linkage probability whereas most species are rare and, consequently, will have a low linkage probability. Significant, positive mutualistic abundance-linkage level relationships are well known (e.g. Elberling and Olesen, 1999). It is not inconceivable that the link-scaling coefficients of <2 that are commonly described for food webs, host–parasitoid networks and mutualistic networks (Ings *et al.*, 2009; Jordano *et al.*, 2003) are at least partly related to this phenomenon, since larger webs will tend to contain long tails of rare species whose interactions are far less likely to be observed than is the case for common species (e.g. Vázquez *et al.*, 2007).

In pollination networks, a longer phenophase of a species translated into more days of potential linkage to new species and an increased linkage level (Olesen *et al.*, 2008). In the Zackenberg network, for example phenophase length explained more than half of the variation in linkage level. Thus, phenology contributed to the skewed frequency distribution of linkage level because it also followed a lognormal frequency distribution. Since linkage probabilities were lower than predicted according to a pure preferential linkage model, Olesen *et al.* (2008) suggested that some species-specific constraints, besides abundance and phenology, limited linkage probability. This seems likely to be true for other networks in which potential consumers and resources might be at least partially separated by being out of phase with one another, such that even species that are (on average) abundant might not necessarily interact, or may do so only rarely: there is evidence, for instance, of looser connections between zooplankton and phytoplankton in Lake Washington, as their respective phenologies have altered at different rates over time as the lake has warmed, so that the consumers no longer peak in abundance at the same time as their resources (Winder and Schindler, 2004a,b).

No detailed (i.e. described with thin time slices) study exists of the temporal dynamics of an ecological network lasting more than a few years: even the four-decade record from Broadstone Stream is punctuated with large gaps (Woodward *et al.*, 2002; Figure 5). Longer term studies often use seasonally pooled data or snapshots taken at a similar time on one or a few occasions per year, over several years (e.g. Yamazaki and Kato, 2003). Olesen *et al.* (unpublished data b), however, examined 12-years of dynamics of a Spanish flower–visitation network between butterflies and their nectar plants, in which the species and their interactions were sampled annually. The study protocol of this multislice network remained exactly the same for all years, that is the same transects, the same observation dates and the same observer. A link was observed whenever a butterfly species visited the flowers of a plant species. Based on these observation data, they constructed a butterfly–plant matrix for each year. They then tracked the appearance ('colonization') and

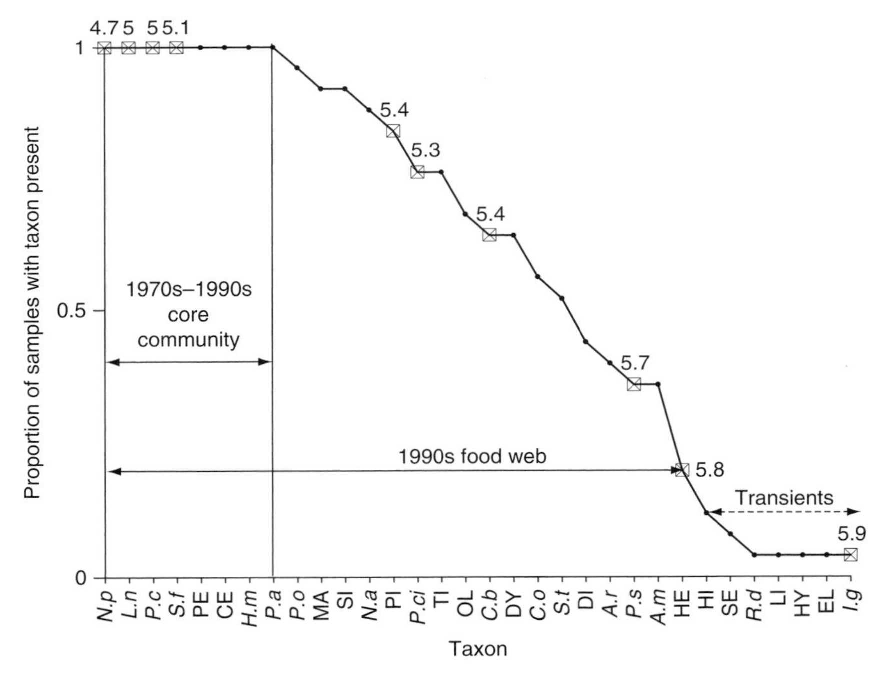

Figure 5 Long-term patterns in the composition of nodes in the Broadstone Stream food web over three decades. Species for which independently derived pH optima were available are highlighted (crossed squares, with respective pH optima in text above each symbol)—note that the progressively less persistent nodes were among the more acid-sensitive taxa (redrawn after Woodward *et al.*, 2002).

disappearance ('extinction') of each species and link between successive years, which enabled them to locate 'hot and cold spots' in the network of high and low temporal dynamics. On average, 59 butterfly species, 50 plant species and 276 links were present in the network each year: annual connectance C was 9.4% consistently among years (for all years pooled: 87 butterflies, 109 plants and 1134 links were noted). Thus, this network also showed global stability, whilst hiding considerable local instability among its species and links. In the butterfly–plant network, individual species and links repeatedly went through colonization or extinction events.

In the butterfly–plant network, linkage level L was used as a simple metric and again all species with $L \leq 2$ and $L > 2$ were defined as specialists and generalists, respectively. These two groups of species were almost equally represented in the network. All 12 annual matrices were nested. Temporal permanence of a species, as defined by the number of years it is present in the network, T, was 8 years for butterflies and 6 years for plants, or 68% and 46% of the 12 year-study period, respectively. Frequency distributions of T had a 2-modal shape, with one mode consisting of temporally stable species observed for 11–12 years, and another of temporally sporadic species observed only for 1–2 years. A 2-modal temporal distribution is also reported for other species communities, for example prairie herbs (Collins and Glenn, 1991). With respect to its long-term behaviour, the network was therefore dominated by two distinct subsets of species, viz. sporadic or low-T species and persistent or high-T species, and this created strongly asynchronous dynamics among species across the network. A British estuarine fish community also consisted of two distinct temporal groups, termed 'persistent' and 'occasional species': persistent species were also often, but not always, common, and occasional species, were often rare (Magurran and Henderson, 2003), as also seen in the Broadstone Stream and Lochnagar food webs (Layer *et al.*, 2010a; Woodward *et al.*, 2002). In a study of a mutualistic network, Carnicer *et al.* (2009) found evidence of marked temporal dynamics in resource use, with two types of frugivore species operating within the web: fast switchers, which were generalists that quickly exploited the new resources when they became abundant, and slow switchers, which maintained their preferred resource links even when these became rare, or more abundant alternative resources became available.

Network science is now a truly cross-disciplinary enterprise, and its toolbox has been assembled from a mix of several, very different, disciplines. Variations over time have been addressed here, but one of the next steps in the study of temporal dynamics of networks is to include different types of link. Mucha *et al.* (2010) have recently developed a framework that enabled them to study the community structure of temporally developing networks, that is by the use of links that connect nodes between time slices. This has now made it possible to study the temporal development of networks with different types of link (multiplexity), and also at different resolution levels, and this work sets the state of the most advanced art with respect to synthesizing temporal dynamics at several scale levels simultaneously. Another promising route is to address space and time simultaneously.

III. SPACE-FOR-TIME SUBSTITUTION

Space and time have long fascinated scientists in general, and biologists in particular. Seemingly endless numbers of papers, ideas and vast amounts of research effort have been devoted to either one of them, but some have also focused upon how they are related and asked: are space and time inherently different? Essentially, is time just one more dimension in a 4D-'space', or are space and time even interchangeable—a uniformitarian thought?

In ecology, one of the most commonly used phrases is 'spatio-temporal variation', although in reality few studies make a real effort to bring the two together (Figure 6). In food-web analysis, for example few studies have been designed explicitly with that purpose in mind (but see Christian *et al.*, 2005; O'Gorman and Emmerson, 2009; Warren, 1989). They often seem to reach similar conclusions, that webs vary substantially along both scales. Closs and Lake (1994), for example assessed the temporal and spatial variation in web structure in an Australian stream. The web was partitioned into three spatial and four temporal slices. The spatial slices were 1.5 km apart but the among-slice variation in web parameters was minor. Thus, if spatial variation did exist, it was manifested at either a coarser or finer scale resolution. The temporal variation, on the other hand, was considerable, even though all slices belonged to the same season. Winemiller (1990) observed the opposite in tropical food webs: considerable among-site variation and much less variation among seasons, although the extent of scales was much wider than in the study of Closs and Lake (1994). In England, Warren (1989) found that food-web complexity increased from the open water of the Skipwith ponds to the pond margins. The greatest variation in web complexity occurred through time, however, with much of this variation due to differences in species life histories. O'Gorman and Emmerson (2010) used spatially replicated food webs to test how food-web properties change through time as a result of targeted extinctions to the webs. They found consistency across the spatial replicates and clear patterns through time for the various targeted extinctions, highlighting the potential for in-depth spatio-temporal analysis of complex food-web structures. Woodward *et al.* (2002) found that the

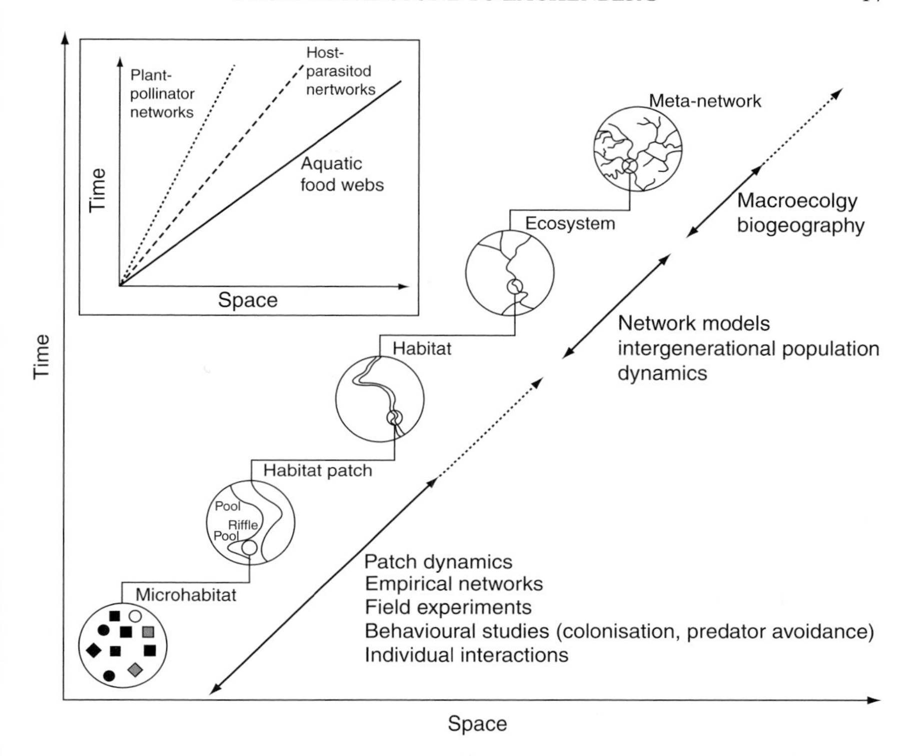

Figure 6 Conceptual representation of ergodicity in a typical fractal-type riverine system (adapted from Woodward and Hildrew, 2002a). Although the fractal nature of temporal and spatial scaling is general, the inset illustrates that ergodicity does not necessarily scale equitably across different systems.

structure of the Broadstone Stream food web varied seasonally, but this was primarily in terms of changes in relative abundance and interaction strengths, rather than extensive turnover in the presence/absence of species or links. However, interannual variation, even over several decades, was often less pronounced, at least in terms of mean annual abundance of the dominant species, than these shorter term seasonal changes (Woodward *et al.*, 2002), and this seems to be true of many temperate and high-latitude freshwaters. Despite this high persistence and the presence of a core of dominant species over three decades, there is evidence of long-term and large-scale covariance within the Broadstone food web, as new (rare) species have colonized the network: over three decades, a progressive rise in pH has mirrored longitudinal changes over about 800 m from the acid source of the stream to its lower more circumneutral reaches. Essentially, the core of acid-tolerant species that predominated in the 1970s have remained largely unchanged in the upper acid reaches, but this group has been added to via invasion and establishment of more acid-sensitive species in the lower reaches, with a general lengthening of food chains and a rise in network complexity: that is there was evidence of time–space interchangeability at the decadal-reach scale (Figure 7).

The studies by Closs and Lake (1994) and many others point towards a long-standing debate in several research disciplines (e.g. in geomorphology and statistical mechanics): are space and time interchangeable or ergodic? The studies of ecological networks mentioned earlier suggest that they are

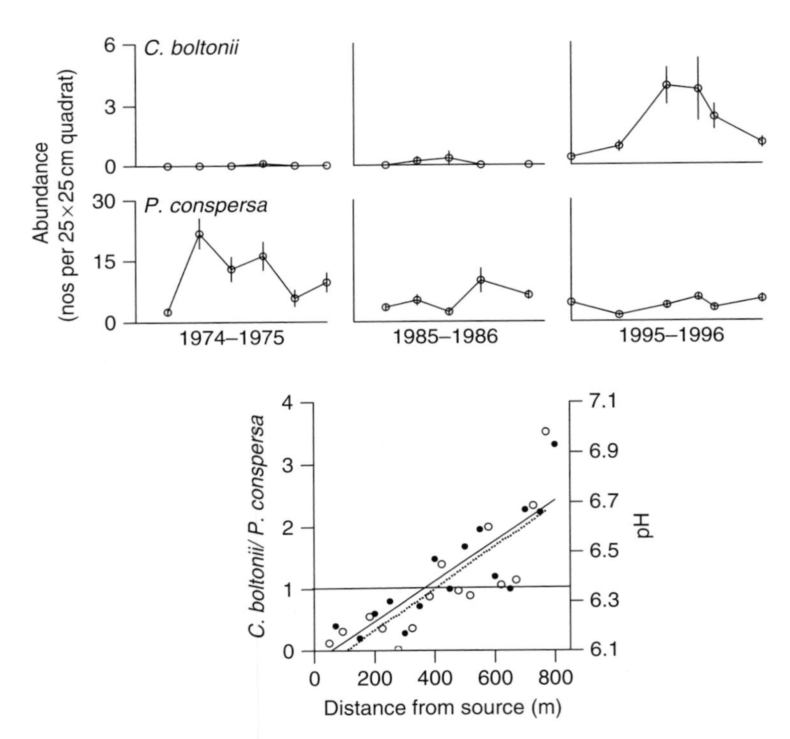

Figure 7 Evidence for ergodicity in the Broadstone Stream food web. Top panels depict abundance of two dominant large predators over 3 years from three decades, as mean summer pH has risen progressively, by about ½ pH unit per decade (Woodward *et al.*, 2002); the lower panel depicts the quotient of numerical abundance of the two predators (solid line) and pH (dashed line) over a spatial gradient across the entire length of the acidified 800 m section of the stream.

often related but not perfectly interchangeable. Ergodic and ergodicity come from Greek: *ergon* and *hodos* meaning 'work/energy' and 'path'. In physics and population ecology (e.g. Åberg, 1992; Cohen, 1979; Maurer, 1999), ergodicity may have more precise definitions than just space–time comparisons. Here, however, we only discuss to what extent spatial and temporal variation are related, which we will call the 'level of ergodicity'. To field ecologists, for example students of natural succession (Southall *et al.*, 2003), it is important to know whether space and time are ergodic or whether the two scales are intrinsically different, that is whether or not it is important to sample along both scales. If time can be interchanged with space (e.g. Figure 6), it allows the researcher much wider flexibility in research design, for example if it is possible to reduce the extent of a field study from, say 3 months to 1 month, if the study area instead is expanded from 1 to 5 ha. It also strengthens the justification for the use of space-for-time substitution (Figures 7 and 8), which remains the *de rigeur* method of inferring impacts of environmental stressors (e.g. Layer *et al.*, 2010b; Rawcliffe *et al.*, 2010), successional development and assembly/disassembly of food webs, given that before-and-after data are rarely available (but see Woodward and Hildrew, 2001). Epistemologically, it is also of profound interest to study the space–time ergodicity of natural phenomena.

Certainly, if we expand along a temporal and spatial scale, both numbers of species and their links increase, but to what extent do they follow similar trajectories? Many food webs may be delimited spatially by their biotope and temporally by the seasons of the year: we speak about lake, forest, or dung pat webs, but rarely about, for example a rainy-season web (Cohen, 1978). All food webs are, of course, to some extent open systems in both time and space, which causes problems with the choice of extent and resolution of scaling of the study (Martinez and Dunne, 1998). Schoenly and Cohen (1991) distinguish between cumulative and time-specific webs, with the former being based on data accumulated during several censuses, whereas the latter are based on data sampled over a single, often short time period, a time slice: the distinction is effectively simply a matter of the 'thickness' of the slice. The same distinction could be applied in a construction of cumulative and space-specific webs. We present some data from Broadstone Stream, which depict how the cumulative or summary annual food web was assembled via cumulative sampling over both space (30 replicate quadrats per date) and time (6 bi-monthly sampling dates), and highlights the potential blurring effects of these commonly used agglomerative approaches (Figure 9). The 'trivariate' structure of the summary network, in which energy fluxes from small, abundant resources to larger, rarer consumers (McLaughlin *et al.*, 2010; Woodward *et al.*, 2005a) is not so clearly evident within a single sampling

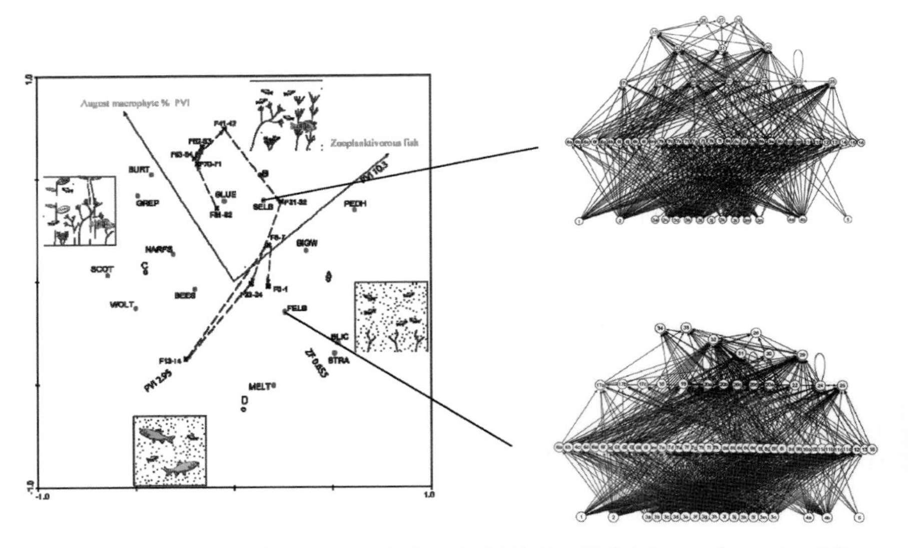

Figure 8 Trajectory of change with time in Felbrigg Hall lake as determined from zooplankton and plant macro-fossil remains and interpretation using a Multivariate Regression Tree Model (Davidson *et al.*, 2010): that is a temporal gradient in food-web structure in responses to long-term eutrophication. Data from food-web nodes from 11 other contemporary lake sites are also shown on this MRT model, which indicates declines in plant abundance and changes in fish community associated with eutrophication: that is a spatial gradient of eutrophication. Two of the sites, Felbrigg and Selbrigg, have had their contemporary food webs constructed, indicating a decline in complexity: the currently 'pristine' Selbrigg site reflects the conditions in the Felbrigg site in the 1920s, prior to the degradation of the food web following several decades of eutrophication (redrawn after Rawcliffe *et al.*, 2010). PVI = percentage volume infested, a measure of the density of rooted plants per unit volume which is an indicator of regime shift as the lake loses its plants and switches into a turbid phytoplankton-dominated state.

occasion, nor within a single habitat patch, which is closer to the scale at which the individual protagonists are actually interacting (Figure 9). Rather, these macro-ecological patterns only emerge when the smaller temporal and spatial scales are combined to form a view of the food web that is averaged, or smoothed, over both time and space. The temporal and spatial extent of a food web must depend to a large extent upon the generation time and home range of the species at the different trophic levels (Thies *et al.*, 2003). Especially from a dynamical context, a study should (ideally) span at least the generation time of the longest lived species in the system to account for potential indirect looping interactions and feedbacks (Montoya *et al.*, 2009; Yodzis, 1988). This makes it virtually impossible to find a satisfactory extent of a study suited to all

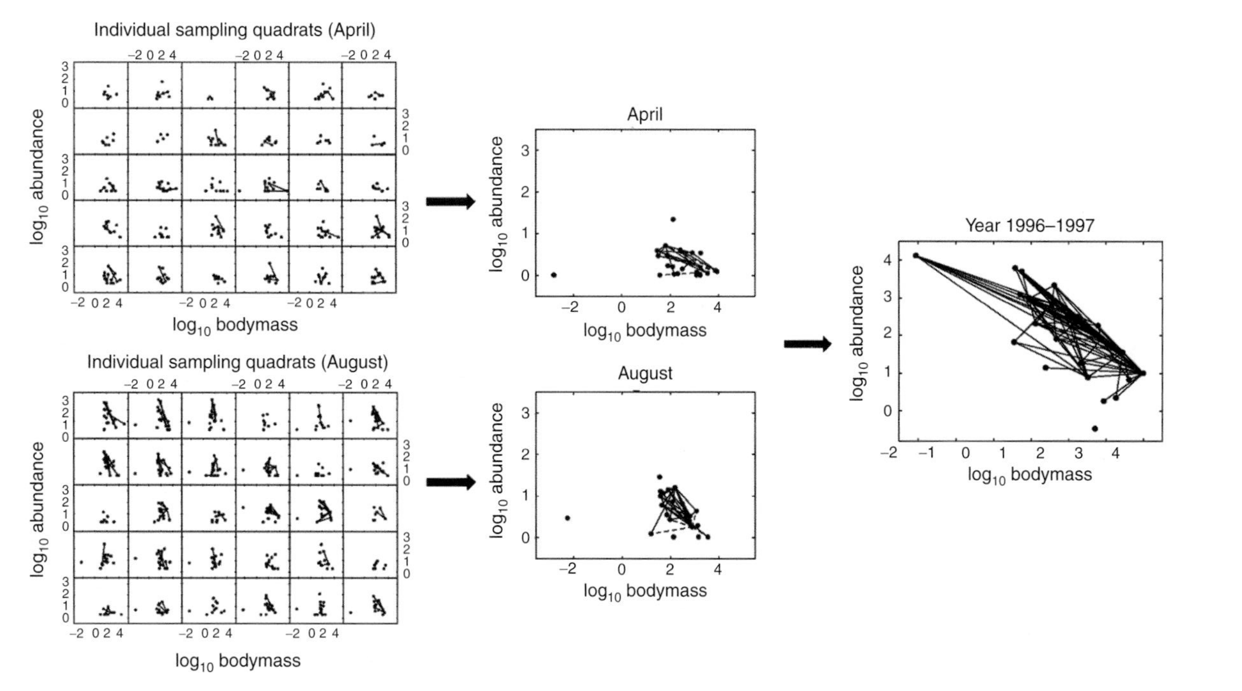

Figure 9 Spatial and temporal agglomeration of an annual summary network (Broadstone Stream 1996–1997 food web). Each data point represents the mass and abundance of a species population on a log–log scale, with links depicting directly observed feeding interactions between predators and their prey. Two (April and August) of the 6 bi-monthly sampling occasions used to build the summary web (far right hand panel) are broken down into food webs constructed for the individual quadrats ($n = 30$, 25 cm × 25 cm basal area) on the left-hand side. These were then used to construct the species averaged web for each date (mid-panels), and ultimately for the 6-date summary web, by combining them with gut contents data from additional qualitative samples of predators whose diets were undersampled in the monthly quantified food webs. See Woodward *et al.* (2005a) for further details of the sampling protocols and site description.

species in a multi-trophic web, so we must almost always accept some form of compromise if we want to study natural systems: however, this is likely to be best achieved if we know which parameters and under which circumstances space and time are ergodic or non-ergodic. At present, these factors are rarely considered, but if we do so at the outset it may give us both more methodological freedom and greater insight into the true spatio-temporal scaling of the network's structure and dynamics.

Bundgaard (2003) analysed the behaviour of a set of pollination-network parameters at increasing spatial and temporal scales. His study system was a 2-mode pollination network consisting of a community of flower-visiting animals and a community of flowering plants. All flowering plant species and all animal-visiting flowers were included, and the study area was a plot of the size of 150×2 m, partitioned into five equally sized space units (1, . . ., 5*S*), each being 30×2 m. The study season was divided into five equally sized time units (1, . . ., 5*T*), each of 1 week of weather suitable for foraging for most flower visitors. This design made it possible to scale up from one (1*S* or 1*T*) to five units (1-5*S* or 1-5*T*) along both the spatial and the temporal dimensions (Figure 10). For each unit, an adjacency matrix, whose elements are either '0' or '1', describing the pollination network was constructed, with '1' indicating presence of a link between two species and '0' that no link was observed. A set of network parameters were calculated as the spatial and temporal scaling increased from 1*S* to 1-5*S* and from 1*T* to 1-5*T*, to give five spatial and five temporal replicates, respectively. Thus, a total of 2×25 matrices (the spatial matrices 1*S*, . . ., 1-5*S* replicated at each of five subsequent weeks; and the temporal matrices 1*T*, . . ., 1-5*T* replicated at each of five sites along the transect) were used in his analysis. Variations in three of the network parameters (species, links and connectance) are shown in Figure 11. The size of the pollinator community, *A*, increased through both time and space (Figure 11A). The trajectories were quite similar and their equations are given in the legend to the figure. Solving these equations for *S* and *T*, Bundgaard (2003) calculated an 'ergodic rule' or 'exchange rate' for *A*: $S_A = 0.55\ T_A^{1.62}$. The same was done for total number of links, *I*, which also increased through both time and space. The trajectories were again similar and an ergodic rule was also suggested for the space–time relationship of *I* (Figure 11B). Connectance, *C*, on the other hand, decreased through space, but was independent of time, so it was not ergodic. Average pollinator linkage level, $\langle L_m \rangle$, was defined as the average number of plant species visited by a pollinator *m*, that is I/A. $\langle L_m \rangle$ increased through both time and space, but slowest through time (Figure 11C). Overall, the exchange rates revealed that one day extra in the field was equivalent to expanding plot area from 5 to 64 m^2, depending on the parameter of interest and the route the researcher selected through the space–time parameter space.

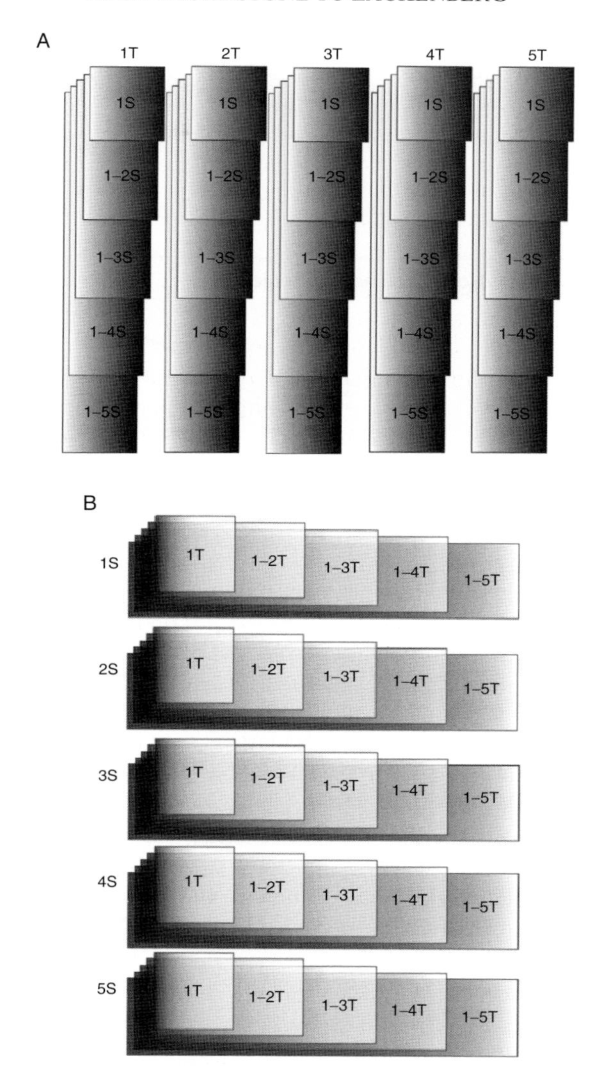

Figure 10 Study design: Each square in the figure has the dimensions 1 week × 60 m^2. A and B, increasing spatial and temporal scale, respectively, from 1 to 5 units (1*S*, …, 1-5*S* or 1*T*, …, 1-5*T*), that is from 60 to 300 m^2 and from 1 to 5 weeks (Bundgaard, 2003).

The parameters behaved broadly as expected, that is scaling up sampling increased both number of species and links. The spatio-temporal variation of $\langle L_m \rangle$ reflected the variation of its components, *A* and *I*. However, *C* tends to decrease as the number of species increases (Olesen and Jordano, 2002;

Petanidou and Potts, 2006) and may fluctuate considerably seasonally (Lundgren and Olesen, 2005; Medan *et al.*, 2006). The spatio-temporal behaviour of C is more difficult to predict because it is a calculatory compromise between numbers of pollinator species, plant species and links.

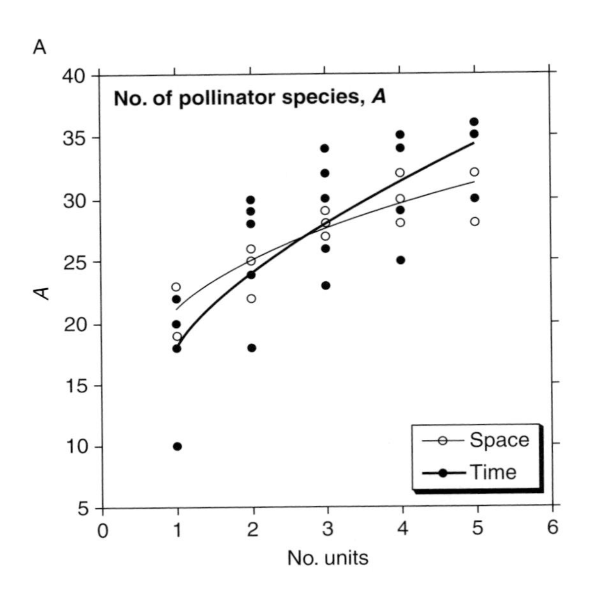

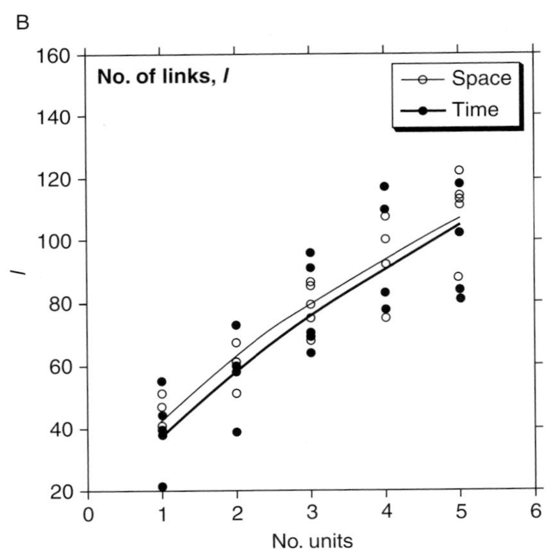

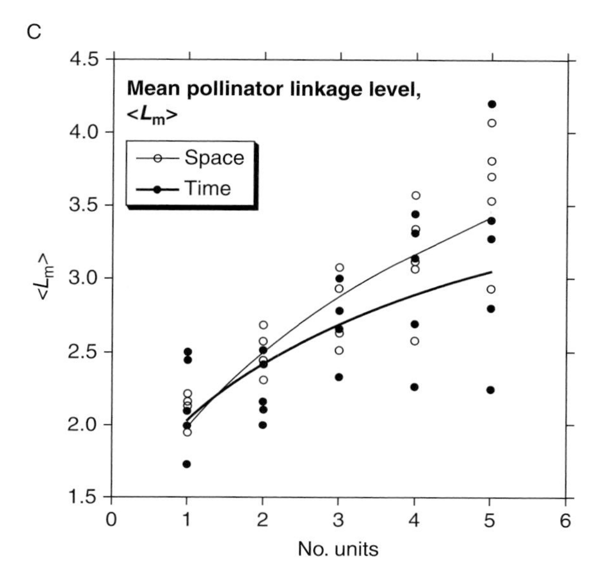

Figure 11 Ergodicity of pollination network parameters: (A), increase in number of pollinator species, A, with increasing scale ($1S$ to $1\text{-}5S$ and $1T$ to $1\text{-}5T$): $A = 21.23S^{0.24}$ and $A = 18.37T^{0.39}$. At $1\text{-}5T \times 1\text{-}5S$, $A = 44$ species; (B) increase in number of links, I, with increasing scale: $I = 42.01S^{0.58}$ and $I = 37.45T^{0.64}$. At $1\text{-}5T \times 1\text{-}5S$, $I = 278$ links; and (C) increase in mean pollinator linkage level, $\langle L_m \rangle$, with increasing scale: $\langle L_m \rangle = 1.98S^{0.34}$ and $\langle L_m \rangle = 2.04T^{0.25}$. At $1\text{-}5T \times 1\text{-}5S$, $\langle L_m \rangle = 6.3$ plant species/pollinator species (Bundgaard, 2003).

Although A was most sensitive to changes in the temporal scale, its development along both dimensions was quite similar, demonstrating that variation in numbers of pollinator species and links, and linkage level along a spatial and a temporal scale, was comparable, that is it was highly ergodic. This may be a key difference between aquatic food webs and pollinator networks. The fact that in many running water food webs, for instance, the medium is constantly mixed and dominated by insect larvae most of the time (adult insects being short lived and mostly aerial) could result in temporal–spatial homogenization to an extent not generally seen in terrestrial systems. This hypothesized difference offers a potential starting point for a more in-depth comparison, for example do plant–pollinator networks occupy a different position on a space–time gradient than stream food webs (Figure 6)? What are the relative rates of spatial and temporal change in the two kinds of network?

In pollination biology, surprisingly few studies have made explicit comparisons of variation in space versus time. Exceptions include investigations of the flower–visitor fauna to single plant species, for example *Lavandula latifolia* (6 years × 4 sites; Herrera, 1988) and *Hormathophylla spinosa* (4 years × 3 sites; Gómez and Zamora, 1999). Herrera (1988) observed weak correlations between spatial and temporal variation in species abundances, but stressed the importance of stochasticity. In future studies, we recommend that an analysis of the variation of parameters in space and time should be more fully integrated and not just, as is *de rigeur* today, analysed separately.

Beyond the intriguing theoretical aspects of ergodicity discussed above, we also have a strong pragmatic reason for reducing sampling effort and improving predictability of cumulative parameter values. The analysis of Bundgaard (2003) demonstrated that a mixed sampling in both space and time captured most of the important parameter variation. Sampling along the space–time diagonal: $(1S, 1T) \rightarrow \cdots \rightarrow (5S, 5T)$ gave much better parameter predictions than only sampling along either the temporal or the spatial dimension ($1T \rightarrow$ 1-5T or $1S \rightarrow$ 1-5S). However, the results reviewed here also suggest that space and time can show fundamental differences and both have to be considered carefully in any ecological study. The more we scale up, the more we risk aggregating species that do not co-occur in networks, that is species become constrained from encountering one another and the potential link between them may be forbidden in time and/or space. Some intriguing and, as yet, largely unanswered questions are raised when considering the spatio-temporal scaling and ergodicity of networks, such as is the large variance in the number of link per species or intraspecific diets often reported simply a question of sampling effects, or does this variance continuously increase regardless of spatio-temporal sampling?

IV. FORBIDDEN LINKS AND LINKAGE CONSTRAINTS

The essence of all network studies is the observation of links among species, but some links are exceedingly difficult to detect and others do not seem to exist at all. Summing all observed links in a network and calculating connectance C, rarely leads to $C > 20\%$, at least not for large and taxonomically well-resolved networks (Dunne *et al.*, 2002a; Olesen and Jordano, 2002; but see Polis, 1991; Warren, 1989; Yodzis, 1998), so often 80% or more of all possible links are not realized. What offers us the most useful insight into network structure—the presence of 20% links or the absence of 80% links? The unexpected absence of a link might have much higher information content than the expected presence of a link.

Very few ecological studies have focused upon these unobserved links (but see Jordano *et al.*, 2003, 2006). In a study of three mutualistic networks, Olesen *et al.* (in press) addressed this issue and suggested that unobserved links are either 'missing' or 'forbidden'. Missing links are essentially methodological yield–effort artefacts and they will eventually be discovered if sampling is extended. The forbidden ones, on the other hand, do not exist and will remain unobserved regardless of sampling effort because of linkage constraints or mismatches, for example temporal, size, reward and physiological–biochemical constraints (Blüthgen *et al.*, 2008; Borrell, 2005; Jordano, 1987; Jordano *et al.*, 2006; Nilsson, 1988; Olesen *et al.*, 2008).

A constraint in time is acting on 'forbidden' links if, for instance, potentially interacting species do not overlap or do so only weakly, for example their phenophases are 'uncoupled' (e.g. Fabina *et al.*, 2009). Temporal uncoupling can therefore mask the expression of other potential constraints. The hypothesis about matches and mismatches of life-cycle events of potentially interacting species was first formulated by Cushing (1975). He focused specifically upon the synchronicity of plankton blooms, but the general idea is more deeply rooted in ecology (Elton, 1927) and also more broadly in other fields (e.g. Bloch and Ritter, 1977, in the concept 'ungleichzeitigkeit'). Size mismatches explain forbidden links between plant species with a long floral corolla tube and a short beak of their pollinating birds (Jordano *et al.*, 2006; Snow and Snow, 1972). Sharks and crabs, for example are physiologically constrained from moving into the cold waters of the Antarctic, but may do so during the next 50 years because of an increase in sea-water temperature, and the consequences to the local fauna are likely to be devastating as the recipient communities have not developed the defence mechanisms to deal with these predators, due to the previous uncoupling of potential encounters over evolutionary time (Aronson *et al.*, 2007). Biochemical constraints are in operation if an insect larva is unable to process the secondary compounds in foliage, for example of the bracken fern *Pteridium aquilinum* (Cooper-Driver *et al.*, 1977). In one of the few studies to quantify the source of forbidden links, Jordano *et al.* (2003) found that 51% of forbidden links in a Mediterranean plant–animal mutualistic network were due to phenological uncoupling, 24% were due to size mismatches and 6% were due to physiological/biochemical constraints (with 19% of possible links left unexplained). Several constraints may even operate in concert, making it exceedingly difficult to delineate their relative contributions and sequence of action. As link biologists and networkers, we want to eliminate all missing links and make ecological-evolutionary explanations of the forbidden ones. For instance, in the Broadstone Stream food web, the likelihood of finding a link was largely related to the product of the abundance of consumers and resources (i.e. a sampling effect related to encounter rate), and also the size of the consumers. There was a range of body sizes over which feeding links were feasible, but upper and, to a lesser extent, lower size refugia represented where the forbidden links were located in the feeding matrix (Figures 12 and 13). The missing links were revealed for interactions between species that could potentially interact (i.e. within the size range of consumer–resource ratios realized among the dominant species) but that were undersampled, that is it

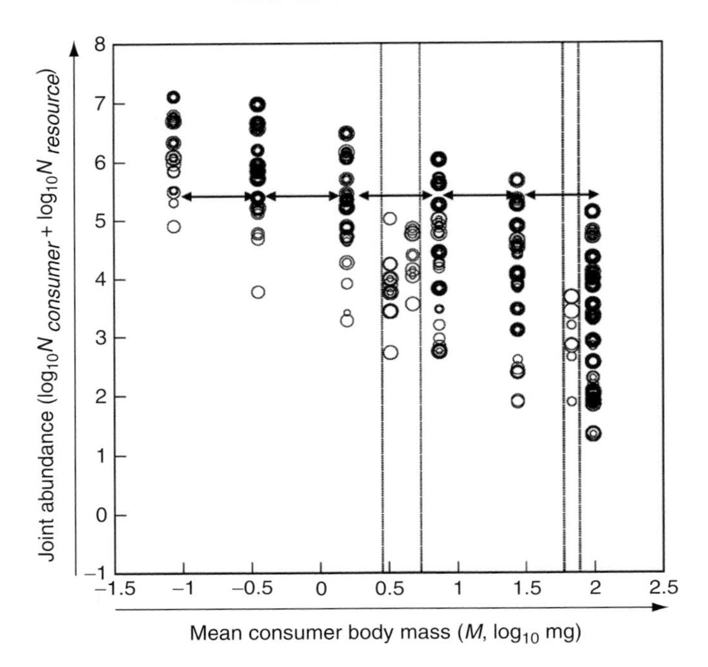

Figure 12 Seasonal feeding matrix for the Broadstone Stream food web, highlighting the location of 'missing' and 'forbidden' links. Each circle represents the presence of a feeding link between consumers (columns) and resources (rows), depicted seasonally such that diameter increases progressively from June 1996, to August, October, December, February and April 1997. Note the three rarest predators (enclosed in dotted lines), with few feeding links, are positioned between the six dominant predators whose mean body masses are evenly spaced on a log-scale, as shown by the double-headed arrows. Note the frequency of link detection increases along the y-axis, such that more links are observed at higher joint abundances of predators and their prey. Similarly, the three rarest predators (highlighted within the vertical dotted lines) have the fewest number of links, due to undersampling of 'missing' links, as revealed by yield–effort curves (see also Woodward *et al.*, 2005a).

was extremely unlikely to find a link between a rare predator and a rare prey species. Among the six dominant predators, on any given sampling date, about 72% of each species' predicted feeding links (using a simple hyperbolic function; after Woodward *et al.*, 2005a) were observed directly, and this was consistent over time and across species (Figure 13): over the entire year, however, saturation of feeding links was observed for all six species (Woodward and Hildrew, 2001). Essentially, the data from Broadstone Stream revealed that body size determined whether links could exist or were 'forbidden' and abundance (=sampling effort) determined whether potential links were detected or 'missing'. The patterning of links within feeding matrices has been successfully predicted recently for a range of complex food webs, including Broadstone,

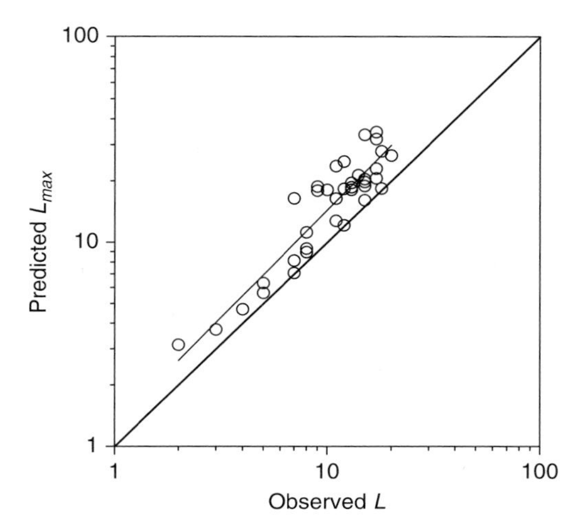

Figure 13 Observed versus predicted number of feeding links for the six dominant predators for the Broadstone Stream food web over six sampling occasions from June 1996, to August, October, December, February and April 1997. Note the scaling exponent = 1, and on average 72% (±2.8SE) of the predicted links (based on a hyperbolic model, after Woodward *et al.*, 2005a,b) were observed for a given consumer on a given date. Over the entire annual cycle, however, the observed feeding links for all six consumers achieved saturation (Woodward *et al.*, 2005a).

using a model based on relatively simple rules related to energetic constraints, foraging theory, and body size (Petchey *et al.*, 2008).

Olesen *et al.* (in press) analysed the Zackenberg pollination network and two Mediterranean seed-dispersal networks to characterize phenological uncoupling during single seasons (Figure 14). The number of missing links was reduced by using two kinds of link currency in each network: flower/fruit-visit data and data on the composition of the pollen load on the body surface of flower visitors and the seeds in faeces of frugivores. Olesen *et al.* (in press) constructed pollination matrices depicting all visitation links observed during the entire season between all pollinators/seed dispersers and their plant species. The species in each plant–animal matrix were sorted in a nested way, that is according to descending linkage level (see Figure 2). Matrix connectance, C, was 15% in the pollination matrix, and 24% and 52% in the seed-dispersal matrices (for comparison, Broadstone Stream food web had a $C = 10\%$ (I/S^2)). Thus, the proportions of unobserved links for the mutualistic networks were $100 - C = 85\%$, 76% and 48%, respectively, and for Broadstone 90%. Olesen *et al.* (in press) subsequently identified all pollen grains on sampled insects to plant species, and this information was added as new links to the visitation matrix: the number of links in the combined visitation-pollen matrix now yielded a C of up to 20%, with these extra 5%

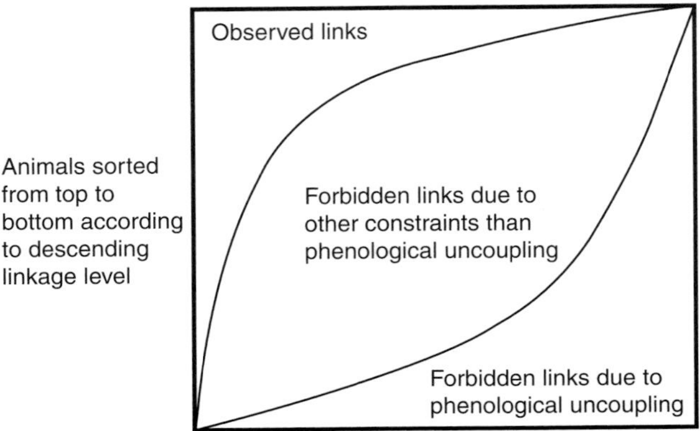

Figure 14 Plant–animal mutualistic interaction matrix sorted in a nested way according to decreasing linkage level from the upper left corner. In a perfectly nested matrix, all links will be inside the area 'Observed links'. Linkage level, abundance and phenophase length are positively correlated. Thus, sorting according to linkage level also, to some extent, means sorting according to phenophase length and abundance. On the other hand, phenophase lengths of an interacting species pair and their level of phenological uncoupling are negatively correlated. Thus, most phenological uncoupling is found in the lower right corner, and both the link pattern and the pattern of phenological uncoupling become nested.

of links representing the estimated number of missing links in the pollination network. The overlap between visit and pollen links was only 33%, underlining the value of using more than one kind of linkage data. In spite of this additional sampling effort, 80% of all potential pollination links still remained unobserved, and most of these were regarded as forbidden.

The reasons for links being forbidden can be ascribed to various constraints, for example phenological uncoupling, morphological mismatches, or handling difficulties. The former alone explained 22–28% of all possible links in the three mutualistic networks described earlier. In the Zackenberg example, most forbidden links due to phenological uncoupling were located in the lower, right corner of the interaction matrix (Figure 14), that is in the link space between rare, short-phenophase specialists. An extinction vortex may arise when a species becomes rarer, has a narrower populational phenophase, becomes more phenologically uncoupled from its mutualists, has reduced fitness, and hence becomes progressively rarer in a positive feedback until ultimately it is lost from the network.

Research on timing in ecology has a long history. Back in 1927, Charles Elton wrote: 'Many of the animals in a community never meet owing to the fact that they become active at different times' (p. 83), that is they are phenologically uncoupled. For example some sea urchin species hide in shelters by day and emerge at night to forage, avoiding overlap with predators that feed only during the day (Crook *et al.*, 2000; Nelson and Vance, 1979). Today, phenological uncoupling is receiving increasing research attention because of its obvious relation to global climate change (e.g. Hegland *et al.*, 2009; Hoye *et al.*, 2007; Inouye *et al.*, 2003; Memmott *et al.*, 2007; Tylianakis *et al.*, 2008; Woodward *et al.*, 2010a,b). For instance, some species respond strongly to climatic change in their onset and length of phenophase, whereas others hardly respond at all (Hoye *et al.*, 2007). Phenophase reshuffling may cause profound changes in network structure and dynamics due to alterations of both the identity and intensity of interactions, and such data will become invaluable for future studies tracking climate-driven changes on network structure, especially in the Arctic, where the climate has already changed the most and where future change is likely to be particularly marked in the coming decades (e.g. Witze, 2008). However, phenological uncoupling also varies with length of season and thus also latitude, so temperature and light availability are often confounded, which makes it difficult to isolate the effects of warming. The proportion of unobserved links explained by phenological uncoupling may range from as much as 28% in the Greenland pollination network (Olesen *et al.*, in press) down to almost nil in networks where one or both of the interacting communities are perennial, such as in a hummingbird–plant network from the island of Trinidad (Jordano *et al.*, 2006; Snow and Snow, 1972), where resident hummingbirds harvest nectar all year round.

Although we often discuss match/mismatch dichotomies, linkage constraints in reality are not always, and perhaps even rarely, such a strong all-or-nothing response. Some links probably remain unobserved because of too narrow a phenological overlap, for example in the Greenland pollination network, 18% of all plant–pollinator species pairs overlapped by only one day. This may be far too insufficient to make linkage likely and thus observed (i.e. it could be difficult in such instances to separate missing from truly forbidden links). In fact, an analysis of the Zackenberg data showed that extensive overlaps were needed to attain a high link probability (Olesen *et al.*, in press). Many links might in fact be weakly constrained, either because of limited opportunities for expression, or because of a wide trait variation of interacting species, allowing only parts of the interacting populations to interact. Examples of these situations can be found in the limited interaction between the most long-tongued individuals of a hawkmoth and orchid individuals with the shortest floral spur (Nilsson, 1988), or optimal foraging by predators that only switch their feeding to an alternative (lower reward) prey species when the density of their preferred prey is low (Baalen *et al.*, 2001). Other constraints were also in operation in the three mutualistic networks, and some links were forbidden because of a size mismatch between, for example a frugivorous bird beak (gape width) and the diameter of a fruit. Other links were forbidden because large frugivores like pigeons and crows had difficulty accessing fruits or because nectarivores like butterflies paid no attention to nectarless 'pollen flowers' (a reward constraint, Olesen *et al.*, 2008). As many as 14–15% of all unobserved links in the seed-dispersal networks were due to these size and accessibility constraints (Olesen *et al.*, in press). Most aquatic food webs are also highly size-structured, typically preventing predators from feeding on species larger (or more than a couple of orders of magnitude smaller) than themselves (Brose *et al.*, 2006; Jennings *et al.*, 2001; Jonsson *et al.*, 2005).

Size-related constraints have been particularly well studied in the context of ecological networks, especially in recent years. Models have been developed and improved over the past three decades, which predict complex food-web structures and properties with increasing accuracy based on simple rules that constrain the interactions between species (e.g. Beckerman *et al.*, 2006; Petchey *et al.*, 2008; Warren, 1996). Many of these models employ simple parameters, such as the number of species, S, and connectance, C. In the early cascade model (Cohen *et al.*, 1990), species are ranked from 1 to S and a consumer only feeds on a resource of lower rank with probability $P=2CS/(S-1)$. This model omits trophic cycles, underestimates interspecific trophic similarity and overestimates food chain length, however. The more recent niche-based models (Warren, 1996; Williams and Martinez, 2000) increased ecological realism by adding contiguity in diets, with consumers feeding on all resources that fall within a particular niche range. The subsequent nested-hierarchy model (Cattin *et al.*, 2004) added a diet discontinuity property, which allowed for the more plausible possibility of gaps in this niche diet range. Interval webs produced by these models are characterized by feeding relationships that correspond to body size hierarchy, with consumers typically feeding on smaller resources. Recently, Beckerman *et al.* (2006) and Petchey *et al.* (2008) have developed the allometric diet breadth model (ADBM), based on foraging theory and an ordering of species according to both body size and diet breadth (derived from energy gained, encounter rate and handling time). This model allows for gaps in the niche space and consumers to feed on a limited amount of resources above them in the ordered set. By including such simple constraints for forbidden links, the ADBM can predict up to 65% of the links in real food webs (Petchey *et al.*, 2008).

Researchers focusing solely on observed interactions may miss important information in their understanding of how species coevolve within complex networks (Thompson, 2005) because the information content in the 'unobserved' is often ignored. Thus, little may make real sense in network analysis except in the light of forbidden links: time, space and body size clearly exert powerful constraints on linkages, and hence on the topological properties and dynamics of ecological networks. In this section, we have discussed the importance of time and size constraints; in the next section, we focus upon spatial constraints, physically manifested as habitat borders.

V. HABITAT BORDERS

A landscape is a patchwork of various kinds of habitat, some of which are clearly distinct, for example a forest and its surrounding arable land, whilst others are less so, for example various kinds of fen. When an individual of a species moves among habitats, it crosses borders or steep transitions in biotic and abiotic conditions, and some borders are more closed or difficult to traverse than others. In addition, borders that are obvious to researchers, for example a forest edge or a coastline, may be porous or non-existent to certain animals, and vice versa. Habitat borders, which despite being ubiquitous natural phenomena, have received surprisingly little research attention in the ecological network literature. For example what is the relationship between penetrability of a habitat border and various traits of a species?

Most network studies are made within a given study plot that has been selected because of its assumed homogenous character. Thus, networks are tacitly treated as closed systems, that is links pointing out from the network are ignored and habitat borders are deliberately excluded in the design of most studies. This inevitably skews our perspective and reaffirms the perception that ecosystems are discrete, closed entities rather than the far more open systems that exist in reality (Polis and Hurd, 1996; Polis *et al.*, 1997; Ward *et al.*, 1998).

Modularity of an ecological network is a signature of a heterogeneous link distribution (Olesen *et al.*, 2007). This property has several causes, of which habitat borders are one, that is we expect an ecological network to increase in modularity if it encompasses habitat borders (Pimm, 2002; Pimm and Lawton, 1980). In the classic study of the arctic food web from Bear Island (Summerhayes and Elton, 1923), for example the web was described as modular because it included three habitats (land, sea and freshwater) (Pimm, 2002). However, sea birds and some fish are transients, blurring borders between habitats and, therefore, modules. In another classic example, every year migratory Pacific salmon species transport significant amounts of marine-derived nutrients into freshwater-food webs during the spawning run, thereby fuelling much of the recipient systems' productivity (Wipfli and Caouette, 1998). Yamazaki and Kato (2003) and a few others have sampled pollination links across habitats, and their networks did indeed have high modularity (Olesen *et al.*, 2007). One would expect plot size to correlate positively with the number of habitats, and indeed after extracting information about plot size and researcher-perceived habitat types from 43 pollination networks, it was apparent that the level of modularity in this study was in fact dependent of plot size (data from Olesen *et al.*, 2007).

A habitat generalist is a species that ignores most borders, which raises the question: does habitat generalization allow specialization along other resource axes without jeopardizing survival, for example with respect to pollination/floral-reward foraging? For example euglossine bees traversing many kilometres to find their unique perfume-producing orchids (Janzen, 1971), or the extremely long-beaked hummingbird *Ensifera ensifera*, which forages over huge mountain stretches to locate very long-tubed nectar flowers (Lindberg and Olesen, 2001). An interesting aspect about generalization and specialization, not yet clarified, is whether levels of generalization to different resources are correlated, that is, is a cost or a benefit associated with a certain generalization level on a given resource axis constraining or facilitating generalization levels in response to other resources? Few empirical studies have addressed this question, and one likely reason for this may be that ecologists themselves only specialize along one research axis. Another insufficiently analyed aspect is to what extent generalization level is a fixed or plastic attribute across space (Fox and Morrow, 1981; Thompson, 2005). In pollination ecology, many studies have demonstrated how the pollinator fauna of a plant species varies geographically (e.g. Armbruster, 1993; Dupont and Skov, 2004; Gómez and Zamora, 1999; Jordano, 1987; Olesen, 1997; Olesen and Jordano, 2002; Thøstesen and Olesen, 1996; Traveset and Sáez, 1997), but even at a local scale the pollinator fauna of a plant species may vary considerably (e.g. Aizen and Feinsinger, 1993; Jennersten, 1988). For example small-scale studies comparing species behaviour in adjacent habitats, varying in disturbance level and successional state (e.g. Feinsinger *et al.*, 1987), and in a series of pollination networks, species were more generalized in early- than in late-successional habitats (Parrish and Bazzaz, 1979). However, many of the early-successional pollinator species were nomadic; that is they were present in several habitats.

The importance of habitat borders to the spatial dynamics of pollination networks and generalization level along different resource axes was analysed by Montero (2005). Her study focused upon the use by pollinator species of two resources: habitat (i.e. physical space) and food (i.e. number of flowering plant species). Habitat width of a species was defined as the number of habitats, H, used by the species (≤ 3 kinds), and pollination-generalization level of a species was the number of flowering plant species visited by the pollinator species within the study site and period, that is L. Montero (2005) studied three adjacent habitats (an interior forest site, a coastal swamp site and a forest light gap in Denmark). The gap was more disturbed than the interior and the coast, whereas the coast and the gap were more open than the interior, and when pooled, the three networks had 105 pollinator species. L and H of pollinators were positively correlated among coast and gap pollinators, but L and H were uncorrelated among interior pollinators. L of species present in both coast and gap was correlated, but not for species present in both interior and gap, and for species in both interior and coast. L of a pollinator thus varied across habitats, that is a pollinator could be a generalist in one habitat and a specialist in an adjacent habitat. This might simply be explained by a difference in abundance: if one of the habitats of a pollinator is less suitable its abundance and, by extension, L may be lower, suggesting that linkage level is both a species property and a local phenomenon (Fox and Morrow, 1981).

In general, both pollination and food-web generalists are expected to dominate in disturbed habitats, whereas specialists are expected to prefer more stable habitats (e.g. Parrish and Bazzaz, 1979; Pickett and Bazzaz, 1978). Vázquez and Simberloff (2002) termed this 'the specialization-disturbance hypothesis', although they did not find supporting evidence in their own study system. In the forest-gap habitat, Montero (2005), however, found a strong positive correlation between pollination and habitat generalization level, lending support to this hypothesis. Species crossing habitats were also generalized in their flower choice, and she hypothesized that gap pollinators were recruited as depicted in Figure 15, so that in the gap, half of the pollinator species are gap specialists, migrating between forest gaps (i.e. they do recognize the existence of the forest edge as a habitat border), whereas the others appear to be colonists, moving in from adjacent, less disturbed habitats (i.e. they do not perceive the forest edge as a border; Parrish and Bazzaz, 1979).

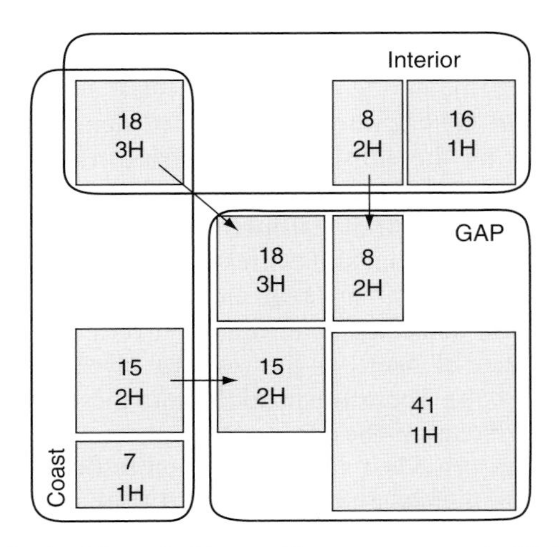

Figure 15 Hypothesized inter-habitat pollinator movements. Eighteen 3H (3 habitats)-species living in both the interior of the forest and on the coast expanded their habitat to include the forest gap. In addition, eight 2H-interior forest species and 15 2H-coast species included the gap in their habitat. Figures are number of pollinator species with a given habitat width, and 1H, 2H and 3H are habitat width of pollinator species.

Porous habitat boundaries are also evident in many food webs. For instance, in the open ocean, standing freshwaters and the lower reaches of rivers, there are often characteristically distinct benthic and pelagic assemblages, which effectively represent spatial food-web modules, but there are also some species and links that connect these two habitats (Krause *et al.*, 2003). This can occur via both bottom-up and top-down effects, with the former being particularly important in benthic systems, which receive a rain of detritus in the form of dead plant material, animal carcases and faecal pellets from the upper surface waters. In addition to these within-ecosystems, across-habitat linkages, there are also a range of across-ecosystem links that can cross borders and connect food webs in seemingly very different environments (Hynes, 1975; Knight *et al.*, 2005; Petersen *et al.*, 1999; Woodward and Hildrew, 2002a,b). The upper reaches of many rivers are fuelled to a large extent by energy inputs derived from riparian leaf litter (Dobson and Hildrew, 1992; Hall *et al.*, 2000; Wallace *et al.*, 1997, 1999), and terrestrial invertebrates that fall into the stream can be sufficiently abundant to support consumer populations at densities often far in excess of what can be sustained solely by in-stream production (e.g. Allen, 1951; Cloe and Garman, 1996; Kawaguchi and Nakano, 2001; Mason and MacDonald, 1982; Nakano *et al.*, 1999). The carcasses of Pacific salmon species, which typically experience 100% mortality after spawning, stimulate both in-stream production (Wipfli and Caouette, 1998) and the growth of terrestrial vegetation, via fertilization from bear faeces (Hilderbrand *et al.*, 1999). Such external subsidies can potentially alter the structure and dynamics of many food webs profoundly (Huxel and McCann, 1998), as demonstrated by the large-scale experimental manipulations of Nakano *et al.* (1999), who revealed the existence of 'apparent trophic cascades' mediated by the subsidy of terrestrial invertebrates to the fish assemblage, and also of Hall *et al.* (2000), who found similarly strong food-web effects of altering terrestrial leaf-litter inputs within a different system. The recent invasion of the Broadstone Stream food web by brown trout (Layer *et al.*, 2010b; Woodward, 2009) has been facilitated by this new apex predator being able to feed on subsidies of terrestrial invertebrates, which are also prominent in the diet of some of the resident invertebrate predators (cf. Woodward *et al.*, 2005b). There are, in addition to these terrestrial-to-aquatic fluxes, considerable fluxes in the other direction, as aquatic insects have an aerial phase that can provide a substantial food resource for many species, including spiders, birds, bats and even humans (Fukui *et al.*, 2006). Consequently, it is important to consider the degree of reciprocity in food-web links, and the magnitude and timing of their manifestation: for instance, leaf fall in temperate systems occurs in a pulse in autumn, whereas insect emergence from freshwaters occurs primarily in the summer. Again, this highlights the need to consider both spatial and temporal aspects that determine the presence and strength of coupling among species within and across networks.

In conclusion, habitat borders differ in their penetrability or porosity to species, nutrients and energy and species in adjacent habitats vary in the extent to which they can and actually do cross borders, that is habitat borders may be perceived asymmetrically by species in neighbouring habitats. However, the complexity of across-habitat or ecosystem links underlines the importance of moving up in spatial scale in ecological studies.

VI. MODULES AND HIERARCHIES

Ecological networks are rich in structural heterogeneity (e.g. Allesina and Pascual, 2009; Allesina *et al.*, 2005; Bascompte *et al.*, 2003; Joppa *et al.*, 2009; Lewinsohn *et al.*, 2006; Melián and Bascompte, 2002, 2004; Olesen *et al.*, 2006, 2007; Olesen *et al.*, submitted for publication; Rezende *et al.*, 2009). They contain motifs, clusters and modules organized in an overall hierarchical structure (Milo *et al.*, 2002). At one extreme, it could even be argued that networks do not exist, and that the biodiversity of species and links is organized into a global modular hierarchy, with networks as artificial descriptions delineated arbitrarily by scientists.

Analysis of patterns has a long tradition in ecology, and biogeographers, community ecologists and networkers have, in particular, explored two kinds of pattern: nestedness and, to a lesser degree, modularity. In general, each pattern has been investigated independently (but see Fortuna *et al.*, 2010), although proof of one pattern does not preclude the presence of others. A network represented as an interaction matrix may have a nested, a gradient or a modular structure (Lewinsohn *et al.*, 2006). Detailed analyses have shown nested patterns to be common in networks of plants and pollinators or seed dispersers, and also in food webs (Bascompte *et al.*, 2003). Ecological networks are often also modular, for example pollination networks (Dupont and Olesen, 2009; Fortuna *et al.*, 2010; Genini *et al.*, 2010; Olesen *et al.*, 2007) and highly resolved plant–herbivore networks (Prado and Lewinsohn, 2004), and this property can also be found in other kinds of network, for example food webs (Krause *et al.*, 2003). Many networks clearly fit one pattern, whereas others fit more than one, and in some, parts of a network may show a different pattern to others (i.e. the network itself is patchy and compartmentalized into different subunits or types of structure, which are nested). In relation to this latter point, according to studies of biochemical and man-made networks, the relevant study unit should be the module, not the network. These networks have a modular structure and individual modules are interpreted as functional units in biochemical networks and geo-political units in the man-made ones, functional analyses should therefore take place at the modular level (Dalsgaard *et al.*, 2008; Olesen *et al.*, 2007). Olesen *et al.* (2007) analysed 51 pollination networks including almost 10,000 species and

20,000 links, and tested for modularity using an algorithm based on simulated annealing (Guimerà and Amaral, 2005). All networks with more than 150 species were modular, whereas networks with less than 50 species were never modular. Both module number and size increased with species number. Some individual modules represented one or a few convergent trait sets. In addition, species played different roles with respect to modularity, and only 15% were structurally important to their network, as either core species or hubs (i.e. highly linked species within their own module), connectors linking different modules, or both. If these key members go extinct, modules and networks may fragment and this can initiate cascades of secondary extinctions (Montoya *et al.*, 2006), suggesting that hubs and connectors should have high conservation value. Modules may be important units in the maintenance of biodiversity because modularity often reflects habitat heterogeneity, divergent selection regimes and phylogenetic clustering.

Pollination networks are small worlds (Jordano *et al.*, 2003; Olesen *et al.*, 2006), but it has been argued that traditional small-world descriptors, such as linkage level, path length and clustering coefficients (see Nooy *et al.*, 2005), are less informative about the overall structure in real networks, especially because these turn out to be modular or even hierarchical (Costa and Sporns, 2005; Guimerà and Amaral, 2005; Guimerà *et al.*, 2007).

Hierarchy theory encompasses the study of macro-patterns of hierarchies, for example by comparing structure and dynamics of different organizational levels (Apollonio, 2002; Waltho and Kolasa, 1994). It makes predictions about phenomena and processes on one hierarchical level inferred from information about adjacent levels, for example given our knowledge about networks of species and their interactions, how would we predict the characteristics of networks at the lower level of interacting individuals? No ready-made hypotheses are available, but from comparative network research we may, as a starting point, predict that networks at adjacent hierarchical levels should behave similarly (Albert and Barabási, 2002). These authors compared characteristics of several real networks, some of which belonged to different hierarchical levels. For instance, a species (e.g. in a food web), a word (e.g. in a synonym network, but not in a specific text) or a chemical compound (e.g. in a metabolic network) are nodes of higher level networks (equivalent to species–species networks), because we can scale down to networks of individuals in nature, single words in a specific text, or single molecules in a specific solution, respectively. For example the word 'network' is a label for the concept of network, which encompasses all specific networks and it has an abundance, frequency or occurrence of four in the previous sentence. Thus, the concept of 'network' is equivalent to 'species' and the four occurrences of the word network in the previous sentence are equivalent to four individuals of this species. In contrast, a homepage (in WWW), a cited scientist (in a citation network) or an airport (in an air-transportation network) is a node of lower level networks, at the level of individuals in an ecological network. If we sort the list of networks of Albert and Barabási (2002) into higher and lower level networks and compare their characteristics, we do not find any significant difference in their most fundamental characteristics. However, an individual-based network studied by Dupont *et al.* (in press), differed in some of its properties from species networks, for example linkage level of individuals was, on average, higher, connectance was higher, and nestedness was lower.

Most complex networks, such as the WWW, and biochemical, gene and cellular networks, are hierarchical and sometimes even self-similar or fractal (Guimerà *et al.*, 2007; Isalan *et al.*, 2008; Sales-Pardo *et al.*, 2007; Song *et al.*, 2005, 2006). A self-similar topology of an observed structure means that it appears the same across a range of scale levels. Thus, a self-similar, modular network is a hierarchy of modular units that look alike, that is modules within modules within modules, etc. and these patterns have also been noted in ecology (Scanlon *et al.*, 2007; Solé and Manrubia, 1995; Storch *et al.*, 2007; Sugihara and May, 1990). Species within groups are expected to have strong ecological-evolutionary impact upon each other, because they are closely linked and thus exert a strong selection pressure upon each other.

In 1974, the French Nobel Prize winner François Jacob wrote: 'Every object that biology studies is a system of systems', that is nature is itself a hierarchical arrangement of complexity. Social, technological and biochemical networks are hierarchically organized into groups or modules of varying size. Olesen *et al.* (submitted for publication) examined if ecological networks have a similar structural organization and discussed the consequences for species and their links. They analysed a pollination network from a phrygana shrubland at Mount Aegaleo in Greece (Petanidou, 1991; Petanidou *et al.*, 2008). Based on 4 years of cumulated observations, a pollination network of 133 plant species, 666 flower–visitor species and 2942 species links was constructed. The network was hierarchically modular with five modules, all of which had a deeper, internal structure, each consisting of 6–7 submodules and most links were within modules or submodules. This was reflected in an increasing connectance *C* with decreasing scale level, that is from network down to module and further down to submodule. For each decreasing scale level, *C* increased 3.5-fold. Forty-three percent of all possible links in a submodule were realized, far higher than in most whole networks. Thus, submodules were indeed highly linked subgroups in the network. In addition, the study revealed a weakly self-similar pattern. Natural phenomena are rarely perfectly self-similar, but their properties seem sufficiently similar over at least a range of scales to make it appropriate to use fractal geometry as an analytical tool (Kallimanis *et al.*, 2002). Since species and links are the most familiar components of biodiversity, demonstrating their hierarchical structuring in mutualistic networks could contribute towards a better understanding of how biodiversity in general is structured. In addition, this hierarchical, modular structure may increase robustness of the network, because disturbances may stay entrenched within submodules or modules and within scale levels (e.g. Maslow and Sneppen, 2002). According to Song *et al.* (2006), a robust modular network may even require a self-similar topology as a prerequisite for stability.

In a sample of 29 modular pollination networks, 254 modules were identified (Olesen *et al.*, 2007), which varied considerably in size, although most with more than 70 species were further submodular. Thus, the finding of hierarchical organization and a weak self-similarity in the large Greek pollination network might be generally applicable to other large pollination networks (Petanidou *et al.*, 2008). Heuristically, a hierarchical organization tells us that it is important when attempting to characterize network structure that we specify an operative scale level and move between scale levels to locate the most informative one (Levin, 1992), although this is rarely done in practice (however, see Kunin, 1998).

Song *et al.* (2006) showed that an important generator of network fractality is multi-scale hub repulsion, that is networks grow through linkage of hubs to peripheral species rather than to other hubs (disassortative node attraction). A hub links to a swarm of peripheral species, which often only have one to two links, a phenomenon also described by the negative connectivity correlation, that is the negative relationship between linkage level of a node and mean linkage level of its nearest neighbours (Melián and Bascompte, 2002). New species colonizing a mutualistic network do indeed most often attach preferentially to hubs (Olesen *et al.*, 2008). However, the link-dense core observed in the nested matrix version of most mutualistic networks also shows some hub attraction, viz. the core of links between hubs (assortative node attraction). Maybe this explains the weak self-similarity observed, because these links are links connecting different modules, and they are therefore also blurring the modularity and hierarchical pattern observed.

A fractal pattern may have another, underlying fractal pattern as its ultimate driver. That is one route towards an understanding of the processes behind the self-similar pattern in mutualistic networks may let us 'step down' to species abundance patterns, sometimes also described as 'fractal' (Mouillot *et al.*, 2000), and one step further down to habitat and physical fractality (e.g. Kallimanis *et al.*, 2002). Thus, a hierarchical distribution of habitat patches may be the ultimate driver of network self-similarity (Storch *et al.*, 2007). This 'scaling of biodiversity' is, of course, very exciting, and opens up many puzzling questions: for example when are ecological networks (like physical fractal systems) near a phase transition between chaos and order (Strogatz, 2005)? What are the dynamic mechanisms behind the appearance of new modules and extinction of old ones? Do new species accumulate or evolve by linking to small modules of species with similar traits or do they evolve from these species, and in this way 'deepening' the fractal structure? To what extent does this happen randomly or through organizational constraints (Olesen *et al.*, 2008)? What importance do connectors and network hubs play with respect to self-similarity? These latter species have several links outside their own module, which show that they simultaneously may link different scale levels together—they may even be an upper class of nodes, belonging to a scale level higher than the rest of their network. Are they also involved in ecological feedback control against perturbations, as is known from biochemical networks and involved in a kind of evolutionary feedback system, as we see in coevolution? These questions remain to be addressed, but they are of potentially huge importance to not just theoretical ecology, but also to applied ecology (e.g. pest management) and conservation biology (e.g. biodiversity management).

VII. INDIVIDUAL-BASED NETWORKS

Networks are organized hierarchically because species are assemblages of individuals and they are themselves parts of increasingly larger assemblages, guilds and trophic levels (e.g. Ravasz and Barabási, 2003). Ecological network analysis most often takes place at species level, that is nodes are species, and links may be trophic (feeding), non-trophic (e.g. competition) or mutualistic (e.g. pollination) interactions between species populations. However, such species–species or *S–S* networks may be scaled down to their constituent individuals, which is the level of organization at which interactions actually occur (Ings *et al.*, 2009; Yvon-Durocher *et al.*, 2010). Thus, every interacting pair of species in a network enfolds a new network at a lower hierarchical level of interacting individuals (Woodward and Warren, 2007).

In ecology, individual–individual or *I–I* networks are known from studies of social interactions among larger animals, for example dolphins (Krause *et al.*, 2007; Lusseau *et al.*, 2008), pollen/gene flow in plant populations (Fortuna *et al.*, 2008), animal-disease dynamics (Perkins *et al.*, 2009), interactions between bats and their roosting trees (Fortuna *et al.*, 2009) and, finally, foraging-niche relationships (Araújo *et al.*, 2008). *I–I* networks are less well developed in the studies of food webs, but recent advances have been made in using individual size-based approaches, rather than species-level characterizations, to determine ecosystem-level properties (Ings *et al.*, 2009; Jennings *et al.*, 2001; Petchey *et al.*, 2008; Woodward and Warren, 2007) and responses to the effects of fishing (Shin *et al.*, 2005).

Most pollination networks studied to date are *S–S* networks, with the exception of Dupont *et al.* (in press), in which a single species pair was extracted from an *S–S* pollination network, and its *I–I* network structure was analysed further. It consisted of 32 individual thistle 'plants' or flowering stems (*Cirsium arvense* (L.) Scop.) and 35 marked honeybees (*Apis mellifera* L.). *C. arvense* attracts many species of flower visitors, but on the days of the experiment, the honeybee was, by far, the most frequent visitor. The study did not discriminate between nectar and pollen foragers and further research may show that the network is atypical because of the social structure of the honeybee. This network had 317 interactions documented by 681 visits. The 2-mode network of individual plants and bees had both a high connectance ($C = 28\%$) and nestedness ($NODF = 0.43$; '**n**estedness metric based on **o**verlap and **d**ecreasing **f**ill') (Bascompte *et al.*, 2003; Dupont *et al.*, 2003; Guimarães Jr. and Guimarães, 2006), but it did not contain modules, probably because of its small size (Dupont and Olesen, 2009; Guimerà and Amaral, 2005; Olesen *et al.*, 2007). Thus, the honeybee–thistle *I–I* network was more densely linked than *S–S* pollination networks in general (60 *S–S* networks, $C = 11 \pm 8\%$, data from Olesen *et al.*, 2007). The two link patterns of nestedness and modularity behaved as in similar-sized *S–S* networks (Bascompte *et al.*, 2003; Olesen *et al.*, 2007). Frequency distributions of *L* fitted best to a truncated power law (Jordano *et al.*, 2003), and interactions were asymmetrical in their frequency (Bascompte *et al.*, 2006). Thus, the shape of the *L* frequency distribution was similar to many *S–S*-pollination networks (Jordano *et al.*, 2003). The constraints behind this need further study, but there is some evidence that forbidden links (Jordano *et al.*, 2003), such as ageing and cost-related constraints (Amaral *et al.*, 2000) might drive truncated linkaging, although there is still much debate surrounding these mechanisms (see Vázquez, 2005).

Dupont *et al.* (in press) constructed a 1-mode network of plants and another one for bees. Both had extremely short average path length between individuals (as low as $\langle l \rangle = 1.03$ for plants and 1.30 for bees) and high average clustering coefficient ($\langle c \rangle = 0.97$ for plants and 0.85 for bees, for definitions of l and c, see Nooy *et al.*, 2005). Some of the parameters (average linkage level, average clustering coefficient) of the thistle–honeybee *I–I* network had values within the range of similar-sized *S–S* pollination networks, others were smaller (average path length) and others again larger (connectance) (Olesen *et al.*, 2006). Thus, *I–I* networks do not seem to behave completely like species networks, but they may be denser and more connected. Pollination *S–S* networks have been shown to be among the smallest networks in the world, but these *I–I* networks may be even smaller (Olesen *et al.*, 2006), suggesting that network patterns might become more pronounced at lower hierarchical levels.

Characteristics of both plants and animals, for example floral display and foraging behaviour (workers, scouts), may be drivers of this described link pattern in *I–I* networks, which again may shape the hierarchical structure and dynamics of *S–S* networks. Larger floral display attracted more bee visits (e.g. Eckhart, 1991). Consequently, the observed link asymmetry may be driven by a skewed floral display-size distribution comprising a few old plants with many flower heads and many young plants with a few flower heads, such that, basic life-history attributes of species can affect the entire topology of the network by cascading upwards to affect higher hierarchical levels (Bianconi *et al.*, 2009). Here, we see a fundamental difference between pollination networks and food webs. In the former, individuals can interact repeatedly, whereas in the latter, the consumer typically destroys the individual resource, at least among predators and prey, although this is not necessarily the case for small herbivores on large plants, or for host–parasite systems (e.g. Lafferty *et al.*, 2008).

I–I pollination networks are detailed road maps of pollen flow in plant populations (Fortuna *et al.*, 2008). The honeybee is probably the world's most generalized flower-visitor species. However, such generalist species may consist of individuals that vary in their level of specialization (Araújo *et al.*, 2008). The ways in which individuals within a species population partition resources may vary markedly. In an opossum species, the diet width of individuals was nested, with specialists consuming food items that were a subset of the diet of more omnivorous individuals (Araújo *et al.*, 2010). The study was about individual opossums foraging on different prey taxa, that is it was at the level of individual-species. However, linkage patterns can be analysed at other levels as well, including species–species, species–individual, and as, in the study of Dupont *et al.* (in press), individual–individual. These hierarchical link patterns need much more study, as variation in pollination and predation specialization at the individual level is far less well known than at the species level. Trap-lining animals follow specific foraging paths, for example hummingbirds and tropical bees (Janzen, 1971) and bumblebees, too, may learn the location of individual plants and follow specific search paths (Manning, 1956). Honeybees, although one of the most generalist pollinators at the species level, may also be patch specific, and have been observed to re-visit nectar-rich flowers on several foraging trips (e.g. McGregor, 1959).

The description by Dupont *et al.* (in press) of an *I–I* pollination network is important for our understanding of *S–S* pollination networks and of ecological networks in general. For example in an Arctic network (Olesen *et al.*, 2008), 23% of all flower–visitor species were only observed once, that is only one individual was observed to visit one plant individual. Thus, a quarter of this *S–S* pollination network is actually based on insect individual–plant individual data. We do not know if this figure is representative for ecological networks in general. Perhaps it is more likely for rare, top predators, and particularly for links that are noted through observational data, for example Wootton's work (1997) on the Tatoosh Island (Washington) food web, the bird links in the Ythan Estuary food web (e.g. Raffaelli and Hall, 1992) or the killer whales in the Californian kelp network (Estes *et al.*, 1998). Interestingly, due to body mass-abundance scaling in food webs, one might expect a skewed distribution in the description of individual versus species-averaged links, such that small, abundant prey species at low trophic levels will have feeding links averaged over many individuals, whereas diets of large, rare predators at high trophic levels may be described to the individual level. This may also contribute to biases in food-web data and potential truncated degree distributions.

We need much more research at this fascinating interface between foraging theory of individuals and ecological network theory (Ings *et al.*, 2009). Particularly lacking are detailed *I–I* descriptions of food webs (but see Woodward and Warren, 2007). Melian *et al.* (in review) have recently analysed a food web with approximately 25,000 individual diets and independent estimations of abundance in several spatio-temporal situations. As in the previous study by Dupont *et al.* (in press), the *I–I* food web does not seem to behave like the species-level network with a highly sensitive species-level connectance to intraspecific variation and sampling effort. Foraging theory suggests that food-web complexity and a high proportion of links can be predicted using individual body sizes (Beckerman *et al.*, 2006; Petchey *et al.*, 2008), but this needs to be tested with empirical *I–I* networks. For example in the Melián *et al.* study, only 35% of the variance in the number of prey per individual predator was accounted by the length of the individual predators. This value was independent of the position and time the sampling was collected. Future research that focuses on collecting individual rather than species-averaged food web should give us far more detailed and ecologically relevant information about interactions, food-web structure and dynamics than is currently the case.

VIII. INVASIONS

To study invasion biology from the perspective of network analysis is to look deep into the spatio-temporal dynamics of nature and, in particular, the range-margin dynamics of species. As species expand their ranges, they invade new habitats and ecological networks. However, the ease with which invaders establish themselves and their ability to persist vary markedly among species, ecosystems, and network types. This raises some key questions of relevance to the study of networks, many of which also have particular resonance within conservation biology. For instance: why do networks differ in their invasibility or resistance to invasion; how do aliens become attached to local networks; and how does network structure change after an invasion? Biological invasions have been intensively studied for decades, but answers to these network-related questions await a shift in the research field that will involve a greater consideration of not just species but of their interactions within complex multispecies systems (Aizen *et al.*, 2008; Carvalheiro *et al.*, 2008;

Lopezaraiza-Mikel *et al.*, 2007; Memmott and Waser, 2002; Olesen *et al.*, 2002). Mutualistic interactions promote—whereas antagonistic interactions often appear to resist—the integration of invasive species into native networks (Elton, 1958; Morales and Aizen, 2006; Simberloff and Holle, 1999). These factors in a network are termed 'biotic facilitation' and 'resistance', respectively. In contrast to biotic resistance, most studies of invasions have ignored facilitative interactions (but see Richardson *et al.*, 2000; Romanuk *et al.*, 2009; Totland *et al.*, 2006; Traveset and Richardson, 2006). Once integrated, the alien may change the structure of the native network, and sometimes it does so in a very dramatic way (Traveset and Richardson, 2006). Unfortunately, we rarely have the opportunity to follow an invasion in detail and census its impact on the native network continuously, particularly because network-level data prior to an invasion are often not available (but see Woodward and Hildrew, 2001).

Native species are old invaders or their descendants, whereas introduced species are young invaders. For many islands, the proportions of endemic, non-endemic native and introduced species are known and an important question is whether an alien forms interactions with native or introduced species, as this could be critical for the recipient biota. Interactions may therefore be categorized into four broad types: endemic interactions (i.e. interactions between endemic species), non-endemic native interactions (i.e. interactions between native species), introduced interactions (i.e. between introduced species) and mixed interactions (e.g. between introduced plants and endemic animals). If introduced species, for example, are more generalized than native ones, that is if they establish or have the capability to establish many interactions with other species, they will more easily interact with newly introduced species and thus facilitate the establishment of the latter. They may then build up large groups of introduced species or 'invader complexes' (D'Antonio and Dudley, 1993) with potentially increasingly detrimental consequences to the native island biota. Invasive plants, for example with generalized pollination biology are expected to achieve more initial reproductive success and therefore a higher establishment probability than specialized pollinated plants (Johnson and Steiner, 2000; Richardson *et al.*, 2000). Simberloff and Holle (1999) suggested the term 'invasional meltdown' to describe the process by which invasive species facilitate one another, and, for example the Hawaiian Islands provide a salutatory example of this, because a large proportion of their ecological networks are now dominated by invasives, which have increased exponentially over time (Gamarra *et al.*, 2005). In the initial establishment phase, alien species involved in highly co-evolved and specialized mutualisms are expected to be at a disadvantage, but if both a plant and a pollinator become established, this may turn into an advantage (Hanley and Goulson, 2003; Richardson *et al.*, 2000): for instance, in Florida, introduced fig tree species started invading when their obligate pollinator wasps were also introduced (Simberloff and Holle, 1999). There is also plenty of evidence from the food-web literature to support the suggestion that trophic generalization facilitates the establishment of exotic invasives: within Europe species such as the topmouth gudgeon fish (*Pseudorasbora parva*), the American signal crayfish (*Pacifastacus leniusculus*), the Chinese mitten crab (*Eriocheir sinensis*), and goldfish (*Carassius auratus*) are classic examples of invasive dietary generalists in freshwaters (Bubb *et al.*, 2004; Herborg *et al.*, 2004; Pinder *et al.*, 2005). There is also evidence that native invaders (i.e. non-exotic species previously not present in a local network) have similar attributes, as suggested by the successive invasions of Broadstone Stream over four decades by progressively larger and more generalist predators at the top of the food web (Woodward, 2009; Woodward and Hildrew, 2001). These native species, however, appear to have less dramatic impacts on their recipient networks than is the case for exotics: for instance, the invasion of the dragonfly *Cordulegaster boltonii* at the top of the Broadstone Stream food web led to marked increases in network complexity, chain length, nestedness, intervality and the number of rigid circuits (Woodward and Hildrew, 2001; Woodward *et al.*, 2005b), but it did not induce any of the strong cascading effects or secondary extinctions that have often accompanied invasions of exotic predators elsewhere (e.g. Flecker and Townsend, 1994).

Olesen *et al.* (2002) explored this in pollination networks on two oceanic islands, the Azorean Flores and the Mauritian Ile aux Aigrettes. At each island site, pollinators and plants were categorized into endemic, non-endemic native and introduced species and all observed permutations of interactions between these groups were mapped. Specifically, they looked for the presence of ‘invader complexes’ of mutualists. On Flores, all pollinator species were insects, whereas the Mauritian site had a pollinating day gecko, *Phelsuma ornata* Gray, as well as a community of insects. Linkage level was higher for endemic species than for non-endemic native and introduced species, whereas for the two latter categories it did not differ. On both islands, observed frequencies of interactions between native (endemic and non-endemic) and introduced pollinators and plants differed from a random association, with introduced pollinators and plants interacting less than would be expected by chance. Thus, the data did not lend support to the existence of invader complexes, rather it suggested that endemic super-generalist species (pollinator or plant species with a very wide pollination niche) included new invaders into their set of food plants or pollinators, thereby facilitating their successful establishment. Reviewing the wider literature, super generalists seem to be a widespread island phenomenon (Olesen *et al.*, 2002), that is island–pollination networks include one or a few species with a very high generalization level compared to co-occurring species. Low density of species on islands may reduce interspecific competition, leading to high abundance (density compensation), wide niches (ecological release) and super generalization (linkage release).

An alternative to the earlier approach is to compare invaded and non-invaded sites, via space-for-time substitution studies. Padrón *et al.* (2009) compared sites invaded by alien *Opuntia* cactus (*O. dillenii* and *O. maxima*) with non-invaded sites on the Balearic and the Canary Islands. The two archipelagos differed considerably in their native network structure, with the Canarian site containing twice as many Diptera species as the Balearic site. *Opuntia* reduced the number of links between native plants and pollinators, and by forging links with a set of the most generalized natives it became integrated into its new networks. In this way, it increased the level of nestedness, although by different mechanisms in the two archipelagos. In highly nested structures, disturbances such as invasions could potentially ramify through the entire network very rapidly (Bascompte *et al.*, 2003; Montoya *et al.*, 2006).

Alien invasions have undoubtedly had more detrimental impacts on islands than anywhere else (Abe, 2006; Kenta *et al.*, 2007; Traveset and Richardson, 2006). On many islands, entire guilds of mutualists have disappeared due to invasions (e.g. Cox and Elmqvist, 2000; Mortensen *et al.*, 2008; Traveset and Richardson, 2006). These island extinctions often have their roots in invasions of the food web, where generalist predators, such as rats, pigs, dogs and brown tree snakes, have extirpated the native pollinators, seed dispersers and frugivores from the mutualistic network (Mortensen *et al.*, 2008). Because of their low species density, islands have small pollination networks and thus a higher connectance, but also endemic super generalists, both among plants and pollinators (Olesen *et al.*, 2002). Such highly connected species often act as keystones by accelerating the rate at which perturbations propagate through the network (Ings *et al.*, 2009; Montoya *et al.*, 2006). In the study of Padrón *et al.* (2009), *Opuntia* interacted with local super generalists. Aliens may even become so dominant that they themselves start to act as hubs, thereby increasing the nestedness of the network and the connectivity among local species. *Opuntia maxima* on the Balearics formed a hub in its invaded network, whereas *O. dillenii* was a peripheral species, only interacting with local super generalists (the honeybee and *Bombus canariensis*). This mechanism is well known: the most generalist native pollinators offer a route ('the nested route to invasion success') to integration for alien plants into native networks (Lopezaraiza-Mikel *et al.*, 2007; Memmott and Waser, 2002; Olesen *et al.*, 2002). When the alien has become linked to the local super generalist through preferential linkage (see Section II), its abundance increases and it begins to link to more and more native species, its abundance increases further and so does its linkage level, until finally it joins the club of super generalists (the upper left corner of a nested interaction matrix, see Figures 2 and 14). The Canarian honeybee–*Opuntia* interaction, on the other hand, could constitute an example of the early stages of a potential invasional meltdown, although it is still not known with certainty if the honeybee is an invader or a native to the Canaries (Bourgeois *et al.*, 2005; O'Dowd *et al.*, 2003; Padrón *et al.*, 2009; Simberloff and Holle, 1999). *Opuntia* flowers during summer, when the spring bloom of most native species has ceased, so only a few late-flowering native species co-flower with the alien (Stald, 2003). Through this unusual flowering ecology, the alien cactus avoids competition with most native plants for pollinators and, on the other hand, its impact on the pollination of natives will also be small. *Opuntia* may even benefit the natives by boosting populations of long-lived pollinators.

Two other studies provide instructive examples of the effects of alien species on invaded mutualistic network structure. Lopezaraiza-Mikel *et al.* (2007) reported an increase in flower–visitor species richness and abundance, and flower visitation when networks become invaded. By analysing 10 pairs of pollination networks with different densities of alien species, Aizen *et al.* (2008) found that connectivity among natives declined in highly invaded networks, because of rewiring of links from native generalists to super-generalist aliens during and after the invasion. An alien may increase nestedness unless it is a specialist linking to native specialists, or if it destroys the native linkage pattern, for instance by stealing ('monopolizing') generalist partners by attacking the core of links among native generalists and 'pulling links out of the core' (Padrón *et al.*, 2009).

Memmott and Waser (2002) studied plant invaders in a mainland pollination network at the regional scale. They found that although alien plant species were visited by fewer, but more generalized flower-visiting animal species than were native plant species, they were nevertheless integrated into the native network. In contrast, in Patagonia, alien species formed a kind of module within the native pollination network (Morales and Aizen, 2006): both these studies focused on networks of natives and aliens where the latter had most likely been present for at least a couple of decades (Memmott and Waser, 2002; Morales and Aizen, 2002).

One of the first steps required for the establishment of an invading plant is its initial reproductive success in the new habitat, so what happens in the first season after the invasion can be critically important. An exotic plant typically leaves its pollinators behind in its home habitat and is initially only faced with those in the recipient–pollination network that include the newcomer into their diet, so the change in the number and frequency of interactions caused by the new arrival may ultimately determine the outcome of the invasion. Besides information about endemicity, various pollination network properties might be expected to affect the reproductive success of the invader (Elberling and Olesen, 1999; Morales and Aizen, 2002, 2006; Olesen, 2000; Olesen *et al.*, 2002; Vázquez *et al.*, 2005). As far as we are aware, nobody has analysed the initial, first-season reproductive behaviour of invading plants and related it to characteristics of not just the invader but also of the recipient habitat and its ecological networks. If legally permissible, it is an easy task, and such an experiment was done by Olesen *et al.* (unpublished data a). Instead of a comparison of invaded and non-invaded networks as in Padrón *et al.* (2009), they experimentally introduced exotic plants into native networks in Denmark and the consequences in the first year after the introduction were scored. Three adjacent habitats were used: a deciduous forest interior, a coastal swamp, and a forest gap. Despite the obvious differences between habitats, they were less than 100 m apart. Three exotic species were 'released': *Gilia achilleaefolia*—a Californian annual, *Vaccaria hispanica*—a S European annual and *Linaria triornithophora*—an Iberian perennial, none of which had any congenerics within the three habitats. The invaders varied in their flower openness (i.e. accessibility of floral resources to insects) and their breeding system. *Gilia* and *Vaccaria* were completely self-compatible, whereas *Linaria* was an obligate out-crosser. Initial abundance of each invader species in a plot was only one flowering plant 16 m^{-2}, similar to that of the rarest native species in the habitats, and this low density was used to mimic the first year of an invasion. In each plot, all interactions between flowering plant species and pollinator species were quantified in terms of the number of visitors per plant per hour.

The three networks (excluding the invaders and their exclusive pollinator species, that is species only visiting the alien plants) had a regional pool of 105 pollinator and 39 flowering plant species. When comparing network structure with and without invaders, the influence of the invaders was minor: they caused a small increase in numbers of pollinator species, network links, and the linkage level of pollinators, and a decrease in connectance and linkage level of plants. On average, an invader species received only half as many pollinator species as a native plant species. Pollinator species visiting both invader and native plant species (1) visited twice as many species and (2) had three times as high a visitation rate as pollinators only visiting natives. In general, alien plants were visited by the most generalized local pollinators and they invaded the network along the nested route, as they invaded the tail in the nested link pattern and then moved towards the generalized core by increasing abundance and linkage level. Based on previous studies, this was expected. However, one observation was not anticipated: the invaders also recruited new ('escort') pollinator species, which were not observed to visit native plants. Thus, invaders were visited less often by fewer but more generalized species and a few novel escort pollinators. In total, 22 pollinator taxa visited the flowers of the three invaders but among these only the syrphid *Episyrphus balteatus*, the bumblebee *Bombus pascuorum*, and the small beetle *Meligethes* sp. were abundant. Species number and composition of pollinator fauna varied substantially among both invaders and networks: the structure of one of the networks is shown in Figure 16. Invaders had a

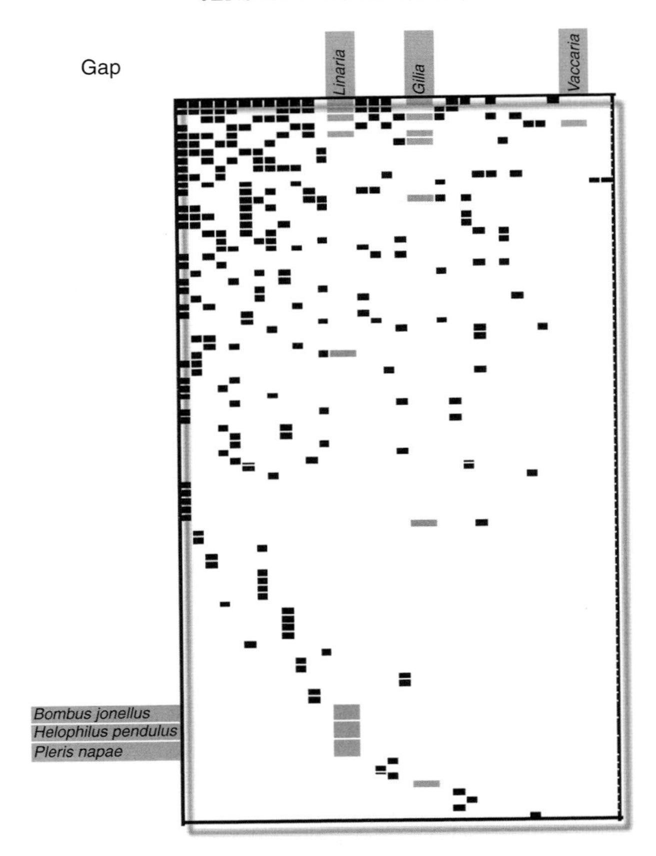

Figure 16 Link matrix of forest gap with the position of the 1-year invaders. Pollinator species are listed in rows, and plant species in columns. The species are sorted according to descending linkage level from the upper left corner, that is in a nested fashion. Names of native insect species only attracted to invader plant species are added and shaded in grey. We envisage that invaders get into the network through one of the tails. They do so mainly because of their low abundance, then they move towards the core, but the speed of progress along this 'nested route' (see text) and the final position of the invaders are not known.

relatively marginal position in the networks, but the more open the flower morphology was and the more self-compatible an invader was, the higher its reproductive success was (measured as seed set). This suggested that selfers should be able to move faster through the tail than outcrossers, but because the former also often has low nectar rewards, it might not be able to penetrate the super-generalist core. Self-incompatible species with a closed flower may be most successful in habitats in which important native pollinators are pre-adapted to the invader's flower morphology. However, although their progression through the tails via the nested route might be slow initially, they may have the potential to join the generalist core. *Gilia* is an example of an invader, which due to its breeding system may reproduce even under low pollinator activity and it may be a successful invader irrespective of insect species density. *Linaria*, on the other hand, produced many seeds in the forest gap but almost none in the other two habitats, most likely because *B. pascuorum* and *E. balteatus* were frequent gap visitors. The former is a long-tongued legitimate pollinator and the latter may either be a pollinator in flowers previously opened by the bumblebees or a pollen scavenger. Thus, the bumblebee is acting as a native 'escort' pollinator in the forest gap and facilitates the establishment of the invader. *Linaria*, which in spite of its closed flower morphology and self-incompatibility, may be a successful invader if suitable local pollinators are available. Native plant species were visited by twice as many pollinator species than was the case for the invasive plant species, in line with the findings of Memmott and Waser's (2002) earlier study. In their data set, native plant species were visited by 34 pollinator species and alien plant species by 24 species. In contrast, Morales and Aizen (2002, 2006) did not find any difference in linkage level to native and alien plants in their study from Patagonia and they also found that alien insects had the same linkage level as native insects (Morales and Aizen, 2002, 2006). Olesen *et al.* (unpublished data a) did not detect any alien insect species at their study site, and the combined results of these different studies suggest that invasion success might be strongly related to the functional attributes and provenance of the recipient–consumer assemblage.

Through comparative and experimental studies of invasives, it is possible to achieve a deeper understanding of the structure and dynamics of ecological networks. There are, of course, legal and ethical considerations to take into account when carrying out experimental manipulations of invasive species: removal experiments are easier to justify than those that involve introductions, but the latter are often the only way to gain an understanding of causal mechanisms that is not confounded by other factors. Future experiments that manipulate propagule size of invader, properties of the invader and structure of receiver networks (e.g. a network with natives that are closely related taxonomically or functionally to the invader) are needed to estimate the importance of phylogenetic relatedness *versus* ecology.

IX. SUPER NETWORKS

In this final section, we move beyond single networks and step up to a higher hierarchical level to focus upon how networks combine to form larger, multi-tiered structures, that is super networks. We also leave ecology temporarily, because, as far as we know, no ecologist has yet taken this step explicitly (however, see Bastolla *et al.*, 2009; Melián *et al.*, 2009). The presence of modules within networks, and networks within networks is a signature of a hierarchical organization. Behind this is a driving physical world shaped by a 'skewed disturbance distribution', that is rare, large disturbances and many, small ones. Palla *et al.* (2005) made a super-network analysis for three kinds of organization, viz. scientific collaboration, word-association and protein interactions. As an introductory example, they used social networks. Each of us, *i*, is member of several networks m_i (job, hobbies, family, political organization, etc.). This multi-membership ties these different social networks together in an exceedingly complex and tiered manner. The overlap size $s_{a,b}{}^{ov}$ between two networks *a* and *b* is the number of shared nodes. The overlap between two networks symbolizes a link between the two. The linkage level or degree of a network *a* becomes $d_a{}^{com}$. The last parameter they introduce is the number of nodes $s_a{}^{com}$ in a network *a*. The distributions of these four parameters m_i, $s_{a,b}{}^{ov}$, $d_a{}^{com}$, and $s_a{}^{com}$ are then used to describe the super networks (Palla *et al.*, 2005). Using this approach, it is possible to target particular parts of the large network and to assign specific functions to these. In addition, the approach makes it also possible to make predictions about the cascading outcome of removing specific nodes, for example in an ecological context, the effect of removing a pollinator species of plant species *a* on a herbivore of plant species *b*, if *a* and *b* both belong to the same super network.

We began by suggesting that ecologists have not taken the step to super networks but, in fact, many food-web ecologists have to some extent, albeit often implicitly and unconsciously. Super networks are embedded in all food webs: for instance, herbivores are members of two 2-mode networks: a plant–herbivore network and a herbivore–predator network. However, such kinds of super network become somewhat trivial when we calculate the four characterizing parameters (*m* for all herbivores are probably 2. $s_{a,b}{}^{ov}$ is probably equal to the size of the herbivore community. d^{com} for the two networks is 1, and finally, s^{com} is equal to the sum of plant and herbivore species, and the sum of herbivore and predator species for the two networks). We could, however, ask far more novel questions, such as: what is the relationship between the linkage levels of herbivores in the two networks? If it is negative, then a herbivore may be both a generalist consumer and a prey to a specialist predator, or vice versa. The next question then arises as to which of these two combinations would impose the highest extinction risk to a herbivore? Each of the two 2-mode antagonistic networks can further be transformed into two 1-mode antagonistic networks, describing the competitive/facilitative interactions within a community or trophic level. The 2-mode plant–herbivore network is transformed into a 1-mode plant network, where links represent shared herbivores, and a 1-mode herbivore network, where links represent shared food-plant species. The 2-mode herbivore–predator network is transformed into a 1-mode herbivore network where links represent shared predators, and a 1-mode predator network, where links represent shared prey species. Thus, we get two kinds of 1-mode herbivore network, in which links represent either shared food plants or predators: these are essentially 'prey overlap' and 'predator overlap' graphs, which are rarely explored in food-web studies (Woodward and Hildrew, 2001, 2002a,b). Are these two networks similar in structure and dynamics?—If not, which one is the most robust to perturbations, for example the invasion of an alien herbivore?

Melián *et al.* (2009) answered some of these super-network questions using a large network containing both mutualistic and antagonistic links from the Doñana Biological Reserve in S Spain. They calculated the ratio between the number of mutualistic to antagonistic links for each plant species in the network. This ratio was very heterogeneously distributed in the plant community. In addition, link strength and this ratio had a strong influence on the species richness of the network.

In a later paper, Palla *et al.* (2007) moved from static to dynamic analysis, and considered the temporal development of such super networks by assessing the change in overlap ($s_{a,b}^{\mathrm{ov}}$) with time. They focused especially on the temporal development of small and large social groups (telephone calls and scientific co-authorships). Small groups persisted if their composition remained unchanged, whereas the persistence of large groups required that they had a strong membership dynamics. They studied in detail the different kinds of development, such as growth or contraction, merging or splitting and birth (colonization) and death (extinction) of groups (modules). To what extent this approach will transfer value to ecology, or food-web ecology in particular, remains to be seen. However, an obvious first step would be to use the concepts and methods to analyse super networks of at least three (partially) interlocking networks, for example plant communities and their pollinators, herbivores, pathogens, etc. Here, it would be interesting to compare the overlap of different plant species, including an increase in the number of networks of which they are members, and, for example relate this approach to the classical niche view. An extension of Palla *et al.* (2007) would then be to conduct a super-network study in relation to the temporal dynamics of niche width or network overlap. Such an approach is also of applied interest in the field of conservation biology: for instance, the effects of invasive species will ramify across more than one network type within a given habitat. Invasive *Opuntia* cactus affects the native seed-dispersal networks as well as the pollination networks in the Canaries and the Balearics (Christensen, 2004; Padrón *et al.*, 2009; Stald, 2003), and the loss of many pollinator species from mutualistic networks on oceanic islands has, for instance, been driven by predatory interactions with exotic species that have invaded the food web (Fritts and Rodda, 1998). Similarly, it is often the diseases and parasites transmitted by exotic species, which operate within the pathogen–host network, that are the main threat to resident species. Island examples include, for example the introduction to the Hawaiian Islands of mosquitoes with the blood parasite *Plasmodium relictum*, causing avian malaria, and avian pox probably introduced with domestic fowl (Parker, 2009). The alien 'entities' in these cases are not just species, but small modules of at least 2–3 interlocked species. In their new environment, they have impacts that can only be understood fully at the level of the super network. In natural systems, most invasions of exotic species are probably such small dangerous packages of 'macrobes' and their attendant microbes.

Another route, which might ultimately prove useful for integrating the analysis of several networks within ecology, was taken by Buldyrev *et al.* (2010). They analysed two interconnected technological networks. On the 28th of September 2003, Italy suffered from an electrical breakdown in its national power-station network, which led to failure of the Internet, which again created feedbacks to the power-station network. Their paper deals with the robustness of interacting networks to such cascading failures. The result was surprising, because the interconnected networks turned out to be very vulnerable to these types of errors, which is diametrically opposed to the perceptions that have been widely held for the last decade about single networks (Albert *et al.*, 2000). Taking the result of Buldyrev *et al.* (2010) to its most extreme, and if there is in effect only one global but multi-tiered ecological network, it may be especially sensitive to perturbations. No ecological network is completely isolated in the long run: even the most isolated oceanic island is connected to the rest of the world's biota through rare colonizations and migrant birds. A fascinating world of complexity is waiting to be studied.

X. CONCLUSIONS

In Rome, 2003, at the conference on 'Growing networks and graphs in statistical physics, finance, biology and social systems', some of the world's leading network researchers from various fields discussed the future for complex network analysis (Anon., 2004). They were particularly fascinated by the large databases becoming available to researchers, the small-world properties of many increasingly larger networks and the robustness of these to errors. They predicted a radical shift from the study of static networks to what they called 'network evolution', or what ecologists would call 'network dynamics' or 'assembly/disassembly'. They suggested 10 novel research areas worth studying: (1) dynamical models for network growth, (2) new statistical distributions to describe the linkage level or degree distribution of networks, (3) why are most networks modular? (4) universal features of network dynamics, (5) how does the dynamics of networks influence their structure? (6) network evolution, and here they meant biological evolution, (7) networks of networks (super networks) and their regulatory mechanisms, (8) methods to study robustness and vulnerability of networks, (9) statistical problems in the analysis of small networks, and (10) why are social networks assortative (i.e. linkage between similar nodes), while biological and technological networks are disassortative? Although probably very few ecologists have read these three pages in the *European Physical Journal*, the last 7 years of network research have indeed been focused upon answering most of the 10 questions by posing new questions, and our paper forms a small part of this gigantic enterprise. We have focused upon network dynamics in space and time, and the hierarchical structure of ecological networks. In addition, we have made an attempt to link the two predominant ecological network types in the current literature: mutualisms and food webs. The next 7 years will most likely be spent on producing better and more sophisticated answers to these 10 big questions from 2003 and new big questions will certainly appear. Ecology will, we believe, turn increasingly towards the study of (1) super networks, (2) genetic techniques to score large ecological networks, (3) disturbance and warning signals in networks, (4) linking network topology and biodiversity/ecosystem functioning, and (5) node properties and individual-based approaches, and (6) network evolution (Bascompte, 2009a,b; Ings *et al.*, 2009; Scheffer *et al.*, 2009; Sugihara and Ye, 2009). Finally, network analysis will become a very important tool in our attempts to formulate more general theories about complexity (e.g. Butts, 2009; Vespignani, 2009) and in many applied projects in ecology in general, including restoration ecology or network management (Memmott, 2009).

One of the founders of modern network theory, Barabási (2009) wrote 'although many fads have come and gone in complexity, one thing is increasingly clear: interconnectivity is so fundamental to the behaviour of complex systems that networks are here to stay'. Research into ecological networks is rapidly gathering speed, and as it finally comes of age in the coming years it will undoubtedly lead us into new and unexpected ways of perceiving the world around us.

ACKNOWLEDGEMENTS

We acknowledge the assistance of J. Bascompte, S. Bek, M. Bundgaard, D. W. Carstensen, B. Dalsgaard, B. K. Ehlers, H. Elberling, M. A. Fortuna, L. Gallop, M. B. García, J. Genini, V. González-Àlvaro, D. M. Hansen, T. T. Ingversen, U. Jacobs, P. Jordano, C. Kaiser-Bunbury, R. Lundgren, A. M. G. Martín, A. Montero, M. Nogales, B. Pádron, T. Petanidou, L. Stald, R. Sweeney, A. Traveset, and A. Valido. The Danish National

Research Council (FNU), the Carlsberg Foundation (CR) and a U.K. Natural Environment Research Council Centre for Population Biology research network grant awarded to GW financed this work.

REFERENCES

Abe, T. (2006). Threatened pollination systems in native flora of the Ogasawara (Bonin) Islands. *Ann. Bot.* **98**, 317–334.

Åberg, P. Size-based demography of the seaweed *Ascophyllum nodosum* in stochastic environment. *Ecology* **73**, 1488–1501.

Aizen, M.A., and Feinsinger, P. (1993). Forest fragmentation, pollination, and plant reproduction in a Chaco dry forest, Argentina. *Ecology* **75**, 330–351.

Aizen, M.A., Morales, C.L., and Morales, J.M. (2008). Invasive mutualists erode native pollination webs. *PLoS Biol.* **6**, 396–403.

Alarcón, R., Waser, N.M., and Ollerton, J. (2008). Year-to-year variation in the topology of a plant–pollinator interaction network. *Oikos* **117**, 1796–1807.

Albert, R., and Barabasi, A.-L. (2002). Statistical mechanics of complex networks. *Rev. Mod. Phys.* **74**, 47–97.

Albert, R., Jeong, H., and Barabasi, A.-L. (2000). Error and attack tolerance of complex networks. *Nature* **406**, 378–382.

Allen, K.R. (1951). The Horokiwi Stream: A study of a trout population. *NZ Mar. Dept. Fish. Bull.* **10–10a**, 1–231.

Allesina, S., and Pascual, M. (2009). Food web models, a plea for groups. *Ecol. Lett.* **12**, 652–662.

Allesina, S., Bodini, A., and Bondavalli, C. (2005). Ecological subsystems via graph theory, the role of strongly connected components. *Oikos* **110**, 164–176.

Amaral, L.A.N., Scala, A., Barthelemy, M., and Stanley, H.E. (2000). Classes of small-world networks. *Proc. Natl. Acad. Sci. USA* **97**, 11149–11152.

Anon. (2004). Virtual round table on ten leading questions for network research. *Eur. Phys. J. B* **38**, 143–145.

Apollonio, S. (2002). Hierarchical Perspectives on Marine Complexities. Columbia University Press, New York.

Araújo, M.S., Guimarães, P.R., Jr., Svanbäck, R., Pinheiro, A., Guimarães, P., Dos Reis, S.F., and Bolnick, D.I. (2008). Network analysis reveals contrasting effects of intraspecific competition on individuals vs. population diets. *Ecology* **89**, 1981–1993.

Araújo, M.S., Martins, E.G., Cruz, L.D., Fernandes, F.R., Linhares, A.X., Reis, S.F. D., and Guimarães, P.R., Jr. (2010). Nested diets: A novel pattern of individual-level resource use. *Oikos* **119**, 81–88.

Armbruster, W.S. (1993). Evolution of plant pollination systems: Hypotheses and tests with the Neotropical vine *Dalechampia*. *Evolution* **47**, 1480–1505.

Aronson, R.B., Thatje, S., Clarke, A., Peck, L.S., Blake, D.B., Wilga, C.D., and Seibel, B.A. (2007). Climate change and invasibility of the Antarctic benthos. *Ann. Rev. Ecol. Evol. Syst.* **38**, 129–154.

Baalen, M.V., Krivan, V., Rijn, P.C.J.V., and Sabelis, M.W. (2001). Alternative food, switching predators, and the persistence of predator–prey systems. *Am. Nat.* **157**, 512–524.

Barabási, A.-L. (2002). Linked. Perseus, Cambridge.

Barabási, A.-L. (2009). Scale-free networks: A decade and beyond. *Science* **325**, 412–413.
Barabási, A.-L., and Albert, R. (1999). Emergence of scaling in random networks. *Science* **286**, 509–512.
Barabási, A.-L., Albert, R., and Jeong, H. (1999). Mean-field theory for scale-free random networks. *Physica A* **272**, 173–187.
Bascompte, J. (2009a). Disentangling the web of life. *Science* **325**, 416–419.
Bascompte, J. (2009b). Mutualistic networks. *Front. Ecol. Environ.* **7**, 429–436.
Bascompte, J., and Jordano, P. (2007). Plant–animal mutualistic networks: The architecture of biodiversity. *Ann. Rev. Ecol. Evol. Syst.* **38**, 567–593.
Bascompte, J., Jordano, P., Melián, C.J., and Olesen, J.M. (2003). The nested assembly of plant–animal mutualistic networks. *Proc. Natl. Acad. Sci. USA* **100**, 9383–9387.
Bascompte, J., Jordano, P., and Olesen, J.M. (2006). Asymmetric coevolutionary networks facilitate biodiversity maintenance. *Science* **312**, 431–433.
Basilio, A.M., Medan, D., Torretta, J.P., and Bartolini, N.J. (2006). A year-long plant–pollinator network. *Austral Ecol.* **31**, 975–983.
Bastolla, U., Fortuna, M.A., Pascual-García, A., Ferrera, A., Luque, B., and Bascompte, J. (2009). The architecture of mutualistic networks minimizes competition and increases biodiversity. *Nature* **458**, 1018–1020.
Beckerman, A.P., Petchey, O.L., and Warren, P.H. (2006). Foraging biology predicts food web complexity. *Proc. Natl. Acad. Sci. USA* **103**, 13745–13749.
Benke, A.C., Wallace, J.B., Harrison, J.W., and Koebel, J.W. (2001). Food web quantification using secondary production analysis: Predaceous invertebrates of the snag habitat in a subtropical river. *Freshw. Biol.* **46**, 329–346.
Bianconi, G., Pin, P., and Marsili, M. (2009). Assessing the relevance of node features for network structure. *Proc. Natl. Acad. Sci. USA* **106**, 11433–11438.
Bloch, E., and Ritter, M. (1977). Nonsynchronism and the obligation to its dialectics. *New German Critique* **11**, 22–38.
Blüthgen, N., Fründ, J., Vázquez, D.P., and Menzel, F. (2008). What do interaction network metrics tell us about specialization and biological traits? *Ecology* **89**, 3387–3399.
Bolnick, D., Yang, L.H., Fordyce, J.A., Davis, J.M., and Svanback, R. (2002). Measuring individual-level resource specialization. *Ecology* **83**, 2936–2941.
Borrell, B.J. (2005). Long tongues and loose niches: Evolution of Euglossine bees and their nectar flowers. *Biotropica* **37**, 664–669.
Bourgeois, K., Suehs, C.M., Vidal, E., and Medail, F. (2005). Invasional meltdown potential: Facilitation between introduced plants and mammals on French Mediterranean islands. *Ecoscience* **12**, 248–256.
Brose, U., Jonsson, T., Berlow, E.L., Warren, P., Banasek-Richter, C., Bersier, L.-F., Blanchard, J.L., Brey, T., Carpenter, S.R., Blandenier, M.-F.C., Cushing, L., Dawah, H.A., *et al.* (2006). Consumer–resource body-size relationships in natural food webs. *Ecology* **87**, 2411–2417.
Bubb, D.H., Thom, T.J., and Lucas, M.C. (2004). Movement and dispersal of the invasive signal crayfish *Pacifastacus leniusculus* in upland rivers. *Freshw. Biol.* **49**, 357–368.
Buldyrev, S.V., Parshani, R., Paul, G., Stanley, H.E., and Havlin, S. (2010). Catastrophic cascade of failures in interdependent networks. *Nature* **464**, 1025–1028.
Bundgaard, M. (2003). Tidslig og Rumlig Variation i et Plante-Bestøvernetværk. MSc. thesis, Aarhus University, Aarhus.

Burgos, E., Ceva, H., Perazzo, R.P.J., Devoto, M., Medan, D., Zimmermann, M., and Delbue, A.M. (2007). Why nestedness in mutualistic networks? *J. Theor. Biol.* **249**, 307–313.

Butts, C.T. (2009). Revisiting the foundations of network analysis. *Science* **325**, 414–416.

Camerano, L. (1880). On the equilibrium of living beings by means of reciprocal destruction. *Acts R. Acad. Sci. Torino* **15**, 393–414.

Carnicer, J., Jordano, P., and Melián, C.J. (2009). The temporal dynamics of resource use by frugivorous birds: A network approach. *Ecology* **90**, 1958–1970.

Carvalheiro, L.G., Barbosa, E.R.M., and Memmott, J. (2008). Pollinator networks, alien species and the conservation of rare plants: *Trinia glauca* as a case study. *J. Appl. Ecol.* **45**, 1419–1427.

Cattin, M.F., Bersier, L.F., Banasek-Richter, C., Baltensperger, R., and Gabriel, J.P. (2004). Phylogenetic constraints and adaptation explain food-web structure. *Nature* **427**, 835–839.

Christensen, W.H. (2004). Invasive Species: Effects of *Opuntia dillenii* on a Natural Ecosystem in the Canary Islands. MSc. thesis, Aarhus University, Aarhus.

Christian, R.R., Baird, D., Luczkovich, J., Johnson, J.C., Scharler, U., and Ulanowicz, R.E. (2005). Role of network analysis in comparative ecosystem ecology of estuaries. In: *Aquatic Food Webs: An Ecosystem Approach* (Ed. by A. Belgrano, U.M. Scharler, J. Dunne and R.E. Ulanowicz), pp. 25–40. Oxford University Press, Oxford.

Clements, R.E., and Long, F.L. (1923). Experimental Pollination. An Outline of the Ecology of Flowers and Insects. Carnegie Institute, Washington.

Cloe, W.W., and Garman, G.C. (1996). The energetic importance of terrestrial arthropod inputs to three warm-water streams. *Freshw. Biol.* **36**, 105–114.

Closs, G.P., and Lake, P.S. (1994). Spatial and temporal variation in the structure of an intermittent-stream food web. *Ecol. Monogr.* **64**, 1–21.

Cohen, J.E. (1978). Food Webs and Niche Space. Princeton University Press, Princeton.

Cohen, J.E. (1979). Ergodic theorems in demography. *Bull. Am. Math. Soc.* **1**, 275–295.

Cohen, J.E., Briand, F., and Newman, C.M. (1990). Community Food Webs: Data and Theory. Springer, Berlin.

Collins, S.L., and Glenn, S.M. (1991). Importance of spatial and temporal dynamics on species regional abundance and distribution. *Ecology* **72**, 654–664.

Cooper-Driver, G., Finch, S., Swain, T., and Bernays, E. (1977). Seasonal variation in secondary plant compounds in relation to the palatability of *Pteridium aquilinum*. *Biochem. Syst. Ecol.* **5**, 117–183.

Costa, L.D.F., and Sporns, O. (2005). Hierarchical features of large-scale cortical connectivity. *Eur. Phys. J. B* **48**, 567–573.

Cox, P.A., and Elmqvist, T. (2000). Pollinator extinction in the Pacific Islands. *Conserv. Biol.* **14**, 1237–1239.

Crook, A.C., Long, M., and Barnes, D.K.A. (2000). Quantifying daily migration in the sea urchin *Paracentrotus lividus*. *J. Mar. Biol. Assoc. UK* **80**, 177–178.

Cushing, D.H. (1975). Marine Ecology and Fisheries. Cambridge University Press, Cambridge.

Dalsgaard, B., Martín, A.M.G., Olesen, J.M., Timmermann, A., Andersen, L.H., and Ollerton, J. (2008). Pollination networks and functional specialization: A test using Lesser Antillean plant-hummingbird assemblages. *Oikos* **117**, 789–793.

D'Antonio, C.M. and Dudley, T.L. (1993) Alien species: The insidious invasion of ecosystems by plants and animals from around the world has become a major environmental problem. *Pacific Discovery* **1993**, summer, 9–11

Darwin, C. (1859). On the Origin of Species by Means of Natural Selection, or the Preservation of Favoured Races in the Struggle for Life. Murray, London.

Davidson, T.A., Sayer, C.D., Langdon, P.G., Burgess, A., and Jackson, M. (2010). Inferring past zooplanktivorous fish and macrophyte density in a shallow lake: Application of a new regression tree model. *Freshw. Biol.* **55**, 584–599.

Dobson, M., and Hildrew, A.G. (1992). A test of resource limitation among shredding detritivores in low order streams in southern England. *J. Anim. Ecol.* **61**, 69–77.

Dunne, J.A., Williams, R.J., and Martinez, N.D. (2002a). Food-web structure and network theory: The role of connectance and size. *Proc. Natl. Acad. Sci. USA* **99**, 12917–12922.

Dunne, J.A., Williams, R.J., and Martinez, N.D. (2002b). Network structure and biodiversity loss in food webs: Robustness increases with connectance. *Ecol. Lett.* **5**, 558–567.

Dupont, Y.L., and Olesen, J.M. (2009). Modules and roles of species in heathland pollination networks. *J. Anim. Ecol.* **78**, 346–353.

Dupont, Y.L., and Skov, C. (2004). Influence of geographical distribution and floral traits on species richness of bees (Hymenoptera: Apoidea) visiting *Echium* species (Boraginaceae) in the Canary Islands. *Int. J. Plant Sci.* **118**, 301–311.

Dupont, Y.L., Hansen, D.M., and Olesen, J.M. (2003). Structure of a plant–flower–visitor network in the high-altitude sub-alpine desert of Tenerife, Canary Islands. *Ecography* **26**, 301–310.

Dupont, Y.L., Trøjelsgaard, K., and Olesen, J.M. (in press). Scaling down from species to individuals, a flower–visitation network between individual honeybees and thistle plants. *Oikos.*

Eckhart, V.M. (1991). The effects of floral display on pollinator visitation vary among populations of *Phacelia linearis* (Hydrophyllaceae). *Evol. Ecol.* **5**, 370–384.

Elberling, H., and Olesen, J.M. (1999). The structure of a high latitude plant–pollinator system: The dominance of flies. *Ecography* **22**, 314–323.

Elton, C.S. (1927). Animal Ecology. Sidgwick & Jackson, London.

Elton, C.S. (1958). The Ecology of Invasions by Plants and Animals. Methuen, London.

Estes, J.A., Tinker, M.T., Williams, T.M., and Doak, D.F. (1998). Killer whale predation on sea otters linking oceanic and nearshore ecosystems. *Science* **282**, 473–476.

Fabina, N.S., Abbott, K.C., and Tucker, R. (2009). Sensitivity of plant–pollinator–herbivore communities to changes in phenology. *Ecol. Model.* **221**, 453–458.

Feinsinger, P., Beach, J.H., Linhart, Y.B., Busby, W.H., and Murray, K.G. (1987). Disturbance, pollinator predictability, and pollination success among Costa Rican cloud forest plants. *Ecology* **68**, 1294–1305.

Flecker, A.S., and Townsend, C.R. (1994). Community-wide consequences of trout introduction in New Zealand streams. *Ecol. Appl.* **4**, 798–807.

Forbes, S.A. (1887). The Lake as a Microcosm. *Bull. Sci. Assoc. (Peoria, IL)* **1**, 77–87.

Fortuna, M.A., García, C., Guimarães, P.R., Jr., and Bascompte, J. (2008). Spatial mating networks in insect-pollinated plants. *Ecol. Lett.* **11**, 490–498.

Fortuna, M.A., Popa-Lisseanu, A.G., Ibáñz, C., and Bascompte, J. (2009). The roosting spatial network of a bird-predator bat. *Ecology* **90**, 934–944.

Fortuna, M.A., Stouffer, D.B., Olesen, J.M., Jordano, P., Mouillot, D., Krasnov, B.R., Poulin, R., and Bascompte, J. (2010). Nestedness versus modularity in ecological networks: Two sides of the same coin? *J. Anim. Ecol.* **79**, 811–817.
Fox, L.R., and Morrow, P.A. (1981). Specialization: Species property or local phenomenon? *Science* **211**, 887–893.
Fritts, T.H., and Rodda, G.H. (1998). The role of introduced species in the degradation of island ecosystems: A case history of Guam. *Ann. Rev. Ecol. Syst.* **29**, 113–140.
Fukui, D., Murakami, M., Nakano, S., and Aoi, T. (2006). Effect of emergent aquatic insects on bat foraging in a riparian forest. *J. Anim. Ecol.* **75**, 1252–1258.
Gamarra, J.G.P., Montoya, J.M., Alonso, D., and Solé, R.V. (2005). Competition and introduction regime shape exotic bird communities in Hawaii. *Biol. Inv.* **7**, 297–307.
Genini, J., Morellato, L.P.C., Guimarães, P.R., Jr., and Olesen, J.M. (2010). Cheaters in mutualism networks. *Biol. Lett* **6**, 494–497.
Gómez, J.M., and Zamora, R. (1999). Generalization vs. specialization in the pollination system of *Hormathophylla spinosa* (Cruciferae). *Ecology* **80**, 796–805.
Guimarães, P.R., Jr., and Guimarães, P.R.J. (2006). Improving the analyses of nestedness for large sets of matrices. *Environ. Model. Softw.* **21**, 1512–1513.
Guimerà, R., and Amaral, L.A.N. (2005). Functional cartography of complex metabolic networks. *Nature* **433**, 895–900.
Guimerà, R., Sales-Prado, M., and Amaral, L.A.N. (2007). Classes of complex networks defined by role-to-role connectivity profiles. *Nat. Phys.* **3**, 63–69.
Hall, S.J., and Raffaelli, D. (1993). Food webs: Theory and reality. *Adv. Ecol. Res.* **24**, 187–239.
Hall, R.O., Wallace, J.B., and Eggert, S.L. (2000). Organic matter flow in stream food webs with reduced detrital resource base. *Ecology* **81**, 3445–3463.
Hanley, M.E., and Goulson, D. (2003). Introduced weeds pollinated by introduced bees: Cause or effect? *Weed Biol. Manag.* **3**, 204–212.
Hegland, S.J., Nielsen, A., Lázaro, A., Bjerknes, A.-L., and Totland, Ø. (2009). How does climate warming affect plant–pollinator interactions? *Ecol. Lett.* **12**, 184–195.
Herborg, L.-M., Rushton, S.P., Clare, A.S., and Bentley, M.G. (2004). Spread of the Chinese mitten crab (*Eriocheir sinensis* H. Milne Edwards) in Continental Europe: Analysis of historical data set. *Hydrobiologia* **503**, 21–28.
Herrera, C.M. (1988). Variation in mutualisms: The spatio-temporal mosaic of a pollinator assemblage. *Biol. J. Linn. Soc.* **35**, 95–125.
Hilderbrand, G.V., Hanley, T.A., Robbins, C.T., and Schwartz, C.C. (1999). Role of brown bears (*Ursus arctos*) in the flow of marine nitrogen into a terrestrial ecosystem. *Oecologia* **121**, 546–550.
Hildrew, A.G. (2009). Sustained research on stream communities: A model system and the comparative approach. *Adv. Ecol. Res.* **41**, 175–312.
Hildrew, A.G., Woodward, G., Winterbottom, J.H., and Orton, S. (2004). Strong density-dependence in a predatory insect: Larger scale experiments in a stream. *J. Anim. Ecol.* **73**, 448–458.
Hoye, T.T., Post, E., Meltofte, H., Schmidt, N.M., and Forchhammer, M.C. (2007). Rapid advancement of spring in the High Arctic. *Curr. Biol.* **17**, R449–R451.
Huxel, G.R., and McCann, K.S. (1998). Food web stability: The influence of trophic flows across habitats. *Am. Nat.* **152**, 460–469.
Hynes, H.B.N. (1975). The stream and its valley. *Verh. Internat. Verein. Theor. Angew. Limnol.* **19**, 1–15.

Ings, T.C., Bascompte, J., Blüthgen, N., Brown, L., Dormann, C.F., Edwards, F., Figueroa, D., Jacob, U., Jones, J.I., Lauridsen, R.B., Ledger, M.E., Lewis, H.M., *et al.* (2009). Ecological networks—Food webs and beyond. *J. Anim. Ecol.* **78**, 253–269.

Inouye, D.W., Saavedra, F., and Lee-Wang, W. (2003). Environmental influences on the phenology and abundance of flowering by *Androsace septentrionalis* (Primulaceae). *Am. J. Bot.* **90**, 905–910.

Isalan, M., Lemerle, C., Michalodimitrakis, K., Horn, C., Beltrao, P., Raineri, E., Garriga-Canut, M., and Serrano, L. (2008). Evolvability and hierarchy in rewired bacterial gene networks. *Nature* **452**, 840–846.

Janzen, D.H. (1971). Euglossine bees as long-distance pollinators of tropical plants. *Science* **171**, 203–205.

Jennersten, O. (1988). Pollination in *Dianthus deltoides* (Caryophyllaceae): Effects of habitat fragmentation on visitation and seed set. *Conserv. Biol.* **2**, 359–366.

Jennings, S., Pinnegar, J.K., Polunin, N.V.C., and Boon, T.W. (2001). Weak cross-species relationships between body size and trophic level belie powerful size-based trophic structuring in fish communities. *J. Anim. Ecol.* **70**, 934–944.

Jeong, H., Néda, Z., and Barabási, A.-L. (2003). Measuring preferential attachment in evolving networks. *Europhys. Lett.* **61**, 567–572.

Johnson, S.D., and Steiner, K.E. (2000). Generalization versus specialization in plant pollination systems. *TREE* **15**, 140–143.

Jonsson, T., Cohen, J.E., and Carpenter, S.R. (2005). Food webs, body size, and species abundance in ecological community description. *Adv. Ecol. Res.* **36**, 1–84.

Joppa, L.N., Bascompte, J., Montoya, J.M., Solé, R., Sanderson, J., and Pimm, S.L. (2009). Reciprocal specialization in ecological networks. *Ecol. Lett.* **12**, 961–969.

Jordano, P. (1987). Patterns of mutualistic interactions in pollination and seed dispersal: Connectance, dependence, and coevolution. *Am. Nat.* **129**, 657–677.

Jordano, P., Bascompte, J., and Olesen, J.M. (2003). Invariant properties in coevolutionary networks of plant–animal interactions. *Ecol. Lett.* **6**, 69–81.

Jordano, P., Bascompte, J., and Olesen, J.M. (2006). The ecological consequences of complex topology and nested structure in pollination webs. In: *Plant–Pollinator Interactions: From Specialization to Generalization* (Ed. by N.M. Waser and J. Ollerton), pp. 173–199. Chicago University Press, Chicago.

Kaiser-Bunbury, C.N., Muff, S., Memmott, J., Müller, C.B., and Caflisch, A. (2010). The robustness of pollination networks to the loss of species and interactions: A quantitative approach incorporating pollinator behaviour. *Ecol. Lett.* **13**, 442–452.

Kallimanis, A.S., Sgardelis, S.P., and Halley, J.M. (2002). Accuracy of fractal dimension estimates for small samples of ecological distributions. *Landscape Ecol.* **17**, 281–297.

Kawaguchi, Y., and Nakano, S. (2001). The contribution of terrestrial invertebrate input to the annual resource budget of salmonid populations in forest and grassland reaches of a headwater stream. *Freshw. Biol.* **46**, 303–316.

Kenta, T., Inari, N., Nagamitsu, T., Goka, K., and Hiura, T. (2007). Commercialized European bumblebee can cause pollination disturbance: An experiment on seven native plant species in Japan. *Biol. Conserv.* **134**, 298–309.

Knight, T.M., McCoy, M.W., Chase, J.M., McCoy, K.A., and Holt, R.D. (2005). Trophic cascades across ecosystems. *Nature* **437**, 880–883.

Krause, A.E., Frank, K.A., Mason, D.M., Ulanowicz, R.E., and Taylor, W.W. (2003). Compartments revealed in food-web structure. *Nature* **426**, 282–285.

Krause, J., Croft, D.P., and James, R. (2007). Social network theory in the behavioural sciences, potential applications. *Behav. Ecol. Sociobiol.* **62**, 15–27.

Kunin, W.E. (1998). Extrapolating species abundance across spatial scales. *Science* **281**, 1513–1515.
Lafferty, K.D., Allesina, S., Arim, M., Briggs, C.J., Leo, G.D., Dobson, A.P., Dunne, J.A., Johnson, P.T.J., Kuris, A.M., Marcogliese, D.J., Martinez, N.D., Memmott, J., *et al.* (2008). Parasites in food webs: The ultimate missing links. *Ecol. Lett.* **11**, 533–546.
Layer, K., Hildrew, A.G., Monteith, D., and Woodward, G. (2010a). Long-term variation in the littoral food web of an acidified mountain lake. *Global Change Biol.*
Layer, K., Riede, J.O., Hildrew, A.G. and Woodward, G. (2010b). Food web structure and stability in 20 streams across a wide pH gradient. *Adv. Ecol. Res.* **42**, 267–301.
Levin, S.A. (1992). The problem of scale in ecology. *Ecology* **73**, 1943–1967.
Lewinsohn, T.M., Prado, P., Jordano, P., Bascompte, J., and Olesen, J.M. (2006). Structure in plant–animal interaction assemblages. *Oikos* **113**, 174–184.
Lewis, O.T., Memmott, J., Lasalle, J., Lyal, C.H.C., Whiteford, C., and Godfray, H.C.J. (2002). Structure of a diverse tropical forest insect–parasitoid community. *J. Anim. Ecol.* **71**, 855–873.
Lindberg, A.B., and Olesen, J.M. (2001). The fragility of extreme specialisation: *Passiflora mixta* and its pollinating hummingbird *Ensifera ensifera*. *J. Trop. Ecol.* **17**, 1–7.
Lopezaraiza-Mikel, M.E., Hayes, R.B., Whalley, M.R., and Memmott, J. (2007). The impact of an alien plant on a native plant–pollinator network: An experimental approach. *Ecol. Lett.* **10**, 539–550.
Lundgren, R., and Olesen, J.M. (2005). The dense and highly connected world of Greenland plants and their pollinators. *Arct. Antarct. Alp. Res.* **37**, 514–520.
Lusseau, D., Whitehead, H., and Gero, S. (2008). Incorporating uncertainty into the study of animal social networks. *Anim. Behav.* **75**, 1809–1815.
Magurran, A.E., and Henderson, P.A. (2003). Explaining the excess of rare species in natural species abundance distributions. *Nature* **422**, 714–716.
Manning, A. (1956). Some aspects of the foraging behaviour of bumblebees. *Behaviour* **9**, 73.
Martinez, N.D., and Dunne, J.A. (1998). Time, space, and beyond: Scale issues in food-web research. In: *Ecological Scale: Theory and Applications* (Ed. by D.L. Peterson and V.T. Parker), pp. 207–226. Columbia University Press, New York.
Maslow, S., and Sneppen, K. (2002). Specificity and stability in topology of protein networks. *Science* **296**, 910–913.
Mason, C.F., and MacDonald, S.M. (1982). The input of terrestrial invertebrates from tree canopies to a stream. *Freshw. Biol.* **12**, 305–312.
Maurer, B.A. (1999). Untangling Ecological Complexity. Chicago University Press, Chicago.
May, R.M. (1972). Will a large complex system be stable? *Nature* **238**, 413–414.
May, R.M. (1973). Stability and Complexity in Model Ecosystems. Princeton University Press, Princeton.
May, R.M., Levin, S.A., and Sugihara, G. (2008). Ecology for bankers. *Nature* **451**, 893–895.
McCann, K.S., Hastings, A., and Huxel, G.R. (1998). Weak trophic interactions and the balance of nature. *Nature* **395**, 794–798.
McCann, K.S., Rasmussen, J., Umbanhowar, J., and Humphries, M. (2005). The role of space, time, and variability in food web dynamics. In: *Dynamic Food Webs: Multispecies Assemblages, Ecosystem Development, and Environmental Change* (Ed.

by P.D. Ruiter, V. Wolters and J.C. Moore), pp. 56–70. Elsevier and AP, Burlington.

McGill, B.J., *et al.* (2007). Species abundance distributions: Moving beyond single prediction theories to integration within an ecological framework. *Ecol. Lett.* **10**, 995–1015.

McGregor, S.E. (1959). Cotton-flowering visitation and pollen distribution by honeybees. *Science* **129**, 97–98.

McLaughlin, O.B., Jonsson, T. and Emmerson, M.C. (2010). Temporal variability in predator–prey relationships of a forest floor food web. *Adv. Ecol. Res.* **42**, 171–264.

Medan, D., Montaldo, N.H., Devoto, M., Mantese, A., Vasellati, V., and Bartoloni, N.H. (2002). Plant–pollinator relationships at two altitudes in the Andes of Mendoza, Argentina. *Arct. Antarct. Alp. Res.* **34**, 233–241.

Medan, D., Basilio, A.M., Devoto, M., Bartoloni, N.J., Torretta, J.P., and Petanidou, T. (2006). Measuring generalization and connectance in temperate, year-long active systems. In: *Plant–Pollinator Interactions: From Specialization to Generalization* (Ed. by N.M. Waser and J. Ollerton), pp. 245–259. Chicago University Press, Chicago.

Melián, C.J., and Bascompte, J. (2002). Complex networks, two ways to be robust? *Ecol. Lett.* **5**, 705–708.

Melián, C.J., and Bascompte, J. (2004). Food web cohesion. *Ecology* **85**, 352–358.

Melián, C.J., Bascompte, J., and Jordano, P. (2005). Spatial structure and dynamics in a marine food web. In: *Complexity in Aquatic Food Webs: An Ecosystem Approach* (Ed. by A. Belgrano, U. Scharler, J. Dunne and R.E. Ulanowicz), pp. 19–24. Oxford University Press, Oxford.

Melián, C.J., Bascompte, J., Jordano, P., and Krivan, V. (2009). Diversity in a complex ecological network with two interaction types. *Oikos* **118**, 122–130.

Melián, C.J., Vilas, C., Baldó, F., Enrique González-Ortegón, E., Drake, P., and Williams, R.J. (in review) A neutral evolutionary and ecological test for individual-based food webs.

Meltofte, H., Christensen, T.R., Elberling, B., Forchhammer, M.C. and Rasch, M. (Eds.) (2008) High-arctic ecosystem dynamics in a changing climate. *Adv. Ecol. Res.* **40,** 1–563

Memmott, J. (1999). The structure of a plant–pollinator food web. *Ecol. Lett.* **2**, 276–280.

Memmott, J. (2009). Foodwebs: A ladder for picking strawberries or a practical tool for practical problems. *Philos. Trans. R. Soc. Lond. B* **364**, 1693–1699.

Memmott, J., and Waser, N.M. (2002). Integration of alien plants into a native flower–pollinator visitation web. *Proc. R. Soc. Lond. B* **269**, 2395–2399.

Memmott, J., Craze, P.G., Waser, N.W., and Price, M.V. (2007). Global warming and the disruption of plant–pollinator interactions. *Ecol. Lett.* **10**, 710–717.

Milo, R., Shen-Orr, S., Itzkovitz, S., Kashtan, N., Chklovskii, D., and Alon, U. (2002). Network motifs: Simple building blocks of complex networks. *Science* **298**, 824–827.

Montero, A. (2005). The Ecology of Three Pollination Networks. MSc thesis, Aarhus University, Aarhus.

Montoya, J.M., and Solé, R.V. (2002). Small world patterns in food webs. *J. Theor. Biol.* **214**, 405–412.

Montoya, J.M., Pimm, S.L., and Solé, R.V. (2006). Ecological networks and their fragility. *Nature* **442**, 259–264.

Montoya, J.M., Woodward, G., Emmerson, M.C., and Solé, R.C. (2009). Indirect effects propagate disturbances in real food webs. *Ecology* **90**, 2426–2433.

Morales, C.L., and Aizen, M.A. (2002). Does invasion of exotic plants promote invasion of exotic flower visitors? A case study from the temperate forests of the southern Andes. *Biol. Invas.* **4**, 87–100.

Morales, C.L., and Aizen, M.A. (2006). Invasive mutualisms and the structure of plant–pollinator interactions in the temperate forests of north-west Patagonia, Argentina. *J. Ecol.* **94**, 171–180.

Mortensen, H.S., Dupont, Y.L., and Olesen, J.M. (2008). A snake in paradise: Disturbance of plant reproduction following extirpation of bird–flower visitors on Guam. *Biol. Conserv.* **141**, 2146–2154.

Mosquin, T., and Martin, J.E. (1967). Observations on the pollination biology of plants on Melville island, N.W.T., Canada. *Can. Field-Nat.* **81**, 201–205.

Mouillot, D., Lepretre, A., Andrei-Ruiz, M.-C., and Viale, D. (2000). The fractal model, a new model to describe the species accumulation process and relative abundance distribution (RAD). *Oikos* **90**, 333–342.

Mucha, P.J., Richardson, T., Macon, K., Porter, M.A., and Onnela, J.-P. (2010). Community structure in time-dependent, multiscale and multiplex networks. *Science* **328**, 876–878.

Nakano, S., Miyasaka, H., and Kuhara, N. (1999). Terrestrial-aquatic linkages: Riparian arthropod inputs alter trophic cascades in a stream food web. *Ecology* **80**, 2435–2441.

Nelson, B.V., and Vance, R.R. (1979). Diel foraging patterns of the sea-urchin *Centrostephanus coronatus* as a predator avoidance strategy. *Mar. Biol.* **51**, 251–258.

Newman, M.E.J. (2001). Scientific collaboration networks. *Proc. Natl. Acad. Sci. USA* **98**, 404–409.

Nielsen, A., and Bascompte, J. (2007). Ecological networks, nestedness and sampling effort. *J. Anim. Ecol.* **95**, 1134–1141.

Nilsson, L.A. (1988). The evolution of flowers with deep corolla tubes. *Nature* **334**, 147–149.

Nooy, W.D., Mrvar, A., and Batagelj, V. (2005). Exploratory Social Network Analysis with Pajek. Cambridge University Press, Cambridge.

O'Dowd, D.J., Green, P.T., and Lake, P.S. (2003). Invasional 'meltdown' on an oceanic island. *Ecol. Lett.* **6**, 812–817.

O'Gorman, E.J., and Emmerson, M.C. (2009). Perturbations to trophic interactions and the stability of complex food webs. *Proc. Natl. Acad. Sci. USA* **106**, 13393–13398.

O'Gorman, E.J. and Emmerson, M.C. (2010). Manipulating interaction strengths and the consequences for trivariate patterns in a marine food web. *Adv. Ecol. Res.* **42**, 301–419.

Olesen, J.M. (1997). From naïveté to experience: Bumblebee queens (*Bombus terrestris*) foraging on *Corydalis cava* (Fumariaceae). *J. Kansas Ent. Soc.* **69**(Suppl.), 274–286.

Olesen, J.M. (2000). Exactly how generalised are pollination interactions? In: *The Scandinavian Association for Pollination Ecology Honours Knut Fægri* (Ed. by Ø. Totland *et al.*), pp. 161–178. Norwegian Acad. Sci. Lett, Oslo.

Olesen, J.M., and Jordano, P. (2002). Geographic patterns in plant–pollinator mutualistic networks. *Ecology* **83**, 2416–2424.

Olesen, J.M., Eskildsen, L.I., and Venkatasamy, S. (2002). Invasion of pollination networks on oceanic islands: Importance of invader complexes and endemic super generalists. *Divers. Distrib.* **8**, 181–192.

Olesen, J.M., Bascompte, J., Dupont, Y.L., and Jordano, P. (2006). The smallest of all worlds, pollination networks. *J. Theor. Biol.* **240**, 270–276.
Olesen, J.M., Bascompte, J., Dupont, Y.L., and Jordano, P. (2007). The modularity of pollination networks. *Proc. Natl. Acad. Sci. USA* **104**, 19891–19896.
Olesen, J.M., Bascompte, J., Elberling, H., and Jordano, P. (2008). Temporal dynamics of a pollination network. *Ecology* **89**, 1573–1582.
Olesen, J.M., Bascompte, J., Dupont, Y.L., Elberling, H., Rasmussen, C. and Jordano, P. Missing and forbidden links in mutualistic networks. *Proc. R. Soc. Lond. B* (in press).
Olesen, J.M., Bascompte, J., Dupont, Y.L., Jordano, P., Petanidou, T. and Sales-Pardo, M. The deep structure of a pollination network, modules, hierarchical organization, and level of self-similarity (submitted for publication).
Olesen, J.M., Dupont, Y.L., Morellato, P. and Genini, J. Seasonal dynamics of pollination networks and their modules (unpublished data a).
Olesen, J.M., Stefanescu, C. and Traveset, A. Strong, long-term dynamics in ecological networks (unpublished data b).
Ollerton, J., and Cranmer, L. (2002). Latitudinal trends in plant–pollinator interactions: Are tropical plants more specialized? *Oikos* **98**, 340–350.
Padrón, B., Traveset, A., Biedenweg, T., Diáz, D., Nogales, M., and Olesen, J.M. (2009). Impact of alien plant invaders on pollination networks from oceanic and continental islands. *PLoS ONE* **4**(7), e6275.
Palla, G., Derényi, I., Farkas, I., and Vicsek, T. (2005). Uncovering the overlapping community structure of complex networks in nature and society. *Nature* **435**, 814–818.
Palla, G., Barabási, A.-L., and Vicsek, T. (2007). Quantifying social group evolution. *Nature* **446**, 664–667.
Parker, P. (2009). Parasites and pathogens. Threats to native birds. In: *Galápagos. Preserving Darwin's Legacy* (Ed. by T.d. Roy), pp. 177–183. D. Bateman & Parque Nacional Galápagos Ecuador, Auckland.
Parrish, J.A.D., and Bazzaz, F.A. (1979). Difference in pollination niche relationships in early and late successional plant communities. *Ecology* **60**, 597–610.
Pascual, M., and Dunne, J.A. (Eds.) (2006). Ecological Networks: Linking Structure to Dynamics in Food Webs Oxford University Press, Oxford.
Perkins, S.E., Cagnacci, F., Stradiotto, A., Arnoldi, D., and Hudson, P.J. (2009). Comparison of social networks derived from ecological data, implications for inferring infectious disease dynamics. *J. Anim. Ecol.* **78**, 1015–1022.
Petanidou, T. (1991). Pollination Ecology in a Phryganic Ecosystem. (in Greek) Dissertation, Aristotelian University, Thessaloniki.
Petanidou, T., and Potts, S.G. (2006). Mutual use of resources in Mediterranean plant–pollinator communities: How specialized are pollination webs? In: *Plant–Pollinator Interactions: From Specialization to Generalization* (Ed. by N.M. Waser and J. Ollerton), pp. 220–244. Chicago University Press, Chicago.
Petanidou, T., Kallimanis, A.S., Tzanopoulos, J., Sgardelis, S.P., and Pantis, J.D. (2008). Long-term observation of a pollination network, fluctuation in species and interactions, relative invariance of network structure and implications for estimates of specialization. *Ecol. Lett.* **11**, 564–575.
Petchey, O.L., Beckerman, A.P., Riede, J.O., and Warren, P.H. (2008). Size, foraging, and food web structure. *Proc. Natl. Acad. Sci. USA* **105**, 4191–4196.
Petersen, I., Winterbottom, J.H., Orton, S., Friberg, N., and Hildrew, A.G. (1999). Emergence and lateral dispersal of adult stoneflies and caddisflies from Broadstone Stream. *Freshw. Biol.* **42**, 401–416.

Pickett, S.T.A., and Bazzaz, F.A. (1978). Organization of an assemblage of early successional species on a soil moisture gradient. *Ecology* **59**, 1248–1255.
Pimm, S.L. (2002). Food Webs. Chicago University Press, Chicago.
Pimm, S.L., and Lawton, J.H. (1980). Are food webs divided into compartments? *J. Anim. Ecol.* **49**, 879–898.
Pinder, A.C., Gozlan, R.E., and Britton, J.R. (2005). Dispersal of the invasive topmouth gudgeon, *Pseudorasbora parva* in the UK: A vector for an emergent infectious disease. *Fish. Manag. Ecol.* **12**, 411–414.
Polis, G.A. (1991). Complex trophic interactions in deserts—An empirical critique of food-web theory. *Am. Nat.* **138**, 123–155.
Polis, G.A., and Hurd, S.D. (1996). Allochthonous inputs across habitats, subsidized consumers and apparent trophic cascades: Examples from the ocean–land interface. In: *Food Webs: Integration of Pattern and Dynamics* (Ed. by G.A. Polis and K.O. Winemiller), pp. 275–285. Chapman and Hall, New York.
Polis, G.A., Anderson, W.B., and Holt, R.D. (1997). Toward an integration of landscape and food web ecology: The dynamics of spatially subsidized food webs. *Ann. Rev. Ecol. Syst.* **28**, 289–316.
Prado, P.I., and Lewinsohn, T.M. (2004). Compartments in insect–plant associations and their consequences for community structure. *J. Anim. Ecol.* **73**, 1168–1178.
Raffaelli, D., and Hall, S.J. (1992). Compartments and predation in an estuarine food web. *J. Anim. Ecol.* **61**, 551–560.
Ravasz, E., and Barabási, A.L. (2003). Hierarchical organization of complex networks. *Phys. Rev. E* **67**, 026112.
Rawcliffe, R., Sayer, C.D., Woodward, G., Grey, J., Davidson, T.A., and Jones, J.I. (2010). Back to the future: Using palaeolimnology to infer long-term changes in shallow lake food webs. *Freshw. Biol.* **55**, 600–613.
Redner, S. (1998). How popular is your paper? *Eur. Phys. J. B* **4**, 131–134.
Reuman, D.C., and Cohen, J.E. (2004). Trophic links' length and slope in the Tuesday Lake food web with species' body mass and numerical abundance. *J. Anim. Ecol.* **73**, 852–866.
Rezende, E.L., Albert, E.A., Fortuna, M.A., and Bascompte, J. (2009). Compartments in a marine food web associated with phylogeny, body mass, and habitat structure. *Ecol. Lett.* **12**, 779–788.
Richardson, D.M., Allsop, N., d'Antonio, C.M., Milton, S.J., and Rejmanek, M. (2000). Plant invasions—the role of mutualisms. *Biol. Rev.* **75**, 65–93.
Romanuk, T.N., Zhou, Y., Brose, U., Berlow, E.L., Williams, R.J., and Martinez, N.D. (2009). Predicting invasion success in complex ecological networks. *Philos. Trans. R. Soc. Lond. B* **364**, 1743–1754.
Rooney, N., McCann, K.S., and Moore, J.C. (2008). A landscape theory for food web architecture. *Ecol. Lett.* **11**, 867–881.
Roughgarden, J. (1972). Evolution of niche width. *Am. Nat.* **106**, 683–718.
Sales-Pardo, M., Guimerà, R., Moreira, A.A., and Amaral, L.A.N. (2007). Extracting the hierarchical organization of complex systems. *Proc. Natl. Acad. Sci. USA* **104**, 15224–15229.
Scanlon, T.M., Caylor, K.K., Levin, S.A., and Rodriguez-Iturbe, I. (2007). Positive feedbacks promote power-law clustering of Kalahari vegetation. *Nature* **449**, 209–213.
Scheffer, M., Bascompte, J., Brock, W.A., Brovkin, V., Carpenter, S.R., Dakos, V., Held, H., Nes, E.H.V., Rietkerk, M., and Sugihara, G. (2009). Early-warning signals for critical transitions. *Nature* **461**, 53–59.

Schoenly, K., and Cohen, J.E. (1991). Temporal variation in food web structure: 16 empirical cases. *Ecol. Monogr.* **61**, 267–298.
Shin, Y.J., Rochet, M.J., Jennings, S., Field, J.G., and Gislason, H. (2005). Using size-based indicators to evaluate the ecosystem effects of fishing. *ICES J. Mar. Sci.* **62**, 384–396.
Simberloff, D., and Holle, B.v. (1999). Positive interactions of nonindigenous species: Invasional meltdown? *Biol. Invas.* **1**, 21–32.
Snow, B.K., and Snow, D.W. (1972). Feeding niches of hummingbirds in a Trinidad valley. *J. Anim. Ecol.* **41**, 471–485.
Solé, R.V., and Manrubia, S.C. (1995). Are rainforests self-organized in a critical state? *J. Theor. Biol.* **173**, 31–40.
Solé, R.V., and Montoya, J.M. (2001). Complexity and fragility in ecological networks. *Proc. R. Soc., Lond. B* **268**, 2039–2045.
Song, C., Halin, S., and Makse, H.A. (2005). Self-similarity of complex networks. *Nature* **433**, 392–395.
Song, C., Havlin, S., and Makse, H.A. (2006). Origins of fractality in the growth of complex networks. *Nat. Phys.* **2**, 275–281.
Southall, E.J., Dale, M.P., and Kent, M. (2003). Spatial and temporal analysis of vegetation mosaics for conservation: Poor fen communities in a Cornish mire. *J. Biogeogr.* **30**, 1427–1443.
Speirs, D.C., Gurney, W.S.C., Winterbottom, J.H., and Hildrew, A.G. (2000). Long-term demographic balance in the Broadstone Stream insect community. *J. Anim. Ecol.* **69**, 45–58.
Stald, L. (2003). Pollination Networks on the Canary Islands. MSc thesis, Aarhus University, Aarhus.
Storch, D., Marquet, P.A., and Brown, J.H. (Eds.) (2007). Scaling Biodiversity. Cambridge University Press, New York.
Strogatz, S.H. (2001). Exploring complex networks. *Nature* **410**, 268–276.
Strogatz, S.H. (2005). Romanesque networks. *Nature* **433**, 365.
Sugihara, G., and May, R.M. (1990). Applications of fractals in ecology. *TREE* **5**, 79–86.
Sugihara, G., and Ye, H. (2009). Cooperative network dynamics. *Nature* **458**, 979–980.
Summerhayes, V.S., and Elton, C.S. (1923). Contributions to the ecology of Spitsbergen and Bear Island. *J. Ecol.* **11**, 214–286.
Tavares-Cromar, A.F., and Williams, D.D. (1996). The importance of temporal resolution in food web analysis—Evidence from a detritus-based stream. *Ecol. Monogr.* **66**, 91–113.
Thies, C., Steffan-Dewenter, I., and Tscharntke, T. (2003). Effects of landscape context on herbivory and parasitism at different spatial scales. *Oikos* **101**, 18–25.
Thompson, J.N. (2005). The Geographic Mosaic of Coevolution. Chicago University Press, Chicago.
Thompson, R.M., and Townsend, C.R. (1999). The effect of seasonal variation on the community structure and food web attributes of two streams: Implications for food web science. *Oikos* **87**, 75–88.
Thøstesen, A.M., and Olesen, J.M. (1996). Pollen removal and deposition by specialist and generalist bumblebees in *Aconitum septentrionale*. *Oikos* **77**, 77–84.
Totland, Ø., Nielsen, A., Bjerknes, A.-L., and Ohlson, M. (2006). Effects of an exotic plant and habitat disturbance on pollinator visitation and reproduction in a boreal forest herb. *Am. J. Bot.* **93**, 868–873.

Traveset, A., and Richardson, D.M. (2006). Biological invasions as disruptors of plant reproductive mutualisms. *TREE* **21**, 208–216.
Traveset, A., and Sáez, E. (1997). Pollination of *Euphorbia dendroides* by lizards and insects: Spatio-temporal variation in patterns of flower visitation. *Oecologia* **111**, 241–248.
Tylianakis, J.M., Didham, R.K., Bascompte, J., and Wardle, D.A. (2008). Global change and species interactions in terrestrial ecosystems. *Ecol. Lett.* **11**, 1351–1363.
Ulrich, W., Almeida-Neto, M., and Gotelli, N.J. (2009). A consumer's guide to nestedness analysis. *Oikos* **118**, 3–17.
Vacher, C., Piou, D., and Desprez-Loustau, M.L. (2008). Architecture of an antagonistic tree/fungus network: The asymmetric influence of past evolutionary history. *PLoS ONE* **3**, e1740.
Vázquez, D.P. (2005). Degree distribution in plant–animal mutualistic networks: Forbidden links or random interactions? *Oikos* **108**, 421–426.
Vázquez, D.P., and Aizen, M.A. (2004). Asymmetric specialization: A pervasive feature of plant–pollinator interactions. *Ecology* **85**, 1251–1257.
Vázquez, D.P., and Simberloff, D. (2002). Ecological specialization and susceptibility to disturbance: Conjectures and refutations. *Am. Nat.* **159**, 606–623.
Vázquez, D.P., Melián, C.J., Williams, N.M., Blüthgen, N., Krasnov, B.R., and Poulin, R. (2007). Species abundance and asymmetric interaction strength in ecological networks. *Oikos* **116**, 1120–1127.
Vespignani, A. (2009). Predicting the behavior of techno-social systems. *Science* **325**, 425–428.
Wallace, J.B., Eggert, S.L., Meyer, J.L., and Webster, J.R. (1997). Multiple trophic levels of a forest stream linked to terrestrial litter inputs. *Science* **277**, 102–104.
Wallace, J.B., Eggert, S.L., Meyer, J.L., and Webster, J.R. (1999). Effects of resource limitation on a detrital-based ecosystem. *Ecol. Monogr.* **69**, 409–442.
Waltho, N., and Kolasa, J. (1994). Organization of instabilities in multispecies systems, a test of hierarchy theory. *Proc. Natl. Acad. Sci. USA* **19**, 1682–1685.
Ward, J.V., Bretschko, G., Brunke, M., Danielopol, D., Gibert, J., Gonser, T., and Hildrew, A.G. (1998). The boundaries of river systems: The metazoan perspective. *Freshw. Biol.* **40**, 531–569.
Warren, P.H. (1989). Spatial and temporal variation in the structure of a freshwater food web. *Oikos* **55**, 299–311.
Warren, P.H. (1996). Structural constraints on food web assembly. In: *Aspects of the Genesis and Maintenance of Biological Diversity* (Ed. by M.E. Hochberg, J. Clobert and R. Barbault), pp. 142–161. Oxford University Press, Oxford.
Waser, N.M., and Ollerton, J. (Eds.) (2006). Plant–Pollinator Interactions: From Specialization to Generalization Chicago University Press, Chicago.
Watts, D.J., and Strogatz, S.H. (1998). Collective dynamics of 'small-world' network. *Nature* **393**, 440–442.
Williams, R.J., and Martinez, N.D. (2000). Simple rules yield complex food webs. *Nature* **404**, 180–183.
Williams, R.J., Berlow, E.L., Dunne, J.A., Barabási, A.-L., and Martinez, N.D. (2002). Two degrees of separation in complex food webs. *Proc. Natl. Acad. Sci. USA* **99**, 12913–12916.
Winder, M., and Schindler, D.E. (2004a). Climatic effects on the phenology of lake processes. *Global Change Biol.* **10**, 1844–1856.
Winder, M., and Schindler, D.E. (2004b). Climate change uncouples trophic interactions in an aquatic ecosystem. *Ecology* **85**, 2100–2106.

Winemiller, K.O. (1990). Spatial and temporal variation in tropical fish trophic networks. *Ecol. Monogr.* **60**, 331–367.
Winemiller, K.O. (1996). Factors driving temporal and spatial variation in aquatic floodplain food webs. In: *Food Webs: Integration of Patterns and Dynamics* (Ed. by G.A. Polis and K.O. Winemiller), pp. 1298–1312. Chapman and Hall, New York.
Winemiller, K.O., and Jepsen, D.B. (1998). Effects of seasonality and fish movement on tropical river food webs. *J. Fish Biol.* **53**, 267–296.
Winemiller, K.O., and Layman, C.A. (2005). Food web science: Moving on the path from abstraction to prediction. In: *Dynamic Food Webs: Multispecies Assemblages, Ecosystem Development, and Environmental Change* (Ed. by P.D. Ruiter, V. Wolters and J.C. Moore), pp. 10–23. Elsevier and AP, Burlington.
Wipfli, M.S., and Caouette, J. (1998). Influence of salmon carcasses on stream productivity: Response of biofilm and benthic macroinvertebrates in southeastern Alaska, USA. *Can. J. Fish. Aquat. Sci.* **55**, 1503–1511.
Witze, A. (2008). Losing Greenland. *Nature* **452**, 798–802.
Woodward, G. (2009). Biodiversity, ecosystem functioning and food webs in freshwaters: Assembling the jigsaw puzzle. *Freshw. Biol.* **54**, 2171–2187.
Woodward, G., and Hildrew, A.G. (2001). Invasion of a stream food web by a new top predator. *J. Anim. Ecol.* **70**, 273–288.
Woodward, G., and Hildrew, A.G. (2002a). Food web structure in riverine landscapes. *Freshw. Biol.* **47**, 777–798.
Woodward, G., and Hildrew, A.G. (2002b). Body-size determinants of niche overlap and intraguild predation within a complex food web. *J. Anim. Ecol.* **71**, 1063–1074.
Woodward, G., and Warren, P.H. (2007). Body size and predatory interactions in freshwaters: Scaling from individuals to communities. In: *Body Size: The Structure and Function of Aquatic Ecosystems* (Ed. by A.G. Hildrew, D. Raffaelli and R. Edmonds-Brown), pp. 98–117. Cambridge University Press, Cambridge.
Woodward, G., Jones, J.I., and Hildrew, A.G. (2002). Community persistence in Broadstone Stream (U.K.) over three decades. *Freshw. Biol.* **47**, 1419–1435.
Woodward, G., Speirs, D.C., and Hildrew, A.G. (2005a). Quantification and temporal resolution of a complex size-structured food web. *Adv. Ecol. Res.* **36**, 85–135.
Woodward, G., Ebenman, B., Emmerson, M., Montoya, J.M., Olesen, J.M., Valido, A., and Warren, P.H. (2005b). Body size in ecological networks. *TREE* **20**, 402–409.
Woodward, G., Perkins, D.M., and Brown, L.E. (2010a). Climate change in freshwater ecosystems: Impacts across multiple levels of organisation. *Philos. Trans. Roy. Soc. B* **365**, 2093–2106.
Woodward, G., Benstead, J.P., Beveridge, O.S., Blanchard, J., Brey, T., Brown, L., Cross, W.F., Friberg, N., Ings, T.C., Jacob, U., Jennings, S., Ledger, M.E., *et al.* (2010b). Ecological networks in a changing climate. *Adv. Ecol. Res.* **42,** 71–138.
Wooton, J.T. (1997). Estimates and tests of per capita interaction strength: Diet, abundance, and impact of intertidally foraging birds. *Ecol. Monogr.* **67**, 45–64.
Yamazaki, K., and Kato, M. (2003). Flowering phenology and anthophilous insect community in a grassland ecosystem at Mt. Yufu, Western Japan. *Contrib. Biol. Lab. Kyoto Univ.* **29**, 255–318.
Yodzis, P. (1988). The indeterminacy of ecological interactions as perceived through perturbation experiments. *Ecology* **69**, 508–515.
Yodzis, P. (1998). Local trophodynamics and the interaction of marine mammals and fisheries in the Benguela ecosystem. *J. Anim. Ecol.* **67**, 635–658.
Yvon-Durocher, G., Reiss, J., Blanchard, J., Ebenman, B., Perkins, D.M., Reuman, D.C., Thierry, A., Woodward, G., and Petchey, O.L. (2010). Across ecosystem comparisons of size structure: Methods, approaches, and prospects. Oikos (in press).

Ecological Networks in a Changing Climate

GUY WOODWARD, JONATHAN P. BENSTEAD, OLIVER S. BEVERIDGE, JULIA BLANCHARD, THOMAS BREY, LEE E. BROWN, WYATT F. CROSS, NIKOLAI FRIBERG, THOMAS C. INGS, UTE JACOB, SIMON JENNINGS, MARK E. LEDGER, ALEXANDER M. MILNER, JOSE M. MONTOYA, EOIN O'GORMAN, JENS M. OLESEN, OWEN L. PETCHEY, DORIS E. PICHLER, DANIEL C. REUMAN, MURRAY S.A. THOMPSON, FRANK J.F. VAN VEEN AND GABRIEL YVON-DUROCHER

ADVANCES IN ECOLOGICAL RESEARCH VOL. 42 0065-2504/10 $35.00

DOI: 10.1016/S0065-2504(10)42002-4

SUMMARY

Attempts to gauge the biological impacts of climate change have typically focussed on the lower levels of organization (individuals to populations), rather than considering more complex multi-species systems, such as entire ecological networks (food webs, mutualistic and host–parasitoid networks). We evaluate the possibility that a few principal drivers underpin network-level responses to climate change, and that these drivers can be studied to develop a more coherent theoretical framework than is currently provided by phenomenological approaches. For instance, warming will elevate individual ectotherm metabolic rates, and direct and indirect effects of changes in atmospheric conditions are expected to alter the stoichiometry of interactions between primary consumers and basal resources; these effects are general and pervasive, and will permeate through the entire networks that they affect. In addition, changes in the density and viscosity of aqueous media could alter interactions among very small organisms and disrupt the pycnoclines that currently compartmentalize many aquatic networks in time and space. We identify a range of approaches and potential model systems that are particularly well suited to network-level studies within the context of climate change. We also highlight potentially fruitful areas of research with a view to improving our predictive power regarding climate change impacts on networks. We focus throughout on mechanistic approaches rooted in first principles that demonstrate potential for application across a wide range of taxa and systems.

I. INTRODUCTION

The Earth's average surface temperature is predicted to rise by 3–5 °C over the next century, far faster than previously experienced by human civilization (Parmesan and Yohe, 2003; Pounds *et al.*, 1999; Thomas *et al.*, 2004). Even larger increases (up to 7.5 °C) are projected for some Arctic regions over the same timeframe based on Global Climate Model simulations (IPCC, 2007). Although there are major uncertainties in these estimates, climate change is nevertheless likely to place considerable environmental stress on many natural systems in the near future (Walther, 2010). Indeed, dramatic changes have already been reported from many ecosystems in recent decades, especially in the Arctic, in the West Antarctic peninsula region and at high altitudes (e.g. Brooks and Birks, 2004; Douglas *et al.*, 1994; Konig *et al.*, 2002; Schofield *et al.*, 2010; Smol *et al.*, 2005) as greenhouse gas emissions and global temperatures have increased. Rising temperatures are just one component of climate change, albeit the one most intensively studied by biologists (Walther, 2010; Walther *et al.*, 2002), and changes in atmospheric composition, the physical properties of aqueous media and weather conditions will also alter the physical environment within which ecosystems operate. The effects of climate change on organisms can therefore be direct (e.g. warming increases ectotherm metabolic rates, or stress on physiological systems), indirect (e.g. shrinking and fragmentation of fresh waters during drought, species effects that are mediated by another species) or a combination of both (Woodward *et al.*, 2010b).

Although the effects of climate change will permeate all levels of biological organization, most research has focused on responses at the lower levels of organization (e.g. range shifts in species populations), and only a few studies have considered community- or ecosystem-level impacts (Hickling *et al.*, 2006; Montoya and Raffaelli, 2010; Parmesan, 2006; Sala *et al.*, 2000; Spooner and Vaughn, 2008; van der Putten *et al.*, 2010). Interspecific interactions within food webs or other types of multi-species ecological networks have been largely ignored within the context of climate change research (but see Emmerson *et al.*, 2005a; Harmon *et al.*, 2009; Ims and Fuglei, 2005; Meerhoff *et al.*, 2007; Woodward *et al.*, 2010a). This represents a critical bottleneck in our predictive ability because network-level responses to stressors cannot simply be extrapolated from studying single species in isolation (Kishi *et al.*, 2005; Raffaelli, 2004; Tylianakis *et al.*, 2008; Woodward, 2009). Part of the reason for this current knowledge gap undoubtedly stems from the perception that ecological networks, which may contain thousands of species and tens of thousands of links, are seemingly too complex to be easily predictable (Montoya *et al.*, 2006; Riede *et al.*, 2010). However, there is increasing evidence that structure and dynamics of even very complex networks might be underpinned by a few relatively simple and predictable rules based on foraging and metabolic constraints, and the distribution of interaction strengths (Beckerman *et al.*, 2006; Berlow *et al.*, 2009; Cohen *et al.*, 2003; McCann *et al.*, 1998; Montoya *et al.*, 2006; Petchey *et al.*, 2008; Reuman and Cohen, 2005; Williams and Martinez, 2000). A new perspective that considers the network level of organization, and its links to other levels, is needed to complement the current phenomenological approaches to develop a more general, mechanistic approach to predicting the impacts of climate change (Ings *et al.*, 2009; Montoya and Raffaelli, 2010; Tylianakis, 2009; Woodward *et al.*, 2010b).

Traditionally, the study of ecological networks has focused on consumer–resource relations in (mostly aquatic) food webs and terrestrial host–parasitoid systems, but recently this has been broadened by the dramatic surge in research on mutualistic networks (e.g. coral-zooxanthellae symbioses; plant–pollinator systems; plant–seed disperser systems) (Ings *et al.*, 2009; Montoya *et al.*, 2006; Olesen and Jordano, 2002; Olesen *et al.*, 2010). In reality, of course, many ecosystems contain all three network types, in addition to competitive interactions, and species may be operating simultaneously within more than one network (e.g. crab spiders that prey upon insect pollinators of flowering plants, which are themselves eaten by herbivores; Ings *et al.*, 2009). The scientific study of these multilayered 'super networks' is still too embryonic to make any firm generalizations at present (Olesen *et al.*, 2010), so the distinctions among the three network types delineated earlier have been retained here for tractability.

Several recent studies and reviews have considered how climate change might affect the higher levels of biological organization (Bascompte and Stouffer, 2009; Wrona *et al.*, 2006), but these have tended to be restricted to considering one type of network (e.g. food webs: Woodward *et al.*, 2010b; plant–pollinator networks: Memmott *et al.*, 2007), one type of system (e.g. terrestrial ecosystems: Tylianakis *et al.*, 2008; running waters: Perkins *et al.*, 2010; marine ecosystems: Hays *et al.*, 2005; Moran *et al.*, 2010; soils: Davidson and Janssens, 2006), one aspect of climate change (e.g. warming: McKee *et al.*, 2003; Yvon-Durocher *et al.*, 2010a,b), or they have emphasized the contingency rather than the potential generality of impacts and responses (e.g. Tylianakis *et al.*, 2008). Here, we aim to improve our currently limited ability to anticipate the effects of climate change on different network types, by attempting to identify emerging components of a research approach that spans multiple levels of organization and is based on general first principles. Consequently, much of the paper concerns how physical and chemical laws (e.g. fluid viscosity, thermodynamics and elemental stoichiometry) that underpin different aspects of climate change act upon individuals and how these effects might be scaled up to predict network-level responses. We also highlight research areas that currently lack general predictive ability at the network level (e.g. species-specific climate envelopes) and discuss to what extent these might be extended and ultimately incorporated into a more predictive, first-principles framework. We identify those systems that are most likely to provide fruitful avenues of further study, and highlight those that seem less promising, in an attempt to help focus future research efforts more efficiently. Finally, we investigate how the different components of climate change (e.g. temperature, atmospheric chemistry), when combined, might act in additive or synergistic ways. Our ultimate objective is to stimulate the development of a novel general theoretical framework to facilitate testable predictions about the likely responses of ecological networks to our changing climate.

Some foundations common to the research approaches we explore are presented in Section II, where we also address the diverse range of methods that will need to be employed. In Sections III–VI, we consider how best to assess the impacts of the different components of climate change we have identified. For each of these components, we identify generalities and remaining contingencies, whilst exploring how effects at lower levels of organization can be understood from general principles and aggregated to produce network-level understanding. Section VII considers how the different components of climate change addressed in Sections III–VI might combine and potentially interact to influence network-level responses in additive or non-additive ways.

II. THE FOUNDATIONS OF A FIRST-PRINCIPLES APPROACH

The general approaches we identify, within which researchers are beginning to understand network-level impacts of climate change as the aggregation of better understood, lower level processes, all rely heavily on quantifying individual variation within species. In particular, many are underpinned by the primacy of body size as an easily measured and unifying property of organisms that is strongly linked to metabolism, behaviour and resource acquisition, all of which are key drivers at the network level of organization. These foundational aspects of the several approaches we identify are elaborated in Section II.A, and some of the key relationships of ecological relevance that will be addressed throughout the paper are highlighted in Table 1. In contrast to the cross-cutting importance of body size and an individual-based perspective, a general understanding of the effects of climate on networks will require the use of a wide range of methodological approaches at different spatial and temporal scales, which are considered in Section II.B.

A. Individuals, Species and Body Size

Ecological networks comprise entities ('nodes') connected to one another by links that represent some form of biological interaction (e.g. herbivory, detritivory, predation, parasitism, pollination) (Ings *et al.*, 2009; Lafferty *et al.*, 2008). Nodes are aggregations of individuals, which are usually lumped together on the basis of taxonomic or functional similarity (Ings *et al.*, 2009; Reiss *et al.*, 2009). The first step towards developing a general framework is to define the relevant entities of interest associated with the nodes and links within a network.

A logical place to start is to use individual organisms, since that is the level of organization at which interactions actually occur, and then to aggregate these entities hierarchically, based on their taxonomic identity or functional roles, to form the network (e.g. Ings *et al.*, 2009; Woodward and Warren, 2007). In contrast to the traditional dichotomy of viewing networks from either a species or size-based perspective, more integrated approaches have

Table 1 Examples of relationships between key biological and environmental parameters that are size- and/or temperature dependent

Ecological phenomenon	Relationship	Network consequence	Examples
Sinking rate versus size (volume, V) in plankton	$aV^{2/3}$	Altered consumer ingestion rates and removal of plankton from euphotic zone to sediment	
Mass-specific metabolic rate ($time^{-1}$) versus size (body mass, M)	$aM^{-1/4}$	Mass-specific metabolic demands increase with size (≈trophic status). Larger species, higher in the food web have higher metabolic demands than smaller species at the lower trophic levels	Peters (1983) Lopez-Urrutia *et al.* (2006)
Nutrient diffusion (mol N (cell time)$^{-1}$) versus size; (D, molecular diffusivity; R, cell radius; ΔC, cell surface nutrient concentration—bulk medium nutrient concentration.	$\propto 4\pi RD\Delta C$	If CO_2 uptake is diffusion-limited, large phytoplankton should be at a competitive advantage as CO_2 levels rise: this could shorten pelagic food chains, if phytoplankton–zooplankton predator–prey mass ratios remain constant	Finkel *et al.* (2010) and references therein.
Population abundance versus size (body mass, M)	(a) $\approx aM^{-0.75}$ (b) $\approx MV^{-1}$	(a) within trophic levels (b) across trophic levels Energy flows from abundant, small species to large rare species within a trophic network	Brown *et al.* (2004), but see Reuman *et al.* (2008).

Size-class abundance versus size; (body mass, M)	$\approx aM^{-0.75}$	Assuming the use of a common energy source, smaller organisms should be favoured as temperatures rise: slopes may be conserved but intercepts and size ranges may shift (e.g. increased prevalence of small phytoplankton in marine systems Moran *et al.*, 2010)	Jennings and Brander (2010); Moran *et al.* (2010), but see Brown *et al.* (2004), Reuman *et al.* (2008)
Metabolic rate (time^{-1}) versus temperature	$\approx e^{-0.64/(k^*T)}$	Energetic demands increase with temperature—this could lead to elevated activity and potential increases in predator–prey encounter rates and interaction strengths	Woodward *et al.* (2002a), Brey (2010)

Independent variables are underlined (see also Finkel *et al.*, 2010; Woodward *et al.*, 2005b for additional relationships based on first principles).

been advocated recently, in which individuals are not only identified to a high level of taxonomic resolution (i.e. species populations) but also described in terms of their functional role, often based on their body size (Barnes *et al.*, 2008; Ings *et al.*, 2009; Woodward and Warren, 2007; Yvon-Durocher *et al.*, 2010b). Body size ($\approx$body mass) is a useful proxy for many functional attributes (especially in aquatic networks), because it captures a large amount of trophic information in a single dimension and, along with temperature, it largely determines an individual's basal metabolic rate and thus its energy requirements (Ings *et al.*, 2009). A key advantage of employing such individual-based networks is that they can be viewed from alternative perspectives simultaneously, allowing, for instance, the relative importance of taxonomic identity versus body size to be assessed (Petchey and Belgrano, 2010; Woodward *et al.*, 2010c). The role of body size in terrestrial networks is perhaps less clear than in aquatic systems, as seemingly idiosyncratic species traits (e.g. ovipositor or proboscis structure) appear to be relatively more important (Ings *et al.*, 2009; Tylianakis *et al.*, 2008). Nonetheless, it can still be an important descriptor of network structure in host–parasitoid networks (Cohen *et al.*, 2005), terrestrial food webs (Petchey *et al.*, 2002), soil food webs (Reuman *et al.*, 2009b) and even plant–pollinator networks (Stang *et al.*, 2006, 2009).

The fundamental links between body size and metabolism underpin a wide range of allometric scaling relationships that operate from the level of the cell, to individuals, to entire ecosystems (Allen *et al.*, 2005; Atkinson, 1994; Brown *et al.*, 2004; Emmerson *et al.*, 2005b; Peters, 1983; Reuman *et al.*, 2008, 2009a; Woodward *et al.*, 2005a,b). The ubiquity of the size-metabolism relationship offers a useful starting point for developing a general theoretical framework, especially because it is also linked directly to two key components of climate change: altered environmental temperature (Brown *et al.*, 2004; Clarke and Fraser, 2004) and consumer–resource CNP ratios (Allen and Gillooly, 2009), both of which are predicted to change significantly as concentrations of carbon-based gases in the atmosphere continue to rise.

B. Experiments and Surveys at Different Scales and Examples

Since it is impossible to carry out a replicated study of climate change on a truly global scale, we need to approach the problem in a layered fashion, using a range of techniques and model systems to build towards a general overview (Figure 1). This is a challenging task, not least because research groups typically act in relative isolation from one another and often use their own bespoke approaches to understanding a particular aspect of climate change. In addition, most ecological research is conducted over short time-scales and at small spatial scales, due largely to logistical and financial constraints, so there is often a mismatch between the scales at which the

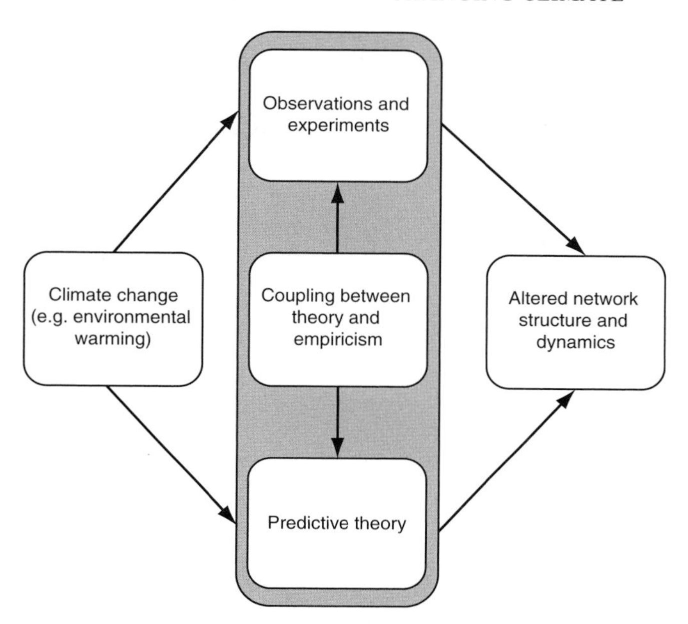

Figure 1 Understanding and predicting the changes in network structure and dynamics that result from climate change (with increasing environmental temperature shown as the example here) require empiricism, theory and close linkage of the two. This link has been made, for example by focusing on the effects of temperature on individuals, their metabolism, their foraging behaviour and other individual level processes. The use of mechanistic models of such effects allows predictions in novel circumstances, which can be validated against observational and experimental studies. Differences between model predictions and empirical results can be used to improve and refine the predictive theory, making the grey shaded region an iterative an continuous process.

environmental drivers operate and the ecological responses are manifested (Kratz *et al.*, 2003; Lane, 1997; Olesen *et al.*, 2010). Ideally, when gauging network-level responses to climate change, studies should run for at least as long as one generation of the longest lived species in the system to take account of as many potential feedback loops as possible because such indirect effects can be critical determinants of how the system responds to perturbations (e.g. Montoya *et al.*, 2009; Yodzis, 1988). The key point here is that time is not absolute *per se*, but a relative measure based on generation times. Unfortunately, it is unrealistic to run such studies in most natural systems as the requisite time could amount to several decades. Shortcuts or alternative approaches are therefore often necessary, such as reconstructing networks by using space-for-time substitutions (Rawcliffe *et al.*, 2010);

hindcasting from preserved samples or palaeoecological cores (e.g. Layer *et al.*, 2010a); exploring the impacts of thermal pollution (e.g. Langford, 1990) or geothermal warming in 'natural experiments' (e.g. Friberg *et al.*, 2009; Woodward *et al.*, 2010a); carrying out manipulative field experiments in mesocosms (e.g. Yvon-Durocher *et al.*, 2010a); the use of experimental microcosms and short-lived microbial and protist taxa as model systems (e.g. Petchey *et al.*, 1999); employing *in silico* simulations (Petchey *et al.*, 2010) or some combination of the above (e.g. Layer *et al.*, 2010b; Montoya *et al.*, 2009).

Space-for-time substitutions represent one of the most widely used approaches in ecological climate change research, but two questions need to be addressed when conducting such studies (e.g. Castella *et al.*, 2001): can spatial snapshots be extrapolated to describe temporal change, and are equilibrial or non-equilibrial conditions being measured? Essentially, both relate to the extent to which ecological responses can keep pace with changing environmental conditions: for instance, transient, non-equilibrial behaviour may be manifested if warming occurs faster than the rate at which the biota can respond, whereas (new) equilibrial conditions may be achieved if the rate of change is relatively slow. Bearing these points in mind, we can identify some idealized criteria for helping to select suitable systems to work with, including: (1) networks should be constructed by direct observation of nodes and links, with body mass, CNP tissue content data and abundance measured, where possible; (2) the system can be perturbed experimentally to simulate some aspect of climate change and to measure its response; (3) measurements of responses can be made at relevant spatial and temporal (generational) scales, across meaningful environmental gradients, and with a minimum of potential confounding effects. In reality, of course, it will often not be possible to meet more than one of these within a given study system. We have identified a sample of several 'model systems' that meet at least some of these criteria and offer potentially promising avenues for future network-based climate change research, which we highlight briefly here and then return to address in more detail later. These systems occupy different portions of the control/realism-replication spectrum, from laboratory-based microcosm experiments to large-scale survey data from close-to-pristine natural ecosystems, such as the Antarctic Weddell Sea food web. This list is not exhaustive and arguably biased towards systems that we are most familiar with, and each has its own advantages and disadvantages, as outlined in Table 2.

At small spatiotemporal scales, laboratory-based experiments with microbial communities provide a high level of control and replication and can be run for many generations over a relatively short time span (weeks to months; Petchey *et al.*, 1999). They have been criticized due to a perceived lack of realism, but nonetheless they represent valuable models for testing cause-and-effect in ways that are not otherwise possible (Daehler and

Table 2 The tool box of approaches used in the study of ecological networks in a changing climate

Approach [usefulness in addressing network responses to climate change]	Strengths	Weaknesses	Examples of community/ecosystem/network-level studies
Ancient fossil data [Limited]	1. True temporal change over deep time (millions of years) 2. Ability to identify recurrent network patterns with contemporary systems despite complete species turnover	1. Inferential and incomplete networks 2. Diets of extinct species unknown 3. Species abundances cannot be quantified accurately	Dunne *et al.* (2008)
Palaeodata [Limited]	1. True temporal change (millennial scales) 2. Good preservation of many key taxa (e.g. bones and scales from vertebrates, chitin from insects and pollen or diatom valves from primary producers) in soils/sediment	1. Potential confounding gradients and additional stressors (e.g. post-1850 eutrophication and acidification in UK) 2. Inferred feeding links; problems with quantification and selective preservation (soft-bodied organisms are often missing from the record) 3. Rates of change faster now than in palaeo record, and reduced potential for adaptation due to recent biodiversity loss	Battarbee (2000), Rawcliffe *et al.* (2010)

(*continued*)

Table 2 *(continued)*

Approach [usefulness in addressing network responses to climate change]	Strengths	Weaknesses	Examples of community/ecosystem/network-level studies
Contemporary temporal surveys [Yes, but correlational]	1. True temporal change (decadal scales) 2. Nodes and links can be observed directly for entire network	1. Very few suitable long-term (decadal) data sets. 2. Potential confounding temporal gradients (e.g. acidification, eutrophication) and other stressors.	Woodward *et al.* (2002), Durance and Ormerod (2007, 2009), Milner *et al.* (2008), Layer *et al.* (2010a)
Space-for-time [Yes, but correlational]	1. 'Proxy stressors' 2. Nodes and links can be observed directly for entire network	1. Biogeographical confounding effects (but this might not be a problem for trait-based/size-based approaches). 2. Not mechanistic. 3. Difficult to distinguish transient and equilibrial dynamics	Lavandier and Décamps (1983), Parker and Huryn (2006), Milner *et al.* (2000)
'Natural experiments' [Yes, but correlational]	1. 'Proxy stressors' 2. No biogeographical confounding effects. 3. Naturally assembled communities 4. High realism 5. Nodes and links can be observed directly for entire network	1. Potentially confounding physical–chemical gradients 2. Lack of control of community composition 3. Limited/no replication 4. Difficult to distinguish transient and equilibrial dynamics	Friberg *et al.* (2009), Woodward *et al.* (2010a)

Field experiments [Yes—partial-whole networks]	1. 'True stressors' 2. Greater degree of control than in surveys—some potential for isolating or combining components of climate change 3. Replication is possible 4. Nodes and links can be observed directly 5. Mechanistic	1. Restricted to small spatiotemporal scales (typically intragenerational) 2. Require rigorous assessment of suitability—for example effect size versus spatial scale × temperature 3. Partial realism	Hogg *et al.* (1995), O'Gorman and Emmerson (2009), Woodward and Hildrew (2002a)
Lab experiments [Yes—partial-whole networks]	1. 'True stressors' 2. High level of control than in surveys—ability to isolate or combine components of climate change 3. High replication is possible 4. Nodes and links can be observed directly 5. Mechanistic	1. Restricted to small spatial scales 2. Require rigorous assessment of suitability—for example effect size versus spatial scale × temperature 3. Limited realism/complexity	Petchey *et al.* (1999) Woodward and Hildrew (2002b)
Models and simulations [Yes]	1. Long-term dynamics and intergenerational change 2. Isolation of key drivers 3. Mechanistic and predictive 4. Inexpensive	1. Limited realism 2. Constrained by gaps in theoretical frameworks and data for parameterization (e. g. phenological matches-mismatches)	Petchey *et al.* (2010)

Strong, 1996). At the opposite end of the control/realism-replication spectrum, long-term surveys of natural systems provide insight into potential responses to climate change in real food webs, but replication is often limited (or non-existent) and there is the ever-present risk that confounding effects (e.g. eutrophication, acidification) could mask potential responses to climate change (e.g. Durance and Ormerod, 2007, 2009). Long-term biomonitoring programmes of natural ecosystems at appropriate scales (decades to centuries) for detecting climatic signals are extremely scarce. In addition, the time scales over which natural networks have assembled may differ substantially from those at which climate change is now operating and, as a result, space-for-time substitutions can also be prone to potentially confounding spatial or bio-geographical effects (e.g. latitude, altitude) (Johnson and Miyanishi, 2008). If these can be minimized, however, such studies can provide important insights with a degree of realism that cannot be captured in experimental studies. Often a combination of approaches can greatly enhance the overall picture: for instance, in empirical surveys of natural systems, Milner *et al.* (2000, 2008) studied glacier retreat through real time over several decades, in addition to using space-for-time substitutions for stream communities in Alaska. Another example of a space-for-time substitution study comes from a catchment of 15 geothermal Icelandic streams that provide a natural experiment for studying the effects of environmental warming in a natural setting, within which field manipulations have also been carried out to disentangle the effects of temperature and nutrient limitation (Friberg *et al.*, 2009; Woodward *et al.*, 2010a). This system is unusual in that all the streams are very close to one another (< 2 km apart) and embedded in the same stream network, yet each has a characteristic temperature regime over its entire length (ranging from 5 to 25 °C, with one 'outlier' at 44 °C), and no additional confounding effects of water chemistry. This allows the effects of temperature to be isolated within a large-scale and long-term 'natural experiment'.

A few intensively studied ecosystems have the dual advantages of possessing extensive time series of data whilst also being amenable to the characterization and experimental manipulation of their ecological networks. Three such model systems include the Ythan Estuary in Scotland (e.g. Emmerson and Raffaelli, 2004), Broadstone Stream in southern England (Hildrew, 2009; Layer *et al.*, 2010b) and Lough Hyne in Southern Ireland (O'Gorman *et al.*, 2008, 2010; O'Gorman and Emmerson, 2010), all of which have been studied for several decades and now have exceptionally well-characterized food webs.

The food web of Broadstone Stream has been studied intensively since the early 1970s: it is strongly size-structured and because encounter rates and hence attack rates increase with temperature, so too do interaction strengths (Hildrew, 2009; Woodward and Hildrew, 2002a; Woodward and Warren, 2007; Woodward *et al.*, 2005a,b). There is some evidence that the food web has altered over four decades in response to an interaction between climatic and pH change. This has been manifested by the invasion and establishment of a new top predator (the dragonfly *Cordulegaster boltonii*), in the 1990s, followed by the more recent establishment of the first vertebrate (brown trout) (Layer *et al.*, 2010b), which has now usurped the dragonfly as the apex predator. These changes, and similar responses seen in long-term data from other acid freshwaters, have been attributed to an interaction between climate change and an amelioration of acidification (Ledger and Hildrew, 2005; Woodward *et al.*, 2002), as prolonged hot, dry summers reduce acid inputs, opening a window of opportunity for new colonists to establish themselves within the food web (Hildrew, 2009). An intriguing aspect of these data is that these invasions occur at the top of the food web. This is mirrored in the increased predominance of trout in geothermally warmed Icelandic streams (Woodward *et al.*, 2010a), which suggests that warming might reduce energetic constraints on secondary production in these food webs.

Lough Hyne is a large, sheltered marine reserve (Figure 2), in which intensive scientific studies have been ongoing since the late 1920s. A large database of information has built up over this period (Wilson, 1984), with extensive time series data available for some of the key species (e.g. Barnes *et al.*, 2002; Figure 2A). Recent attempts have also been made to quantify the benthic compartment of the food web, combined with investigations of ecosystem functioning and metabolic theory (O'Gorman *et al.*, 2008; Yvon-Durocher *et al.*, 2008; Figure 2B). Experimental manipulations of experimental food webs within the Lough have shown that larger species have weaker interaction strengths per unit biomass (as opposed to numerical *per capita* effects) than do smaller species, so the loss of the former taxa, which is often predicted as a consequence of climate change, should result in increased mean interaction strengths and has the potential to destabilize the trophic network (O'Gorman and Emmerson, 2010).

In addition to these 'traditional' food webs that describe interactions between predators and prey and between primary consumers and basal resources, insect host–parasitoid networks also represent useful model systems, particularly because their links can be readily observed and quantified (Ings *et al.*, 2009; van Veen *et al.*, 2006). The typically short generation times of the species involved mean that responses to environmental change can happen quickly, and these systems have already proven useful to measure community-level impacts of land use (MacFadyen *et al.*, 2009; Tylianakis *et al.*, 2007). The population dynamics of host–parasitoid interactions have been studied extensively (Hassell, 2000) and models of pair-wise species interactions have been extended to describe the dynamics of at least simple experimental communities very successfully (Bonsall and Hassell, 1998; van Veen *et al.*, 2005). In principle then, host–parasitoid networks have the potential to provide important insights into how climate change might affect multi-species communities, and to test theoretical predictions in the real

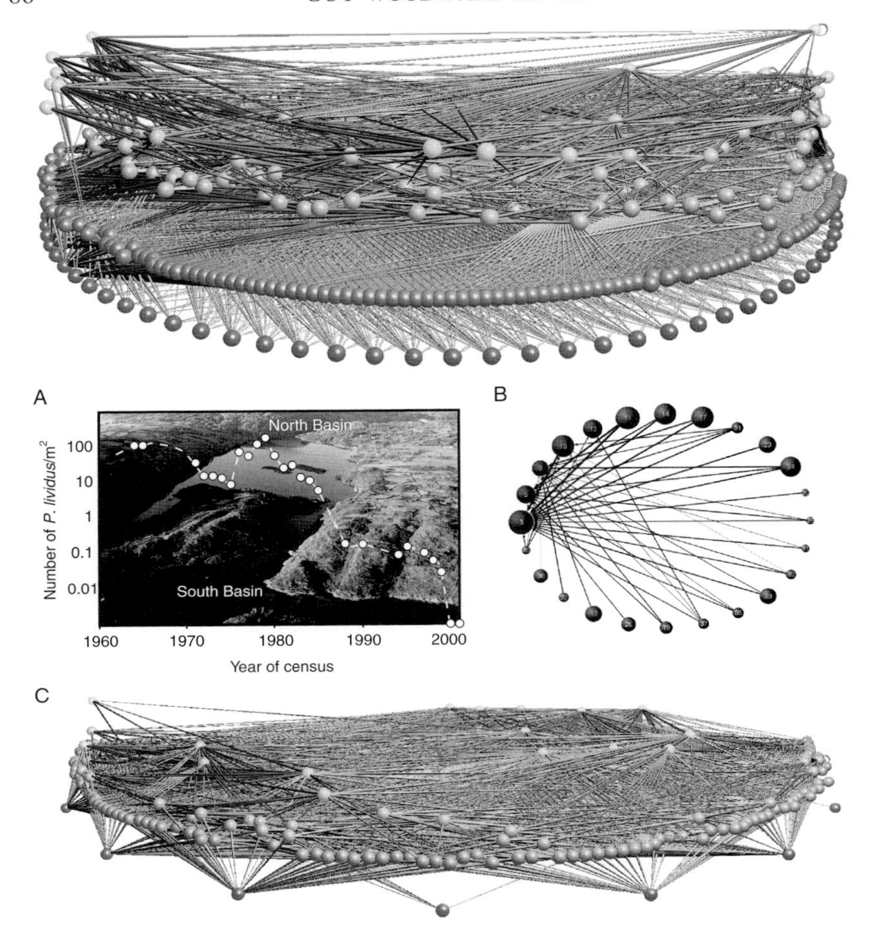

Figure 2 The greater Lough Hyne food web (U. Jacob, unpublished data), highlighting studies using survey-based data versus experimental field trials. (A) Long-term population change in a keystone species within the food web. Population fluctuation of *Paracentrotus lividus* abundance in the South Basin of Lough Hyne (after Barnes *et al.*, 2002). Insert is the position of Lough Hyne in Ireland. (B) A quantitative portion of the food web from the sublittoral zone is depicted for a guild of predatory fishes and their prey, with the width of arrows representing the strength of the interaction (Yvon-Durocher *et al.*, 2008 for full details) and the size of the nodes representing total population biomass. (C) A subset of 10 consumer species were manipulated in naturally assembling experimental communities as part of a long-term field trial (O'Gorman and Emmerson, 2009). Food webs such as the one shown represent replicable communities that mirror the complexity of the greater Lough Hyne food web.

world within a feasible timeframe. It must, however, also be borne in mind that there are some important differences between host–parasitoid and predator–prey networks, especially in the degree of diet specialization and hence connectance, due to the nature of the interactions involved, at least in terms of the *per capita* interaction strengths between consumers and resources (Ings *et al.*, 2009; van Veen *et al.*, 2005), so some caution should be exercised when making wider generalizations based on these systems.

III. NETWORK RESPONSES TO CLIMATE CHANGE COMPONENTS: IMPACTS OF WARMING ON ORGANISMS

Environmental warming is probably the most familiar manifestation of global climate change (e.g. Burgmer *et al.*, 2007; Deutsch *et al.*, 2008; Tylianakis, 2009) and its effects have been documented at all levels of biological organization, at least in controlled systems, although most of the research to date has focused on the lower levels. Because of the strong and consistent temperature dependence of many physiological rates (Bystrom *et al.*, 2006; Charnov, 2003; Gillooly *et al.*, 2001; Ings *et al.*, 2009), studying the effects of warming offers promise for developing a systematic approach within which individual-level effects can be scaled up to the network level. For instance, within a network, temperature can have both direct effects (e.g. on the sum of metabolic activity of all individuals in the population) and indirect effects, mediated by their influences on the attributes of organisms and their interactions. Here, we review the current evidence for the effects of warming on individual-level physiological rates, interactions among individuals and populations, and the mostly phenomenological information on its influence at the higher levels of organization. In parallel, we also assess the potential for integration into a general approach and identify some important remaining contingencies that merit further study. These relate mostly to indirect effects of warming through other physical processes, the implications of which, though likely far reaching, are still too poorly understood to be incorporated into any general framework at present.

A. Individual-Level Effects

Body temperature affects a plethora of individual-level physiological rates, including metabolic rate (Brey, 2010; Clarke and Johnston, 1999; Gillooly *et al.*, 2001), life history milestones (Charnov, 2003; Charnov and Gillooly, 2004), growth (Gillooly *et al.*, 2001), and even rates of nucleotide substitution in molecular evolution (Gillooly *et al.*, 2005; but see Lanfear *et al.*, 2007). The effects of warming have now been documented at all levels of biological organization, in laboratory-based microcosms (e.g. Figure 3) as well as in metastudies, and at the individual-level environmental temperature is a strong determinant of reproductive rates (Muller and Geller, 1993), cell size (Atkinson *et al.*, 2003; Richardson and Schoeman, 2004), and movement rates (Petchey *et al.*, unpublished data; Woodward and Hildrew, 2002a). These individual-level responses appear broadly consistent with the effects of temperature on biochemical reactions, as typically described by the Arrhenius relationship (e.g. Brey, 2010; Clarke and Johnston, 1999; Gillooly *et al.*, 2001). The literature on the effects of temperature on individual-level and species-level physiological rates and life-history parameters is large (e.g. Beverton and Holt, 1959; Clarke and Johnston, 1999; Dunham *et al.*, 1989), and we provide only a cursory overview here. In addition, many of the well-known allometric body-size scaling relationships (e.g. Peters, 1983; Yodzis and Innes, 1992) merit augmentation by fitting temperature as an additional component to predict ecological responses (Clarke, 2006). These

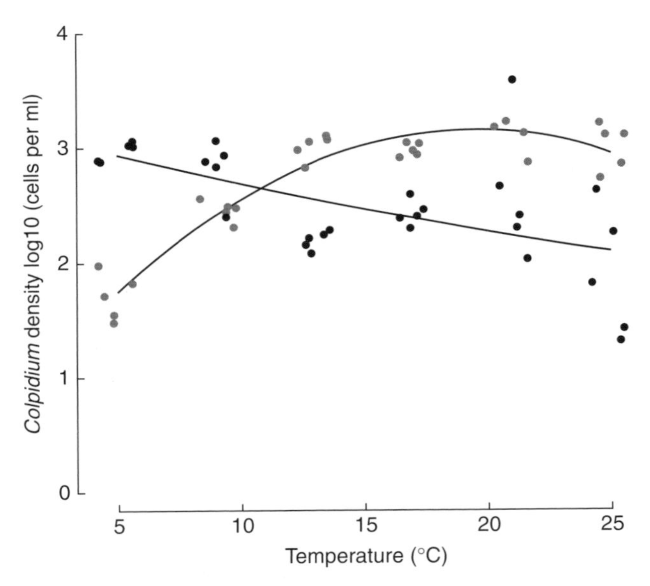

Figure 3 Interspecific interactions fundamentally change the impact of temperature on population dynamics. Grey circles are *Colpidium* cultured with only bacteria. Black are *Colpidium* cultured also in the presence of their predator *Didinium*. Circles are the mean population densities over 2 weeks, lines are statistically significant models (at $p < 0.05$). An artificial x-axis offset has been applied to aid clarity of interpretation. Reproduced with permission from Beveridge *et al.* (2010) in which full details of the work can be found.

individual-based relationships have ramifications that filter upwards to the higher levels of organization: for example elevated metabolic rates could increase interaction strengths and hence reduce the dynamic stability of the network as a whole (Kokkoris *et al.*, 2002).

B. Interaction-Level Effects

Feeding and other interactions between individuals are the final manifestation of a series of discrete steps, each of which is related to physiological rates that can depend on temperature. For feeding to occur, and hence for a trophic link to be manifested within a network, the sequence of these steps ranges from detection, to capture, to consumption of resources, to digestion, each of which is often strongly temperature and size dependent (e.g. Woodward and Hildrew, 2002a). This provides a useful starting point for developing firmer predictions as to how networks as a whole might respond to warming. For instance, because warming tends to increase mobility and activity (especially among ectotherms), attack rates should rise with increasing temperature and handling times should shorten, which will in turn affect ingestion rates and *per capita* interaction strengths within the network (e.g. Woodward and Hildrew, 2002a,c; Figure 4). Estimates of the activation energies of parasitism rate, attack rate, feeding rate, grazing rate and ingestion rate range from 0.46 to 0.81 eV (Brown *et al.*, 2004; Vasseur and McCann, 2005), and the few data that are currently available suggest that handling times may have activation energies that range from 0.13 to 0.71 eV (Petchey *et al.*, 2010). At present, there is still no firm consensus among the still too-few studies that have estimated the temperature dependence of ingestion (Kingsolver and Woods, 1998; Rall *et al.*, 2010; Sarmento *et al.*, 2010; Thompson, 1978; Woodward and Hildrew, 2002a) and assimilation rates (e.g. Giguere, 1981; Short and Ward, 1981a,b; Zhang and Li, 2004).

C. Population and Community-Level Effects

A key question that needs to be addressed at the level of organization of interactions between populations (or size classes) is: does the efficiency of the trophic transfer of biomass from prey to consumer assemblages change with temperature? For instance, the size spectrum of a community allows inferences to be made about trophic transfer efficiency within a food web without the need for direct observations of feeding links, as long as information on predator–prey mass ratios is available (Jennings and Brander, 2010). There is a wealth of size-spectrum data from marine systems, within which size-based (i.e. ‘individual size distributions’) rather than species-based (i.e. after Layer

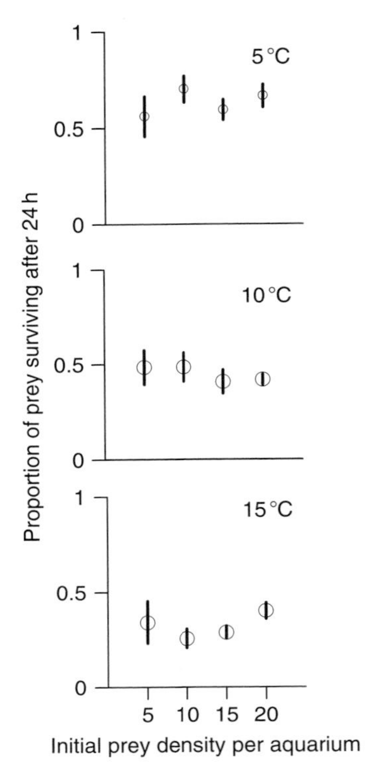

Figure 4 Temperature-dependent changes in interaction strengths between the top predator (the dragonfly *C. boltonii*) and a dominant prey species (the stonefly *Nemurella pictetii*) within the Broadstone Stream food web. Replicated laboratory experiments were carried out using aquaria maintained at three temperatures: prey survival declined as temperatures increased, due to elevated encounter rates resulting from increased activity levels (redrawn after Woodward *et al.*, 2002a).

et al., 2010b; McLaughlin *et al.*, 2010; O'Gorman and Emmerson, 2010) Mass-abundance (*MN*) scaling relations have typically been used to infer food web structure and energetics. An in-depth analysis of how mass-abundance scaling relationships and transfer efficiencies might change across temperature gradients (*v* across latitude or over long-term series) could offer important insights here, with potentially far-reaching implications for fisheries management (Jennings and Brander, 2010; Perry *et al.*, 2010). There is evidence from correlational studies that *MN* slopes are surprisingly consistent, even in networks that vary fundamentally in species composition and are found at different temperatures. Boudreau and Dickie (1992), for instance, compared size spectra in a range of aquatic ecosystems and demonstrated that slopes were independent of temperature, whereas changes in intercepts reflected differences in primary production. Marquet *et al.* (1990)

and Jonsson *et al.* (2005) have also provided evidence that *MN* slopes may be conserved within a narrow range in natural communities, despite considerable turnover in species composition, and that this conservatism appears to be maintained by density compensation coupled with body size shifts. Current evidence seems to suggest that, for a given level of primary production, the size structure, abundance, trophic level and consumer production might be broadly predictable and not strongly temperature dependent (Jennings and Brander, 2010). This implies that if climate-induced changes in primary production can be predicted, and if climate change does not fundamentally alter the rules of community assembly, responses in some properties of consumer assemblages (e.g. changes in the intercept of size-based *MN* relationships) can be predicted. This size-based approach could provide useful general null models with which to gauge the biological impacts of climate change in multi-species networks, even in the absence of explicitly characterized species populations (Jennings and Brander, 2010), because feeding links can be generalized from predator–prey body mass ratios that do not vary systematically with temperature or primary production at oceanic scales (Barnes *et al.*, 2010).

Moran *et al.* (2010) have recently combined allometric *MN* scaling relationships of such individual-based size distributions with the temperature-size rule to predict an increased prevalence of very small picophytoplankton within the overall marine phytoplankton assemblage as temperatures rise. They found that temperature alone accounted for 73% of the variance in the relative contribution of these small cells to total phytoplankton biomass, irrespective of differences in inorganic nutrient loading (Moran *et al.*, 2010), suggesting that environmental warming might be the primary driver among the different components of climate change.

D. Individual-to-Network Level Effects

At present, the available results at the network-based levels of organization, where nodes and the links between them are described explicitly (rather than implicitly, as in size-spectra approaches), are still largely phenomenological and less unified by theory at present than those at the lower levels. As such, they serve primarily as guides for developing theoretical frameworks that can integrate across levels: ultimately, emerging theory needs to be able to explain observed network-level properties as the outcome of the better-understood phenomena manifested at the lower levels of organization. Vermaat *et al.* (2009) identified connectance, richness and productivity as the three principal components of networks, each of which may be influenced (directly or indirectly) by temperature change. Higher level properties of networks, including the topological or dynamic stability of the system and the role of indirect effects, whereby nodes influence one another ‘at a distance' (Montoya *et al.*, 2009) are thus likely to be strongly temperature dependent, as are the lower levels of organization that underpin them.

If, for instance, the separate components of the predation sequence do indeed have different activation energies, as suggested earlier, this has implications for the network as a whole. If temperature dependencies vary between the different steps, warming could alter both food web structure (e.g. Petchey *et al.*, 2010) and interaction strengths (e.g. Woodward and Hildrew, 2002a), which are two key determinants of network stability (May, 1972, 1973; McCann, 2000). Recent theory predicts, for example that differential scaling in the temperature dependence of attack rates and handling times will alter the frequency of feeding interactions and, potentially, connectance within a network (Petchey *et al.*, 2010). Such changes have the potential to alter network dynamics, but the current dearth of detailed, standardized empirical data on these temperature dependencies and their effects on network structure makes testing these theoretical predictions impossible at present. This rather surprising gap in our understanding of two fundamental processes (handling times and attack rates) that underpin ecological networks clearly needs to be addressed systematically across a range of systems.

In an experimental study of protist food webs that included information on body size, species identity and feeding links, population extinction caused by gradual warming was much more frequent for larger species than for smaller ones (Petchey *et al.*, 1999), as has also often been predicted for larger metazoans (Raffaelli, 2004; Woodward *et al.*, 2005b). The rate of response and capacity to do so is related to the body size (and thus is also often linked to the trophic level) of the species involved, because larger species have slower turnover times (Woodward *et al.*, 2005b). Experimental manipulations of temperature and food chain length have revealed that the effect of warming on population size depends strongly on the number of trophic levels (Beveridge *et al.*, 2010) and interspecific competition (Jiang and Morin, 2004). Fox and Morin (2001), in contrast, found little evidence of temperature effects on population sizes, let alone interaction-dependent temperature effects. At the network and ecosystem level data are still scant, and few experiments bar that of Petchey *et al.* (1999) have characterized the responses of even moderately complex communities exposed to temperature manipulations. The increased vulnerability to warming among larger organisms at the higher trophic levels, as predicted by theory and reported by Petchey *et al.* (1999), appears to contrast with observations from Icelandic geothermal streams, where there was a lengthening of food chains and increased mean individual size and abundance of the largest predators (brown trout) as temperatures increased (Woodward *et al.*, 2010a; Figure 5). This is suggestive that the top predators in cold high latitude and/or altitude systems might be under physiological cold stress (cf. Hari *et al.*, 2006), and as a result they might benefit from bottom-up effects of warming via increased

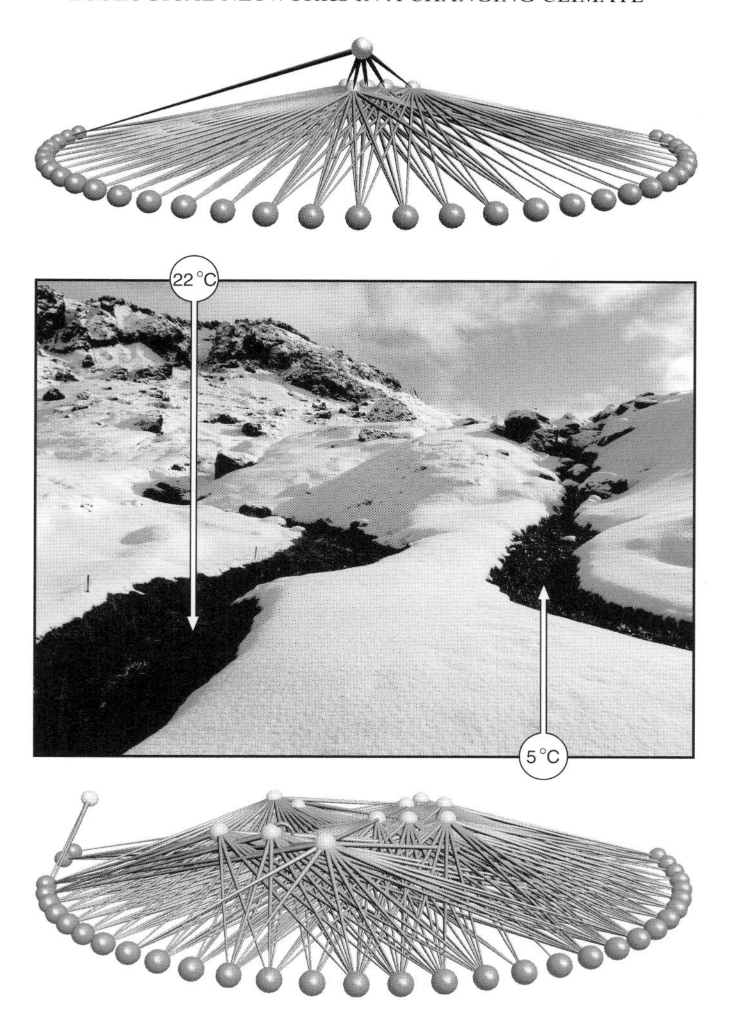

Figure 5 Icelandic stream food webs under ambient conditions (5 °C) and geothermal warming (22 °C), in April 2009, from within a catchment of 15 streams spanning a wide natural thermal range without additional confounding chemical gradients or dispersal constraints (Friberg *et al.*, 2009; Woodward *et al.*, 2010b). As temperatures increase across streams, the abundance and mean body size of brown trout, the apex predator, increase. The pair of streams exhibit this phenomenon: the cold stream (lower food web) is fishless and omnivorous invertebrates occupy the highest trophic level, whereas trout are the top predators in the adjacent warm stream (upper food web), less than 2 m away. Food web data: D. E. Pichler *et al.* (unpublished data).

primary production at the base of the food web (Friberg *et al.*, 2009). The seemingly different network-level responses seen in laboratory microcosms versus those observed in the field clearly merit further exploration from an individual-based perspective that can account for the true metabolic costs of operating at different temperatures.

E. Evolutionary and Biogeographical Effects of Warming

Most ecosystem-level studies have used 'static' temperature differences as a proxy measure of environmental warming, in which systems are maintained at a fixed level above ambient and compared with reference conditions (e.g. Liboriussen *et al.*, 2005; Yvon-Durocher *et al.*, 2010a; Figure 6). Notable exceptions include Petchey *et al.* (1999), who used a 2 °C warming per week over 7 weeks in laboratory microcosms, where one week was equivalent to ca. 100 generations of many of the protists within their assembled food webs. Long-term ecological responses to warming will also be accompanied by evolutionary processes (Harmon *et al.*, 2009; Tylianakis, 2009), and there are suggestions that the capacity for adaptation might be impaired at colder temperatures. The suggested reason for this is the link between individual metabolism, generation time and DNA evolution, which has been proposed as a possible explanation of latitudinal gradients in species richness (Allen and Gillooly, 2009; Stegen *et al.*, 2009) and hence the size of local ecological networks. If generation times are too long, this may hinder the ability of populations to turn over and for adaptations to spread fast enough to keep pace with a rapidly changing environment: once again, this should make large species in higher latitude networks particularly vulnerable to warming. It might also partially explain the rapid declines in Arctic char in Lake Windermere in the UK, a cold, deep lake that has seen increases in more eurythermic predatory fish in recent decades (Winfield *et al.*, 2010). If this is a particular example of a more general phenomenon related to an interaction between environmental temperature and long generation times in large, cold-stenothermic species, we might expect to see especially rapid turn-over in these higher trophic levels at higher latitudes and altitudes.

IV. NETWORK RESPONSES TO THE COMPONENTS OF CLIMATE CHANGE: IMPACTS OF WARMING ON AQUEOUS MEDIA

All ecological networks operate within a physical landscape that is itself influenced by temperature: consequently, warming can affect biota both directly (i.e. via metabolic constraints) and indirectly (i.e. via environmental constraints). For instance, climate change will not only influence the viscosity of body fluids and water, but also the form in which the latter occurs in the environment (i.e. vapour, rain, ice) and the amount and timing of its flux

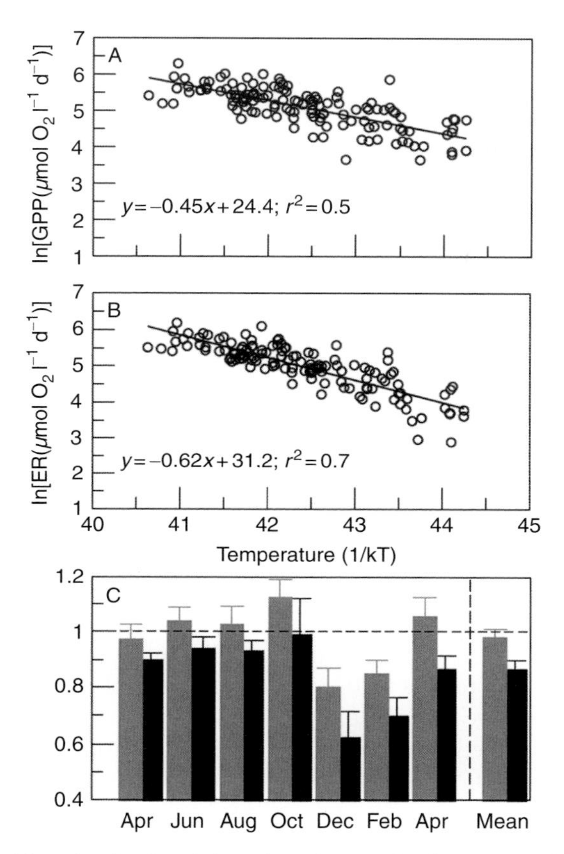

Figure 6 Watching the food web breathe: changes in the 'metabolic balance' (i.e. ecosystem respiration (ER)/gross primary production (GPP)) of pond mesocosms during a one-year warming experiment. The temperature dependences of GPP (A) and ER (B) were significantly different and constrained by the activation energies of photosynthesis and respiration, respectively. As a result, ER increased more rapidly with increases in temperature than did GPP. This shifted the metabolic balance of the warmed mesocosms (grey bars depict warmed treatments and their standard errors, and black bars depict the ambient treatments C), resulting in a 13% reduction in their carbon sequestration capacity that could be predicted from models derived from the MTE. Figure redrawn after Yvon-Durocher *et al.* (2010a) Note: in (A) and (B) temperature increases from right to left.

through ecosystems (e.g. droughts, floods, glacial meltwaters). Here, we assess temperature-related changes in aqueous media, in terms of its impacts on fluid viscosity and the form and distribution of the occurrence of water in the environment, and the implications of these more indirect consequences of warming for different levels of biological organization. We leave aside other important effects of climate change in aquatic systems, in particular ocean acidification resulting from increasing dissolved CO_2, whose effects even at the population level are still poorly understood (Pelejero *et al.*, 2010).

A. Fluid Viscosity: Impacts on Individuals, Interactions and Networks

Fluid viscosity and solute solubility are temperature dependent and, since all living organisms are composed primarily of aqueous fluids, warming will affect the supply and removal rate of key nutrients and waste products to and from body tissues (at least for ectotherms), unless biological mechanisms can countermand these physical changes to an individual's internal environment. By extension, environmental warming also has important implications for the physical properties of the external media within which ecological networks operate, especially in aquatic systems, where the viscosity of the medium will affect buoyancy and sinking rates—and hence encounter rates between consumers and resources. The sinking velocity of phytoplankton, for instance, increases with cell volume according to Stoke's law, described as $\propto aV^{2/3}$ (Finkel *et al.*, 2010; Table 1). A change in water temperature from 5 to 10 °C is associated with a change in viscosity (dynamic) from 1.52×10^{-3} N s m^{-2} to 1.31×10^{-3} N s m^{-2}. The implications of this depend on the organism's size and speed, which can be summarized by the Reynolds number, *Re* (a ratio between the inertia of water being displaced and the viscosity of the environment) (Purcell, 1977). Large, fast-moving organisms ($Re > 450$) operate in the inertial regime and are relatively unaffected by changes in viscosity, whereas small, slow-moving organisms ($Re < 1$) operate in the viscous regime and are highly susceptible to changes in viscosity (Sommer, 1996). The diel vertical migration of plankton, and hence consumer–resource encounter rates could be disrupted in pelagic food webs, especially for the smallest organisms at the base of the food web with the lowest Reynolds numbers. Reduced viscosity will reduce the costs of swimming even as it supports faster rates of sinking, so the balance between the effects of viscosity on passive sinking versus active swimming could be critical in determining the strength of interactions among the planktonic organisms at the base of many aquatic food webs. Because changes in viscosity and resultant changes in sinking rates and in the energetic costs of swimming follow known formulas, these effects on aquatic systems could and should be addressed via a systematic theory.

Aquatic networks often comprise basal resource species of phytoplankton or microbial decomposers, and small actively swimming consumers, which have an *Re* less than one (Sommer, 1996). Diel vertical migration of phytoplankton and the motility of bacteria (Schneider and Doetsch, 1974), protists (Winet, 1976), copepods (Larsen *et al.*, 2008), rotifers (Hagiwara *et al.*, 1998), echinoderms (Podolsky and Emlet, 1993) and even fishes (Hunt Von Herbing and Keating, 2003) all depend upon the viscosity of water. Many lower trophic-level organisms use cilia or flagellae for locomotion or feeding, the beat frequency and displacement of which are determined by density of the surrounding water (Machmer, 1972; Riisgård and Larsen, 2007; Sleigh, 1956), adding to the potential impacts of temperature-dependent changes in viscosity on aquatic communities. Additionally, the interference caused to *Daphnia* feeding on cyanobacteria by the latter's filaments declines with increasing viscosity (Abrusàn, 2004). Hence, consumer–resource encounter rates and dynamics may change in response to changes in viscosity, in addition to the direct effects of temperature *per se*. Indeed, the feeding or ingestion rates of several aquatic organisms are known to decrease with increasing viscosity (Abrusàn, 2004; Bolton and Havenhand, 1998, 2005; Loiterton *et al.*, 2004; Podolsky, 1994), and these effects may even be sufficient to qualitatively alter population dynamics (Harrison, 1995; Luckinbill, 1973; Seuront and Vincent, 2008).

Laboratory experiments have manipulated the viscosity of water via the use of thickening agents such as methyl cellulose, polyvinylpyrrolidone (PVP), Dextran, and Ficoll, which has enabled the viscosity-based effects of warming to be isolated from those associated with temperature *per se* (e.g. changes in metabolic rate). For example Podolsky (1994) quantified the proportion of feeding rates of echinoderm larvae (*Dendraster excentricus*) that were attributable to temperature-dependent changes in viscosity at 22 °C (viscosity = 1.02 cP), 12 °C (viscosity = 1.30 cP) and 22 °C with the addition of Dextran, resulting in a viscosity 1.30 cP. This study revealed that the 10 °C reduction in temperature was associated with a 67% reduction in feeding rate, 61% of which was directly attributed to changes in viscosity. In running waters, the ability to resist shear stress, particularly among small organisms, will decline if the viscosity of water increases (Statzner *et al.*, 1988), and in terrestrial plant–pollinator networks nectar might become too 'sticky' for some smaller insects to handle effectively. In flowers with an open morphology, nectar concentration and hence viscosity increase with temperature, and this may change the composition of the visitor fauna, in addition to lowering harvesting rates (Borrell, 2006).

B. The Form and Distribution of Water in the Environment

The rates and extent of warming predicted under the IPCC (2007) climate change scenarios will not be uniform over space and time, and ecological responses will therefore depend to a large extent upon the physical environment within which they are played out. At the whole-ecosystem scale, the nature and size of the medium within which the ecological network operates will determine its response to rising atmospheric temperature. For instance, large water bodies, such as the open oceans, have sufficiently high heat capacity and thermal inertia to smooth out short-term temperature fluctuations, which are also predicted to increase with warming (i.e. both the mean and variability of temperature are likely to rise in many areas, IPCC, 2007). In contrast, small water bodies exposed to direct insolation can fluctuate by tens of degrees Celsius over the course of a single day. For instance, water temperatures in streams fed from small springs or surface run-off (Figure 7)

Figure 7 Mountain and spring-fed stream food webs for tributaries of the Ivishak River in arctic Alaska (69°1′N, 147°43′W) (redrawn after Parker and Huryn, 2006). Mean food chain length was 3.04 in the spring stream and 1.83 in the mountain stream. A fish species, the Dolly Varden char (*Salvelinus malma*), was the top predator in the mountain stream, whereas an invertivorous riverine bird species, the American dipper (*Cinclus mexicanus*), was the top predator in the spring stream. Difference in mean food chain length between streams was due largely to the presence of *C. mexicanus* at the spring stream.

are far more variable than those supplied by large reserves of groundwater. The latter typically have very stable thermal regimes that reflect their subterranean provenance (e.g. cf. Friberg *et al.*, 2009; Parker and Huryn, 2006; Woodward *et al.*, 2010a). Non-geothermally influenced streams rarely exceed 29 °C because at elevated air temperatures the vapour pressure deficit above the water surface increases drastically, causing strong evaporative cooling (Mohseni and Stefan, 1999) that sets an upper limit to the ultimate temperature they can attain. In contrast, because maximum temperature is not limited in terrestrial ecosystems in this way, we might expect to see stronger metabolic constraints operating on individuals within terrestrial networks, relative to their aquatic counterparts, as atmospheric temperatures rise in the Tropics.

One profound effect of warming on the physical environment is the 'habitat fragmentation' that can arise within waterbodies where pycnoclines delineate sharp gradients of water density. A classic example of this is the widespread phenomenon of thermoclines in aquatic systems, which separate the warm surface waters in the euphotic zone from the cooler and less productive deeper waters in lakes and oceans, with the latter also commonly exhibiting haloclines of differing density and hence viscosity (Gnanadesikan, 1999). Many standing waters and oceanic regions are likely to become more strongly stratified as temperatures rise, and the development and persistence of strong thermoclines could impair the recycling of nutrients to the photic zone from deeper, colder waters (Finkel *et al.*, 2010) by preventing vertical mixing in the water column. The thermal imbalances resulting from predicted surface water warming in the coming decades might ultimately lead to large-scale changes in the existing ocean circulation currents, such as the posited change in the Pacific Ocean from today's ENSO (El Niño Southern Oscillation) to a permanent El Nino or La Nina state (Timmermann *et al.*, 1999; but see Sarmiento *et al.*, 2004). At present, on a global scale, pycnoclines tend to disappear in the open ocean at around 50–60° latitude, due to reduced salinity and/or temperature change, whereas further from the poles they prevent extensive mixing and help to retain the relative biological distinctness of the food webs in different waterbodies. The physical vertical segregation of waters of different density and temperature created by pycnoclines has implications for individual organisms and their interactions within networks. As sinking rates of plankton are positively correlated with body size (which also determines which consumer species exploit the plankton), and since warming should favour smaller phytoplankton (e.g. Moran *et al.*, 2010; Strecker *et al.*, 2004), changes in the vertical depth and horizontal distribution of pycnoclines due to warming could alter food chain lengths, network structure and the flux of nutrients in many lakes and regions of the open oceans. We return to consider these potential interactions between temperature, viscosity and nutrient fluxes within food webs in more detail in Sections V and VI.

At high latitudes and altitudes, ice sheets and glaciers are in rapid retreat in many parts of the world, leading to the changes in total runoff volume, the timing of spring/summer runoff peaks and the contributions from different water sources (Milner *et al.*, 2009). Changes in rainfall patterns will alter many terrestrial and freshwater ecological networks, but these effects will be far from uniform across the globe (Boulton, 2003). For instance, the hotter, drier summers and wetter winters predicted for many temperate areas in the near future will create more intense and frequent floods and droughts (Milly *et al.*, 2006), whereas at higher latitudes and altitudes floods could increase in spring and early summer due to more rapid snowmelt (Barnett *et al.*, 2005; Hannah *et al.*, 2007). New hydrologic disturbance regimes have the potential to alter the rate and magnitude of supply of basal resources to primary consumers to eliminate sensitive species and, hence, to restructure entire networks. Some recent long-term experiments have shown that drought episodes can have powerful effects on the organization and functioning of stream food webs (e.g. Harris *et al.*, 2007; Ledger *et al.*, 2009; Walters and Post, 2008). For example in a two-year field manipulation, repeated droughts reduced the availability of basal resources (Ledger *et al.*, 2008), eliminated members of the consumer assemblages (especially large, rare species high in the food web: Figure 8) and thereby suppressed total secondary production by more than 50% (Ledger *et al.*, unpublished data). The fragmentation of freshwaters during droughts can break network linkages in both space and time (Zwick, 1992), with large species being affected disproportionately more strongly than smaller, more *r*-selected species, which may benefit, at least temporarily, from the release of top-down or competitive control as suggested by population irruptions during recolonization when the water returns (e.g. Ledger and Hildrew, 2001). Although it may be possible to make some generalizations, because of these apparent system-specific contingencies in their effects on different species and other network components, it is difficult with our current understanding to incorporate hydrological changes into any general or predictive theory at present.

V. NETWORK RESPONSES TO THE COMPONENTS OF CLIMATE CHANGE: ATMOSPHERIC COMPOSITION AND ECOLOGICAL STOICHIOMETRY

In addition to the direct and indirect effects of warming, elevated environmental CO_2 concentrations associated with climate change could also alter the structure and dynamics of ecological networks fundamentally. One of the key mechanisms of relevance here is the way in which increasing CO_2 will

Figure 8 Control (right-hand channel, B, in the photograph) and monthly dewatered (left-hand channel, A) food webs for two experimental stream channels, constructed at the end of a 22-month hydrological manipulation conducted by Ledger *et al.* (2009). Taxa that became locally extinct tended to be rare and high in the food web, and the network became simpler as species and links were lost in response to drought stress (food web data: M. Ledger *et al.*, unpublished data).

influence the cellular composition of primary producers, which typically have CNP ratios that reflect ambient environmental concentrations (Elser and Hessen, 2005; Kominoski *et al.*, 2007; Redfield, 1934). Changes in environmental CO_2 concentrations will alter the elemental composition of C:nutrient ratios in autotrophs (Hessen *et al.*, 2004, 2005; Urabe *et al.*, 2003), which possess considerable stoichiometric plasticity relative to their more homeostatic consumers (Sterner and Elser, 2002). RuBisCO (Ribulose-1,5-bisphosphate carboxylase oxygenase), the enzyme that catalyses CO_2 fixation, is typically less than half saturated at current global atmospheric CO_2 concentrations for most phytoplankton (Badger *et al.*, 1998; Giordano *et al.*, 2005). This is typically the rate-limiting step in the Calvin cycle, so increases in this greenhouse gas have the potential to alter rates of CO_2 assimilation dramatically, particularly in the global ocean where the lion's share of the planet's carbon fixation takes place (Cox *et al.*, 2000). Experiments have revealed that algae grown at a range of dissolved CO_2 concentrations differ markedly in their stoichiometry, with those grown at high CO_2 and low P exhibiting faster photosynthetic rates but high atomic C:P ratios, which reduced the efficiency of energy and mass transfer to their consumers (Urabe *et al.*, 2003). These stoichiometric shifts among primary producers can be explained by changes in sub-cellular metabolism, as under high CO_2 photosynthesis, carbon fixation and assimilation are stimulated, resulting in greater allocation to storage of C-rich sugars relative to P-rich ribosomes required for rapid growth (Urabe *et al.*, 2003).

Heterotrophs are typically far more homeostatic in their stoichiometry than is the case for autotrophs (Sterner and Elser, 2002). As a result, changes in the food quality of basal resources (as defined by their C: nutrient ratios) can exert strong influences on consumer production and abundance within the food web (Norby *et al.*, 2001). For example Urabe *et al.* (2002, 2003) and Hessen *et al.* (2002) have shown that growth rates of herbivorous *Daphnia* decline markedly when fed on high C:P versus low C:P algae. These studies highlight how stoichiometric consumer–resource imbalances resulting from increased CO_2 could ramify to the network level of organization, as a consequence of changes in autotrophic stoichiometry at the base of the food web. Similarly, elevated CO_2 in a 9-year field experiment in a terrestrial system revealed suppressed herbivore abundance, but increased ingestion rates, reflecting the need of consumers to eat more plant material to extract a similar level of nutriment when C:nutrient levels increase (Stiling and Cornelissen, 2007). Large shifts in the stoichiometry of basal resources with climate change are thus likely to alter the distribution and fluxes of biomass within many food webs, in both terrestrial and aquatic systems. Because the effects of stoichiometry on individuals and interactions have been much studied (e.g. Sterner and Elser, 2002) and the field continues to expand rapidly, there is considerable potential for integrating these effects more deeply into an overall theory to explain network effects of climate change based on lower level principles.

Body size is once again revealed as a potentially important determinant of responses to climate change, as it also influences uptake rates and the stoichiometric effects of increased CO_2. For instance, phytoplankton cell size, elemental requirements and composition constrain resource uptake and processing rates (see Finkel *et al.*, 2010 and references therein). In many marine systems, phytoplankton body size tends to increase with nutrient availability, so elevated CO_2 should favour larger organisms (the opposite of the case for warmer temperatures; Daufresne *et al.*, 2009; Strecker *et al.*, 2007). In turn, increased phytoplankton cell size under elevated CO_2 will increase sinking rates and hence could reduce the ability for consumers to recycle carbon before it is lost to the sediment: because large phytoplankton tend to be grazed by large zooplankton (e.g. Cyr and Curtis, 1999) the resultant food webs should be shorter, as fewer trophic steps are required to reach consumers of a given body size, but a higher proportion of primary production could be lost before it is consumed, due to faster sinking rates. Since most of the organic carbon exported to the deep sea is derived from larger, denser cells that sink more rapidly (Laws *et al.*, 2000), the size structure of planktonic assemblages has important consequences for how climate change influences consumer–resource stoichiometry, the flux of energy through the food web and, ultimately, the ability of the global ocean to absorb and store atmospheric CO_2 (Kohfeld *et al.*, 2005; Sarmiento and Wofsy, 1999; Watson and Orr, 2003). Scaling relationships between phytoplankton cell size and key ecological and physiological processes have recently been used to model primary production, carbon pools and rates of export to the deep ocean (Finkel *et al.*, 2010), highlighting the importance of understanding these fundamental links between body size, metabolism and elemental fluxes within ecological networks.

In addition to the trophic links between primary producers and herbivores, many ecosystems rely heavily on inputs of energy and nutrients via detrital pathways (e.g. soil food webs), often via external subsidies (Moore *et al.*, 2004). Consequently, the C:nutrient content of detrital resources, which are largely derived from dead primary producers, should also increase as atmospheric CO_2 levels rise. These ratios are key drivers of many ecosystem processes associated with decomposition (e.g. Hladyz *et al.*, 2009; Norby *et al.*, 2001; Tuchman *et al.*, 2002) and, ultimately, the fluxes of energy and nutrients to the higher trophic levels within food webs. Carbon-rich detrital resources can also serve to modulate the potentially destabilizing effects on network structure of 'fast' autochthonous food chains based on primary production (Rooney *et al.*, 2006). It remains to be seen how the interplay between these two types of trophic pathway will respond to climate change, but some initial studies suggest that elevated CO_2 can shift the balance from the relative importance of algae versus detritus to the higher trophic levels (e.g. Kominoski *et al.*, 2007).

VI. NETWORK ASSEMBLY AND DISASSEMBLY: SPATIAL AND TEMPORAL MATCHES AND MISMATCHES

A. Climate Envelope Models, Invasions and Extinctions: Spatial Rewiring of Ecological Networks

To date, the effects of environmental change on species distributions have been predicted primarily using 'habitat modelling' and 'climate envelope' approaches. In both cases, a species' future distribution is predicted from knowledge about its current distribution, the environmental conditions within that distribution and predictions about future environmental conditions (Pearson and Dawson, 2003). The majority of these modelling approaches do not, however, consider species as interacting components of a wider ecological network (Kissling *et al.*, 2010): rather, they assume that each species experiences completely independent effects of environmental change in isolation. This represents a major shortcoming, given the powerful effects network structure and dynamics have on the ability of organisms to colonize and establish themselves within an ecosystem. Network-level responses are more than simply the aggregate sum of all their component species-level responses and, as such, the latter cannot simply be extrapolated to predict the former (Tylianakis, 2009; Woodward, 2009).

Making predictions about network assembly based on a species-by-species perspective will be challenging given that the recipient network governs the likelihood that incoming species and life history stages will become established, whilst the incoming organisms also rewire and hence shape the network. Consequently, understanding the patterns and processes that govern network assembly will be crucial in determining the potential impacts of climate change. The key driving forces behind network assembly and disassembly are immigration, speciation and environmental filtering (Weiher and Keddy, 1999). As new habitats are created or old habitats are altered through climatic events, such as environmental warming, changing ocean currents, ice scouring, droughts or flooding, these factors will determine how species and links are gained or lost from the network. Immigration and speciation will be governed by the spatial and temporal dynamics of the system. If species adapted to new environmental conditions already exist nearby, the primary mechanism for assembly will be colonization. Dispersal ability plays a key role here, and certain taxa in particular systems (e.g. aquatic organisms without flying adult stages in landlocked lakes, East–West oriented river networks or freshwaters in small oceanic islands) will have little opportunity to track poleward warming by invading new networks (Ings *et al.*, 2009; Woodward and Hildrew, 2002b). If the environment is sufficiently isolated from source populations of potential new species, network assembly is more likely to be fulfilled by evolution or adaptation of existing species through time (Emerson and Gillespie, 2008). Environmental filtering may lead to network disassembly, if certain species can no longer tolerate altered biotic or abiotic factors in the locale, and this could overpower the effects of stochastic processes predicted by neutral theory, for example birth, death, colonization, speciation (Chase, 2007).

In the absence of barriers to migration, range shifts and expansions will lead to the arrival of new species in local communities as climate changes (Rahel and Olden, 2008). Through the links that these species forge in the network, they can have direct and indirect effects on many other species in the community (Henneman and Memmott, 2001) and hence the structure and thereby the stability of the entire web (Romanuk *et al.*, 2009). Non-independent responses of species to environmental change provide the potential for complex effects on patterns of community assembly and disassembly, and evidence from small-scale experiments suggests that indirect effects of temperature change can be substantial (Davis *et al.*, 1998). This growing body of evidence makes it increasingly clear that it is no longer tenable to make simple predictions about networks based only on species' climate envelopes; rather, the species that can invade or go extinct are dependent not only on changes in environmental conditions, but also on the other species that have invaded already, or have gone extinct already.

In addition, new species assemblages may emerge due to the differential rates of range shifts by species within ecosystems: present assemblages of interacting populations will not simply shift further north or to the west or to higher altitudes. Some species will move faster and further than others. Short-lived species with high dispersal abilities will re-assemble differently within networks from those which are long-lived and have low dispersal potentials (Montoya and Raffaelli, 2010). Spatial dislocations will have important impacts for above-ground versus below-ground terrestrial assemblages, which, although intimately linked, are characterized by quite different rate processes, and could become dislocated as ranges shift through climate change. An example is the effects of such dislocations on future plant distributions and diversity, which might explain why some plants may be lost, whereas others may become more abundant, in their native versus new ranges owing to climate change (Van der Putten *et al.*, 2010).

Despite the strong effects interspecific interactions and network structure can have on community assembly (e.g. Warren, 1996), there are still few studies of how environmental change affects the rate and trajectory of assembly, and fewer still of disassembly (González, 2000), to be able to make reliable generalizations at present. Although not explicitly linked to climate change effects *per se*, a recent whole-network study was carried out in a series of large subtidal mesocosms in Lough Hyne to explore empirical patterns and processes related to food web assembly and disassembly

(O'Gorman and Emmerson, 2009, 2010). Identical core assemblages of predatory fish, decapods and echinoderms were established within the cages. Smaller species from lower trophic levels were allowed to recruit naturally, creating multiple replicates of complex food webs consisting of over 100 species. Targeted species extinctions carried out on subsets of the mesocosms highlighted the importance of both strong and weak interactors for food web stability (O'Gorman and Emmerson, 2009; Figure 2C), with cascading effects of strong interactors on ecosystem processes being dampened in the presence of weakly interacting species. Further, the loss of either strong or weak interactors increased the variability of ecosystem process rates and reduced the resistance of the communities to secondary extinctions and invasions. Given that the strong and weak interactors manipulated in this study were the largest species present in the webs, occurring at high trophic levels, these targeted extinctions mirror the loss of large species predicted by climate change scenarios and thus highlight the potential impacts of such losses on network stability (e.g. Raffaelli, 2004).

Some additional evidence for potential changes to freshwater networks under future climatic scenarios can be drawn from a range of studies conducted along temporal and spatial (altitudinal or latitudinal) environmental gradients as proxies for climate change (e.g. Friberg *et al.*, 2009; Lavandier and Décamps, 1983; Parker and Huryn, 2006; Woodward *et al.*, 2010a). In the Alaskan study of Parker and Huryn (2006) (Figure 7), higher streambed disturbance, due to more frequent and intense peak flows arising precipitation-based flow (rather than spring-derived), reduced mean food chain length, altered the identity of top predators and the proportion of biomass at different trophic levels, and significant differences in material and energy flow. Some glacially influenced streams will warm as ice masses shrink in response to climate change, which is generally predicted to increase overall invertebrate diversity whilst simultaneously facilitating the local extinction of cold-stenotherm taxa. However, Flory and Milner (1999) showed that the extinction of an early-colonizing chironomid genus (*Diamesa*), however, was not due to increased water temperature *per se*, but to competitive exclusion by another chironomid, *Pagasia partica*, which colonized later, indicating that warming strengthened biotic interactions, thereby influencing network assembly indirectly. Nevertheless, Milner *et al.* (2008) showed in a long-term study of a stream where water temperature had increased from 2 to 18 °C that the mechanism of community assembly was typically tolerance, and that only a few taxa (<15%) were lost, suggesting that in these kinds of systems networks will increase in both size and complexity as glacial influences wane.

In a study of biomass production in quantified stream food webs along an altitudinal gradient in the French Pyrénées, Lavandier and Décamps (1983) characterized changes in the distribution and magnitude of energy flows, together with increases in taxonomic richness, along a maximum water temperature gradient from 4.5 to 13 °C. The larvae of small taxa (Chironomidae) were replaced as the dominant primary consumer of periphyton and detritus by larger taxa (*Baetis* mayflies and *Allogamus auricollis* caddisflies) along the thermal gradient. However, the Chironomidae still formed the major energy source for invertebrate predators across all sites, perhaps owing to their smaller body size not affording them an upper body-size refugium from gape-limited predators (cf. Woodward *et al.*, 2005a). Such taxonomic and functional shifts in mountain stream food webs might be expected more widely across the world, owing to global reductions in glacier and snowpack extent (e.g. Brown *et al.*, 2007; Milner *et al.*, 2009).

Highly resolved studies of natural systems that can be used for detecting climate change impacts on ecological networks are still scarce, but several model systems, including those we identified in Section II, offer promise within this context. The biogeographical isolation of Antarctica provides a close-to-pristine environment for studying intact ecosystems (Arntz *et al.*, 1994; Chown and Gaston, 2000; Clarke, 1983; Dayton, 1990; Gray, 2001; Hedgpeth, 1971) and the recently characterized food web of the eastern Weddell Sea shelf and slope region (Figure 9) contains 488 species and over 16,000 feeding links (Brose *et al.*, 2005; Jacob, 2005). Many species are opportunistic trophic generalists, resulting in high omnivory and linkage density within the network (Brenner *et al.*, 2001; Jacob *et al.*, 2003). Rising temperatures and shifts in the position of warm and cold waters as the ice sheets melt, however, could open the door to colonization of a suite of new apex predators: invasions of shark species not previously found in the Antarctic could potentially exert powerful top-down effects upon the slow-growing and poorly defended resident fauna in the food web (Aronson *et al.*, 2007; Clarke *et al.*, 2004), and this is likely to lead to extensive rewiring of the network if these species are lost (Figure 9). In addition to potential increases in the rate and number of these biological perturbations (i.e. invasions), physical disturbances are also likely to increase in the Antarctic, where icebergs are being calved from glaciers and ice sheets at an accelerating rate, which can alter large areas of benthic marine food webs by scouring the sea bed and effectively resetting the clock of food web assembly at more local scales (Teixidó *et al.*, 2007; Figure 10).

We need a mechanistic framework that allows us to predict where a given species is likely to forge links in a given network: several approaches based on body size have been used successfully in predator–prey networks (e.g. Petchey *et al.*, 2002), and other approaches have also been developed recently in the plant–pollinator and host–parasitoid literature. These need to be refined to account for changing environmental conditions and then tested for their ability to predict network effects of climate change. Ives and Zhu (2006)have proposed a method that could be applied to host–parasitoid webs, based on the phylogenetic structure of the network. Essentially, if the

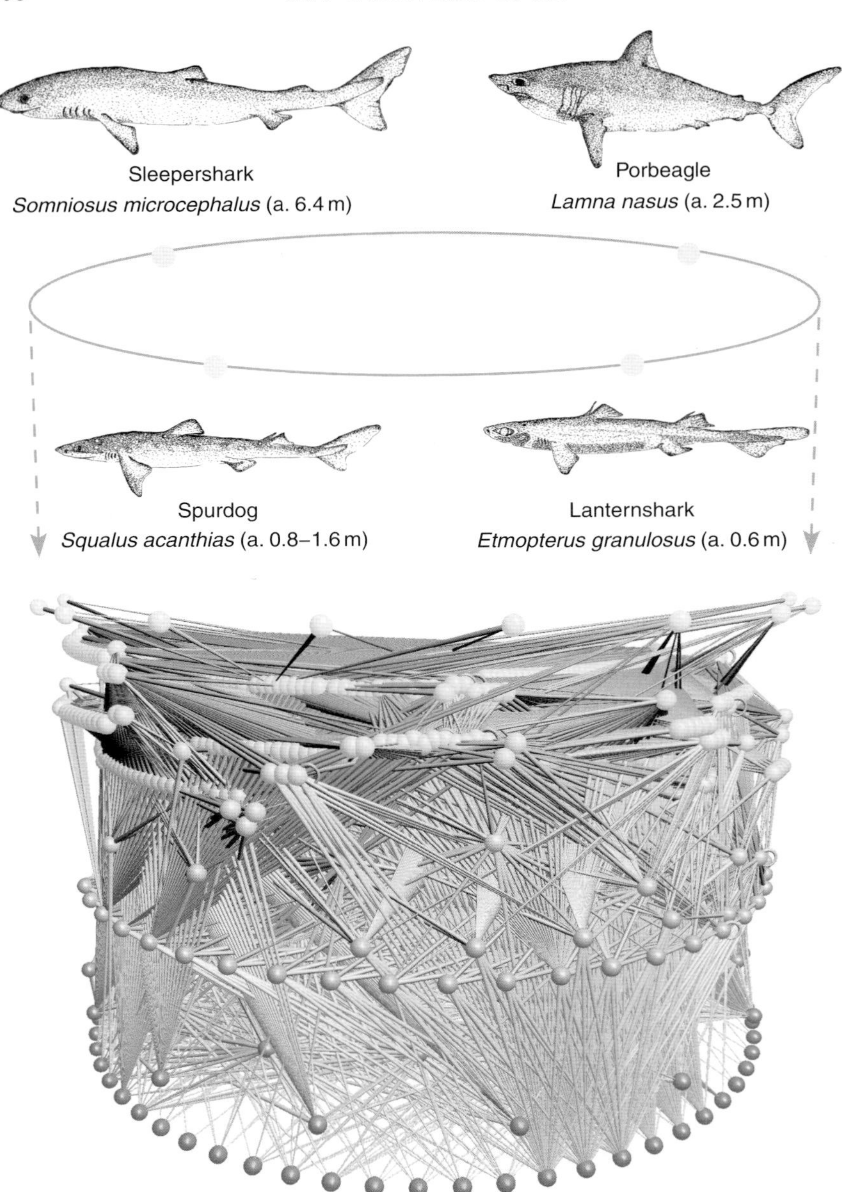

Figure 9 Directional change in the Weddell Sea food web, Antarctica. The pre-invasion food web contains 492 species, 16,137 links and no 'true' top predators; the post-invasion food web contains 497 species, 16,344 links and true top predators. The post-invasion web here is a hypothetical web, constructed by simply including the sharks and links with prey items they would likely take according to expert opinion. Cascading effects including possible extinctions are not considered.

ERRATUM

Chapter 2 — Ecological Networks in a Changing Climate

Page 109: Figure 10 - A portion of the figure is missing in the chapter. Please refer the correct figure below.

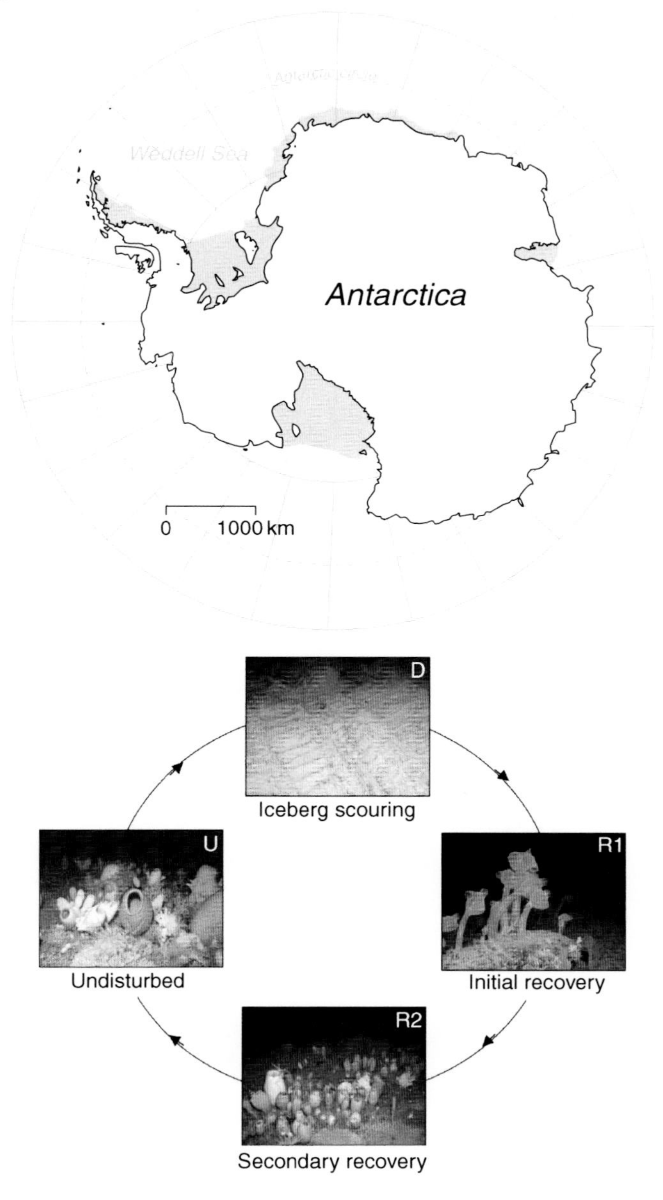

Figure 10 Weddell Sea food web, Antarctica, depicting cycles of ice-scouring of the sea bed and subsequent recovery of the benthic food web. Photographs © supplied by Julian Gutt at the Alfred Wegener Institute for Polar and Marine Research, 27568 Bremerhaven, Germany.

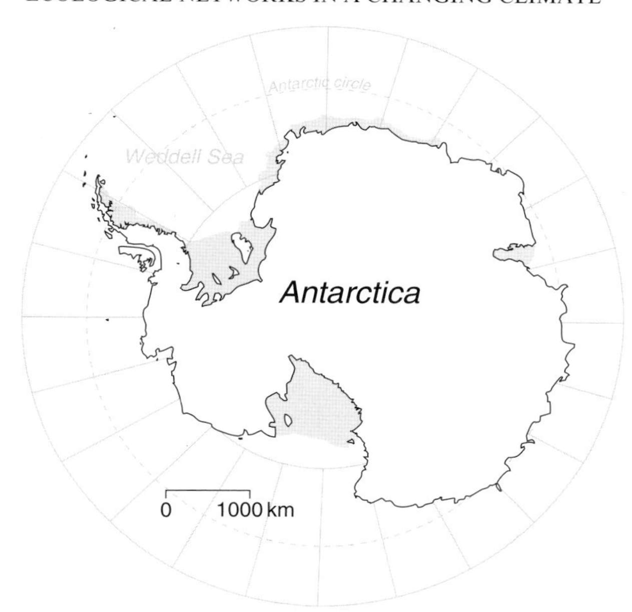

Figure 10 Weddell Sea food web, Antarctica, depicting cycles of ice-scouring of the sea bed and subsequent recovery of the benthic food web. Photographs © supplied by Julian Gutt at the Alfred Wegener Institute for Polar and Marine Research, 27568 Bremerhaven, Germany.

host range of a parasitoid is constrained by host phylogeny (i.e. hosts within the range are more closely related to each other than they are to hosts outside the parasitoid's range), and likewise the 'parasitoid range' of a host is constrained by parasitoid phylogeny, then it should be possible to calculate the probability that an invading species with a given phylogenetic position will interact with each of the species already in the network. Similar approaches might be usefully extended to other systems, particularly plant–pollinator networks where modular structures seem to be extremely common, and which might reflect analogous phylogenetic constraints within closely related taxa and coevolutionary links between consumers and resources (Jordano *et al.*, 2010). Unlike food webs and host–parasitoid systems, however, the interactions within these networks are primarily mutually beneficial and also typically more temporally dynamic (at least at higher latitudes) (Olesen *et al.*, 2010). In addition to the current range of approaches and modelling techniques used to predict network assembly, more recent work has identified a possible link between ecological stoichiometry and invasion success (González *et al.*, 2010). This suggests another possible means to employ a first-principles approach, in line with the general aims of this chapter, as a predictive tool that could help to reduce the high levels of contingency which typify many of the current phenomenological approaches used to assess species range shifts (e.g. bioclimatic envelope models).

B. Phenological Matches and Mismatches: Temporal Rewiring of Ecological Networks

The impacts of climate change on ecological networks are likely to act via both spatial (previous section) and temporal/phenological coupling and decoupling of interactions (Winder and Schindler, 2004). Direct, species-level climate change effects that may precipitate phenological rewiring of networks could include decreased development time, and increased survival and a longer growing season, leading to more generations per season. These changes may also impact population dynamics, as observed for Arctic insect herbivores (Strathdee *et al.*, 1993). Changes in generation time may be relatively slow to respond to a changing climate (Hulle *et al.*, 2008), especially if photoperiod (which is independent of climate change), rather than temperature, is a key driver of life-history events (such as diapause and reproduction). The important role of phenology, that is the timing of key life history events, in ecological communities was recognized many decades ago by Charles Elton, who devoted an entire chapter to 'Time and animal communities' in his influential book *Animal Ecology* (1927). He wrote: 'Many of the animals in a community never meet owing to the fact that they become active at different times' (p. 83). In modern ecological parlance, Elton was highlighting the important role of phenology in controlling the links that are realized in ecological networks (cf. Durant *et al.*, 2007). The so-called match/mismatch hypothesis, a term first coined by Cushing (1975), originally focused on the extent of coincidence between the time of hatching of fish larvae and the time when their planktonic prey become available. This was later extended by Pope *et al.* (1994), who discussed how larval and juvenile fishes needed to 'ride' the seasonal wave of plankton production to grow and survive: if the phenology of consumers and resources becomes decoupled due to differential responses (which again, may be related to correlates between body size, individual metabolism, and life history and developmental rates), this could break or weaken many of the key links, especially those at the base of the food web.

In host–parasitoid and other insect-dominated networks in temperate zones, for example insects enter an overwintering state, either as eggs, larvae, pupae or adults, depending on the taxon. The effects of warming could simply result in all insects becoming active earlier in the season to the same degree (e.g. Harper and Peckarsky, 2006), but there is also evidence that increasing day length, which is independent of climate change, plays an important role for some taxa. This could lead to a mismatch in early season emergence of species that rely differently on light versus temperature cues, with powerful repercussions for community dynamics, due partly to priority effects. For example an early emerging aphid species can support a population of a parasitoid that prevents the establishment of a later-arriving aphid species through apparent competition, while synchronized arrival leads to exclusion of the former aphid and the parasitoid, due to resource competition and unstable dynamics, respectively (Jones *et al.*, 2009). Although it is currently difficult to make general predictions about these kinds of effect, given the limited availability of suitable data, it is likely that the relative importance of day length and temperature for different taxa depends on the developmental stage in which they overwinter. It seems reasonable to suggest that species that overwinter as adults might simply need to reach a minimum temperature for activity to commence, whereas those that overwinter as pupae, for example may need external cues such as day length to trigger the onset or completion of metamorphosis. If this is the case, it gives us some identifiable species traits that could potentially be used in future work to formulate more general theories and to improve our ability to predict how and when phenological decoupling might arise within certain types of ecological networks.

Within most ecological networks, a large proportion of possible links remain unobserved. Some of these are missing from available data because of insufficient sampling, as revealed by yield–effort curves (e.g. Ings *et al.*, 2009; Woodward *et al.*, 2005a), whereas others are truly absent in nature and will not be observed, irrespective of sampling effort (Jordano *et al.*, 2003; Olesen *et al.*, 2010). Truly absent links are accounted for by biological phenomena, including phenological and spatial uncoupling, size- or reward-mismatching and foraging constraints (Jordano, 1987; Jordano *et al.*, 2006; Nilsson, 1988; Olesen *et al.*, 2008). Phenological uncoupling (Figure 11) occurs when the phenophases (activity periods) of species in a network do not overlap (Cushing, 1975; Jordano, 1987), and such overlaps often have to be substantial for links to be both manifested and detected by an observer (Jordano, 1987).

Olesen *et al.* (2010) have demonstrated the importance of phenological uncoupling (weak or no overlap in phenophases) in a Greenland pollination network from the high Arctic (74°30′N, 21°00′W), where connectance was only 15%, that is the proportion of unobserved potential links was 85%. They found that phenological uncoupling accounted for nearly one-third of all unobserved links in this network. However, the importance of phenological uncoupling varies among networks and is likely to be more extreme at the edge of species ranges and at high latitudes, due to a severely constrained growing season, whereas it may be insignificant where one or both of the interacting entities are perennial. Examples of the latter scenario include the

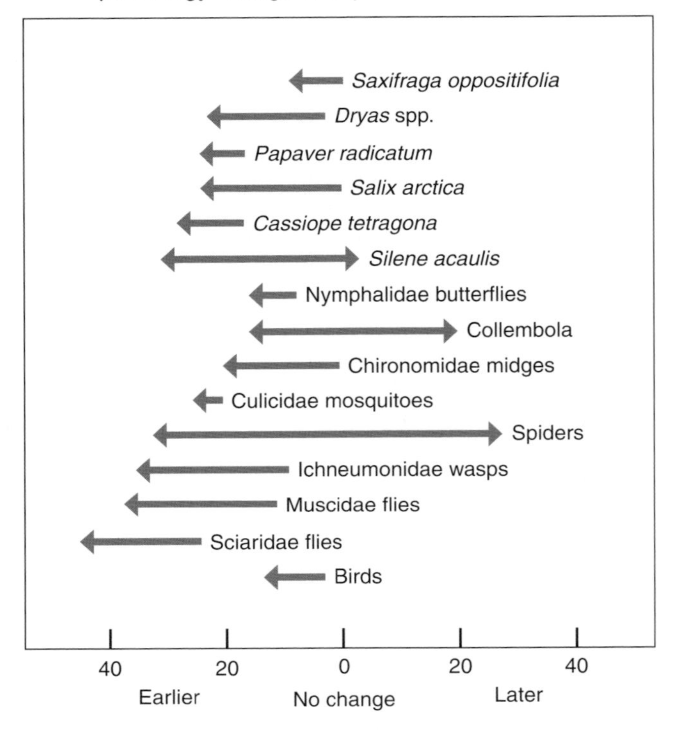

Figure 11 Phenological matches and mismatches in plant–pollinator networks in Greenland (after Olesen *et al.*, 2008). See also Olesen *et al.* (2010) for further details of the study system and network properties.

tropical Trinidad hummingbird–plant network studied by Snow and Snow (1972), where resident hummingbirds harvest nectar all year round, and tropical domatia ant–plant networks (Fonseca and Ganade, 1996), where both mutualists are intimately linked to each other for extended periods.

There is increasing evidence that significant changes in the phenology of plants and animals have occurred in response to recent climatic changes (Forrest *et al.*, 2010) and uncoupling has now been documented in a number of systems (herbivorous insects/insectivorous birds/raptors: Both *et al.*, 2009; Thackeray *et al.*, 2010), highlighting potential for climate-driven reshuffling of phenologies to alter network structure and dynamics dramatically (e.g. Hegland *et al.*, 2009; Høye *et al.*, 2007; Inouye, 2008; Inouye *et al.*, 2003; Memmott *et al.*, 2007; Tylianakis *et al.*, 2008). A recent review of phenological change in terrestrial and aquatic systems in the UK has highlighted that higher trophic levels (i.e. secondary consumers) are slower to respond to climatic changes than are the lower trophic levels, making the former especially vulnerable to phenological uncoupling (Thackaray *et al.*, 2010).

Additional challenges to predicting responses to climatic change arise if the onset and length of phenophase are species specific (Høye *et al.*, 2007). Further, differential responses may even occur among populations of the same species: for example phenological uncoupling has been reported for great tits and the winter moth larvae they feed their young on in Denmark (Visser *et al.*, 2009), whereas another population in the UK has been able to match the earlier emergence of winter moth caterpillars (Charmantier *et al.*, 2008). These differences are likely to reflect both genetic variation and phenotypic plasticity.

Because species responses are embedded within the wider network, the latter's degree of modularity or compartmentalization has important consequences for phenological matches/mismatches (Olesen *et al.*, 2010). Large mutualistic networks are typically nested, modular and hierarchical, with a dense core of links shared by small club of generalists connected by two tails to a swarm of specialists. Consequently, the latter may have their relatively short phenophase displaced considerably before they become uncoupled from their generalized linkage partner (Olesen *et al.*, 2010). The hierarchical nature of these nested link patterns adds a strong element of robustness against perturbations and helps to facilitate species coexistence, at least in mutualistic networks: most species in modules are short-lived specialists linked to local module hubs of intermediate lifespan, which are further linked to network hubs of greater longevity (Olesen *et al.*, 2010). Consequently, the importance of phenophase displacement among specialists can be mitigated by module hubs, and displacement among these may be buffered by network hubs, such that nestedness and modularity can counteract the effects of phenophase disturbance. Climate change could potentially trigger extinction vortices within mutualistic networks: for instance, as a species becomes rarer its population phenophase shortens, which increases the chances of it becoming phenologically uncoupled from its resources, which in turn can reduce mean fitness and abundance, and so on. However, phenological coupling/uncoupling appears to be a gradual phenomenon, and Olesen *et al.* (unpublished data) have shown that many unobserved links in a network may be due to limited temporal overlap, which needs to be substantial to achieve a high linkage probability (Olesen *et al.*, 2010).

Whilst these models give invaluable insights into the importance of phenological constraints in plant–pollinator networks, they do have limitations. In particular, they still lack a general mechanistic basis, and thus predictive power, and they have to rely upon making fundamental assumptions about pollinator responses to global warming due to the current lack of data. It is therefore vital that future studies begin to characterize species responses to elevated temperature directly (e.g. see studies on causal effects of temperature on bird reproduction; Visser *et al.*, 2009), and also to test more rigorously for presence of easily measured traits (e.g. body size) that could improve our ability to generalize. Within food webs, there are suggestions, for instance that the extent of phenological uncoupling varies with trophic status and/or body size, with faster responses occurring among smaller organism lower in the web, which can release them from top-down control as their more slowly responding consumers become progressively shifted out of phase (e.g. Thackeray *et al.*, 2010; Winder and Schindler, 2004). There is also evidence that community assembly over biogeographic scales is related to food web structure, in a seemingly analogous manner to that observed in mutualistic networks, with webs initially being composed of a core of a few trophic generalist species to which more specialist taxa attach themselves progressively over time (Piechnik *et al.*, 2008).

Although many of the studies highlighted earlier have demonstrated the importance of phenological change in ecological networks, a more mechanistic viewpoint is needed to begin to predict the potential impacts of climate change on ecological networks through phenological coupling/uncoupling. Detailed phenophase data will become essential baseline information for future studies estimating climate effects on network structure, especially in Arctic networks, where climate changes are most marked (e.g. Witze, 2008). High-quality phenophase data are available for some network types, and models incorporating such data have suggested that plant–pollinator interactions are likely to be severely disrupted, with consumers potentially encountering periods without food and plants encountering periods without pollinators (Memmott *et al.*, 2007). The development of this understanding is currently hindered by the population level focus of most phenophase research to date (i.e. from when the first individual in a population initiates its activity until the last ceases). By implication, more abundant species are likely to have a longer population-level phenophase at a given site, but from a network perspective, interactions between individuals are crucial, so in future we will need to shift our focus to consider changes in individual phenophase, or at least mean individual phenophase (Olesen *et al.*, 2010).

VII. MULTIPLE AND INTERACTING COMPONENTS OF CLIMATE CHANGE

So far, we have subdivided climate change into several seemingly neatly delimited components (e.g. changes in environmental temperature, atmospheric composition and fluid viscosity). In reality, though, many of these will be occurring simultaneously, and may act additively or synergistically (e.g. Feuchtmayr *et al.*, 2009; Moss *et al.*, 2003). Synergies, which might either exacerbate or ameliorate the overall effects of climate change, have been largely overlooked from a network-level perspective. In this final section, we will consider how the components of climate change combine, in terms of their impacts on the higher levels of biological organization, in either an additive or synergistic manner.

A. Combined Impacts of Warming and Atmospheric Change on Metabolism and Stoichiometry Within Ecological Networks

Metabolic theory and ecological stoichiometry provide a useful roadmap for accomplishing the task of assessing the combined effects of two of the most obvious aspects of climate change, warming and elevated CO_2 levels in the environment, because they draw clear linkages between easily measured characteristics (e.g. body size, temperature) and higher level ecological dynamics (e.g. metabolism, population growth, energy and material flux). Moreover, these unifying frameworks are rooted in first principles of thermodynamics and mass conservation and can be applied across multiple levels of organization, from individuals to whole ecosystems (Brown *et al.*, 2004; Sterner and Elser, 2002).

Metabolism and rates of ecological processes are clearly inseparable from the stoichiometric requirements of organisms for biologically important elements such as nitrogen (N) and phosphorus (P), and these may differ markedly between consumers and resources within a network (Allen and Gillooly, 2009; Gillooly *et al.*, 2005; Reich and Oleksyn, 2004; Sterner and Elser, 2002). For instance, N:P ratios of marine primary producers increase with ambient temperature, whereas this does not seem to be the case for animals, which show a virtually negligible response (Figure 12). Building new biomass and maintaining elemental homeostasis (as in the case of consumers) necessitate the uptake or assimilation of N or P in some proportion to the amount of carbon fixed or assimilated. Although metabolic theory may explain significant variability in metabolism and energy flux across broad ranges in temperature and body size, important additional variation may be related to resource stoichiometry (Brown *et al.*, 2004; Jeyasingh, 2007; Sterner, 2004; Sweeney *et al.*, 1986). Thus, theoretical and empirical advances that combine metabolic and stoichiometric approaches should strongly improve our ability scale-up lower level effects to predict interactive effects of climate change on food webs and ecosystems (Allen and Gillooly, 2009).

For many ecosystems, the components needed to develop a systematic approach to understanding these additive or interactive effects of environmental warming and changes in CO_2 concentrations may already be available to a

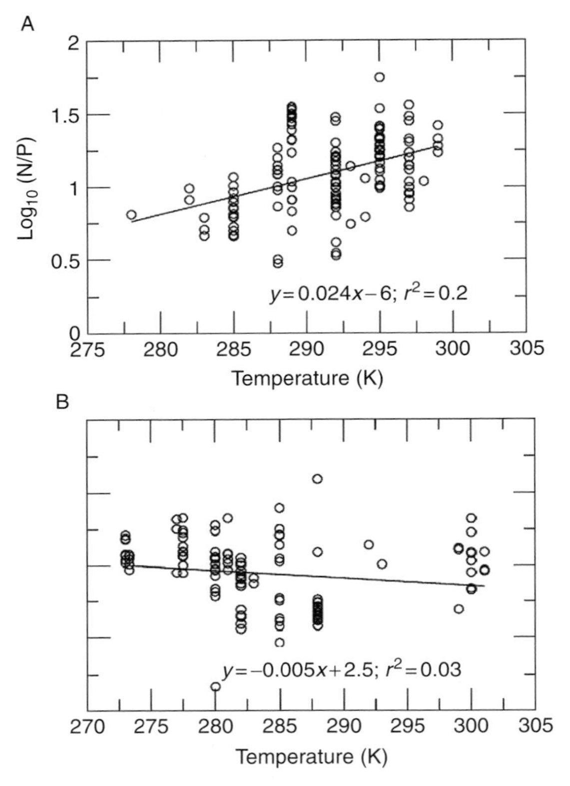

Figure 12 The effects of temperature on NP ratios in marine autotrophs (A) and heterotrophs (B). Data are redrawn from Brey *et al.* (2010). Notice that the Log10 (NP) of autotrophs is strongly, positively related to the mean annual temperature (K) of the organisms environment, while the Log10 (NP) of heterotrophs is relatively independent of environmental temperature. These patterns likely reflect the relative stoichiometric plasticity of autotrophs relative to the homeostatic nature of heterotrophs. However, a mechanistic explanation for the apparent increase in plant NP with increasing temperature is currently lacking in the literature.

large extent but have not yet been integrated into a network-level perspective. Studies into the interacting effects of the different components of climate change on the growth of primary producers at the base of the network provide a useful starting point for gauging the potential joint effects of elevated CO_2 and temperature, especially in aquatic systems, for which both theory and experimental evidence are better developed. For instance, phytoplankton growth physiology has been extensively studied using chemostats and modelled using a range of generalizations (e.g. Klausmeier *et al.*, 2004a,b; Litchman *et al.*, 2004; White and Zhao, 2009) of the classic Droop model (Droop, 1983) of nutrient uptake and growth. Importantly, parameters of this model and its variants are related to body size (Litchman *et al.*, 2007, 2009) and temperature (Aksnes and Egge, 1991; Geider *et al.*, 1998). The influence of the availability of multiple nutrients has also been extensively modelled and examined experimentally (e.g. Geider *et al.*, 1998; Urabe and Sterner, 1996). Recent work has used allometric parameterization of models to show quantitatively that the optimal size of competing phytoplankton cells increases with the typical period between nutrient upwelling events (Litchman *et al.*, 2007, 2009), and a similar approach could clearly be used to model the separate and combined effects of increased temperature and CO_2 on this competitive balance.

In a study of lake plankton, Hessen *et al.* (2004) showed algal C:P ratios were significantly higher in summer than in winter, which they attributed to seasonal shifts in the thermal stimulation of photosynthesis. Both temperature and CO_2 concentrations affect autotroph stoichiometry in ways that can ramify through the food web, and theory suggests a competitive trade-off exists between growth and nutrient storage in phytoplankton along an axis of individual body size. Essentially, larger phytoplankton can store nutrients to survive periods of resource limitation, whereas smaller phytoplankton can grow faster when nutrients are abundant; however, if CO_2 uptake is diffusion-limited larger cells should be favoured as concentrations rise (Aksnes and Egge, 1991; Finkel *et al.*, 2010; Litchman *et al.*, 2009). This raises the intriguing suggestion that elevated CO_2 levels and rising temperatures might have opposite (i.e. synergistic) effects on body size. Since temperature affects rates of both nutrient uptake (Aksnes and Egge, 1991) and growth (Geider *et al.*, 1998), it should alter this competitive balance, potentially changing the size structure of marine phytoplankton assemblages at the base of the food web. Stratification within the water column of many lakes and oceans is also likely to increase as temperatures rise (unless mitigated by increased mixing due to more intense storm events), reducing availability of limiting nutrients (i.e. low N:P ratios) to phytoplankton in the photic zone (Richardson and Schoeman, 2004; Winder *et al.*, 2009). In addition to the potential impacts on size-selective herbivores, changes in phytoplankton body size will alter the rate of sinking through pycnoclines, which has additional implications for encounter rates with zooplankton at the next trophic level, and hence the rates and magnitude of energy transfer and nutrient cycling with the food web. If, for instance, temperature, rather than CO_2 availability, is the primary driver, smaller phytoplankton should be favoured, and there is some evidence to suggest that this might be the case (e.g. Moran *et al.*, 2010).

One exciting avenue towards blending metabolic and stoichiometric approaches that has great potential for predicting climate change effects on ecological networks is the application of threshold elemental ratio (TER) models (Frost *et al.*, 2006; Sterner, 1997; Sterner and Hessen, 1994). These incorporate information about consumer elemental composition (CNP ratios) and physiology (assimilation efficiencies, respiration) to quantify the resource elemental ratios at which the consumer switches from limitation by one element (e.g. C) to another (e.g. P; Frost *et al.*, 2006). In addition, these models provide quantitative estimates of consumer–resource elemental imbalances, which have the potential to influence community assembly, material flux in food webs and interaction strengths. Such models can be easily modified to incorporate changes in temperature (via effects on individual respiration or ingestion; Frost and Elser, 2002) or food quality (via effects on CNP ratios of prey items), and, when applied to entire food webs, may provide a powerful tool for predicting climate-induced shifts in food web structure and dynamics.

The TER approach incorporates important stoichiometric differences between autotrophs and heterotrophs that may lead to contrasting responses to climate warming and emergent effects on network patterns and processes (Sterner and Elser, 2002). For example autotrophs show considerable variation in C:nutrient ratios (even within a single species) and can vary strongly with regard to their nutrient-use efficiencies (Cross *et al.*, 2005; Rhee and Gotham, 1981; Sterner *et al.*, 1997; Vitousek, 1982). This flexibility means that a broad range of carbon fixation rates can be maintained with the same rate of nutrient supply. Thus, holding nutrient concentrations constant, both autotrophic production and C:nutrient ratios should increase with warming.

In contrast to autotrophs, elemental stoichiometry of heterotrophic consumers is relatively fixed or homeostatic (Elser *et al.*, 2000; Frost *et al.*, 2003, but see Cross *et al.*, 2003; Demott *et al.*, 2001). Although consumer metabolic rate should respond positively to increases in temperature (Gillooly *et al.*, 2001), this increased metabolism must be matched with increased assimilation of nutrients to maintain positive growth or homeostasis (Frost *et al.*, 2005). In the short term, a combination of flexible (and increasing) autotroph C:nutrient ratios and fixed consumer C:nutrient ratios should result in elevated elemental imbalances with increased temperature. Despite immediate increases in consumer metabolic rates, these increased imbalances should negatively affect growth and production of consumers, particularly those taxa with high nutrient requirements and low TERs (Figure 13).

Long-term food web responses to elevated temperature should, however, be fundamentally different because of time lags between short-term physiological responses of individual taxa and taxonomic shifts in the community. Over these longer time scales, elevated temperatures and low nutrient availability (Figure 13) should select for consumers with relatively low nutrient requirements (Frost *et al.*, 2006), high body C:nutrient ratios (Woods *et al.*, 2003) and high TERs. These warming-induced shifts in food web structure and stoichiometry should therefore alter the identity of the nodes (i.e. species turnover and/or altered size-classes), as well as the strength of interactions and the rates of elemental fluxes within trophic networks.

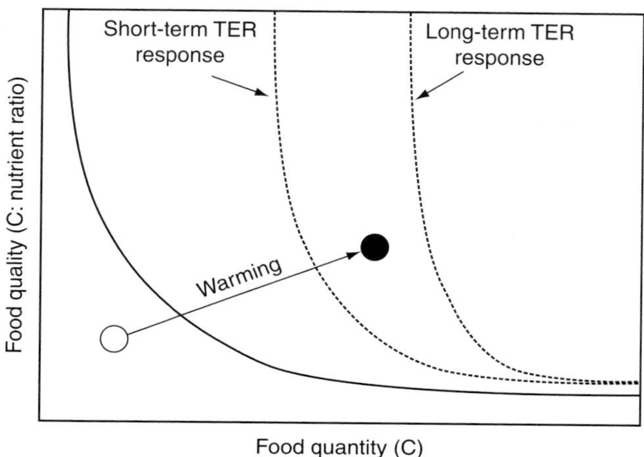

Figure 13 Predicted changes in food quantity and quality with increased temperature (from open circle in cold streams to filled circle in warm streams) and consumer threshold elemental ratio isoclines (TERs; solid, dashed, and dotted lines). Both food quantity and C:nutrient ratios are predicted to increase with warming. The solid line represents the TER of cold-adapted taxa in cold streams. If the combination of food quantity and quality falls below this line (as in the open circle), the consumer is predicted to be quantity (or C) limited. In contrast, if the combination of food quantity and quality falls above this line, the consumer is predicted to be nutrient limited (Sterner, 1997). The dashed line represents short- to mid-term response of the TER of this taxon in warm streams; the shift in TER is due to increased consumer respiration at higher temperatures (Frost and Elser, 2002). In this case, declines in food quality (increased C:nutrient ratios) and shifts in TER (driven by increased consumer respiration at higher temperatures) lead to nutrient limitation of consumer production and reduced flows of carbon and nutrients through this consumer. The dotted line represents predicted TER isoclines of consumers in temperature-acclimated communities (i.e. long-term evolutionary responses or shifts in community structure). Taxa in these communities are predicted to have high TERs and consequently should not experience warming-induced nutrient limitation.

Although use of TER models is still in its infancy (i.e. most studies have focused on single consumer–resource interactions; Anderson and Hessen, 2005; Frost and Elser, 2002), there is great potential for expansion to examine the dynamics of whole food webs. Importantly, recent theoretical advances (Allen and Gillooly, 2009; Gillooly *et al.*, 2005) that integrate metabolic theory and ecological stoichiometry should aid in application of this approach to the study of complex food webs. For example Allen and Gillooly (2009) derived an elegant model that predicts TERs based on the densities and individual mass of fundamental metabolic units (e.g. protein complexes) and organelles (e.g. ribosomes and mitochondria) responsible for metabolism, which scale predictably with body size and temperature. These authors demonstrated a reasonable fit between their theoretically based TERs and those parameterized from literature values (Frost *et al.*, 2006) for >30 aquatic animal taxa spanning a broad array of body size and phylogenetic history. These results are encouraging and suggest that scaling theory may indeed help guide predictions about food web responses to climate change. Nonetheless, as we progress towards using these models, more rigorous validation against empirical food web data will be essential.

B. Other Additive and Synergistic Effects of the Components of Climate Change

The potential synergies among the components of climate change other than temperature and elevated CO_2 are currently far less well known. For instance, although a few studies have highlighted the potential for viscosity-derived effects in ecological networks, little is known about their importance relative to other temperature-dependent mechanisms (e.g. metabolic rates). The metabolic costs of predator-induced phenotypic defences, which are prevalent in many zooplankters, may, for instance be determined by viscosity, which might account for their reduced expression at lower temperatures (Lagergren *et al.*, 2000). Whilst we have quantitative evidence for the importance of temperature-dependent viscosity for individual responses that affect encounter and ingestion rates (e.g. via altered swimming speed and feeding rates), our understanding of the implications for population and community dynamics is still in its infancy.

Given the potential for a complex array of synergies between the different components of climate change to affect ecological networks, a key priority is to distinguish the likely main effects and interactions from those that are subsidiary or trivial. The powerful overarching effects across multiple levels of organization (Brown *et al.*, 2004) of body size and temperature (and their combined effect on metabolism) are strong contenders for being principal drivers. As such, the temperature–metabolism relationship might be viewed as the first axis of interest, with others representing the residual variation about the line. Similarly, body size has well-known effects on individual foraging behaviour, such as allometric relationships with handling time and attack rates (Petchey *et al.*, 2002). Thus, quantifying body sizes of consumers and resources (and recording the environmental temperature) should capture a relatively large portion of the relevant characteristics of an ecological network. It is encouraging that this appears to the case for the small but increasing number of food webs that have been described in this way (e.g. Jonsson *et al.*, 2005; Layer *et al.*, 2010b; McLaughlin *et al.*, 2010; Mulder and Elser, 2009; O'Gorman and Emmerson, 2010; Petchey *et al.*, 2002; Woodward *et al.*, 2005a), but more data are still needed before we can make robust generalizations.

Among the subsidiary (but potentially interacting) axes, stoichiometric constraints may well be the next major determinant of the strength of interactions in many ecological networks, at least between the basal resources and the primary consumers that represent the first links in the food chain (e.g. Jeyasingh, 2007). The outline of a predictive framework for gauging likely responses of ecological networks to climate change, based upon well-described and relatively simple relationships related to metabolic scaling, foraging theory, and ecological stoichiometry, seems to be edging slowly towards being within our grasp (e.g. Petchey *et al.*, 2010). This general framework could then be overlain with the seemingly more contingent effects of species-specific bioclimatic envelopes, phenological effects and community assembly/disassembly models. Indeed, some recent studies have started to forge links from first principles to these higher-level responses, such as the posited relationship between ecological stoichiometry and invasion success (i.e. chemical-metabolic constraints on network assembly) suggested by González *et al.* (2010).

VIII. CONCLUSIONS

One of the key challenges for predicting effects of climate change on ecological networks is developing theory that explicitly incorporates both the separate components of climate change and their potential interactions among them (e.g. increased temperature × atmospheric CO_2). In an ideal world, ecological networks would be constructed from individual-based data, with directly observed nodes and links, over multiple generations, and at spatial and temporal scales that capture the essential properties of the study system, because this provides the most relevant and reliable data with which to assess the likely responses of these complex systems to climate change. Further, the relevant components of climate change under investigation should be imposed on the system as dynamic stressors, to identify causal relationships and to differentiate between potentially transient and equilibrial responses. Once these criteria have been met, a range of more specific hypotheses can be tested rigorously within a general predictive framework. Of course, in reality compromises must be made, but even if these goals are only partially met we will be better able to develop a more coherent and less contingent view of how Earth's ecosystems might respond to climate change. Given that one commonly recurring theme is that larger organisms seem to be those most adversely affected by many of the components of climate change (e.g. Jeppesen *et al.*, 2003; Meerhoff *et al.*, 2007; Moran *et al.*, 2010), and that metabolic and stoichiometric constraints underpin so many key ecological processes, focusing on these drivers and responses seems to be an obvious place to start developing the network-level perspective we need to help predict the impacts of future climate change. In particular, we need to start constructing, monitoring and manipulating networks from as wide a range of natural systems as possible over extended time frames, and across multiple levels of organization. Disentangling the responses to climate signals in natural multi-species systems is a major challenge that will require considerable logistic commitments on a truly global scale and over many years, but initiating such research at this still early stage will undoubtedly prove invaluable for developing our predictive capacity in the near future, as our climate continues to change apace.

ACKNOWLEDGEMENTS

We would like to thank Stephanie Parker and Alex Huryn for providing additional food web data, Haley Cohen, Graham Wall, Gisli Gislason and Jon Olafsson for assistance with sample collection for the Icelandic food webs, and Sarah Pottinger and Jacques Deere for helping to organize the workshops. Network images were produced with FoodWeb3D, written by R.J. Williams and provided by the Pacific Ecoinformatics and Computational Ecology Lab (www.foodwebs.org, Yoon *et al.*, 2004). This project was supported by a Natural Environment Research Council Centre for Population Biology grant awarded to G.W.

REFERENCES

Abrusàn, G. (2004). Filamentous cyanobacteria, temperature and *Daphnia* growth: The role of fluid mechanics. *Oecologia* **141**, 395–401.

Aksnes, D.L., and Egge, J.K. (1991). A theoretical model for nutrient uptake in phytoplankton. *Mar. Ecol. Prog. Ser. Oldendorf* **70**, 65–72.

Allen, A.P., and Gillooly, J.F. (2009). Towards an integration of ecological stoichiometry and the metabolic theory of ecology to better understand nutrient cycling. *Ecol. Lett.* **12**, 369–384.

Allen, A.P., Gillooly, J.F., and Brown, J.H. (2005). Linking the global carbon cycle to individual metabolism. *Funct. Ecol.* **19**, 202–213.

Anderson, T.R., and Hessen, D.O. (2005). Threshold elemental ratios for carbon versus phosphorus limitation in *Daphnia*. *Freshw. Biol.* **50**, 2063–2075.

Arntz, W.E., Brey, T., and Gallardo, V.A. (1994). Antarctic marine zoobenthos. *Oceanogr. Mar. Biol. Annu. Rev.* **32**, 241–304.

Aronson, R.B., Thatje, S., Clarke, A., Peck, L.S., Blake, D.B., Wilga, C.D., and Seibel, B.A. (2007). Climate change and the invasibility of the Antarctic benthos. *Annu. Rev. Ecol. Evol. Syst.* **38**, 129–154.

Atkinson, D. (1994). Temperature and organism size—A biological law for ectotherms. *Adv. Ecol. Res.* **25**, 1–58.

Atkinson, D., Ciotti, B.J., and Montagnes, D.J.S. (2003). Protists decrease in size linearly with temperature: ca. 2.5% C-1. *Philos. Trans. R. Soc. Lond. B* **270**, 2605–2611.

Badger, M.R., Andrews, T.J., Whitney, S.M., Ludwig, M., Yellowlees, D.C., Leggat, W., and Price, G.D. (1998). The diversity and coevolution of Rubisco, plastids, pyrenoids, and chloroplast-based CO2-concentrating mechanisms in algae. *Can. J. Bot.-Rev. Can. Bot.* **76**, 1052–1071.

Barnes, D.K.A., Verling, E., Crook, A., Davidson, I., and O'Mahoney, M. (2002). Local population disappearance follows (20 yr after) cycle collapse in a pivotal ecological species. *Mar. Ecol. Prog. Ser.* **226**, 311–313.

Barnes, C., Bethea, D.M., Brodeur, R.D., Spitz, J., Ridoux, V., Pusineri, C., Chase, B.C., Hunsicker, M.E., Juanes, F., Kellermann, A., Lancaster, J., Ménard, F., *et al.* (2008). Predator and body sizes in marine food webs. *Ecology* **89**, 881.

Barnes, C., Maxwell, D.L., Reuman, D.C., and Jennings, S. (2010). Global patterns in predator–prey size relationships reveal size-dependency of trophic transfer efficiency. *Ecology* **91**, 222–232.

Barnett, T.P., Adam, J.C., and Lettenmaier, D.P. (2005). Potential impacts of a warming climate on water availability in snow-dominated regions. *Nature* **438**, 303–309.

Bascompte, J., and Stouffer, D.B. (2009). The assembly and disassembly of ecological networks. *Philos. Trans. R. Soc. Lond. B* **364**, 1781–1787.

Battarbee, R.W. (2000). Palaeolimnological approaches to climate change, with special regard to the biological record. *Quatern. Sci. Rev.* **19**, 107–124.

Beckerman, A.P., Petchey, O.L., and Warren, P.H. (2006). Foraging biology predicts food web complexity. *Proc. Natl. Acad. Sci. USA* **103**, 13745–13749.

Berlow, E.A., Dunne, J.A., Martinez, N.D., Starke, P.B., Williams, R.J., and Brose, U. (2009). Simple prediction of interaction strengths in complex food webs. *Proc. Natl. Acad. Sci. USA* **106**, 187–191.

Beveridge, O.S., Humphries, S., and Petchey, O.L. (2010). The interacting effects of temperature and food chain length on trophic abundance and ecosystem function. *J. Anim. Ecol.* **79**, 693–700.

Beverton, R.J.H., and Holt, S.J. (1959). A review of the lifespans and mortality rates of fish in nature, and their relation to growth and other physiological characteristics. *CIBA Found. Symp, Lifespan Animals* **5**, 142–177.

Bolton, T.F., and Havenhand, J.N. (1998). Physiological versus viscosity-induced effects of an acute reduction in water temperature on microsphere ingestion by trochophore larvae of the serpulid polychaete Galeolaria caespitose. *J. Plankton Res.* **20**, 2153–2164.

Bolton, T.F., and Havenhand, J.N. (2005). Physiological acclimation to decreased water temperature and the relative importance of water viscosity in determining the feeding performance of larvae of a serpulid polychaete. *J. Plankton Res.* **27**, 875–879.

Bonsall, M.B., and Hassell, M.P. (1998). Population dynamics of apparent competition in a host–parasitoid assemblage. *J. Anim. Ecol.* **67**, 918–929.

Borrell, B.J. (2006). Mechanics of nectar-feeding in the orchid bee *Euglossa imperialis*: Pressure, viscosity and flow. *J. Exp. Biol.* **209**, 4901–4907.

Both, C., van Asch, M., Bijlsma, R.G., van den Burg, A.B., and Visser, M.E. (2009). Climate change and unequal phenological changes across four trophic levels: Constraints or adaptations? *J. Anim. Ecol.* **78**, 73–83.

Boudreau, P.R., and Dickie, L.M. (1992). Biomass spectra of aquatic ecosystems in relation to fisheries yield. *Can. J. Fish. Aquat. Sci.* **49**, 1528–1538.

Boulton, A.J. (2003). Parallels and contrasts in the effects of drought on stream macroinvertebrate assemblages. *Freshw. Biol.* **48**, 1173–1185.

Brenner, M., Buck, B.H., Cordes, S., Dietrich, L., Jacob, U., Mintenbeck, K., Schröder, A., Brey, T., Knust, R., and Arntz, W. (2001). The role of iceberg scours in niche separation within the Antarctic fish genus *Trematomus*. *Polar Biol.* **24**, 502–507.

Brey, T. (2010). An empirical model for estimating aquatic invertebrate respiration. *Methods Ecol. Evol.* 92–101.

Brooks, S.J., and Birks, H.J.B. (2004). The dynamics of Chironomidae (Insecta: Diptera) assemblages in response to environmental during the past 700 years on Svalbard. *J. Paleolimnol.* **31**, 483–498.

Brose, U., Cushing, L., Banasek-Richter, C., Berlow, E., Bersier, L.F., Blanchard, J.L., Brey, T., Carpenter, S.R., Cattin-Blandenier, M.F., Cohen, J.E., Dell, T., Edwards, F., *et al.* (2005). Empirical consumer–resource body size ratios. *Ecology* **86**, 2545.

Brown, J.H., Gillooly, J.F., Allen, A.P., Savage, V.M., and West, G.B. (2004). Toward a metabolic theory of ecology. *Ecology* **85**, 1771–1789.

Brown, L.E., Hannah, D.M., and Milner, A.M. (2007). Vulnerability of alpine stream biodiversity to shrinking glaciers and snowpacks. *Global Change Biol.* **13**, 958–966.

Burgmer, T., Hillebrand, H., and Pfenninger, M. (2007). Effects of climate-driven temperature changes on the diversity of freshwater macroinvertebrates. *Oecologia* **151**, 93–100.

Bystrom, P., Andersson, J., Kiessling, A., and Eriksson, L.O. (2006). Size and temperature dependent foraging capacities and metabolism: Consequences for winter starvation mortality in fish. *Oikos* **115**, 43–52.

Castella, E., Adalsteinsson, H., Brittain, J.E., Gislason, G.M., Lehmann, A., Lencioni, V., Lods-Crozet, B., Maiolini, B., Milner, A.M., Olafsson, J.S., Saltveit, S.J., and Snook, D.L. (2001). Macrobenthic invertebrate richness and composition along a latitudinal gradient of European glacier-fed streams. *Freshw. Biol.* **46**, 1811–1831.

Charmantier, A., McCleery, R.H., Cole, L.R., Perrins, C., Kruuk, L.E.B., and Sheldon, B.C. (2008). Adaptive phenotypic plasticity in response to climate change in a wild bird population. *Science* **320**, 800–803.

Charnov, E.L. (2003). Life history invariants: Some explorations of symmetry in evolutionary ecology. Oxford Series in Ecology and Evolution Oxford University Press, Oxford, UK, 168 pp.

Chase, J.M. (2007). Drought mediates the importance of stochastic community assembly. *Proc. Natl. Acad. Sci. USA* **104**, 17430–17434.

Chown, S.L., and Gaston, K.J. (2000). Island hopping invaders hitch a ride with tourists in the Southern Ocean. *Nature* **408**, 637.

Clarke, A. (1983). Life in cold water: The physiological ecology of polar marine ectotherms. *Oceanogr. Mar. Biol. Annu. Rev.* **21**, 341–453.

Clarke, A. (2006). Temperature and the metabolic theory of ecology. *Funct. Ecol.* **20**, 405–412.

Clarke, A., and Fraser, K.P.P. (2004). Why does metabolism scale with temperature? *Funct. Ecol.* **18**, 243–251.

Clarke, A., and Johnston, N.M. (1999). Scaling of metabolic rate with body mass and temperature in teleost fish. *J. Anim. Ecol.* **68**, 893–905.

Clarke, A., Aronson, R.B., Crame, J.A., Gili, J.M., and Blake, D.B. (2004). Evolution and diversity of the benthic fauna of the Southern Ocean Continental Shelf. *Antarct. Sci.* **16**, 559–568.

Cohen, J.E., Jonsson, T., and Carpenter, S.R. (2003). Ecological community description using the food web, species abundance, and body size. *Proc. Natl. Acad. Sci. USA* **100**, 1781–1786.

Cohen, J.E., Jonsson, T., Muller, C.B., Godfray, H.C.J., and Savage, V.M. (2005). Body sizes of hosts and parasitoids in individual feeding relationships. *Proc. Natl. Acad. Sci. USA* **102**, 684–689.

Cox, P.M., Betts, R.A., Jones, C.D., Spall, S.A., and Totterdell, I.J. (2000). Acceleration of global warming due to carbon-cycle feedbacks in a coupled climate model. *Nature* **408**, 184–187.

Cross, W.F., Benstead, J.P., Rosemond, A.D., and Wallace, J.B. (2003). Consumer–resource stoichiometry in detritus-based streams. *Ecol. Lett.* **6**, 721–732.

Cross, W.F., Benstead, J.P., Frost, P.C., and Thomas, S.A. (2005). Ecological stoichiometry in freshwater benthic systems: Recent progress and perspectives. *Freshw. Biol.* **50**, 1895–1912.

Cushing, D.H. (1975). Marine Ecology and Fisheries. Cambridge University Press, Archive, 279 pp.

Cyr, H., and Curtis, J.M. (1999). Zooplankton community size structure and taxonomic composition affects size-selective grazing in natural communities. *Oecologia* **118**, 306–315.

Daehler, C.C., and Strong, D.R. (1996). Can you bottle nature? The roles of microcosms in ecological research. *Ecology* **77**, 663–664.

Daufresne, M., Lengfellner, K., and Sommer, U. (2009). Global warming benefits the small in aquatic ecosystems. *Proc. Natl. Acad. Sci. USA* **106**, 12788–12793.

Davidson, E.A., and Janssens, I.A. (2006). Temperature sensitivity of soil carbon decomposition and feedbacks to climate change. *Nature* **440**, 165–173.

Davis, A.J., Jenkinson, L.S., Lawton, J.H., Shorrocks, B., and Wood, S. (1998). Making mistakes when predicting shifts in species range in response to global warming. *Nature* **391**, 783–786.

Dayton, P.K. (1990). Polar benthos. In: *Polar Oceanography, Part B: Chemistry, Ecology and Geology* (Ed. by W.O. Smith, Jr.), pp. 631–685. Academic Press, London.

Demott, W.R., Gulati, R.D., and Van Donk, E. (2001). Effects of dietary phosphorus deficiency on the abundance, phosphorus balance, and growth of *Daphnia cucullata* in three hypereutrophic Dutch lakes. *Limnol. Oceanogr.* **46**, 1871–1880.

Deutsch, C.A., Tewksbury, J.J., Huey, R.B., Sheldon, K.S., Ghalambor, C.K., Haak, D.C., and Martin, P.R. (2008). Impacts of climate warming on terrestrial ectotherms across latitude. *Proc. Natl. Acad. Sci. USA* **105**, 6668–6672.

Douglas, M.S.V., Smol, J.P., and Blake, W. (1994). Marked post-18th century environmental change in high-Arctic ecosystems. *Science* **266**, 416–419.

Droop, M.R. (1983). 25 years of algal growth kinetics: A personal view. *Bot. Mar.* **26**, 99–112.

Dunham, A.E., Grant, B.W., and Overall, K.L. (1989). Interfaces between biophysical and physiological ecology and the population ecology of terrestrial vertebrate ectotherms. *Physiol. Zool.* **62**, 335–355.

Dunne, J.A., Williams, R.J., Martinez, N.D., Wood, R.A., and Erwin, D.E. (2008). Compilation and network analyses of Cambrian food webs. *PLoS Biol.* **6**, 693–708.

Durance, I., and Ormerod, S.J. (2007). Effects of climatic variation on upland stream invertebrates over a 25 year period. *Global Change Biol.* **13**, 942–957.

Durance, I., and Ormerod, S. (2009). Trends in water quality and discharge confound long-term warming effects on river macroinvertebrates. *Freshw. Biol.* **54**, 388–405.

Durant, J.M., Hjermann, D.O., Ottersen, G., and Stenseth, N.C. (2007). Climate and the match or mismatch between predator requirements and resource availability. *Climate Res.* **33**, 271–283.

Elser, J.J., and Hessen, D.O. (2005). Biosimplicity via stochiometry; The evolution of food web structure and processes. In: *Aquatic Food Webs: An Ecosystem Approach* (Ed. by A. Belgrano, U.M. Scharler, J. Dunne and R.E. Ulanowicz), pp. 7–18. Oxford University Press, Oxford, UK.

Elser, J.J., Fagan, W.F., Denno, R.F., Dobberfuhl, D.R., Folarin, A., Huberty, A., Interlandi, S., Kilham, S.S., McCauley, E., Schulz, K.L., Siemann, E.H., and Sterner, R.W. (2000). Nutritional constraints in terrestrial and freshwater food webs. *Nature* **408**, 578–580.

Emerson, B.C., and Gillespie, R.G. (2008). Phylogenetic analysis of community assembly and structure over space and time. *Trends Ecol. Evol.* **23**, 619–630.

Emmerson, M., and Raffaelli, D. (2004). Predator–prey body size, interaction strength and the stability of a real food web. *J. Anim. Ecol.* **73**, 399–409.

Emmerson, M., Bezemer, T.M., Hunter, M.D., and Jones, T.H. (2005a). Global change alters the stability of food webs. *Global Change Biol.* **11**, 490–501.

Emmerson, M.E., Montoya, J.M., and Woodward, G. (2005b). Allometric scaling and body-size constraints in complex food webs. In: *Dynamic Food Webs: Multi-species Assemblages, Ecosystem Development, and Environmental Change* (Ed. by P.C. de Ruiter, V. Wolters and J.C. Moore). Academic Press, San Diego, USA.

Feuchtmayr, H., Moran, R., Hatton, K., Connor, L., Heyes, T., Moss, B., Harvey, I., and Atkinson, D. (2009). Global warming and eutrophication: Effects on water chemistry and autotrophic communities in experimental hypertrophic shallow lake mesocosms. *J. Appl. Ecol.* **46**, 713–723.

Finkel, Z.V., Beardall, J., Flynn, K.J., Quigg, A., Rees, T.A.V., and Raven, J.A. (2010). Phytoplankton in a changing world: Cell size and elemental stoichiometry. *J. Plankton Res.* **32**, 119–137.

Flory, E.A., and Milner, A.M. (1999). The role of competition in invertebrate community development in a recently formed stream in Glacier Bay National Park, Alaska. *Aquat. Ecol.* **33**, 175–184.

Fonseca, C.R., and Ganade, G. (1996). Asymmetries, compartments and null interactions in an Amazonian ant–plant community. *J. Anim. Ecol.* **65**, 339–347.

Forrest, J., Inouye, D.W., and Thomson, J.D. (2010). Flowering phenology in subalpine meadows: Does climate variation influence community co-flowering patterns? *Ecology* **91**, 431–440.

Fox, J.W., and Morin, P.J. (2001). Effects of intra- and interspecific interactions on species responses to environmental change. *J. Anim. Ecol.* **70**, 80–90.

Friberg, N., Christensen, J.B., Olafsson, J.S., Gislason, G.M., Larsen, S.E., and Lauridsen, T.L. (2009). Relationship between structure and function in streams contrasting in temperature: Possible impacts of climate change on running water ecosystems. *Freshw. Biol.* **54**, 2051–2068.

Frost, P.C., and Elser, J.J. (2002). Growth responses of littoral mayflies to the phosphorus content of their food. *Ecol. Lett.* **5**, 232–240.

Frost, P.C., Tank, S.E., Turner, M.A., and Elser, J.J. (2003). Elemental composition of littoral invertebrates from oligotrophic and eutrophic Canadian lakes. *J. N. Am. Benthol. Soc.* **22**, 51–62.

Frost, P.C., Evans-White, M.A., Finkel, Z.V., Jensen, T.C., and Matzek, V. (2005). Are you what you eat? Physiological constraints on organismal stoichiometry in an elementally imbalanced world. *Oikos* **109**, 18–28.

Frost, P.C., Benstead, J.P., Cross, W.F., Hillebrand, H., Larson, J.H., Xenopoulos, M.A., and Yoshida, T. (2006). Threshold elemental ratios of carbon and phosphorus in aquatic consumers. *Ecol. Lett.* **9**, 774–779.

Geider, R.J., Macintyre, H.L., and Kana, T.M. (1998). A dynamic regulatory model of phytoplanktonic acclimation to light, nutrients, and temperature. *Limnol. Oceanogr.* **43**, 679–694.

Giguere, L.A. (1981). Food assimilation efficiency as a function of temperature and meal size in larvae of *Chaoborus trivittatus* (Diptera: Chaoboridae). *J. Anim. Ecol.* **50**, 103–109.

Gillooly, J., Brown, J.H., West, G.B., Savage, V.M., and Charnov, E.L. (2001). Effects of size and temperature on metabolic rate. *Science* **293**, 2248–2251.

Gillooly, J.F., Allen, A.P., Brown, J.H., Elser, J.J., Martinez del Rio, C., Savage, V.M., West, G.B., Woodruff, W.H., and Woods, H.A. (2005). The metabolic basis of whole-organism RNA and phosphorus content. *Proc. Natl. Acad. Sci. USA* **102**, 11923–11927.

Giordano, M., Beardall, J., and Raven, J.A. (2005). CO_2 concentrating mechanisms in algae: Mechanisms, environmental modulation, and evolution. *Annu. Rev. Plant Biol.* **56**, 99–131.

Gnanadesikan, A. (1999). A simple predictive model for the structure of the oceanic pycnocline. *Science* **283**, 2077–2079.

González, A. (2000). Community relaxation in fragmented landscapes: The relation between species richness, area and age. *Ecol. Lett.* **3**, 441–448.

González, A., Kominoski, J.S., Danger, M., Ishida, S., Iwai, N., and Rubach, A. (2010). Can ecological stoichiometry help explain patterns of biological invasions? *Oikos* **119**, 779–790.

Gray, J.S. (2001). Antarctic marine benthic biodiversity in a world-wide latitudinal context. *Polar Biol.* **24**, 633–641.

Hagiwara, A., Yamamiya, N., and Belem de Araujo, A. (1998). Effect of water viscosity on the population growth of the rotifer *Brachionus plicatilis* Muller. *Hydrobiology* **387–388**, 489–494.

Hannah, D.M., Brown, L.E., Milner, A.M., Gurnell, A.M., McGregor, G.R., Petts, G.E., Smith, B.P.G., and Snook, D.L. (2007). Integrating climate–hydrology–ecology for alpine river systems. *Aquat. Conserv.: Mar Freshw. Ecosyst.* **17**, 636–656.

Hari, R.E., Livingstone, D.M., Siber, R., Burkhardt-Holm, P., and Guttinger, H. (2006). Consequences of climate change for water temperature and brown trout populations in Alpine rivers and streams. *Global Change Biol.* **12**, 10–26.

Harmon, J.P., Moran, N.A., and Ives, A.R. (2009). Species response to environmental change: Impacts of food web interactions and evolution. *Science* **323**, 1347–1350.

Harper, M.P., and Peckarsky, B.L. (2006). Emergence cues of a mayfly in a high-altitude stream ecosystem: Potential response to climate change. *Ecol. Appl.* **16**, 612–621.

Harris, R.M.L., Milner, A.M.M., Armitage, P.D., and Ledger, M.E. (2007). Replicability of physicochemistry and macroinvertebrate assemblages in stream mesocosms: Implications for experimental research. *Freshw. Biol.* **52**, 2434–2443.

Harrison, G.W. (1995). Comparing predator prey models to Lucknbills's experiment with *Didinium* and *Paramecium*. *Ecology* **76**, 357–374.

Hassell, M.P. (2000). The spatial and temporal dynamics of host–parasitoid interactions. Oxford University Press, 154 pp.

Hays, G.C., Richardson, A.J., and Robinson, C. (2005). Climate change and marine plankton. *Trends Ecol. Evol.* **20**, 337–344.

Hedgpeth, J.W. (1971). Perspectives of benthic ecology in Antarctica. In: *Research in the Antarctic* (Ed. by L.O. Quam), pp. 93–136. American Association for the Advancement of Science, Washington.

Hegland, S.J., Nielsen, A., Lázaro, A., Bjerknes, A.-L., and Totland, Ø. (2009). How does climate warming affect plant–pollinator interactions? *Ecol. Lett.* **12**, 184–195.

Henneman, M.L., and Memmott, J. (2001). Infiltration of a Hawaiian community by introduced biological control agents. *Science* **293**, 1314–1316.

Hessen, D.O., Faerovig, P.J., and Andersen, T. (2002). Light, nutrients, and P:C ratios in algae: Grazer performance related to food quality and quantity. *Ecology* **83**, 1886–1898.

Hessen, D.O., Agren, G.I., Anderson, T.R., Elser, J.J., and De Ruiter, P.C. (2004). Carbon, sequestration in ecosystems: The role of stoichiometry. *Ecology* **85**, 1179–1192.

Hessen, D.O., van Donk, E., and Gulati, R. (2005). Seasonal seston stoichiometry: Effects on zooplankton in cyanobacteria-dominated lakes. *J. Plankton Res.* **27**, 449–460.

Hickling, R., Roy, D.B., Hill, J.K., Fox, R., and Thomas, C.D. (2006). The distributions of a wide range of taxonomic groups are expanding polewards. *Global Change Biol.* **12**, 450–455.

Hildrew, A.G. (2009). Sustained research on stream communities: A model system and the comparative approach. *Adv. Ecol. Res.* **41**, 175–312.

Hladyz, S., Gessner, M.O., Giller, P.S., Pozo, J., and Woodward, G. (2009). Resource quality and stoichiometric constraints in a stream food web. *Freshw. Biol.* **54**, 957–970.

Hogg, I.D., Williams, D.D., Eadie, J.M., and Butt, S.A. (1995). The consequences of global warming for stream invertebrates: A field simulation. *J. Thermal Biol.* **20**, 199–206.

Høye, T.T., Post, E., Meltofte, H., Schmidt, N.M., and Forchhammer, M.C. (2007). Rapid advancement of spring in the High Arctic. *Curr. Biol.* **17**, R449–R451.

Hulle, M., Bonhomme, J., Maurice, D., and Simon, J.C. (2008). Is the life cycle of high arctic aphids adapted to climate change? *Polar Biol.* **31**, 1037–1042.

Hunt von Herbing, I., and Keating, K. (2003). Temperature-induced changes in viscosity and its effects on swimming speed in larval haddock. The Big Fish Bang, Proceedings of the 26th Annual Larval Fish Conference, 23–34.

Ims, R.A., and Fuglei, E. (2005). Trophic interaction cycles in tundra ecosystems and the impact of climate change. *Bioscience* **55**, 311–322.

Ings, T.C., Montoya, J.M., Bascompte, J., Bluthgen, N., Brown, L., Dormann, C.F., Edwards, F., Figueroa, D., Jacob, U., Jones, J.I., Lauridsen, R.B., Ledger, M.E., *et al.* (2009). Ecological networks—Beyond food webs. *J. Anim. Ecol.* **78**, 253–269.

Inouye, D.W. (2008). Effects of climate change on phenology, frost damage, and floral abundance of montane wildflowers. *Ecology* **89**, 353–362.

Inouye, D.W., Saavedra, F., and Lee-Wang, W. (2003). Environmental influences on the phenology and abundance of flowering by *Androsace septentrionalis* (Primulaceae). *Am. J. Bot.* **90**, 905–910.

IPCC Climate Change (2007). The Physical Sciences Basis. In: *Contribution of Working Group I to the Fourth Assessment Report of the Intergovernmental Panel on Climate Change* (Ed. by M. Parry, O. Canziani, J. Palutkof, P. Van der Linden and C. Hanson). Cambridge University Press, Cambridge, UK.

Ives, A.R., and Zhu, J. (2006). Statistics for correlated data: phylogenies, space, and time. *Ecol. Appl.* **16**, 20–32.

Jacob, U. (2005). Trophic Dynamics of Antarctic Shelf Ecosystems—Food Webs and Energy Flow Budgets. University of Bremen, Germany, PhD Thesis, 125 pp.

Jacob, U., Terpstra, S., and Brey, T. (2003). The role of depth and feeding in regular sea urchins niche separation—An example from the high Antarctic Weddell Sea. *Polar Biol.* **26**, 99–104.

Jennings, S.J., and Brander (2010). Predicting the effects of climate change on marine communities and the consequences for fisheries. *J. Mar. Syst.* **79**, 418–426.

Jeppesen, E., Søndergaard, M., and Jensen, J.P. (2003). Climatic warming and regime shifts in lake food webs: Some comments. *Limnol. Oceanogr.* **48**, 1346–1349.

Jeyasingh, P.D. (2007). Plasticity in metabolic allometry: The role of dietary stoichiometry. *Ecol. Lett.* **10**, 282–289.

Jiang, L., and Morin, P.J. (2004). Temperature-dependent interactions explain unexpected responses to environmental warming in communities of competitors. *J. Anim. Ecol.* **73**, 569–576.

Johnson, E.A., and Miyanishi, K. (2008). Testing the assumptions of chronosequences in succession. *Ecol. Lett.* **11**, 419–431.

Jones, T.S., Godfray, H.C.J., and van Veen, F.J.F. (2009). Resource competition and shared natural enemies in experimental insect communities. *Oecologia* **159**, 627–635.

Jonsson, T., Cohen, J.E., and Carpenter, S.R. (2005). Food webs, body size, and species abundance in ecological community description. *Adv. Ecol. Res.* **36**, 1–84.

Jordano, P. (1987). Patterns of mutualistic interactions in pollination and seed dispersal: Connectance, dependence asymmetries, and coevolution. *Am. Nat.* 657–677.

Jordano, P., Bascompte, J., and Olesen, J.M. (2003). Invariant properties in coevolutionary networks of plant–animal interactions. *Ecol. Lett.* **6**, 69–81.

Jordano, P., Bascompte, J., and Olesen, J.M. (2006). The ecological consequences of complex topology and nested structure in pollination webs. In: *Plant–pollinator interactions: From specialization to generalization* (Ed. by N.M. Waser and J. Ollerton), pp. 173–199. University of Chicago Press, Chicago.

Jordano, P., Vázquez, D.P., and Bascompte, J. (2010). Redes complejas de interaccions mutualistas planta–animal. In: *Ecología y evolución de interacciones planta–animal: conceptos y applicaciones* (Ed. by R. Medel, M.A. Aizen and R. Zamora), pp. 17–41. Edotorial Universitaria, Santiago de Chile.

Kingsolver, J., and Woods, H. (1998). Interactions of temperature and dietary protein concentration in growth and feeding of *Manduca sexta* caterpillars. *Physiol. Entomol.* **23**, 354–359.

Kishi, D., Murakami, M., Nakano, S., and Maekawa, K. (2005). Water temperature determines strength of top-down control in a stream food web. *Freshw. Biol.* **50**, 1315–1322.

Kissling, W.D., Field, R., Korntheuer, H., Heyder, U., and Böhning-Gaese, K. (2010). Woody plants and the prediction of climate-change impacts on bird diversity. *Philos. Trans. R. Soc. Lond. B* **365**, 2035–2045.

Klausmeier, C.A., Litchman, E., and Levin, S.A. (2004a). Phytoplankton growth and stoichiometry under multiple nutrient limitation. *Limnol. Oceanogr.* **49**, 1463–1470.

Klausmeier, C.A., Litchman, E., Daufresne, T., and Levin, S.A. (2004b). Optimal nitrogen-to-phosphorus stoichiometry of phytoplankton. *Nature* **429**, 171–174.

Kohfeld, K.E., Quere, C.L., Harrison, S.P., and Anderson, R.F. (2005). Role of marine biology in glacial–interglacial CO_2 cycles. *Science* **308**, 74–78.

Kokkoris, G.D., Troumbis, A.Y., and Lawton, J.H. (2002). Patterns of species interaction strength in assembled theoretical competition communities. *Ecol. Lett.* **2**, 70–74.

Kominoski, J.S., Moore, P.A., Wetzel, R.G., and Tuchman, N.C. (2007). Elevated CO_2 alters leaf-litter-derived dissolved organic carbon: Effects on stream periphyton and crayfish feeding preference. *J. N. Am. Benthol. Soc.* **26**, 663–672.

Konig, K.A., Kamenik, C., Schmidt, R., Agustí-Panareda, A., Appleby, P., Lami, A., Prazakova, M., Rose, N., Schnell, O.A., Tessadri, R., Thompson, R., and Psenner, R. (2002). Environmental changes in an alpine lake (Gossenköllesee, Austria) over the last two centuries—The influence of air temperature on biological parameters. *J. Paleolimnol.* **28**, 147–160.

Kratz, T.K., Deegan, L.A., Harmon, M.E., and Lauenroth, W.K. (2003). Ecological variability in space and time: Insights gained from the US LTER Program. *Bioscience* **53**, 57–67.

Lafferty, K.D., Allesina, S., Arim, M., Briggs, C.J., DeLeo, G., Dobson, A., Dunne, J.A., Johnson, P.T.J., Kuris, A.M., Marcogliese, D.J., Martinez, N.D., Memmott, J., *et al.* (2008). Parasites in food webs: The ultimate missing links. *Ecol. Lett.* **11**, 533–546.

Lagergren, R., Lord, H., and Stenson, J.A.E. (2000). Influence of temperature on hydrodynamic costs of morphological defences in zooplankton: Experiments on models of *Eubomina* (Cladocera). *Funct. Ecol.* **14**, 380–387.

Lane, A.M.J. (1997). The U.K. Environmental Change Network database: An integrated information resource for long-term monitoring and research. *J. Environ. Manage.* **51**, 87–105.

Lanfear, R., Thomas, J.A., Welch, J.J., Brey, T., and Bronham, L. (2007). Metabolic rate does not calibrate the molecular clock. *Proc. Natl. Acad. Sci. USA* **104**, 15388–15393.

Langford, T.E. (1990). Ecological effects of thermal discharges. Pollution Monitoring Series. Elsevier Applied Science, 468 pp. UK.

Larsen, P.S., Madsen, C.V., and Riisgard, H.U. (2008). Effect of temperature and viscosity on swimming velocity of the copepod *Acartia tonsa*, brine shrimp *Artemia salina* and rotifer *Brachionus plicatilis*. *Aquat. Biol.* **4**, 47–54.

Lavandier, P., and Décamps, H. (1983). Un torrent d'altitude dans les Pyrénées: l'Estaragne. In: *Ecosystèmes Limniques* (Ed. by F. Bourlière and M. Lamotte), pp. 81–111. Masson, Paris.

Laws, E.A., Falkowski, P.G., Smith, W.O.J., *et al.* (2000). Temperature effects on export production in the open ocean. *Global Biogeochem. Cycle* **14**, 1231–1246.

Layer, K., Hildrew, A.G., Monteith, D., and Woodward, G. (2010a). Long-term variation in the littoral food web of an acidified mountain lake. *Global Change Biol.* 10.1111/j.1365-2486.2010.02195.x (in press).

Layer, K., Riede, J.O., Hildrew, A.G., and Woodward, G. (2010b). Food web structure and stability in 20 streams across a wide pH gradient. *Adv. Ecol. Res.* **42.**

Ledger, M.E., and Hildrew, A.G. (2001). Recolonization by the benthos of an acid stream following a drought. *Arch. Hydrobiol.* **152**, 1–17.

Ledger, M.E., and Hildrew, A.G. (2005). The ecology of acidification: Patterns in stream food webs across a pH gradient. *Environ. Pollut.* **137**, 103–118.

Ledger, M.E., Harris, R.M.L., Armitage, P.D., and Milner, A.M. (2008). Disturbance frequency influences patch dynamics in stream benthic algal communities. *Oecologia* **155**, 809–819.

Ledger, M.E., Harris, R.M.L., Armitage, P.D., and Milner, A.M. (2009). Realism of model ecosystems: An evaluation of physicochemistry and macroinvertebrate assemblages in artificial streams. *Hydrobiologia* **617**, 91–99.

Liboriussen, L., Landkildehus, F., Meerhoff, M., Bramm, M.E., Søndergaard, M., Christoffersen, K., Richardson, K., Søndergaard, M., Lauridsen, T.L., and Jeppesen, E. (2005). Global warming: Design of a flow-through shallow lake mesocosm climate experiment. *Limnol. Oceanogr.: Methods* **3**, 1–9.

Litchman, E., Klausmeier, C.A., and Bossard, P. (2004). Phytoplankton nutrient competition under dynamic light regimes. *Limnol. Oceanogr.* **49**, 1457–1462.

Litchman, E., Klausmeier, C.A., Schofield, O.M., and Falkowski, P.G. (2007). The role of functional traits and trade-offs in structuring phytoplankton communities: Scaling from cellular to ecosystem level. *Ecol. Lett.* **10**, 1170–1181.

Litchman, E., Klausmeier, C.A., and Yoshiyama, K. (2009). Contrasting size evolution in marine and freshwater diatoms. *Proc. Natl. Acad. Sci. USA* **106**, 2665–2670.

Loiterton, B., Sundbom, M., and Vrede, T. (2004). Separating physical and physiological effects of temperature on zooplankton feeding rate. *Aquat. Sci.* **66**, 13–129.

Lopez-Urrutia, A., San Martin, E., Harris, R.P., and Irigoien, X. (2006). Scaling the metabolic balance of the oceans. *Proc. Natl. Acad. Sci. USA* **103**, 8739–8744.

Luckinbill, L.S. (1973). Coexistence in laboratory populations of *Paramecium aurelia* and its predator *Didinium nasutum*. *Ecology* **54**, 1320–1327.

Macfadyen, S., Gibson, R., Raso, L., Sint, D., Traugott, M., and Memmott, J. (2009). Parasitoid control of aphids in organic and conventional farming systems. *Agr. Ecosyst. Environ.* **133**, 14–18.

Machmer, H. (1972). Ciliary activity and the origin of metachrony in *Paramecium*: Effects of increased viscosity. *J. Exp. Biol.* **57**, 239–259.

Marquet, P.A., Navarrete, S.A., and Castilla, J.C. (1990). Scaling population density to body size in rocky intertidal communities. *Science* **250**, 1125–1127.

May, R.M. (1972). Will a large complex system be stable? *Nature* **238**, 413–414.

May, R.M. (1973). Stability and Complexity in Model Ecosystems. Princeton University Press, Princeton.

McCann, K., Hastings, A., and Huxel, G.R. (1998). Weak trophic interactions and the balance of nature. *Nature* **395**, 794–798.

McKee, D., Atkinson, D., Collings, S.E., Eaton, J.W., Gill, A.B., Harvey, I., Hatton, K., Heryes, T., Wilson, D., and Moss, B. (2003). Response of freshwater microcosm communities to nutrients, fish, and elevated temperature during winter and summer. *Limnol. Oceanogr.* **48**, 707–722.

McLaughlin, O.B., Jonsson, T., and Emmerson, M.C. (2010). Temporal variability in predator–prey relationships of a forest floor food web. *Adv. Ecol. Res.* **42**, 171–264.

Meerhoff, M., Clemente, J.M., De Mello, F.T., Iglesias, C., Pedersen, A.R., and Jeppesen, E. (2007). Can warm climate-related structure of littoral predator assemblies weaken the clear water state in shallow lakes? *Global Change Biol.* **13**, 1888–1897.

Memmott, J., Craze, P.G., Waser, N.M., and Price, M.V. (2007). Global warming and the disruption of plant–pollinator interactions. *Ecol. Lett.* **10**, 710–717.

Milly, P.C.D., Dunne, K.A., and Vecchia, A.V. (2006). Global pattern of trends in streamflow and water availability in a changing climate. *Nature* **438**, 347–350.

Milner, A.M., Knudsen, E.E., Soiseth, C., Robertson, A.L., Schell, D., Phillips, I.T., and Magnusson, K. (2000). Colonization and development of stream communities across a 200-year gradient in Glacier Bay National Park, Alaska. *Can. J. Fish. Aquat. Sci.* **57**, 2319–2335.

Milner, A.M., Robertson, A.E., Monaghan, K., Veal, A.J., and Flory, E.A. (2008). Colonization and development of a stream community over 28 years; Wolf Point Creek in Glacier Bay, Alaska. *Front. Ecol. Environ.* **6**, 413–419.

Milner, A.M., Brown, L.E., and Hannah, D.M. (2009). Hydroecological effects of shrinking glaciers. *Hydrol. Process.* **23**, 62–77.

Mohseni, O., and Stefan, H.G. (1999). Stream temperature/air temperature relationship: A physical interpretation. *J. Hydrol.* **218**, 128–141.

Montoya, J.M., and Raffaelli, D. (2010). Climate change, biotic interactions and ecosystem services. *Philos. Trans. R. Soc. Lond. B* **365**, 2013–2018.

Montoya, J.M., Pimm, S.L., and Solé, R.V. (2006). Ecological networks and their fragility. *Nature* **442**, 259–264.

Montoya, J.M., Woodward, G., Emmerson, M.C., and Sole, R.C. (2009). Indirect effects propagate disturbances in real food webs. *Ecology* **90**, 2426–2433.

Moore, J.C., Berlow, E.L., Coleman, D.C., de Ruiter, P.C., Dong, Q., Hastings, A., Johnson, N.C., McCann, K.S., Melville, K., Morin, P.J., Nadelhoffer, K., Rosemond, A.D., *et al.* (2004). Detritus, trophic dynamics and biodiversity. *Ecol. Lett.* **7**, 584–600.

Moran, X.A., López-Urrutia, Á., Calvo-Díaz, A., and Li, W.K.W. (2010). Increasing importance of small phytoplankton in a warmer ocean. *Global Change Biol.* **16**, 1137–1144.

Moss, B., McKee, D., Atkinson, D., Collings, S.E., Eaton, J.W., Gill, A.B., Harvey, I., Hatton, K., Heyes, T., and Wilson, D. (2003). How important is climate? Effects of warming, nutrient addition and fish on phytoplankton in shallow lake microcosms. *J. Appl. Ecol.* **40**, 782–792.

Mulder, C., and Elser, J.J. (2009). Soil acidity, ecological stoichiometry and allometric scaling in grassland food webs. *Global Change Biol.* **15**, 2730–2738.

Muller, H., and Geller, W. (1993). Maximum growth-rates of aquatic ciliated protozoa—The dependence on body size and temperature reconsidered. *Arch. Hydrobiol.* **126**, 315–327.

Nilsson, L.A. (1988). The evolution of flowers with deep corolla tubes. *Nature* **334**, 147–149.

Norby, R.J., Cotrufo, M.F., Ineson, P., O'Neill, E.G., and Canadell, J.G. (2001). Elevated CO_2, litter chemistry, and decomposition: A synthesis. *Oecologia* **127**, 153–165.

O'Gorman, E.J., and Emmerson, M.C. (2009). Perturbations to trophic interactions and the stability of complex food webs. *Proc. Natl. Acad. Sci. USA* **106**, 13393–13398.

O'Gorman, E.J., and Emmerson, M.C. (2010). Manipulating interaction strengths and the consequences for trivariate patterns in a marine food web. *Adv. Ecol. Res.* **42**, 301–419.

O'Gorman, E.J., Enright, R.A., and Emmerson, M.C. (2008). Predator diversity enhances secondary production and decreases the likelihood of trophic cascades. *Oecologia* **158**, 557–567.

Olesen, J.M., and Jordano, P. (2002). Geographic patterns in plant–pollinator mutualistic networks. *Ecology* **83**, 2416–2424.

Olesen, J.M., Bascompte, J., Elberling, H., and Jordano, P. (2008). Temporal dynamics of a pollination network. *Ecology* **89**, 1573–1582.

Olesen, J.M., Dupont, Y.L., O'Gorman, E.J., Ings, T.C., Layer, K., Melián, C.J., Troejelsgaard, K., Pichler, D.E., Rasmussen, C., and Woodward, G. (2010). From Broadstone to Zackenberg: Space, time and hierarchies in ecological networks. *Adv. Ecol. Res.* **42**, 1–69.

Parker, S.M., and Huryn, A.D. (2006). Food web structure and function in two arctic streams with contrasting disturbance regimes. *Freshw. Biol.* **51**, 1249–1263.
Parmesan, C. (2006). Ecological and evolutionary responses to recent climate change. *Ann. Rev. Ecol. Evol. Syst.* **37**, 637–669.
Parmesan, C., and Yohe, G. (2003). A globally coherent fingerprint of climate change impacts across natural systems. *Nature* **421**, 37–42.
Pearson, R.G., and Dawson, T.P. (2003). Predicting the impacts of climate change on the distribution of species: Are bioclimate envelope models useful? *Global Ecol. Biogeogr.* **12**, 361–371.
Pelejero, C., Calvo, E., and Hoegh-Guldberg, O. (2010). Paleo-perspectives on ocean acidification. *Trends Ecol. Evol.* **25**, 332–344.
Perkins, D.M., Reiss, J., Yvon-Durocher, G., and Woodward, G. (2010). Global change and food webs in running waters. *Hydrobiologia* (in press).
Perry, R.I., Cury, P., Brander, K., Jennings, S., Möllmann, C., and Planque, B. (2010). Sensitivity of marine systems to climate and fishing: Concepts, issues and management responses. *J. Mar. Syst.* **79**, 427–435.
Petchey, O.L., and Belgrano, A. (2010). Body-size distributions and size-spectra: universal indicators of ecological status? *Biol. Lett.* **6**, 434–437.
Petchey, O.L., Beckerman, A.P., Riede, J.O., and Warren, P.H. (2008). Size, foraging, and food web structure. *Proc. Natl. Acad. Sci. USA* **105**, 4191–4196.
Petchey, O.L., McPhearson, P.T., Casey, T.M., and Morin, P.J. (1999). Environmental warming alters food-web structure and ecosystem function. *Nature* **402**, 69–72.
Petchey, O.L., Casey, T.J., Jiang, L., McPhearson, P.T., and Price, J. (2002). Species richness, environmental fluctuations, and temporal change in total community biomass. *Oikos* **99**, 231–240.
Petchey, O.L., Brose, U., and Rall, B.C. (2010). Predicting the effects of temperature on food web connectance. *Philos. Trans. R. Soc. Lond. B* **365**, 2081–2091.
Peters, R.H. (1983). In: *The Ecological Implications of Body Size* (Ed. by E. Beck, H.J.B. Birks and E.F. Connor), pp. 24–44. Cambridge University Press, Cambridge, UK.
Piechnik, D.A., Lawler, S.P., and Martinez, N.D. (2008). Food-web assembly during a classic biogeographic study: Species' "trophic breadth" corresponds to colonization order. *Oikos* **117**, 665–674.
Podolsky, R.D. (1994). Temperature and water viscosity: Physiological versus mechanical effects on suspension feeding. *Science* **265**, 100–103.
Podolsky, R.D., and Emlet, R.B. (1993). Separating the effects of temperature and viscosity on swimming and water movement by Sand Dollar larvae (*Dendraster excentricus*). *J. Exp. Biol.* **176**, 207–221.
Pope, J.G., Shepherd, J.G., and Webb, J. (1994). Successful surf-riding on size spectra: The secret of survival in the sea. *Philos. Trans. R. Soc. Lond. B* **343**, 41–49.
Pounds, J.A., Fogden, M.P.L., and Campbell, J.H. (1999). Biological response to climate change on a tropical mountain. *Nature* **398**, 611–615.
Purcell (1977). Life at low Reynolds number. *Am. J. Phys.* **45**, 1–11.
Raffaelli, D. (2004). How extinction patterns affect ecosystems. *Science* **306**, 1141–1142.
Rahel, F.J., and Olden, J.D. (2008). Assessing the effects of climate change on aquatic invasive species. *Conserv. Biol.* **22**, 521–533.
Rall, B.C., Vucic-Pestic, O., Ehnes, R.B., Emmerson, M.C., and Brose, U. (2010). Temperature, predator–prey interaction strength and population stability. *Global Change Biol.* 10.1111/j.1365-2486.2009.02124.x.

Rawcliffe, R., Sayer, C.D., Woodward, G., Grey, J., Davidson, T.A., and Jones, J.I. (2010). Back to the future: Using palaeolimnology to infer long-term changes in shallow lake food webs. *Freshw. Biol.* **55**, 600–613.

Redfield, A.C. (1934). On the proportions of organic derivatives in sea water and their relation to the composition of plankton. In: *James Johnstone Memorial Volume* (Ed. by R.J. Daniel), pp. 176–192. Liverpool University Press, Liverpool.

Reich, P.B., and Oleksyn, J. (2004). Global patterns of plant leaf N and P in relation to temperature and latitude. *Proc. Natl. Acad. Sci. USA* **101**, 11001–11006.

Reiss, J., Bridle, J.R., Montoya, J.M., and Woodward, G. (2009). Emerging horizons in biodiversity and ecosystem functioning research. *Trends Ecol. Evol.* **24**, 505–514.

Reuman, D.C., and Cohen, J.E. (2005). Estimating relative energy fluxes using the food web, species abundance, and body size. *Adv. Ecol. Res.* **36**, 137–182.

Reuman, D.C., Mulder, C., Raffaelli, D., and Cohen, J.E. (2008). Three allometric relations of population density to body mass: Theoretical integration and empirical tests in 149 food webs. *Ecol. Lett.* **11**, 1216–1228.

Reuman, D.C., Mulder, C., Banašek-Richter, C., Cattin Blandenier, M.-F., Breure, A.M., Den Hollander, H.A., Kneitel, J.M., Raffaelli, D., Woodward, G., and Cohen, J.E. (2009a). Allometry of body size and abundance in 166 food webs. *Adv. Ecol. Res.* **41**, 1–44.

Reuman, D.C., Cohen, J.E., and Mulder, C. (2009b). Human and environmental factors influence soil food webs' abundance-mass allometry and structure. *Adv. Ecol. Res.* **41**, 45–85.

Rhee, G.-Y., and Gotham, I.J. (1981). The effect of environmental factors on phytoplankton growth: Temperature and the interactions of temperature with nutrient limitation. *Limnol. Oceanogr.* **26**, 635–648.

Richardson, A.J., and Schoeman, D.S. (2004). Climate impact on plankton ecosystems in the Northeast Atlantic. *Science* **305**, 1609–1612.

Riede, J.O., Rall, B.C., Banasek-Richter, C., Navarrete, S.A., Wieters, E.A., and Brose, U. (2010). Scaling of food-web properties with diversity and complexity across ecosystems. *Adv. Ecol. Res.* **42**, 139–170.

Riisgård, H.U., and Larsen, P.S. (2007). Viscosity of seawater controls beat frequency of water-pumping cilia and filtration rate of mussels *Mytilus edulis*. *Mar. Ecol. Prog. Ser.* 141–150.

Romanuk, T.N., Zhou, Y., Brose, U., Berlow, E.L., Williams, R.J., and Martinez, N. D. (2009). Predicting invasion success in complex ecological networks. *Philos. Trans. R. Soc. Lond. B* **364**, 1743–1754.

Rooney, N., McCann, K., Gellner, G., and Moore, J.C. (2006). Structural asymmetry and the stability of diverse food webs. *Nature* **442**, 265–269.

Sala, O.E., Chapin Iii, F.S., Armesto, J.J., Berlow, E., Bloomfield, J., Dirzo, R., Huber-Sanwald, E., Huenneke, L.F., Jackson, R.B., Kinzig, A., *et al.* (2000). Global biodiversity scenarios for the year 2100. *Science* **287**, 1770–1774.

Sarmento, H., Montoya, J.M., Vázquez-Domínguez, E., Vaqué, D., and Gasol, J.M. (2010). Warming effects on marine microbial food web processes: How far can we go when it comes to predictions? *Philos. Trans. R. Soc. Lond. B* **365**, 2137–2149.

Sarmiento, J.L. and Wofsy, S.C. (1999). A U.S. carbon cycle science plan: Report of the Carbon and Climate Working Group, *U.S. Global Change Res. Program*, Washington, DC

Sarmiento, J.L., Slater, R., Barber, R., Bopp, L., Doney, S.C., Hirst, A.C., Kleypas, J., Matear, R., Mikolajewicz, U., Monfray, P., Soldatov, V., Spall, S.A., *et al.* (2004). Response of ocean ecosystems to climate warming. *Glob. Biogeochem. Cycles* **18**, 1–23.

Schneider, W.R., and Doetsch, R.N. (1974). Effect of viscosity on bacterial motility. *Bacteriology* **117**, 696–701.
Schofield, O., Ducklow, H.W., Martinson, D.C., Meredith, M.P., Moline, M.A., and Frazer, W.R. (2010). How do polar marine ecosystems respond to rapid climate change? *Science* **328**, 1520–1523.
Seuront, L., and Vincent, D. (2008). Increased seawater viscosity, *Phaeocystis globosa* spring bloom and Temora longicornis feeding and swimming behaviours. *Mar. Ecol. Prog. Ser* **363**, 131–145.
Short and Ward (1981b). Trophic ecology of three winter stoneflies (Plecoptera). *Am. Mid Naturalist* **105**, 341–347.
Short, R.A., and Ward, J.V. (1981a). Benthic detritus dynamics in a mountain stream. *Holarctic Ecol.* **4**, 32–35.
Sleigh, M.A. (1956). Metachronism and frequency of beat in the peristomial cilia of *Stentor*. *J. Exp. Biol.* **33**, 15–28.
Smol, J.P., Wolfe, A.P., Birks, H.J.B., Douglas, M.S.V., Jones, V.J., Korhola, A., Pienitz, R., Rühland, K., Sorvari, S., Antoniades, D., *et al.* (2005). Climate-driven regime shifts in the biological communities of arctic lakes. *Proc. Natl. Acad. Sci. USA* **102**, 4397–4402.
Snow, B.K., and Snow, D.W. (1972). Feeding niches of hummingbirds in a Trinidad valley. *J. Anim. Ecol.* **41**, 471–485.
Sommer, U. (1996). Plankton ecology: The past two decades of progress. *Naturwissenschaften* **83**, 293–301.
Spooner, D.E., and Vaughn, C.C. (2008). A trait-based approach to species' roles in stream ecosystems: Climate change, community structure, and material cycling. *Oecologia* **158**, 307–317.
Stang, M., Klinkhamer, P.G.L., and van der Meijden, E. (2006). Size constraints and flower abundance determine the number of interactions in a plant–flower visitor web. *Oikos* **112**, 111–121.
Stang, M., Klinkhamer, P.G.L., Waser, N.M., Stang, I., and van der Mejden, E. (2009). Size-specific interaction patterns and size matching in a plant–pollinator interaction web. *Ann. Bot. Lond.* **103**, 1459–1469.
Statzner, B., Gore, J.A., and Resh, V.H. (1988). Hydraulic stream ecology: Observed patterns and potential applications. *J. N. Am. Benthol. Soc.* **7**, 307–360.
Stegen, J.C., Enquist, B.J., and Ferriere, R. (2009). Advancing the metabolic theory of biodiversity. *Ecol. Lett.* **12**, 1001–1015.
Sterner, R.W. (1997). Modelling interactions of food quality and quantity in homeostatic consumers. *Freshw. Biol.* **38**, 473–481.
Sterner, R.W. (2004). A one-resource "stoichiometry"? *Ecology* **85**, 1813–1816.
Sterner, R.W., and Elser, J.J. (2002). Ecological Stoichiometry: The Biology of Elements from Molecules to the Biosphere. Princeton University Press, Princeton.
Sterner, R.W., and Hessen, D.O. (1994). Algal nutrient limitation and the nutrition of aquatic herbivores. *Annu. Rev. Ecol. Syst.* **25**, 1–29.
Sterner, R.W., Elser, J.J., Fee, E.J., Guildford, S.J., and Chrzanowski, T.H. (1997). The light:nutrient ratio in lakes: The balance of energy and materials affects ecosystem structure. *Am. Nat.* **150**, 663–684.
Stiling, and Cornelissen (2007). How does elevated carbon dioxide (CO_2) affect plant–herbivore interactions? A field experiment and meta-analysis of CO_2-mediated changes on plant chemistry and herbivore performance. *Global Change Biol.* **9**, 1823–1842.
Strathdee, A.T., Bale, J.S., Block, W.C., Coulson, S.J., Hodkinson, I.D., and Webb, N.R. (1993). Effects of temperature elevation on a field population of

Acyrthosiphon svalbardicum (Hemiptera, Aphididae) on Spitsbergen. *Oecologia* **96**, 457–465.

Strecker, A.L., Cobb, T.P., and Vinebrooke, R.D. (2004). Effects of experimental greenhouse warming on phytoplankton and zooplankton communities in fishless alpine ponds. *Limnol. Oceanogr.* **49**, 1182–1190.

Sweeney, B.W., Vannote, R.L., and Dodds, P.J. (1986). Effects of temperature and food quality on growth and development of a mayfly, *Leptophlebia intermedia*. *Can. J. Fish. Aquat. Sci.* **43**, 12–18.

Teixidó, N., Garrabou, J., Gutt, J., and Arntz, W.E. (2007). Iceberg disturbance and successional spatial patterns: The case of the shelf Antarctic benthic communities. *Ecosystems* **10**, 143–158.

Thackeray, S.J., Sparks, T.H., Frederiksen, M., Burthe, S., Bacon, P.J., Bell, J.R., Botham, M.S., Brereton, T.M., Bright, P.W., Carvalho, L., Clutton-Brock, T., and Dawson, A. (2010). Trophic level asynchrony in rates of phenological change for marine, freshwater and terrestrial environments. *Global Change Biol.* 10.1111/ j.1365-2486.2010.02165.x(in press).

Thomas, C.D., Cameron, A., Green, R.E., Bakkenes, M., Beaumont, L.J., Collingham, Y.C., Erasmus, B.F.N., De Siqueira, M.F., Grainger, A., and Hannah, L. (2004). Extinction risk from climate change. *Nature* **427**, 145–148.

Thompson, D.J. (1978). Towards a realistic predator prey model effect of temperature on functional response and life history of larvae of damselfly, *Ischnura elegans*. *J. Anim. Ecol.* **47**, 757–767.

Timmermann, A., Oberhuber, J., Bacher, A., Esch, M., Latif, M., and Roeckner, E. (1999). Increased El Niño frequency in a climate model forced by future greenhouse warming. *Nature* **398**, 694–697.

Tuchman, N.C., Wetzel, R.G., Rier, S.T., Wahtera, K.A., and Teeri, J.A. (2002). Elevated atmospheric CO_2 lowers leaf litter nutritional quality for stream ecosystem food webs. *Global Change Biol.* **8**, 163–170.

Tylianakis, J. (2009). Warming up food webs. *Science* **323**, 1300–1301.

Tylianakis, J.M., Tscharntke, T., and Lewis, O.T. (2007). Habitat modification alters the structure of tropical host–parasitoid food webs. *Nature* **445**, 202–205.

Tylianakis, J., Didham, R.K., Bascompte, J., and Wardle, D.A. (2008). Global change and species interactions in terrestrial ecosystems. *Ecol. Lett.* **11**, 1351–1363.

Urabe, J., and Sterner, R.W. (1996). Regulation of herbivore growth by the balance of light and nutrients. *Proc. Natl. Acad. Sci. USA* **93**, 8465–8469.

Urabe, J., Kyle, M., Makino, W., Yoshida, T., Andersen, T., and Elser, J.J. (2002). Reduced light increases herbivore production due to stoichiometric effects of light/ nutrient balance. *Ecology* **83**, 619–627.

Urabe, J., Togari, J., and Elser, J.J. (2003). Stoichiometric impacts of increased carbon dioxide on a planktonic herbivore. *Global Change Biol.* **9**, 818–825.

Van der Putten, W.H., Macel, M., and Visser, M.E. (2010). Predicting species distribution and abundance responses to climate change: Why it is essential to include biotic interactions across trophic levels. *Philos. Trans. R. Soc. Lond. B* **365**, 2025–2034.

Van Veen, F.J.F., Van Holland, P.D., and Godfray, H.C.J. (2005). Stable coexistence in insect communities due to density- and trait-mediated indirect effects. *Ecology* **86**, 1382–1389.

Van Veen, F.J.F., Memmott, J. and Godfray, H.C.J. (2006). Indirect effects, apparent competition and biological control. In: *Trophic and Guild Interactions in Biological Control*, pp. 145–169.

Vasseur, D.A., and McCann, K.S. (2005). A mechanistic approach for modeling temperature dependent consumer–resource dynamics. *Am. Nat.* **166**, 184–198.

Vermaat, J.E., Dunne, J.A., and Gilbert, A.J. (2009). Major dimensions in food-web structure properties. *Ecology* **90**, 278–282.

Visser, M.E., Holleman, L.J.M., and Caro, S.P. (2009). Temperature has a causal effect on avian timing of reproduction. *Proc. R. Soc. B-Biol. Sci.* **276**, 2323–2331.

Vitousek, P.M. (1982). Nutrient cycling and nutrient use efficiency. *Am. Nat.* **119**, 553–572.

Walters, A.W., and Post, D.M. (2008). An experimental disturbance alters fish size structure but not food chain length in streams. *Ecology* **89**, 3261–3267.

Walther, G.R. (2010). Community and ecosystem responses to recent climate change. *Philos. Trans. R. Soc. Lond. B* **365**, 2019–2024.

Walther, G.-R., Post, E., Convey, P., Menzel, A., Parmesan, C., Beebee, T.J.C., Fromentin, J.-M., Hoegh-Guldberg, O., and Bairlein, F. (2002). Ecological responses to recent climate change. *Nature* **416**, 389–395.

Warren, P.H. (1996). Structural constraints on food web assembly. In: *Aspects of the Genesis and Maintenance of Biological Diversity* (Ed. by M.E. Hochberg, J. Clobert and R. Barbault), pp. 142–161. Oxford University Press, Oxford, UK.

Watson, A.J., and Orr, J.C. (2003). Carbon dioxide fluxes in the global ocean. In: *Ocean Biogeochemistry: The Role of the Ocean Carbon Cycle in Global Change* (Ed. by M.J.R. Fasham), Global Change—The IGBP Seriespp. 123–143.

Weiher, E., and Keddy, P.A. (1999). Relative abundance and evenness patterns along diversity and biomass gradients. *Oikos* 355–361.

White, M.C., and Zhao, X.Q. (2009). A periodic Droop model for two species competition in a chemostat. *Bull. Math. Biol.* **71**, 145–161.

Williams, R.J., and Martinez, N.D. (2000). Simple rules yield complex food webs. *Nature* **404**, 180–183.

Wilson, K. (1984). A bibliography of Lough Hyne (Ine) 1687-1982. *J. Life Sci. R. Dubl. S* **5**, 1–11.

Winder, M., and Schindler, D.E. (2004). Climate change uncouples trophic interactions in an aquatic ecosystem. *Ecology* **85**, 2100–2106.

Winder, M., Reuter, J.E., and Schladow, S.G. (2009). Lake warming favours small-sized planktonic diatom species. *Philos. Trans. R. Soc. Lond. B* **276**, 427–435.

Winet, H. (1976). Ciliary propulsion of objects in tubes—Wall drag on swimming *Tetrahymena* (Ciliata) in presence of mucin and other long-chain polymers. *J. Exp. Biol.* **64**, 283–302.

Winfield, I.J., Hateley, J., Fletcher, J.M., James, J.B., Bean, C.W., and Clabburn, P. (2010). Population trends of Arctic charr (*Salvelinus alpinus*) in the UK: Assessing the evidence for a widespread decline in response to climate change. *Hydrobiologia* **650**, 55–65.

Witze, A. (2008). Losing Greenland. *Nature* **452**, 798–802.

Woods, H.A., Makino, W., Cotner, J.B., Hobbie, S.E., Harrison, J.F., Acharya, K., and Elser, J.J. (2003). Temperature and the chemical composition of poikilothermic organisms. *Funct. Ecol.* **17**, 237–245.

Woodward, G. (2009). Biodiversity, ecosystem functioning and food webs in fresh waters: Assembling the jigsaw puzzle. *Freshw. Biol.* **54**, 2171–2187.

Woodward, G., and Hildrew, A.G. (2002a). Differential vulnerability of prey to an invading top predator: Integrating field surveys and laboratory experiments. *Ecol. Entomol.* **27**, 732–744.

Woodward, G., and Hildrew, A.G. (2002b). Food web structure in riverine landscapes. *Freshw. Biol.* **47**, 777–798.

Woodward, G., and Hildrew, A.G. (2002c). The impact of a sit-and-wait predator: Separating consumption and prey emigration. *Oikos* **99**, 409–418.

Woodward, G., and Warren, P.H. (2007). Body size and predatory interactions in freshwaters: Scaling from individuals to communities. In: *Body Size: The Structure and Function of Aquatic Ecosystems* (Ed. by A.G. Hildrew, D. Raffaelli and R. Edmonds-Brown), pp. 98–117. Cambridge University Press, Cambridge.

Woodward, G., Jones, J.I., and Hildrew, A.G. (2002). Community persistence in Broadstone Stream (U.K.) over three decades. *Freshw. Biol.* **47**, 1419–1435.

Woodward, G., Speirs, D.C., and Hildrew, A.G. (2005a). Quantification and resolution of a complex, size-structured food web. *Adv. Ecol. Res.* **36**, 85–135.

Woodward, G., Ebenman, B., Emmerson, M., Montoya, J.M., Olesen, J.M., Valido, A., and Warren, P.H. (2005b). Body-size in ecological networks. *Trends Ecol. Evol.* **20**, 402–409.

Woodward, G., Perkins, D.M., and Brown, L. (2010a). Climate change in freshwater ecosystems: Impacts across multiple levels of organisation. *Philos. Trans. R. Soc. Lond. B* **365**, 2093–2106.

Woodward, G., Christensen, J.B., Olafsson, J.S., Gislason, G.M., Hannesdottir, E. R., and Friberg, N. (2010b). Sentinel systems on the razor's edge: Effects of warming on Arctic stream ecosystems. *Global Change Biol.* **16**, 1979–1991.

Woodward, G., Blanchard, J., Lauridsen, R.B., Edwards, F.K., Jones, J.I., Figueroa, D., Warren, P.H., and Petchey, O.L. (2010). Individual-based food webs: species identity, body size and sampling effects. *Adv. Ecol. Res.* **43** (in press).

Wrona, F.J., Prowse, T.D., Reist, J.D., Hobbie, J.E., Lévesque, L.M.J., and Vincent, W.F. (2006). Climate change effects on aquatic biota, ecosystem structure and function. *Ambio* **35**, 359–369.

Yodzis, P. (1988). The indeterminacy of ecological interactions as perceived through perturbation experiments. *Ecology* **69**, 508–515.

Yodzis, P., and Innes, S. (1992). Body size and consumer–resource dynamics. *Am. Nat.* **139**, 1151–1175.

Yoon, I., Williams, R.J., Levine, E., Yoon, S., Dunne, J.A., and Martinez, N.D. (2004). Webs on the Web (WoW): 3D visualization of ecological networks on the WWW for collaborative research and education. *Proc. ISandT/SPIE Symp. Electron. Imag., Vis. Data Anal.* **5295**, 124–132.

Yvon-Durocher, G., Montoya, J.M., Emmerson, M.C., and Woodward, G. (2008). Macroecological patterns and niche structure in a new marine food web. *Cent. Eur. J. Biol.* **3**, 91–103.

Yvon-Durocher, G., Woodward, G., Jones, J.I., Trimmer, M., and Montoya, J.M. (2010a). Warming alters the metabolic balance of ecosystems. *Philos. Trans. R. Soc. Lond. B* **365**, 2117–2126.

Yvon-Durocher, G., Reiss, J., Blanchard, J., Ebenman, B., Perkins, D.M., Reuman, D.C., Thierry, A., Woodward, G., and Petchey, O.L. (2010b). Across ecosystem comparisons of size structure: Methods, approaches, and prospects. *Oikos* (in press).

Zhang, Y.-P., and Li, X. (2004). The thermal dependence of food assimilation and locomotor performance in southern grass lizards, *Takydromus sexlineatus* (Lacertidae). *J. Thermal. Biol.* **29**, 45–53.

Zwick, P. (1992). Stream habitat fragmentation—A threat to biodiversity. *Biodivers. Conserv.* **1**, 80–97.

Scaling of Food-Web Properties with Diversity and Complexity Across Ecosystems

JENS O. RIEDE, BJÖRN C. RALL, CAROLIN BANASEK-RICHTER, SERGIO A. NAVARRETE, EVIE A. WIETERS, MARK C. EMMERSON, UTE JACOB AND ULRICH BROSE

SUMMARY

Trophic scaling models describe how topological food-web properties such as the number of predator–prey links scale with species richness of the community. Early models predicted that either the link density (i.e. the number of links per species) or the connectance (i.e. the linkage probability between any pair of species) is constant across communities. More recent analyses, however, suggest that both these scaling models have to be rejected, and we discuss several hypotheses that aim to explain the scale dependence of these complexity parameters. Based on a recent, highly resolved food-web compilation, we analysed the scaling behaviour of 16 topological parameters and found significant power–law scaling relationships with diversity (i.e. species richness) and

ADVANCES IN ECOLOGICAL RESEARCH VOL. 42 0065-2504/10 $35.00

DOI: 10.1016/S0065-2504(10)42003-6

complexity (i.e. connectance) for most of them. These results illustrate the lack of universal constants in food-web ecology as a function of diversity or complexity. Nonetheless, our power–law scaling relationships suggest that fundamental processes determine food-web topology, and subsequent analyses demonstrated that ecosystem-specific differences in these relationships were of minor importance. As such, these newly described scaling relationships provide robust and testable cornerstones for future structural food-web models.

I. INTRODUCTION

Over the last several centuries, physicists have developed a variety of scaling laws, such as Newton's law of universal gravitation, which holds that the gravitational force between two bodies is proportional to the product of their masses and the inverse of their squared distance. The change in gravitational force with distance is well described by a scaling law, where the gravitational constant and the exponent (negative square) are constant with respect to distance. Scaling laws thus indicate, but do not prove, the fundamental process that governs the relationship between variables. In search of analogues of the grand laws of physics, ecologists have been searching for ecological scaling models that can be generalized across organisms, populations and even entire ecosystems (Lange, 2005; O'Hara, 2005). Among the most promising approaches, trophic scaling models predict relationships between topological food-web properties, such as the number of predator–prey feeding interactions (links, L) and the species richness (S, hereafter: diversity) of the community (Dunne, 2006). In diversity–topology relationships, scale refers to the number of species; and ecologists have searched for universal food-web constants that equally apply to species-poor and species-rich ecosystems. Early trophic scaling models suggested that link density—the number of links per species (L/S)—is constant across food webs of varying species richness (Cohen and Briand, 1984). This 'link-species scaling law' is in agreement with the classical stability criterion of random networks, which holds that local population stability is maintained if link density remains below a critical threshold that, in turn, depends on the average interaction strength (May, 1972). Subsequent early trophic scaling models proposed constancy of additional food-web properties, including the proportions of top species (T, species consuming other species whilst they have no consumers), intermediate species (I, species that consume and are consumed by other species) and basal species (B, species without resource species below them within a food chain, e.g. plants or detritivores) (Cohen and Briand, 1984), and constant proportions of links between these trophic groups: T–I, T–B, I–I and I–B links (Cohen and Briand, 1984). Empirical tests using early food-web data rendered support to these scaling laws (Briand and Cohen, 1984; Cohen and Briand, 1984; Cole *et al.*, 2006), but the quality of the data employed has cast doubt on the validity of these findings, largely due to poor taxonomic resolution, limited sampling effort and the presence of biological impossibilities (e.g. birds included as basal species) (Hall and Raffaelli, 1993; Ings *et al.*, 2009; Paine, 1988; Polis, 1991).

Other studies based on data of higher quality demonstrated that link density, the proportions of top, intermediate and basal species, and the proportions of *T–I*, *T–B*, *I–I* and *I–B* links are not constant across the diversity scale (Hall and Raffaelli, 1991; Martinez, 1991, 1993a; Schoener, 1989; Warren, 1989; Winemiller, 1990). Earlier findings of scale invariance were consequently ascribed to a range of methodological artefacts arising from inadequate sampling, strong species aggregation and poor data resolution (Bersier *et al.*, 1999; Goldwasser and Roughgarden, 1997; Hall and Raffaelli, 1991; Martinez, 1991, 1993b; Martinez *et al.*, 1999).

While the improved data demonstrated scale dependence of link density, an alternative hypothesis proposed that connectance (C)—the linkage probability of any pair of species in the food web ($C = L/S$)—should be constant across ecosystems of variable species richness (Martinez, 1992). Models with constant link density assume that any species can consume a fixed number of the coexisting species, whereas the constant–connectance model holds that any species can consume a fixed fraction of the coexisting species. The latter hypothesis initially received some empirical support (Martinez, 1992, 1993a, b; Spencer and Warren, 1996), but further analyses of more recent food-web data suggest that neither link density nor connectance are constant across the diversity scale (Brose and Martinez, 2004; Dunne, 2006; Montoya and Sole, 2003; Schmid-Araya *et al.*, 2002).

Much of this trophic scaling debate has focused on parameters of food-web complexity, such as the link density or connectance (Dunne, 2006). Other recent approaches that have addressed the scaling of additional topological food-web parameters have been inspired by physicists' scaling laws and introduced scale-dependent properties, but with constant scaling exponents (Camacho *et al.*, 2002a,b; Garlaschelli *et al.*, 2003). This implies that the studied food-web properties vary with the diversity of the communities, but this variance is described by universally constant exponents. For instance, they found significant scaling relationships of food-web properties such as the fractions of top, intermediate and basal species and the number of links among them (Martinez, 1994), the clustering coefficient (Camacho *et al.*, 2002a,b; Dunne *et al.*, 2002) and the average path length between any pair of species in a food web (Camacho *et al.*, 2002a,b; Williams *et al.*, 2002). However, all these studies still suffered from data limitation by either being based on older food-web collections of poor resolution or new compilations of high-quality data that included fewer than 20 food webs. Over the last decade or so, additional collections of higher quality food webs have become available (Brose *et al.*, 2006a; Brose and Martinez, 2004; Townsend *et al.*, 1998), but systematic and comprehensive analyses of scaling relationships in these data have yet to be undertaken.

In the present study, we attempt to fill this void by analysing the scaling of 19 food-web properties (see Section II for a description) with species richness (diversity) and connectance (complexity) using a collection of 65 food webs from terrestrial, lake, stream, estuarine and marine ecosystems (see Appendix for a detailed overview). Additionally, we tested for significant differences in scaling relationships among these five ecosystem types. This approach extends prior studies testing for significant deviations of marine (Dunne *et al.*, 2004) or Cambrian food-web topology (Dunne *et al.*, 2008) from those of other ecosystems. Our analyses also address whether the different ecosystem types included possess specific topologies, or whether there are consistent scaling relationships that hold across ecosystems, which would indicate the existence of general constraints upon the structure of ecological networks.

II. METHODS

A. The Food-Web Data Set

We illustrate trophic scaling relationships using a data set of 65 food webs from a variety of habitats (see Appendix for an overview of the food webs). This compilation includes 13 food webs that have been used in prior meta-studies (Cattin *et al.*, 2004; Dunne, 2006; Dunne *et al.*, 2004; Montoya and Sole, 2003; Stouffer *et al.*, 2005; Williams and Martinez, 2000) and six food webs from a meta-study on natural consumer–resource body-mass ratios (Brose *et al.*, 2006a,b). They are complemented by four further webs from the banks of Lake Neuchâtel (Cattin, 2004), nine of the largest webs from a study of 50 lakes in the Adirondack Mountains of New York State (Havens, 1992), eight stream webs from the collection of 10 New Zealand webs (Townsend *et al.*, 1998) and five terrestrial island food webs (Piechnik *et al.*, 2008). We did not consider food webs dominated by parasitoid or parasitic interactions, because the physical constraints by which they are governed differ from those that govern predator–prey interactions (Brose *et al.*, 2006a,b), thus modifying complexity patterns (Lafferty *et al.*, 2006). This choice is not meant to imply that such interactions are not of importance for the structure and function of the food webs, rather that maintaining a focus on free-living predator–prey interactions in a consistent, standardized manner helps elucidate the underlying processes. Overall, the data compilation analysed here includes food webs from 14 lakes or ponds, 25 streams or rivers, five brackish water of estuaries and salt marshes, six marine and 15 terrestrial ecosystems. The number of taxonomic species in these food webs ranges between 27 and 492, and the number of links ranges from 60 to 16,136.

B. Food-Web Topology

Nineteen food-web properties were calculated for each of the 65 taxonomic food webs studied (see Figures 1 and 3 for an overview). The properties analysed were: (1) the total number of links in the food webs, L; (2) the number of links per species, L/S; (3) connectance, C; the fractions of (4) top

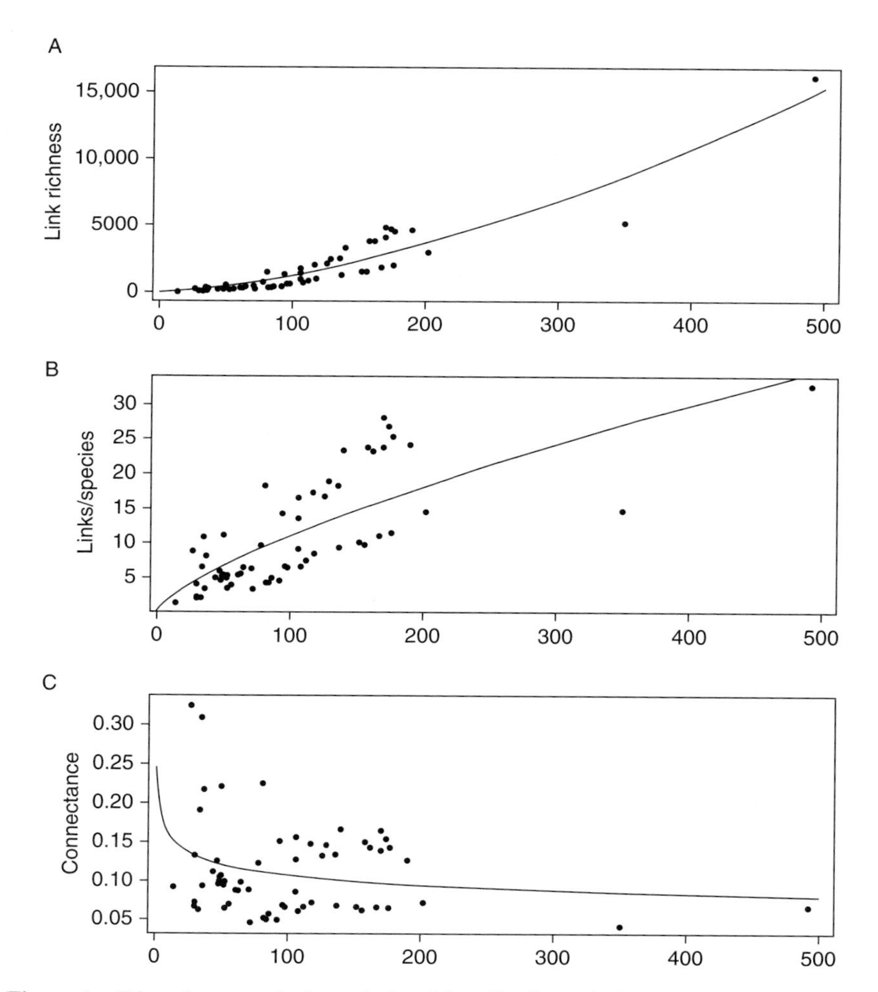

Figure 1 Diversity–complexity relationships. Scaling of (A) trophic link richness (exponent = 1.57 ± 0.07, $p < 0.001$), (B) links per species (exponent = 0.71 ± 0.08, $p < 0.001$) and (C) connectance (exponent = -0.18 ± 0.09, $p = 0.057$) with species richness of the food webs.

species (species with resources but without consumers), (5) intermediate species (species with resources and consumers), (6) basal species (species with consumers but without resources); (7) herbivores (species that consume basal species); (8) omnivores (species consuming resources across more than one trophic level); (9) cannibals (species partially feeding on con-specifics); (10) species in loops (circular link structures originating and ending at the same species); (11 and 12) the standard deviations of the species' generality (the number of resources) and vulnerability (the number of consumers); (13) linkedness (the total number of links to resources and consumers); (14) the average short-weighted trophic level (SWTL, average of the prey-averaged trophic level and the shortest chain for each species) across all species in the food web; (15) the average shortest chain length (shortest chain of trophic links from a species to a basal species) across all species in the food web; (16) the species' mean similarity (mean of the maximum trophic similarity of each species to any other species in the same food web); (17) the characteristic path length (mean over all shortest trophic paths between any pair of species in a food web); (18) the mean clustering coefficient (probability that two interacting species are linked to a third species); (19) and the diet discontinuity (the proportion of triplets of taxa with an irreducible gap in feeding links over the number of possible triplets), which have also been used in prior studies (Cattin *et al.*, 2004; Dunne *et al.*, 2004).

C. Statistical Analyses

First, we analysed the power–law scaling of the complexity parameters (1–3, as listed earlier): L, L/S and C, as a function of species richness, S:

$$L = aS^b \tag{1a}$$

$$L/S = aS^b \tag{1b}$$

$$C = aS^b \tag{1c}$$

where a and b are constants. For the remaining 16 topological food-web properties, P, we analysed the power–law scaling with species richness, S, and connectance, C:

$$P = aS^b C^c \tag{2}$$

where a, b and c are constants. We used a non-linear least-squares regression method to fit Eqs. 1a, 1b, 1c and 2 to the data (function 'nls' within the standard package 'stats' provided by the statistical software **R** 2.9.0, R Development Core Team, 2009). Significant scaling exponents b and c were interpreted as an indication of diversity and complexity scaling, respectively.

Two prominent scaling models predict that the scaling exponent, b, of the relationship between the number of links and species richness should be one ('the link-species scaling law', Cohen and Briand, 1984) or two ('the constant connectance hypothesis', Martinez, 1992). We tested these specific predictions by calculating the normally distributed probabilities of the z scores:

$$p\left(z = \frac{b - \mu}{\sigma}\right) \tag{3}$$

where b is the estimated exponent, σ its standard error, and μ represents the expected prediction.

Subsequently, we used the residuals of the fitted power–law models (Eq. 2) to test for signatures of the five ecosystem types (lake, stream, estuarine, marine and terrestrial ecosystems) in the scaling relationships. Our first analyses addressed significant differences in the overall food-web structure between these ecosystem types by a cluster analysis based on Euclidean distances (function 'hclust' within the package 'stats' provided by the statistical software **R** 2.9.0, R Development Core Team, 2009). Prior to this analysis, the residuals were normalized to zero mean (which they already had before) and unit variance. In a second analysis, we tested for significant differences in specific topological properties among the ecosystem types. For each of the 16 food-web properties, we carried out independent ANOVAs with the residuals as dependent variables and the ecosystem type as the factorial independent variable in one-way ANOVAs. Significant ANOVAs were followed by Tukey HSD post hoc tests.

III. RESULTS

The food webs in our data collection comprise between 27 and 492 species with 60–16,136 trophic links reported, 2–32.8 links per species, and connectance values that ranged from 0.04 to 0.33 (see Appendix for an overview of the food webs). In our first analysis, we illustrate the scaling of link richness, link density (links per species) and connectance with diversity (Figure 1).

A. Complexity–Diversity Relationships

The scaling of link richness (i.e. the number of trophic links, L) with species richness (S) should follow a power law with a slope of one ($\mu = 1$, Eq. 3) according to the 'link-species scaling law' (Cohen and Briand, 1984), or a slope of two ($\mu = 2$, Eq. 3) according to the 'constant connectance hypothesis' (Martinez, 1992). The power–law model (Eqs. 1a, 1b, 1c) fitted to our data

Table 1 Fit of power–law scaling models (Eqs. 1a, 1b, 1c) for link richness, links/species and connectance depending on species richness

	Intercept			Slope		
	Estimate	Std. error	*p*-Value	Estimate	Std. error	*p*-Value
Link richness	0.86715	0.34271	0.0139	1.57372	0.06843	<0.001
Link/species	0.4002	0.15731	0.0134	0.71976	0.07731	<0.001
Connectance	0.24588	0.09872	0.0154	−0.18032	0.094	0.0596

yielded an exponent, b, of 1.57 ± 0.07 (mean ± s.e., Figure 1A, Table 1), which differed significantly from one ($z = 8.1429$, $p < 0.001$) and two ($z = -6.1429$, $p < 0.001$). Further analyses suggested that link density increased significantly ($p < 0.001$), whereas connectance tended to decrease with diversity ($p = 0.06$), (Figure 1B and C, Table 1). This implied that more diverse food webs were characterized by more links per species, but a lower connectance, than food webs of low diversity. Two of the food webs in our data set contained considerably more species than the other food webs: these were those of the Weddell Sea, with 492 species, and Lough Hyne, with 350 species. It is possible that these data points might have exerted an undue influence on our conclusions, but repeating the analyses while excluding these two data points yielded similar results.

B. Ecosystem Types and Complexity–Stability Relationships

The distribution of the data in Figure 1 suggested two groups of food webs below and above the power–law model: one group of food webs with more links, a higher link density and a higher connectance and a second group of food webs with fewer links, a lower link density and a lower connectance than predicted by the power–law regression models. This resulted in positive and negative residuals of the first and second data group, respectively. We hypothesized that these differences might be driven by characteristics of specific ecosystems, such as the higher connectance and link density that appears to generally occur in marine ecosystems (Dunne *et al.*, 2004). Hence, we employed three independent one-way ANOVAs to test for significant effects of the ecosystem type on the residuals of link richness (Figure 2A), links per species (Figure 2B) and connectance (Figure 2C). We found significant effects of the ecosystem types on link richness and links per species. Subsequent post hoc tests revealed that this pattern was driven by higher link richness ($p = 0.048$) and links per species ($p = 0.021$) in river ecosystems relative to terrestrial ecosystems (Figure 2A and B). All other differences

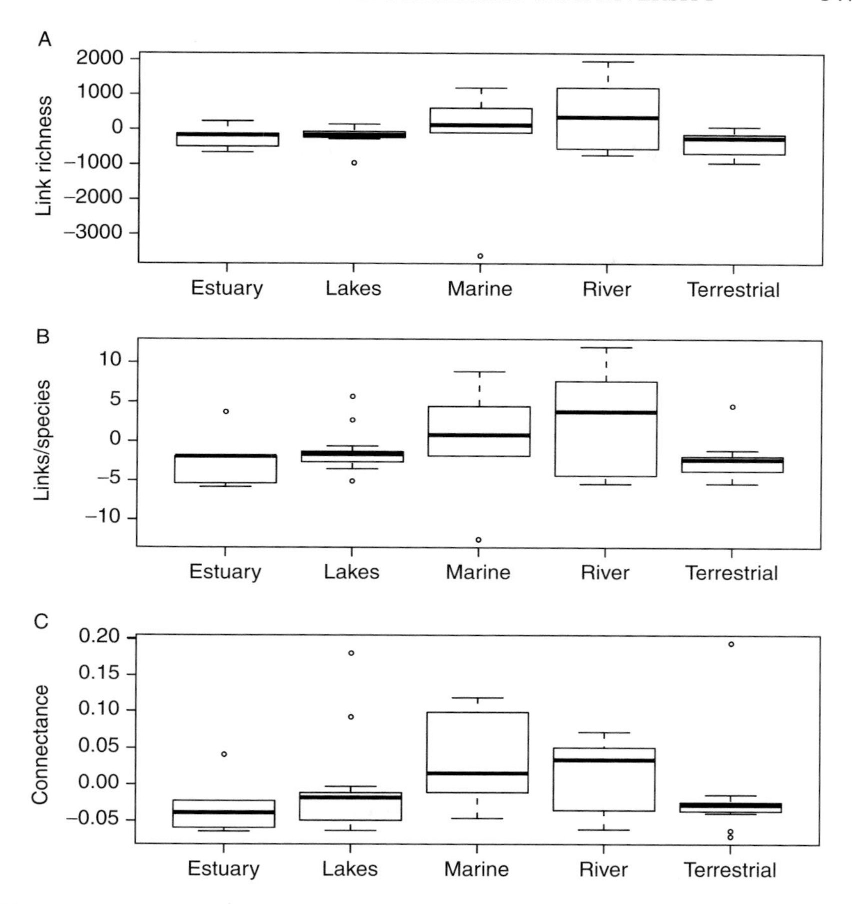

Figure 2 Ecosystem-type specific differences in diversity–complexity scaling: differences in the residuals of the diversity–complexity relationships (shown in Figure 1) for (A) trophic link richness (note that only the relationship between terrestrial and river ecosystems is significantly different; $p=0.048$), (B) links per species (note that only the relationship between terrestrial and river ecosystems is significantly different; $p=0.022$) and (C) connectance depending on the ecosystem types.

among pairs of ecosystem types were not significant. Interestingly, our analyses did not reveal any significant difference in connectance between the ecosystem types (Figure 2C). This also revealed that the grouping of ecosystems in types employed in our study did not explain the residual pattern apparent in Figure 1. Together, these results suggested that river ecosystems were richer in links and links per species than terrestrial ecosystems, whereas otherwise complexity was similar across ecosystem types.

C. Topology–Diversity Relationships

We subsequently analysed the scaling of 16 additional food-web properties with diversity (Figure 3) and connectance (Figure 4) by fitting the power–law regression model (Eq. 2) to the data. We found that 10 food-web properties were significantly correlated with species richness in the way predicted by a power function (Table 2): the fractions of top, intermediate, basal, omnivore and cannibalistic species, the number of species in loops, the standard

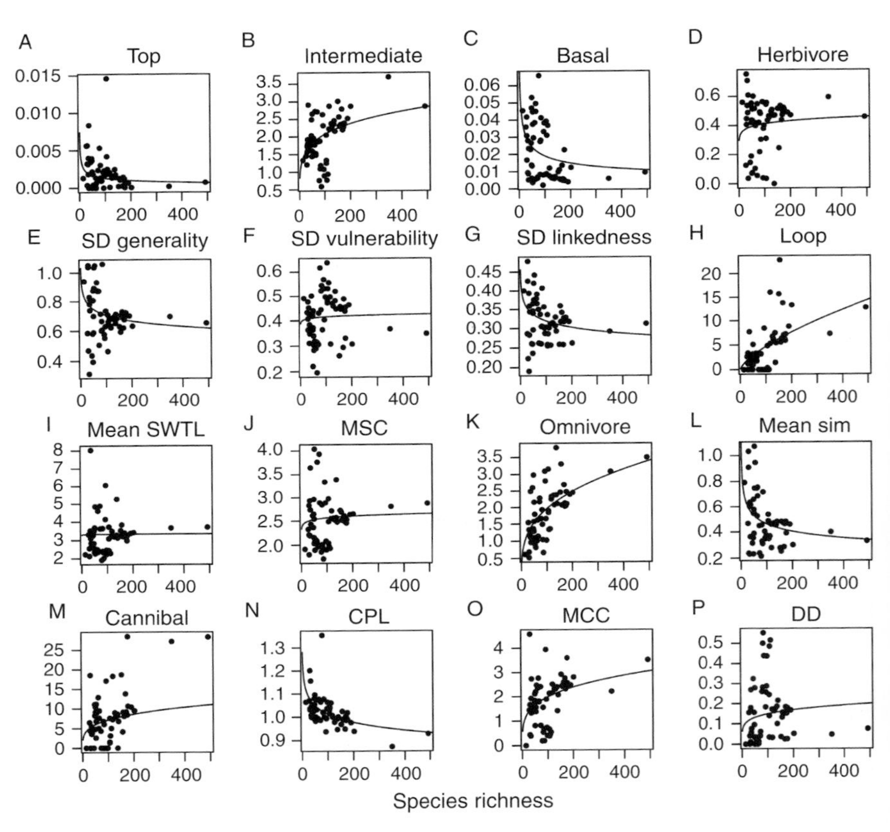

Figure 3 The diversity scaling (Eq. 2) of 16 food-web properties: fraction of (A) top, (B) intermediate, (C) basal, (D) herbivore, (K) omnivore and (M) cannibalistic species, (H) species in loops, the standard deviations of (E) generality, (F) vulnerability and (G) linkedness, (I) the mean short trophic level (SWTL), (J) the mean shortest chain length (MSC), (L) the mean similarity, (N) the characteristic path length (CPL), (O) the mean clustering coefficient (MCC) and (P) the diet discontinuity (DD). See Table 2 for fitted model parameters. For the *y*-axis, we used the normalized residuals for each food-web property and ecosystem type (see Section II for details).

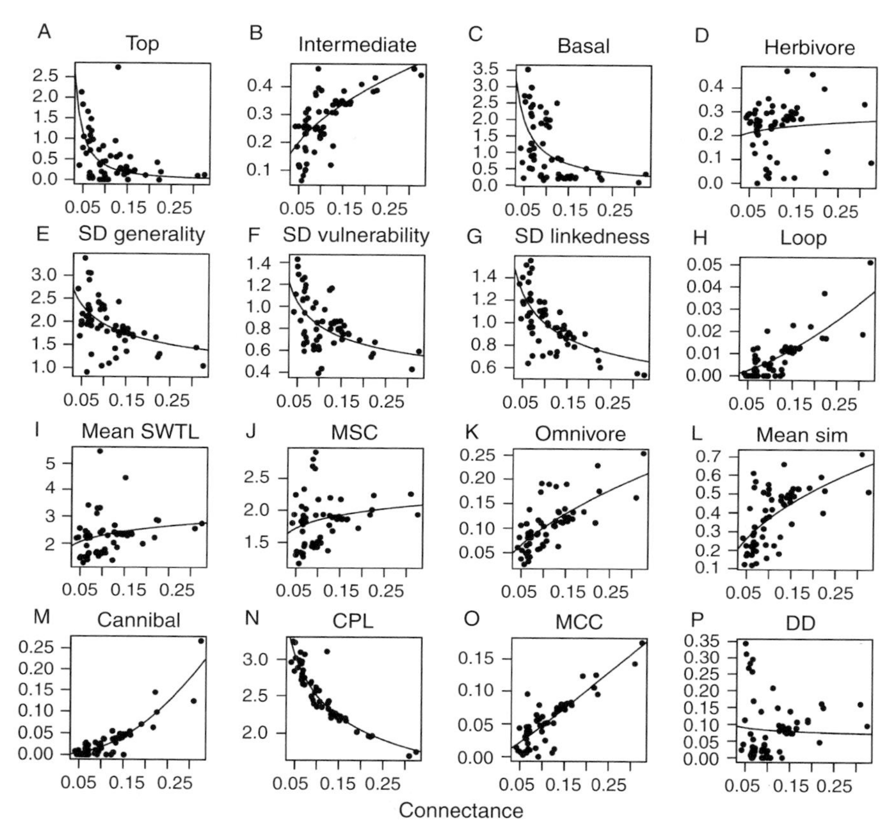

Figure 4 The complexity scaling (Eq. 2) of 16 food-web properties: fraction of (A) top, (B) intermediate, (C) basal, (D) herbivore, (K) omnivore and (M) cannibalistic species, (H) species in loops, the standard deviations of (E) generality, (F) vulnerability and (G) linkedness, (I) the mean short trophic level (SWTL), (J) the mean shortest chain length (MSC), (L) the mean similarity, (N) the characteristic path length (CPL), (O) the mean clustering coefficient (MCC) and (P) the diet discontinuity (DD). See Table 2 for fitted model parameters. For the *y*-axis, we used the normalized residuals for each food-web property and ecosystem type (see Section II for details).

deviation of linkedness, the mean similarity, the characteristic path length and the mean clustering coefficient.

Moreover, 12 food-web properties were significantly correlated with connectance in the way predicted by the power–law model of Eq. 2 (Table 2): the fractions of top, intermediate, basal, omnivore and cannibalistic species, the number of species in loops, the standard deviations of generality, vulnerability and linkedness, the mean similarity, the characteristic path length and the mean clustering coefficient.

Table 2 Fit of a power–law scaling model (Eq. 2) of different topological food-web properties depending on connectance and species richness

	a			*b* (species-richness)			*c* (connectance)		
Variables	Estimate	Std. deviation	*p*	Estimate	Std. deviation	*p*	Estimate	Std. deviation	*p*
Fraction. Top	0.007	0.008	0.368	−0.382	0.182	0.04	−1.68	0.414	<0.001
Frac. Intermediate	0.832	0.192	<0.001	0.198	0.049	<0.001	0.474	0.071	<0.001
Frac. Basal	0.097	0.072	0.187	−0.353	0.135	0.011	−1.004	0.257	<0.001
Frac. herbivore	0.298	0.132	0.028	0.071	0.09	0.438	0.112	0.13	0.392
SD generality	1.032	0.229	<0.001	−0.083	0.043	0.061	−0.276	0.066	<0.001
SD vulnerability	0.388	0.09	<0.001	0.015	0.044	0.735	−0.328	0.07	<0.001
SD linkedness	0.456	0.078	<0.001	−0.077	0.033	0.022	−0.339	0.051	<0.001
Frac. in loop	0.185	0.081	0.026	0.701	0.114	<0.001	1.445	0.192	<0.001
Mean SWTL	3.288	1.053	0.003	0.003	0.067	0.969	0.157	0.095	0.101
Mean short chain	2.335	0.458	<0.001	0.021	0.04	0.611	0.102	0.058	0.082
Frac omnivore	0.414	0.108	<0.001	0.339	0.056	<0.001	0.612	0.082	<0.001
Mean similarity	1.132	0.299	<0.001	−0.194	0.062	0.003	0.487	0.081	<0.001
Frac cannibal	2.171	0.589	<0.001	0.26	0.102	0.014	2.078	0.185	<0.001
Char path length	1.281	0.071	<0.001	−0.051	0.011	<0.001	−0.291	0.017	<0.001
Mean cluster coef.	0.588	0.147	<0.001	0.265	0.063	<0.001	1.098	0.098	<0.001
Diet discontinuity	0.064	0.064	0.317	0.183	0.192	0.344	−0.113	0.289	0.697

The food-web parameters are: fraction of top, intermediate, basal, herbivore, cannibalistic species, species in loops, the standard deviations of generality, vulnerability and linkedness, the mean short-weighted trophic level, the mean shortest chain length, the mean similarity, the characteristic path length, the mean clustering coefficient and the diet discontinuity (see Section II for details).

With increasing species richness the fractions of top and basal species decreased, whereas the fractions of intermediate and omnivorous species increased (Figure 3A–C and K). Additionally, the standard deviation of the species' linkedness and their mean trophic similarity decreased (Figure 3G and L). This implied that in the more diverse food webs the predominantly increasing number of intermediate species yielded a more similar distribution of the number of links across species but less similarity in who consumes whom. Moreover, the characteristic path length between any pair of species decreased (Figure 3N) and the mean clustering coefficient increased (Figure 3O) with increasing species richness. Thus, in more diverse food webs, the species were assembled in clusters of sub-webs and consequently, the average length of the trophic paths between pairs of species decreased.

With increasing food-web connectance, our analyses indicated that the fractions of top and basal species, the standard deviations of generality, vulnerability and linkedness, and the mean clustering coefficient decreased, whereas the fraction of intermediate, omnivore and cannibalistic species, the proportions of species in loops, the mean trophic similarity among species and the mean cluster coefficient increased with connectance (Figure 4, Table 2).

D. Ecosystem Types and Topology–Diversity Relationships

While these analyses suggested a significant fit of the general diversity and complexity scaling model for topological food-web properties (Eq. 2), we also addressed significant differences in trophic scaling among the five ecosystem types (lake, stream, estuarine, marine and terrestrial ecosystems). The cluster analysis of the normalized residuals of the fitted trophic scaling model illustrated whether the overall similarity or dissimilarity of the food-web topologies could be ascribed to the ecosystem types (Figure 5). This analysis revealed that despite some topological similarities in food webs located closely to each other, no systematic grouping of food webs according to the ecosystem types emerged. This residual analysis suggested that (1) the trophic scaling models (Figures 3 and 4) held across ecosystem types without systematic deviations, and (2) despite variance in species richness and connectance between ecosystem types, natural food webs possessed similar overall topologies.

Subsequently, we carried out more detailed analyses of variance (ANOVA) of the residuals of the 16 food-web properties among the five ecosystem types (Figures 6 and 7). We found significant signatures of the ecosystem types in the fractions of intermediate (Figure 6B), basal (Figure 6C) and herbivorous species (Figure 6D), the standard deviations

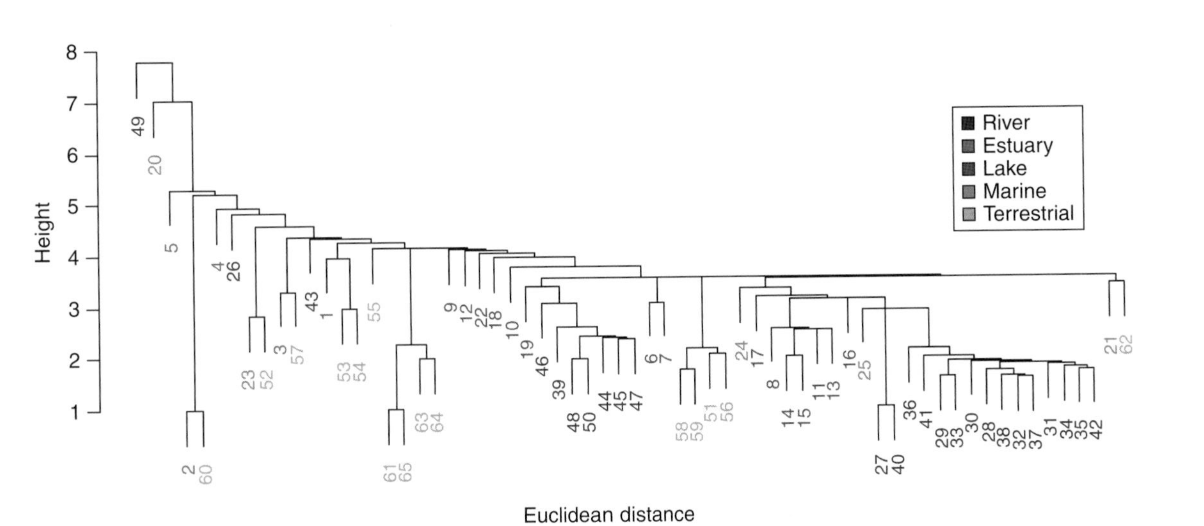

Figure 5 Clustering of food webs according to the residual variation in the 16 food-web properties (see Figures 4 and 5). Euclidian distances were calculated with species richness and connectance corrected values. Colour code for the different ecosystems: blue = rivers, red = estuaries, green = lakes, pink = marine ecosystems, orange = terrestrial ecosystems. The web numbers are given in the Appendix. (For interpretation of the references to colour in this figure legend, the reader is referred to the Web version of this chapter.)

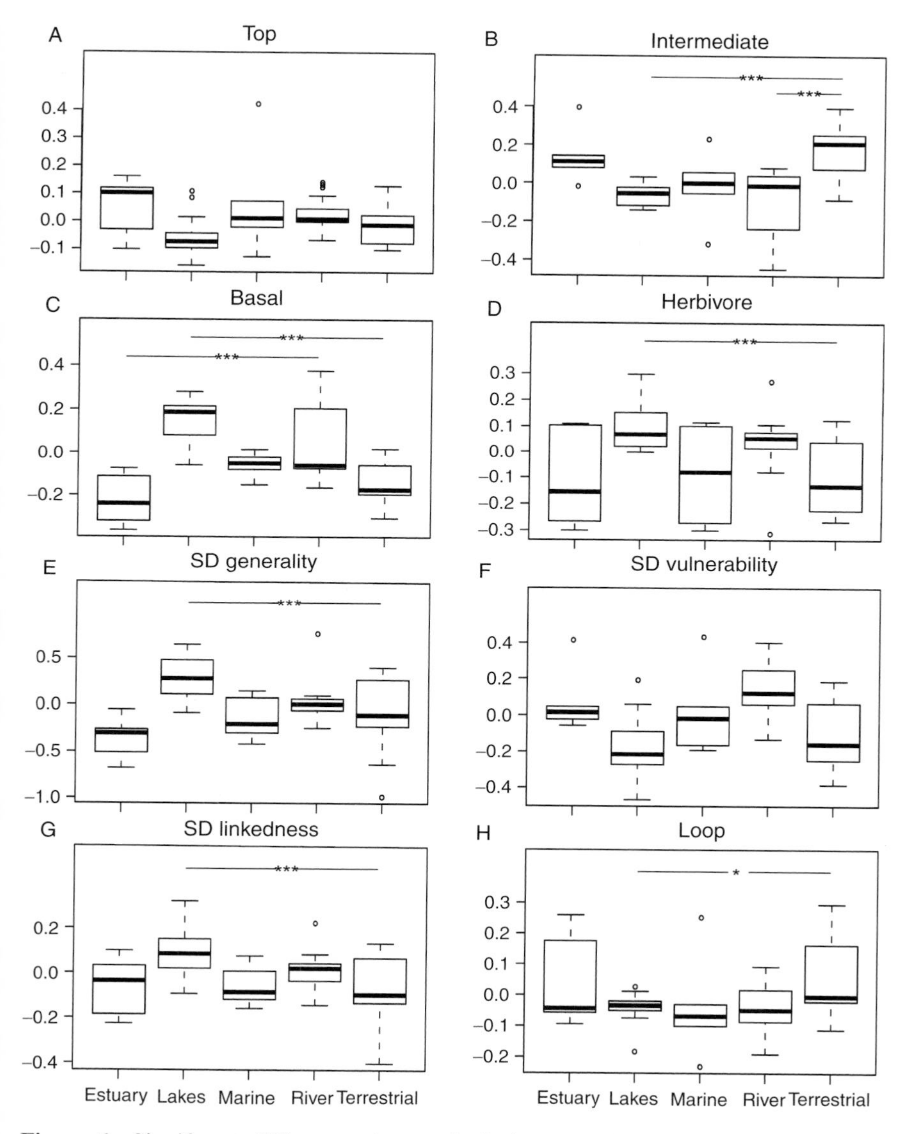

Figure 6 Significant differences in topological properties (the residuals of their complexity and diversity scaling according to Eq. 2—see Figures 3 and 4) between the ecosystem types: the fractions of (A) top, (B) intermediate, (C) basal, and (D) herbivore species, the standard deviations of (E) generality, (F) vulnerability and (G) linkedness, (H) the fraction of species in loops. Significant differences between ecosystems types were calculated by Tukey HSD post hoc tests (*$p < 0.05$; ***$p < 0.001$).

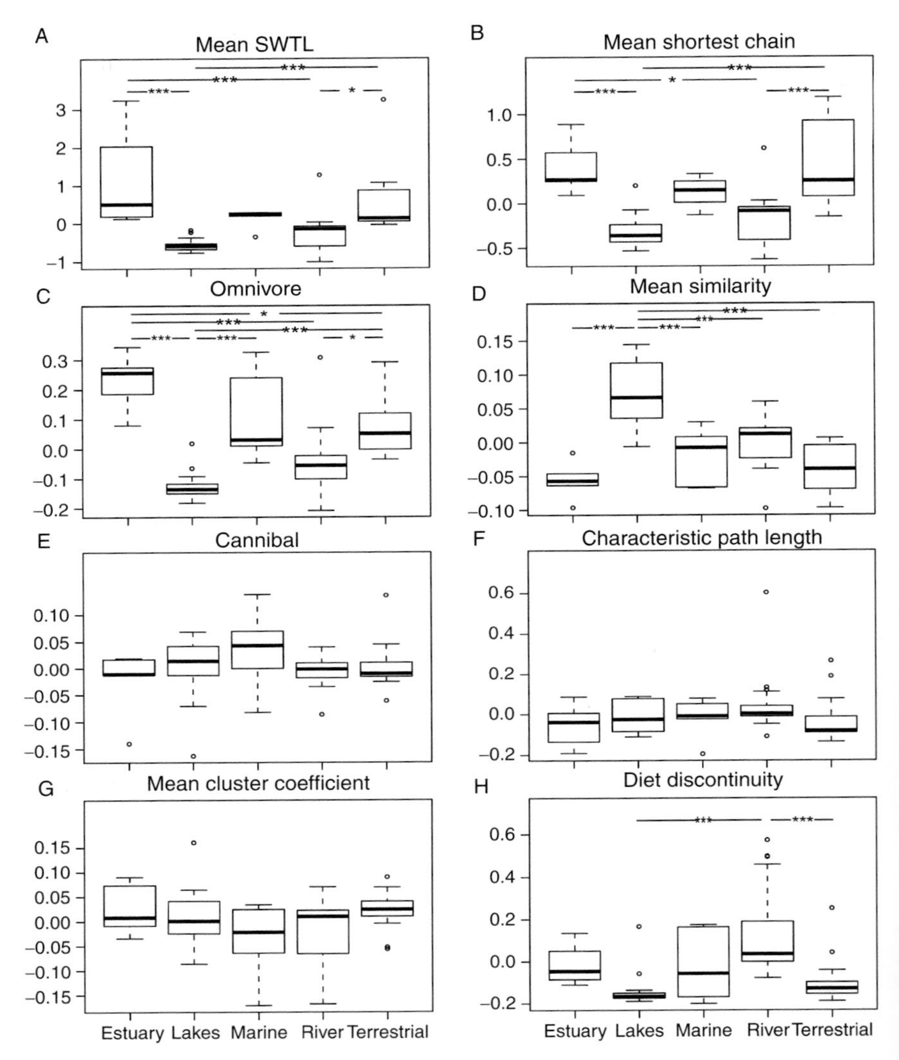

Figure 7 Significant differences in topological properties (the residuals of their complexity and diversity scaling according to Eq. 2—see Figures 3 and 4) between the ecosystem types: (A) the mean short-weighted trophic level, (B) the mean shortest chain length, (C) the fraction of omnivores, (D) the mean similarity, (E) the fraction of cannibals, (F) the characteristic path length, (G) the mean clustering coefficient and (H) the diet discontinuity. Significant differences between ecosystems types were calculated by Tukey HSD post hoc tests (*$p < 0.05$; ***$p < 0.001$).

of generality (Figure 6E) and linkedness (Figure 6G), the fraction of species in trophic loops (Figure 6H), the mean short-weighted trophic levels (Figure 7A), the mean shortest chain lengths (Figure 7B), the fraction of omnivores (Figure 7C), the mean trophic similarity of species (Figure 7D) and the diet discontinuity (Figure 7H). These differences were subsequently explored in more detail using post hoc tests (Figures 6 and 7). Across the 16 food web properties, this included a total of 160 pairwise comparisons of ecosystem types. Overall, 26% (43 out of 160 combinations) of these possible combinations were significantly different from each other (see levels of significance indicated in Figures 6 and 7 for details). This suggested that the majority of topological food-web properties followed similar diversity–complexity scaling models across ecosystems. Some systematic differences emerged between lake and terrestrial ecosystems in terms of the fractions of intermediate, basal, herbivorous and omnivorous species, the mean short-weighted trophic level, the mean shortest chain length, the mean similarity and the fraction of species in loops (Figures 6 and 7). Additionally, some estuaries and lakes differed in the mean short-weighted trophic level, the mean shortest chain length, the fraction of omnivores and the mean similarity. Overall, these analyses suggest that most differences in food-web topology occurred when comparing terrestrial and lake ecosystems, whereas terrestrial and marine ecosystems were most similar.

IV. DISCUSSION

A. Complexity–Diversity Relationships

Over several decades, the search for universal food-web constants that equally apply to species-poor and species-rich ecosystems has focused on complexity–diversity relationships. While the link-species scaling law predicts constancy of the number of links per species (Cohen and Briand, 1984), the alternative constant–connectance hypothesis holds that network connectance should remain constant as species richness increases (Martinez, 1992). Based on a new collection of novel food webs, our results suggest that both scaling models, the link-species scaling law and the constant–connectance hypothesis, have to be rejected. Instead, we found that link richness and the number of links per species increase, whereas connectance decreases along the diversity scale. Caution must be exerted, however, regarding the latter conclusion as the result was only marginally significant, and large variation in connectance was observed among food webs of similar species richness, particularly at poor to intermediate richness levels. Additional food-web data are necessary to provide a more robust evaluation of this hypothesis. Moreover, our results suggest that there is a stronger constraint to the maximum connectance possible for different food webs, which defines a clear upper boundary of the diversity–connectance relationship, while no lower boundary is apparent. Overall, in contrast to classic models, our results support the recent paradigm shift from the previously held perception that connectance is constant to one that it is scale dependent (Brose *et al.*, 2004; Dunne, 2006; Montoya and Sole, 2003; Schmid-Araya *et al.*, 2002).

Interestingly, the link-species scaling law is in agreement with the classical stability criterion of random networks, which holds that local population stability is maintained if link density falls below a critical threshold which, in turn, depends on the average interaction strength (May, 1972). The lack of constancy in the number of links per species in our and other recent empirical analyses indicates that this topological stability criterion cannot be responsible for the stability of complex natural food webs. Instead, recent work has demonstrated that the specific body-mass structure of natural food webs may provide the critically important dynamic stability of the trophic networks (Brose, 2010; Brose *et al.*, 2006b; Otto *et al.*, 2007). In particular, negative diversity–stability relationships in complex food webs without body-mass structure are converted into neutral to slightly positive relationships if the natural body-mass structure is accounted for (Brose, 2010; Brose *et al.*, 2006a,b). Moreover, implementing natural body-mass distributions in models of complex food webs also yields positive relationships between complexity and stability (Rall *et al.*, 2008). Together, these dynamic model analyses provide a potential explanation for the stability of highly diverse food webs, despite the high number of links per species.

B. Explanations for the Scale Dependence of Complexity

Several potential explanations for the scale dependence of links per species and connectance can be identified. First, in communities with many interacting species, the decrease of connectance with diversity might result from a methodological artefact (Paine, 1988), namely that the difficulty of identifying trophic links among a large number of species increases with species richness. This yields a potentially lower sampling intensity of links in more diverse food webs, which could account for a decrease in connectance with species richness (Bersier *et al.*, 1999; Goldwasser and Roughgarden, 1997; Ings *et al.*, 2009; Martinez *et al.*, 1999). Ultimately, an adequate sampling effort can only be guaranteed if yield–effort curves demonstrate saturation in link richness with sampling effort for every food web (Ings *et al.*, 2009; Woodward and Hildrew, 2001) or if extrapolation methods suggest a high sampling coverage (Brose and Martinez, 2004; Brose *et al.*, 2003). While this is certainly desirable for future food-web compilations, the currently available data lack this information and we cannot entirely rule out that sampling effect contributes to the decrease in connectance with species richness.

Second, the increase in links per species with species richness could be primarily driven by an increasing number of weak links (i.e. links with a low energy flux), whereas the number of strong links per species might be constant. Empirical studies have indeed found interaction strengths to be highly skewed towards many weak and a few strong links (Fagan and Hurd, 1994; Goldwasser and Roughgarden, 1993; O'Gorman and Emmerson, in press; Paine, 1992; Woodward *et al.*, 2005b; Wootton, 1997). Taking the variability in energy flux between links into consideration, initial tests found the overall number of links per species increased with species richness, whereas quantitative versions of link density, weighting the links according to their energy flux, remained scale invariant (Banasek-Richter *et al.*, 2005). Thus, the distribution of energy fluxes becomes more unequal as systems accrue species, possibly due to the increase in weak links. This implies that species can have strong interactions with only a limited number of the coexisting species, whilst the number of weak interactions increases continuously with species richness. While the former 'sampling effect' suggests that the number of links could be under-reported in more diverse food webs, the use of quantitative information on individual trophic links along with their corresponding descriptors (Banasek-Richter *et al.*, 2004; Bersier *et al.*, 2002) suggests that most of these links in diverse food webs are weak, and therefore, may be unimportant for calculating 'ecologically meaningful' measures of connectance or link density. However, this implication needs to be reconciled with recent theoretical work stressing the importance of weak links for the organization and dynamics of natural food webs (Berlow, 1999; McCann *et al.*, 1998; Navarrete and Berlow, 2006; O'Gorman and Emmerson, 2009).

Third, food-web stability might require that during community assembly diversity is negatively correlated with complexity. This argument is based on the finding that species-poor communities exhibit Poissonian degree distributions (i.e. the frequency of species with a specific number of links), whereas species-rich communities have more skewed distributions (Montoya and Sole, 2003). Thus, increasing diversity primarily leads to an increase in species with few links, which decreases connectance. Classic stability analyses have shown that population stability decreases with increases in both diversity and connectance (May, 1972). When natural food webs assemble, the destabilizing effect of increasing diversity needs to be balanced by a resulting decrease in connectance to avoid instability (Montoya and Sole, 2003). This stability argument links variation in species diversity and community complexity in a mechanistic manner.

Fourth, processes that increase diversity might reduce species' ability to co-exist, which would also decrease connectance. However, the constant–connectance and link-species scaling models assume that species consume a fixed fraction or a fixed number, respectively, of the co-existing species (Cohen and Briand, 1984; Martinez, 1992). Thus, these models predict constancy in the scaling exponents only if co-existence does not change with diversity. However, potential consumer and resource species do not necessarily co-exist in meta-communities at larger spatial scales (Brose *et al.*, 2004; Olesen *et al.*, in press). If species richness across food webs increases with the spatial extent of the habitats, connectance will decrease with species richness due to a decrease in predator–prey co-occurrence. Link-area models based on this argument have successfully predicted the number of links, link density and connectance of aquatic food webs ranging in spatial scale from local habitats to landscapes (Brose *et al.*, 2004). Interestingly, the exponent of the power–law link-species model at the scale of local habitats was close to two, as predicted by the constant–connectance model, whereas it decreased to lower values when larger spatial scales were included, where species' co-existence is expected to break down (Brose *et al.*, 2004). Similarly, predator–prey co-existence may also break down with increasing habitat complexity (Keitt, 1997). Increasing habitat complexity or architectural complexity of the vegetation leads to higher species richness as many predators can specialize on specific sub-habitats, such as distinct vegetation layers (Brose, 2003; Olesen *et al.*, in press; Tews *et al.*, 2004). The localized occurrence of these predators in sub-habitats may yield reduced connectance as the predators do not co-exist with all prey species that fall within their feeding niche. Interestingly, strong support for the constant–connectance hypothesis comes from the pelagic food webs of 50 lakes (Martinez, 1993a) and aquatic microcosms (Spencer and Warren, 1996). In these relatively homogeneous habitats, increases in habitat complexity play little or no role in increasing species richness, which allows for constant connectance. In contrast, increasing species richness in stream communities was correlated with decreases in connectance, which might be partially explained by variations in habitat complexity (Schmid-Araya *et al.*, 2002).

Fifth, predator specialization might decrease connectance in the more diverse food webs. The feeding ranges of consumers are limited to specific body-size ranges of potential resource species (Brose *et al.*, 2008; Vucic-Pestic *et al.*, 2010). With increasing species richness, the likelihood of that particularly small or large species occur also increases, which should result in an increase in the body-size range with increasing species richness. If the body-size range increases with the species richness of the community, connectance will decrease with community diversity due to physical feeding constraints (i.e. upper and lower size refugia from predation). Moreover, the possibility to select from multiple prey species increases for any predator with increasing species richness. Therefore, predators in more diverse communities may specialize on a subset of their potential feeding niche that includes prey species that are easier to exploit or are less defended. In addition, uneven abundances of potential prey within the feeding range may induce predator switching behaviour that creates temporally unexploited prey of low abundances. This hypothesis suggests that the prey abundance of the unrealized links should be lower than that for the realized links (e.g. Olesen *et al.*, in press). In concordance with these arguments, Beckerman *et al.* (2006) offer a mechanistic explanation for how patterns in connectance arise within food webs. Based on foraging theory, they assume that predators feed preferentially on the energetically most rewarding prey. Their 'diet breadth model' relates food-web complexity to the species' foraging biology and does well in predicting the scaling of connectance with species richness (Beckerman *et al.*, 2006). The optimization constraints regarding the species' foraging behaviour thus determine the complexity of their predicted food webs. Interestingly, recent extensions of this optimal foraging approach in the allometric diet breadth model also provide successful predictions of the topology of complex food webs: that is not simply how many links are expressed, but also where they are located within the feeding matrix (Petchey *et al.*, 2008). Together, these models stress the key role of the natural body-mass structure in determining food-web topology, which also supports the hypothesis that connectance may decrease with increasing species richness, as a result of extended body-mass ranges within the community size spectrum.

Each of the aforementioned hypotheses may offer only a partial explanation of the variance of connectance with species richness, and they are not mutually exclusive. Most likely, the mechanisms underlying the observed patterns are multi-causal and vary with the spatial scale. The 'sampling hypothesis' suggests that mere sampling artefacts are responsible for the decrease in connectance with species richness, whereas all other hypotheses invoke ecological processes as the drivers behind this pattern. In addition to the empirical patterns, these biological hypotheses support the conclusion that connectance decreases with diversity and lend weight to the emerging paradigm that connectance is scale dependent in community food webs.

C. Topology–Diversity Relationships

Analyses of topological scaling relationships in our new data collection revealed significant diversity scaling of 10 food-web properties and complexity scaling for 12 out of the 16 properties. Our results support the conclusion of prior studies that the fractions of top and basal species decrease with diversity and that the fraction of intermediate species increases (Hall and Raffaelli, 1991; Martinez, 1991, 1993a; Schoener, 1989; Warren, 1989; Winemiller, 1990), thus supporting the classic scaling relationships. They also indicate that species-rich food webs exhibit greater variability in species' linkedness (i.e. the overall number of links) than do species-poor webs. This finding is consistent with an earlier study (Montoya and Sole, 2003) that suggests that species-rich food webs have a more uneven distribution of links among species, a feature that may increase population stability. Moreover, we found that the mean clustering coefficient (the likelihood that two species that are linked to the same species are also linked to each other) increases with diversity. Surprisingly, this empirical result contradicts a prior analytical result based on niche-model food webs (Camacho *et al.*, 2002b). It does, however, lend support to another analytical finding that the mean shortest path length between species decreases with diversity (Williams *et al.*, 2002). Together, these findings suggest that species-rich food webs are more compartmentalized and have shorter average path length between pairs of species than species-poor ones, which suggests that food webs of high diversity are organized by combining sub-web compartments, within which species are closely linked to each other.

D. Ecosystem Types and Trophic Scaling Relationships

While our analyses indicated robust scaling models of topological food-web properties with diversity and complexity, substantial residual variation around these trends remained. In line with earlier approaches for marine (Dunne *et al.*, 2004) and Cambrian food-web topology (Dunne *et al.*, 2008), we aimed to examine significant signatures of ecosystem types in this residual variation. However, in contrast to these previous studies, we did not employ the niche model (Williams and Martinez, 2000) as the null model to remove the dominant effects of species richness and connectance on food-web topology. Analyses based on the niche model share its assumptions whilst ignoring those of alternative topological models and their predictions (Allesina *et al.*, 2008; Cattin *et al.*, 2004; Stouffer *et al.*, 2005; Williams and Martinez, 2008). To avoid such pre-assumptions of the analyses, we used power–law scaling models of food-web topology against species richness and connectance to test for effects of ecosystem type on the residuals of the scaling relationships. These residuals were thus independent of the effects of diversity and complexity on food-web topology.

As in the studies of Dunne *et al.* (2004, 2008), we did not find strong effects of ecosystem types on food-web topology. While our cluster analysis indicated that food-web topologies were not generally grouped according to their ecosystem type, the more detailed analyses of variance for individual food-web properties suggested the presence of more specific differences, as revealed by pairwise comparisons. Our comparison of 160 pairwise combinations in post hoc tests could be criticized for multiple comparisons, since, on average, chance events should result in significant differences ($p < 0.05$) for at least 5% of the combinations. Family-wise corrections of Type I error probability for such a large number of comparisons (Peres-Neto, 1999) render the power of the tests uninformatively low (García, 2004; Moran, 2003). From a 'false-rejection-rate' perspective (García, 2004), however, our analyses identified significant differences among pairs for 26% (43 out of 160) of the combinations, which shows that differences among ecosystem types were more than just random events. To avoid over-interpreting potential statistical artefacts, however, we refrain from a detailed interpretation of individual pairwise combinations of food-web topologies from different ecosystems, although there were indications of some systematic differences between lake and terrestrial ecosystems (eight significant differences in topological properties), and between estuaries and lakes (four significant differences in topological properties). A cautious interpretation of these differences suggests that pelagic ecosystems such as lakes might possess a somewhat different network topology than terrestrial and estuarine (mainly benthic) ecosystems (cf. Yvon-Durocher *et al.*, 2010). Interestingly, this is consistent with differences in the body-mass structure between these ecosystems, which may be related to systematic effects of hard surfaces employed by predators while consuming carcasses (Brose *et al.*, 2006a,b). Overall, however, our analyses indicated that the majority of topological food-web properties followed similar diversity–complexity scaling rules across ecosystems. These robust scaling models are certainly suggestive of the presence of general building rules for ecological networks, even across seemingly very different ecosystem types.

V. CONCLUSIONS

As in some previous studies (Brose *et al.*, 2004; Dunne, 2006; Montoya and Sole, 2003; Schmid-Araya *et al.*, 2002), our results suggest that neither links per species nor connectance are scale-invariant constants across communities of varying species richness. In the same vein, our results also illustrated that most food-web properties scaled with the diversity and complexity of the communities. After several decades of debate surrounding trophic scaling theories, it now seems unlikely that there are universal scale-independent constants in natural food webs that hold true for all communities, from those that are relatively depauperate to those that are species rich. Nevertheless, recent works support trophic scaling models that predict relationships between parameters of food-web topology and diversity with constant scaling exponents (Camacho *et al.*, 2002a; Garlaschelli *et al.*, 2003). While these scaling relationships are certainly not as easy to explain as may be desired, they provide a means of understanding the inevitable interrelation of the different parameters within complex food webs. A mechanistic understanding of why complex food webs appear to share a fundamental network structure mediated by species richness and connectance is yet to be achieved—perhaps we share this aspect with physicists, who are still lacking a mechanistic explanation with rigorous empirical support for the gravitational force several centuries after Newton phrased the universal law of gravitation.

Despite this lack of universal constants in food-web ecology, many theoretical aspects of food-web ecology have been clarified and substantially refined in the last decade. Recent structural food-web models (Allesina *et al.*, 2008; Cattin *et al.*, 2004; Stouffer *et al.*, 2005; Williams and Martinez, 2000) include dependence of network topology on species richness and connectance, and predict food-web properties depending on contiguous feeding ranges within an ordered set of species' niches (Williams and Martinez, 2000), phylogenetic constraints on feeding interactions (Cattin *et al.*, 2004) and exponential degree distributions (Stouffer *et al.*, 2005). The integration of such research with core concepts from other research areas, such as the role of body-size constraints on predator–prey interactions (Beckerman *et al.*, 2006; Brose, 2010; Brose *et al.*, 2006a,b, 2008; Layer *et al.*, 2010; McLaughlin *et al.*, in press; Petchey *et al.*, 2008; Vucic-Pestic *et al.*, 2010; Woodward *et al.*, 2005a,b, 2010;Wootton and Emmerson, 2005), offers a potentially promising way in which to start to develop a more mechanistic basis for explaining the trophic patterns we see in nature.

APPENDIX

Table A1 Overview of all food webs sources used in this study, with references of predation matrix, number of interactions, number of links, proportion of links/species, connectance and the general habitat

Webnr.	Common web name	Taxonomic species	Links	Links/species	conn*	Predation matrix source
	Estuary/Saltmarsh					
1	Carpinteria	72	238	3.31	0.05	Lafferty *et al.* (2006)
2	Chesapeake Bay	36	121	3.36	0.09	Baird and Ulanowicz (1989)
3	St. Mark's	48	220	4.58	0.1	Christian and Luczkovich (1999)
4	Ythan2010	92	416	4.52	0.05	Cohen *et al.* (2009)
5	Mangrove Estuary wet season	94	1339	14.24	0.15	Heymans *et al.* (2002)
	Lakes					
6	Alford Lake	56	219	3.91	0.07	Havens (1992)
7	Balsam Lake	53	182	3.43	0.06	Havens (1992)
8	Beaver Lake	61	327	5.36	0.09	Havens (1992)
9	Big Hope Lake	30	120	4	0.13	Havens (1992)
10	Bridge Brook Lake	75	552	7.36	0.1	Havens (1992)
11	Chub Pond	65	417	6.42	0.1	Havens (1992)
12	Connery Lake	30	60	2	0.07	Havens (1992)
13	Hoel Lake	49	254	5.18	0.11	Havens (1992)
14	Long Lake	65	416	6.4	0.1	Havens (1992)
15	Stink Lake	53	280	5.28	0.1	Havens (1992)
16	Little Rock Lake	176	2009	11.41	0.06	Havens (1992)
17	Sierra Lakes	37	298	8.05	0.22	Harper-Smith *et al.* (2005)
18	Skipwith Pond	35	379	10.83	0.31	Warren (1989)

(continued)

Table A1 (*continued*)

Webnr.	Common web name	Taxonomic species	Links	Links/species	conn*	Predation matrix source
19	Tuesday Lake 1984	50	268	5.36	0.11	Jonsson *et al.* (2005)
	Marine					
20	Chile Food web	106	1436	13.55	0.13	Navarette and Wieters (unpublished data)
21	Lough Hyne	350	5131	14.66	0.04	Jacob (unpublished data)
22	Mondego Zostera Meadows	47	278	5.91	0.13	Patricio and Marques (2006)
23	Caribbean Reef, small	50	555	11.1	0.22	Opitz (1996)
24	NE US Shelf	81	1482	18.3	0.23	Link (2002)
25	Weddell Sea	492	16136	32.8	0.07	Brose *et al.* (2006a,b) and Jacob (2005)
	River/streams					
26	Bere Stream	137	1276	9.31	0.07	Woodward *et al.* (2008)
27	Broadstone Stream	34	221	6.5	0.19	Woodward *et al.* (2005b)
28	Alamitos Creek	162	3756	23.19	0.14	Harrison (2003)
29	Caldero Creek	126	2109	16.74	0.13	Harrison (2003)
30	Corde Matre Creek	106	1757	16.58	0.16	Harrison (2003)
31	Coyote	190	4583	24.12	0.13	Harrison (2003)
32	Guadeloupe Creek	174	4662	26.79	0.15	Harrison (2003)
33	Guadeloupe River	136	2487	18.29	0.13	Harrison (2003)
34	Los Gatos Creek	177	4480	25.31	0.14	Harrison (2003)
35	Los Trancos Creek	129	2440	18.91	0.15	Harrison (2003)
36	San Francisquito Creek	140	3266	23.33	0.17	Harrison (2003)

37	Saratoga Creek,	158	3754	23.76	0.15	Harrison (2003)
38	Steverson Creek	170	4776	28.09	0.17	Harrison (2003)
39	Blackrock	82	348	4.24	0.05	Harrison (2003)
40	Broad	34	221	6.5	0.19	Townsend *et al.* (1998)
41	Ross	117	2024	17.3	0.15	Townsend *et al.* (1998)
42	Penetetia creek	170	4037	23.75	0.14	Harrison (2003)
43	Sutton	86	422	4.91	0.06	Townsend *et al.* (1998)
44	Canton	108	707	6.55	0.06	Townsend *et al.* (1998)
45	Dempster	106	965	9.1	0.09	Townsend *et al.* (1998)
46	German	84	352	4.19	0.05	Townsend *et al.* (1998)
47	Healy	96	633	6.59	0.07	Townsend *et al.* (1998)
48	Kyeburn	98	628	6.41	0.07	Townsend *et al.* (1998)
49	Little Kyeburn	78	749	9.6	0.12	Townsend *et al.* (1998)
50	Stony	112	831	7.42	0.07	Townsend *et al.* (1998)
	Terrestrial					
51	Grand Caricaie Cl C1	118	997	8.45	0.07	Cattin Blandenier (2004)
52	Coachella	27	237	8.78	0.33	Polis (1991)
53	EcoWeb 59	30	65	2.17	0.07	Cohen (1989)
54	EcoWeb 60	33	68	2.06	0.06	Cohen (1989)
55	El Verde	156	1509	9.67	0.06	Waide and Reagan (1996)
56	Grand Caricaie Sn C2	152	1525	10.03	0.07	Cattin Blandenier (2004)
57	St. Martin	44	217	4.93	0.11	Goldwasser and Roughgarden (1993)
58	Grand Caricaie Cm C1	202	2930	14.5	0.07	Cattin Blandenier (2004)
59	Grand Caricaie Cm M2	167	1830	10.96	0.07	Cattin Blandenier (2004)
60	Simberloff_E1	36	121	3.36	0.09	Simberloff and Abele (1975)
61	Simberloff_E2	63	347	5.51	0.09	Simberloff and Abele (1975)
62	Simberloff_E3	49	242	4.94	0.1	Simberloff and Abele (1975)
63	Simberloff_E7	52	255	4.9	0.09	Simberloff and Abele (1975)
64	Simberloff_E9	71	446	6.28	0.09	Simberloff and Abele (1975)
65	Simberloff_ST2	63	347	5.51	0.09	Simberloff and Abele (1975)

*conn = connectance.

Binary link structures of these 65 food webs are available upon request by 'Ulrich Brose'.

ACKNOWLEDGEMENTS

The manuscript was greatly improved by comments and suggestions from Neo D. Martinez, Jennifer Dunne and Stefan Scheu. U. B., B. C. R. and J. O. R. are supported by the German Research Foundation (BR 2315/4-1, 9-1).

REFERENCES

Allesina, S., Alonso, D., and Pascual, M. (2008). A general model for food web structure. *Science* **320**, 658–661.

Baird, D., and Ulanowicz, R.E. (1989). The seasonal dynamics of the Chesapeake Bay ecosystem. *Ecol. Monogr.* **59**, 329–364.

Banasek-Richter, C., Cattin, M.F., and Bersier, L.F. (2004). Sampling effects and the robustness of quantitative and qualitative food-web descriptors. *J. Theor. Biol.* **226**, 23–32.

Banasek-Richter, C., Cattin Blandenier, M.-F., and Bersier, L.-F. (2005). Food web structure: from scale invariance to scale dependence, and back again? In: *Dynamic Food Webs* (Ed. by P.C. De Ruiter, V. Wolters and J.C. Moore), pp. 48–55. Academic Press, Burlington, USA.

Beckerman, A.P., Petchey, O.L., and Warren, P.H. (2006). Foraging biology predicts food web complexity. *Proc. Natl. Acad. Sci. USA* **103**, 13745–13749.

Berlow, E.L. (1999). Strong effects of weak interactions in ecological communities. *Nature* **398**, 330–334.

Bersier, L.F., Dixon, P., and Sugihara, G. (1999). Scale-invariant or scale-dependent behavior of the link density property in food webs: a matter of sampling effort? *Am. Nat.* **153**, 676–682.

Bersier, L.F., Banasek-Richter, C., and Cattin, M.F. (2002). Quantitative descriptors of food-web matrices. *Ecology* **83**, 2394–2407.

Briand, F., and Cohen, J.E. (1984). Community food webs have scale-invariant structure. *Nature* **307**, 264–267.

Brose, U. (2003). Bottom-up control of carabid beetle communities in early successional wetlands: mediated by vegetation structure or plant diversity? *Oecologia* **135**, 407–413.

Brose, U. (2010). Body-mass constraints on foraging behaviour determine population and food-web dynamics. *Funct. Ecol.* **24**, 28–34.

Brose, U., Martinez, N.D., and Williams, R.J. (2003). Estimating species richness: Sensitivity to sample coverage and insensitivity to spatial patterns. *Ecology* **84**, 2364–2377.

Brose, U. & Martinez, N. (2004). Estimating the richness of species with variable mobility. *Oikos* **105**, 292–300.

Brose, U., Ostling, A., Harrison, K., and Martinez, N.D. (2004). Unified spatial scaling of species and their trophic interactions. *Nature* **428**, 167–171.

Brose, U., Jonsson, T., Berlow, E.L., Warren, P., Banasek-Richter, C., Bersier, L.F., Blanchard, J.L., Brey, T., Carpenter, S.R., Cattin Blandenier, M.-F., Cushing, L., Dawah, H.A., *et al.* (2006a). Consumer–resource body-size relationships in natural food webs. *Ecology* **87**, 2411–2417.

Brose, U., Williams, R.J., and Martinez, N.D. (2006b). Allometric scaling enhances stability in complex food webs. *Ecol. Lett.* **9**, 1228–1236.

Brose, U., Ehnes, R.B., Rall, B.C., Vucic-Pestic, O., Berlow, E.L., and Scheu, S. (2008). Foraging theory predicts predator–prey energy fluxes. *J. Anim. Ecol.* **77**, 1072–1078.

Camacho, J., Guimera, R., and Amaral, L.A.N. (2002a). Analytical solution of a model for complex food webs. *Phys. Rev. E* **65**, 030901.

Camacho, J., Guimera, R., and Amaral, L.A.N. (2002b). Robust patterns in food web structure. *Phys. Rev. Lett.* **88**, 228102.

Cattin Blandenier, M. (2004). *Food web ecology: models and application to conservation*, thesis, University of Neuchatel, Switzerland.

Cattin Blandenier, M.-F. (2004). Food Web Ecology: Models and Application to Conservation. Thesis, Université de Neuchâtel (Suisse).

Cattin, M.F., Bersier, L.F., Banasek-Richter, C., Baltensperger, R., and Gabriel, J.P. (2004). Phylogenetic constraints and adaptation explain food-web structure. *Nature* **427**, 835–839.

Christian, R. & Luczkovich, J. (1999). Organizing and understanding a winter's seagrass foodweb network through effective trophic levels. *Ecological Modelling* **117**, 99–124.

Cohen, J.E. (1989). Ecologists Co-operative Web Bank (ECOWebTM).

Cohen, J.E., and Briand, F. (1984). Trophic links of community food webs. *Proc. Natl. Acad. Sci. USA* **81**, 4105–4109.

Cohen, J.E., Schittler, D.N., Raffaelli, D.G., and Reuman, D.C. (2009). Food webs are more than the sum of their tritrophic parts. *Proc. Natl. Acad. Sci. USA* **106**, 22335–22340.

Cole, J.J., Carpenter, S.R., Pace, M.L., Van de Bogert, M.C., Kitchell, J.L., and Hodgson, J.R. (2006). Differential support of lake food webs by three types of terrestrial organic carbon. *Ecol. Lett.* **9**, 558–568.

Dunne, J.A. (2006). The network structure of food webs. In: *Ecological Networks: Linking Structure to Dynamics in Food Webs* (Ed. by M. Pascual and J.A. Dunne), pp. 27–86. Oxford University Press.

Dunne, J.A., Williams, R.J., and Martinez, N.D. (2002). Network structure and biodiversity loss in food webs: Robustness increases with connectance. *Ecol. Lett.* **5**, 558–567.

Dunne, J.A., Williams, R.J., and Martinez, N.D. (2004). Network structure and robustness of marine food webs. *Mar. Ecol. Prog. Ser.* **273**, 291–302.

Dunne, J.A., Williams, R.J., Martinez, N.D., Wood, R.A., and Erwin, D.H. (2008). Compilation and network analyses of Cambrian food webs. *PLoS Biol.* **6**, e102.

Fagan, W.F., and Hurd, L.E. (1994). Hatch density variation of a generalist arthropod predator: Population consequences and community impact. *Ecology* **75**, 2022–2032.

García, L. (2004). Escaping the Bonferroni iron claw in ecological studies. *Oikos* **105**, 657–663.

Garlaschelli, D., Caldarelli, G., and Pietronero, L. (2003). Universal scaling relations in food webs. *Nature* **423**, 165–168.

Goldwasser, L., and Roughgarden, J. (1993). Construction and analysis of a large Caribbean food web. *Ecology* **74**, 1216–1233.

Goldwasser, L., and Roughgarden, J. (1997). Sampling effects and the estimation of food-web properties. *Ecology* **78**, 41–54.

Hall, S.J., and Raffaelli, D. (1991). Food-web patterns—Lessons from a species-rich web. *J. Anim. Ecol.* **60**, 823–842.

Hall, S.J., and Raffaelli, D.G. (1993). Food webs—Theory and reality. In: *Adv. Ecol. Res.* (Ed. by M. Begon and A.H. Fitter), Vol. 24, pp. 187–239. Academic Press, London.

Harper-Smith, S., Berlow, E., Knapp, R., Williams, R., and Martinez, N. (2005). Communicating ecology through food webs: Visualizing and quantifying the effects of stocking alpine lakes with fish. In: *Dynamic Food Webs: Multispecies assemblages, ecosystem development, and environmental change*. Elsevier/Academic Press.

Harrison, K. (2003). Effects of Land Use and Dams on Stream Food Web Ecology in Santa Clara Valley, California 127. Thesis, San Francisco State University.

Havens, K. (1992). Scale and structure in natural food webs. *Science* **257**, 1107–1109.

Heymans, J.J., Ulanowicz, R.E., and Bondavalli, C. (2002). Network analysis of the South Florida Everglades graminoid marshes and comparison with nearby cypress ecosystems. *Ecol. Model.* **149**, 5–23.

Ings, T.C., Montoya, J.M., Bascompte, J., Bluthgen, N., Brown, L., Dormann, C.F., Edwards, F., Figueroa, D., Jacob, U., Jones, J.I., Lauridsen, R.B., Ledger, M.E., *et al.* (2009). Ecological networks—Beyond food webs. *J. Anim. Ecol.* **78**, 253–269.

Jacob, U. (2005). Trophic Dynamics of Antarctic Shelf Ecosystems—Food Webs and Energy Flow Budgets. Thesis, University of Bremen.

Jonsson, T., Cohen, J.E., and Carpenter, S.R. (2005). Food webs, body size, and species abundance in ecological community description. *Adv. Ecol. Res.* **36**, 1–84.

Keitt, T.H. (1997). Stability and complexity on a lattice: Coexistence of species in an individual-based food web model. *Ecol. Model.* **102**, 243–258.

Lafferty, K.D., Dobson, A.P., and Kuris, A.M. (2006). Parasites dominate food web links. *Proc. Natl. Acad. Sci. USA* **103**, 11211–11216.

Lange, M. (2005). Ecological laws: What would they be and why would they matter? *Oikos* **110**, 394–403.

Layer, K., Riede, J.O., Hildrew, A.G., and Woodward, G. (2010). Food web structure and stability in 20 streams across a wide pH gradient. *Adv. Ecol. Res.* **42**, 265–299.

Link, J. (2002). Does food web theory work for marine ecosystems. *Mar. Ecol. Prog. Ser.* **230**, 1–9.

Martinez, N.D. (1991). Artifacts or attributes—Effects of resolution on the Little Rock Lake food web. *Ecol. Monogr.* **61**, 367–392.

Martinez, N.D. (1992). Constant connectance in community food webs. *Am. Nat.* **139**, 1208–1218.

Martinez, N.D. (1993a). Effect of scale on food web structure. *Science* **260**, 242–243.

Martinez, N.D. (1993b). Effects of resolution on food web structure. *Oikos* **66**, 403–412.

Martinez, N.D. (1994). Scale-dependent constraints on food-web structure. *Am. Nat.* **144**, 935–953.

Martinez, N.D., Hawkins, B.A., Dawah, H.A., and Feifarek, B.P. (1999). Effects of sampling effort on characterization of food-web structure. *Ecology* **80**, 1044–1055.

May, R.M. (1972). Will a large complex system be stable? *Nature* **238**, 413–414.

McCann, K., Hastings, A., and Huxel, G.R. (1998). Weak trophic interactions and the balance of nature. *Nature* **395**, 794–798.

McLaughlin, O.B., Jonsson, T., and Emmerson, M.C. (2010). Temporal variability in predator–prey relationships of a forest floor food web. *Adv. Ecol. Res.* **42**, 171–264.

Montoya, J.M., and Sole, R.V. (2003). Topological properties of food webs: From real data to community assembly models. *Oikos* **102**, 614–622.

Moran, M.D. (2003). Arguments for rejecting the sequential Bonferroni in ecological studies. *Oikos* **100**, 403–405.

Navarrete, S.A., and Berlow, E.L. (2006). Variable interaction strengths stabilize marine community pattern. *Ecol. Lett.* **9**, 526–536.

O'Gorman, E.J., and Emmerson, M.C. (2010). Manipulating interaction strengths and the consequences for trivariate patterns in a marine food web. *Adv. Ecol. Res.* **42**, 301–419.

O'Gorman, E.J., and Emmerson, M.C. (2009). Perturbations to trophic interactions and the stability of complex food webs. *Proc. Natl. Acad. Sci. USA* **106**, 13393–13398.

O'Hara, R.B. (2005). The anarchist's guide to ecological theory. Or, we don't need no stinkin' laws. *Oikos* **110**, 390–393.

Olesen, J.M., Dupont, Y.L., O'Gorman, E., Ings, T.C., Layer, K., Melián, C.J., Troejelsgaard, K., Pichler, D.E., Rasmussen, C., and Woodward, G. (2010). From Broadstone to Zackenberg: Space, time and hierarchies in ecological networks. *Adv. Ecol. Res.* **42**, 1–69.

Opitz, S. (1996). *Trophic interactions in Caribbean coral reefs. Technical Report 43.* ICLARM, Manily.

Otto, S.B., Rall, B.C., and Brose, U. (2007). Allometric degree distributions facilitate food-web stability. *Nature* **450**, 1226–1229.

Paine, R.T. (1988). Food Webs—Road maps of interactions or grist for theoretical development. *Ecology* **69**, 1648–1654.

Paine, R.T. (1992). Food-web analysis through field measurement of per capita interaction strength. *Nature* **355**, 73–75.

Patricio, J., and Marques, J.C. (2006). Mass balanced models of the food web in three areas along a gradient of eutrophication symptoms in the south arm of the Mondego estuary (Portugal). *Ecol. Model.* **197**, 21–34.

Peres-Neto, P.R. (1999). How many statistical tests are too many? The problem of conducting multiple ecological inferences revisited. *Mar. Ecol. Prog. Ser.* **176**, 303–306.

Petchey, O.L., Beckerman, A.P., Riede, J.O., and Warren, P.H. (2008). Size, foraging, and food web structure. *Proc. Natl. Acad. Sci. USA* **105**, 4191–4196.

Piechnik, D.A., Lawler, S.P., and Martinez, N.D. (2008). Food-web assembly during a classic biogeographic study: Species' "trophic breadth" corresponds to colonization order. *Oikos* **117**, 665–674.

Polis, G.A. (1991). Complex trophic interactions in deserts: An empirical critique of food-web theory. *Am. Nat.* **138**, 123–155.

R Development Core Team (2009). R: A Language and Environment for Statistical Computing. 3-900051-07-0, Austria, Vienna.

Rall, B.C., Guill, C., and Brose, U. (2008). Food-web connectance and predator interference dampen the paradox of enrichment. *Oikos* **117**, 202–213.

Schmid-Araya, J.M., Schmid, P.E., Robertson, A., Winterbottom, J., Gjerløv, C., and Hildrew, A.G. (2002). Connectance in stream food webs. *J. Anim. Ecol.* **71**, 1056–1062.

Schoener, T.W. (1989). Food webs from the small to the large. *Ecology* **70**, 1559–1589.

Simberloff, D.S., and Abele, L.G. (1975). Island biogeography theory and conservation practice. *Science* **191**, 285–286.

Spencer, M., and Warren, P.H. (1996). The effects of energy input, immigration and habitat size on food web structure: A microcosm experiment. *Oecologia* **108**, 764–770.

Stouffer, D.B., Camacho, J., Guimera, R., Ng, C.A., and Amaral, L.A.N. (2005). Quantitative patterns in the structure of model and empirical food webs. *Ecology* **86**, 1301–1311.

Tews, J., Brose, U., Grimm, V., Tielborger, K., Wichmann, M.C., Schwager, M., and Jeltsch, F. (2004). Animal species diversity driven by habitat heterogeneity/diversity: The importance of keystone structures. *J. Biogeogr.* **31**, 79–92.

Townsend, C.R., Thompson, R.M., McIntosh, A.R., Kilroy, C., Edwards, E., and Scarsbrook, M.R. (1998). Disturbance, resource supply, and food-web architecture in streams. *Ecol. Lett.* **1**, 200–209.

Vucic-Pestic, O., Rall, B.C., Kalinkat, G., and Brose, U. (2010). Allometric functional response model: Body masses constrain interaction strengths. *J. Anim. Ecol.* **79**, 249–256.

Waide, R., and Reagan, W. (1996). *The food web of a tropical rainforest*. University of Chicago Press, Chicago.

Warren, P.H. (1989). Spatial and temporal variation in the structure of a fresh-water food web. *Oikos* **55**, 299–311.

Williams, R.J., and Martinez, N.D. (2000). Simple rules yield complex food webs. *Nature* **404**, 180–183.

Williams, R.J., and Martinez, N.D. (2008). Success and its limits among structural models of complex food webs. *J. Anim. Ecol.* **77**, 512–519.

Williams, R.J., Martinez, N.D., Berlow, E.L., Dunne, J.A., and Barabási, A.-L. (2002). Two degrees of separation in complex food webs. *Proc. Natl. Acad. Sci. USA* **99**, 12913–12916.

Winemiller, K.O. (1990). Spatial and temporal variation in tropical fish trophic networks. *Ecol. Monogr.* **60**, 331–367.

Woodward, G., and Hildrew, A.G. (2001). Invasion of a stream food web by a new top predator. *J. Anim. Ecol.* **70**, 273–288.

Woodward, G., Ebenman, B., Emmerson, M., Montoya, J.M., Olesen, J.M., Valido, A., and Warren, P.H. (2005a). Body size in ecological networks. *Trends Ecol. Evol.* **20**, 402–409.

Woodward, G., Speirs, D.C., and Hildrew, A.G. (2005b). Quantification and temporal resolution of a complex size-structured food web. *Adv. Ecol. Res.* **36**, 85–135.

Woodward, G., Papantoniou, G., Edwards, F., and Lauridsen, R.B. (2008). Trophic trickles and cascades in a complex food web: Impacts of a keystone predator on stream community structure and ecosystem processes. *Oikos* **117**, 683–692.

Woodward, G., Benstead, J.P., Beveridge, O.S., Blanchard, J., Brey, T., Brown, L., Cross, W.F., Friberg, N., Ings, T.C., Jacob, U., Jennings, S., Ledger, M.E., *et al.* (2010). Ecological networks in a changing climate. *Adv. Ecol. Res.* **42**, 71–138.

Wootton, J.T. (1997). Estimates and tests of per-capita interaction strength: Diet, abundance, and impact of intertidally-foraging birds. *Ecol. Monogr.* **67**, 45–64.

Wootton, J. & Emmerson, M. (2005). Measurement of interaction strength in nature. *Annual Reviews Ecol. Evol. Systems* **36**, 419–444.

Yvon-Durocher, G., Reiss, J., Blanchard, J., Ebenman, B., Perkins, D.M., Reuman, D.C., Thierry, A., Woodward, G., and Petchey, O.L. (2010). Across ecosystem comparisons of size structure: Methods, approaches, and prospects. *Oikos* (in press).

Temporal Variability in Predator–Prey Relationships of a Forest Floor Food Web

ÓRLA B. McLAUGHLIN, TOMAS JONSSON
AND MARK C. EMMERSON

ADVANCES IN ECOLOGICAL RESEARCH VOL. 42

0065-2504/10 $35.00
DOI: 10.1016/S0065-2504(10)42004-8

SUMMARY

Connectance webs represent the standard data description in food web ecology, but their usefulness is often limited in understanding the patterns and processes within ecosystems. Increasingly, efforts have been made to incorporate additional, biologically meaningful, data into food web descriptions, including the construction of food webs using data describing the body size and abundance of each species. Here, data from a terrestrial forest floor food web, sampled seasonally over a 1-year period, were analysed to investigate (i) how stable the body size–abundance and predator–prey relationships of an ecosystem are through time and (ii) whether there are system-specific differences in body size–abundance and predator–prey relationships between ecosystem types.

I. INTRODUCTION

The loss of biodiversity and the consequences for ecosystem functioning have been extensively debated (Hooper *et al.*, 2005; Loreau *et al.*, 2001; Naeem *et al.*, 1994; Wardle, 1999). In the last decade, a large number of biodiversity–ecosystem functioning (BEF) studies have been conducted in a wide range of systems, and many have highlighted the negative consequences of species loss on ecosystems (Petermann *et al.*, 2010; Srivastava and Bell, 2009; Zavaleta *et al.*, 2010). However, such studies have been largely limited to a consideration of horizontal biodiversity loss within single trophic levels (Hector *et al.*, 1999; Naeem and Li, 1997). Empirically predicting the consequences of biodiversity loss from complex multitrophic systems is often limited by the feasibility of replicating large-scale factorial experiments that examine the effects of biodiversity on ecosystem processes, such as primary productivity, secondary production and decomposition rates (Hector *et al.*, 1999; Jonsson and Malmqvist, 2000; Worm *et al.*, 2006). The few studies that have examined the consequences of biodiversity loss across multiple trophic levels (vertical biodiversity) have done so using factorial experiments of simple systems using either additive or substitutive designs (e.g. Cardinale *et al.*, 2006; Jonsson and Malmqvist, 2003; Snyder *et al.*, 2008). Predicting the consequences of biodiversity loss in 'real' multitrophic level systems is thus still compromised by our limited knowledge of interactions within and across multiple trophic levels in natural food webs (Reiss *et al.*, 2009).

In order to understand how communities are organized, it is essential to quantify the strength of interactions between species and within the food web as whole. Estimating the strength of trophic interactions has been a rate-limiting step in predicting the likely consequences of biodiversity loss from complex food webs (Kaiser-Bunbury *et al.*, 2010; Maron *et al.*, 2010). In recent years, it has become increasingly evident that food webs are often composed of a few strong and many weak trophic interactions, and this distribution is important in conferring stability upon the system as a whole (Emmerson and Raffaelli, 2004; McCann *et al.*, 1998; Neutel *et al.*, 2002). Theoretical studies have shown that the inclusion of weak interactors tends to suppress the destabilizing effect of strong interactors within a community (McCann *et al.*, 1998), whilst studies that integrate empirical and theoretical work have shown that asymmetries in the pattern of interaction strengths (Pimm, 1984) and the presence of weak links in omnivorous food web loops can also promote stability (Neutel *et al.*, 2002). Increasingly, a range of studies have explored the importance of predator and prey body mass M for the determination of interaction strength between predators and prey (Brose *et al.*, 2005; Emmerson and Raffaelli, 2004; Emmerson *et al.*, 2005; Jonsson and Ebenman, 1998; Otto *et al.*, 2007). These approaches have used the ratio M_j/M_i to estimate the *per capita* interaction strength between predators j and prey i, where M is a measure of (usually species-averaged) body mass. The usefulness of such theoretical approaches for predicting the consequences of biodiversity loss for ecosystem functioning in real systems is, however, still restricted by the scarcity of appropriate empirical data with which to test and validate models. There is therefore a need for high-quality data that describe the composition (abundance and body mass of each species), structure (body mass–abundance relationships and food web topology) and functioning (energy flow and associated system process rates) of real food webs.

The relationship between body mass and abundance has been examined empirically (Damuth, 1981; Kerr and Dickie, 2001; Layer *et al.*, 2010a,b; Leaper and Raffaelli, 1999; O'Gorman and Emmerson, 2010), but the exact allometric relationship is disputed (Jennings and Mackinson, 2003). Characteristics of body size can affect a range of ecological characteristics of individuals and species, including resource use and partitioning (Arim *et al.*, 2010, Schmid and Schmid-Araya, 2007), which in turn can affect the structuring of the community itself. The incorporation of body mass–abundance data into descriptions of food web structure allows greater insight into how these metrics are interrelated with other bivariate relationships, and to understand mechanistically or functionally the empirical community patterns that emerge (Woodward *et al.*, 2005a). To date, the availability of such data remains limited (but see Jonsson *et al.*, 2005; Woodward *et al.*, 2005b), which constrains the extent to which generalizations can be made to test theoretical predictions of the likely consequences of biodiversity change and how perturbations will spread through real systems (Emmerson and Raffaelli, 2004; Montoya *et al.*, 2009).

Connectance webs (i.e. binary feeding matrices of who eats whom) have been the standard data description in food web ecology for decades, but their usefulness is limited in understanding the ecological consequences of predator–prey interactions due to a lack of information on interaction strengths and species abundances. There are growing efforts to incorporate additional, biologically meaningful data into food web descriptions to better understand and describe general patterns. Recent approaches include the construction of '*MN* webs', whereby feeding links are overlain on biplots of body mass (*M*) and abundance (*N*) of species (Jonsson *et al.*, 2005), and trophochemical webs, which include stoichiometric data, usually in the form of carbon, nitrogen and phosphorus ratios, on the species within the network (Mülder *et al.*, 2006; Sterner and Elser, 2002). Cohen *et al.* (2003) coined the term 'trivariate patterns' to describe the combination of body mass, abundance and food web topology data in *MN* webs. Through the analysis of the Tuesday Lake web, Jonsson *et al.* (2005) illustrated the interrelationship between the structural properties and energy fluxes within a food web and the distributions of body mass and abundance. They stressed that many patterns in the structure of ecological communities that have traditionally been treated as independent are, in fact, connected. One perhaps surprising finding was that some of the trivariate patterns appeared to be relatively robust after a major ecosystem perturbation in the form of introduction of the piscivorous fish largemouth bass (*Micropterus salmoides*) to Tuesday Lake. This implies the existence of general, recurrent patterns in ecosystems that might be predictably consistent from one community to another. Few studies have been able to repeat this analysis due to a lack of appropriate datasets (but see Layer *et al.*, 2010b; Reuman *et al.*, 2009; Woodward *et al.*, 2005b), and there is a need to test the ideas proposed by Jonsson *et al.* (2005) with new data from additional ecosystems to ascertain whether these seemingly recurrent, predictable patterns hold true.

Temporally resolved body size and abundance data are needed to address these questions. Here, we expand the database of available *MN* webs and test the generality of previously described *MN* relationships, using new data from a terrestrial forest floor food web (the Gearagh alluvial forest ecosystem, Ireland) (Figure 1), sampled seasonally over a 1-year period, to characterize (i) the seasonal persistence of the body size–abundance and predator–prey relationships within the food web and to (ii) identify whether there are system-specific differences in body size–abundance and predator–prey relationships between aquatic and terrestrial systems. Lindeman (1942) suggested that there are systematic contrasts in trophic efficiency and energy flow between terrestrial and aquatic ecosystems (Shurin *et al.*, 2006). We hypothesize that at the scale studied, that is on a density basis (m^2), there will be no difference between the body size–abundance and predator–prey relationships found in the Gearagh and aquatic ecosystems. Previous studies have indicated that fundamental differences might exist in the size structure of terrestrial and aquatic systems at large spatial scales (Cyr *et al.*, 1997; Pace *et al.*, 1999;

Figure 1 Photograph of a typical island sampled during this study in the Gearagh.

Shurin *et al.*, 2006). For instance, aquatic systems are dominated by microscopic plankton, zooplankton and large fish, where ontogenic shifts in body mass are observed, whereas, terrestrial systems are characterized by large primary producers, which are fed on by either small, herbivorous arthropods, or large, grazing mammals (and adult body mass is asymptotic). This new food web is described in terms of network topology and the body mass and abundance of component species, and measurements of detrital and vegetative biomass were also made to complement and develop the *MN* web approach of Cohen *et al.* (2003, 2009). The new Gearagh data thus contributes to the small but growing number of other well-described complex food webs that have been constructed using this approach (e.g. Amundsen *et al.*, 2009; Jonsson *et al.*, 2005; Layer *et al.*, 2010a,b; O'Gorman and Emmerson, 2010; Woodward *et al.*, 2005b). This predator-dominated terrestrial invertebrate food web describes the feeding interactions between above-ground predators and their above- and below-ground prey. This study also characterizes a range of empirical relationships between predator and prey body size, and uses these relationships to analyse whether there are differences between aquatic and terrestrial food webs (see Section II for a detailed field site description).

A. Predictions for the Gearagh Food Web Based on Empirical Patterns of Food Webs

If consumers are mostly larger than their resources, in a bivariate plot of consumer body mass (M_C, *x*-axis) and resource body mass (M_R, *y*-axis), the data points will be located in the upper right triangle of the plot (Warren and

Lawton, 1987; Woodward and Hildrew, 2002a). If this is a general ecological phenomenon, when plotting consumer body mass as a function of resource body mass the majority of the data points for the Gearagh should also populate the upper right triangle of the plot. Furthermore, if consumers are larger than their resources, then body mass (M) should increase with trophic height (T), where T is the average trophic position of a species in all food chains of which it is a member (Jonsson *et al.*, 2005). In the studies that have examined such relationships quantitatively (Jennings and Mackinson, 2003; Jennings *et al.*, 2001; Jonsson *et al.*, 2005; Woodward and Hildrew, 2002a), consumer trophic height (T_C) typically scales exponentially with consumer body mass (M_C) (but see Layman *et al.*, 2005). This scaling relationship is important because it produces asymmetries in the likelihood of trophic interactions and may in turn have important consequences for population dynamics and resource partitioning among predators (Chase, 1999; Woodward and Hildrew, 2002b).

In many systems, consumers (excluding parasites) will not only be larger but also rarer than their resources. If this is true, then within a bivariate plot of the numerical abundance of consumers (N_C, x-axis) and resources (N_R, y-axis), the majority of the data points should be located in the upper left triangle of the plot. Empirically, many studies (Damuth, 1981; Jennings and Mackinson, 2003; Kerr and Dickie, 2001) have shown that numerical abundance (N) decreases allometrically with body size (i.e. with constants α and β: $N = \alpha M^{\beta}$). Both empirical patterns and theoretical predictions typically approximate that the slope (β) of the body mass–abundance relationship on log–log scales will lie between -0.75 and -1.2 (White *et al.*, 2007). When the relationship is examined at local (the relationship between the average body size of a species (M_{sp}) and its population density (N) where all population densities are taken from a single region), rather than global (the relationship between the average body size of a species (M_{sp}) and its average population density compiled from the literature (N_{comp}), where densities can be taken from any point on the globe) scales (White *et al.*, 2007), some studies have found this slope to be much shallower, average -0.25 (Blackburn and Gaston, 1997; but see Jonsson *et al.*, 2005). Again, if this is a widespread and common phenomenon, we would also expect to find that (i) the majority of data points will be found in the upper left triangle in a plot of resource numerical abundance as a function of consumer numerical abundance; (ii) the relationship between numerical abundance and body size for the Gearagh is allometric and (iii) since the Gearagh is a local community, the slope will be shallower than -0.75.

If consumer body mass is only slightly greater than resource body mass, while consumer numerical abundance consistently tends to be much smaller than resource numerical abundance, then resource biomass abundance (i.e. $M_R \times N_R = B_R$) should exceed consumer biomass abundance (i.e. $M_C \times N_C = B_C$). We should expect, then, that in the bivariate plot of consumer biomass abundance (B_C, x-axis) and resource biomass abundance (B_R, y-axis), the majority of data points will be located in the lower right triangle of the plot.

Within Elton's (1927) pyramid of numerical and biomass abundances, larger and rarer organisms are typically found at the top of the food chain. Lindeman (1942) predicted that these larger organisms that occupy higher trophic levels would tend be more omnivorous and that they would be more likely to feed on a wider range of prey items, that is an increase in trophic generality, and this in general is supported in several empirical studies (e.g. Woodward and Hildrew, 2002b). This is in broad accordance with the assumptions of a range of models that have been proposed to explain food web topological properties, for example the cascade model (Cohen and Newman, 1985), the niche model (Williams and Martinez, 2000), the nested hierarchy model (Cattin *et al.*, 2004) and the generalized cascade model (Stouffer *et al.*, 2005). These models generally predict that the potential number of prey of a predator increases with trophic rank or height. An accompanying prediction, also in accordance with the cascade model, suggests that species at high trophic levels will themselves have fewer predators, and hence reduced trophic vulnerability (Cohen and Newman, 1985; Schoener, 1989). Consequently, we expect that trophic height in the Gearagh food web will increase with trophic generality but decrease with vulnerability.

To facilitate comparisons amongst food webs, we used the trivariate approach proposed by Cohen *et al.* (2003, 2009) to examine the size structure of a community and explore a range of food web patterns that are typically considered independent. To date, the full complement of univariate, bivariate and trivariate patterns have been described and analysed for Tuesday Lake (Jonsson *et al.*, 2005) only, while a slightly smaller set of relationships have been described for Broadstone Stream (Woodward *et al.*, 2005b), for the Ythan estuary (Woodward *et al.*, 2005a) a set of 20 streams and for the Lochnagar lake food web (Layer *et al.*, 2010a,b). Finally, we tested the hypothesis that seasonal changes in productivity (i.e. standing stock of plant biomass) drive dynamical shifts in food web topology and reflect the importance of seasonally emerging consumers (from larval to adult stages) on the resulting food web structure.

II. METHODS

A. Field Site Description

The Gearagh woodland is a mixed deciduous alluvial forest in the floodplain of the River Lee, South-West Cork, Ireland (N 51°52′09″, W 9°01′00″) (Figure 1). This ecosystem is one of the last semi-natural forested floodplains in Europe (Brown, 1997), and was designated a statutory national nature reserve in 1987. Historically, the Gearagh extended approximately 5.5 km in length (Cross and Kelly, 2003), but almost 60% of the woodland was felled and flooded between 1955 and 1956 (White, 1985) to accommodate the River Lee hydroelectric scheme. The remaining 200 ha today is a complicated braided river system composed of approximately 13 channels, each 1–7 m wide. The main channels are stabilized by tree roots, creating a mosaic of small islands. Over time, the accumulation of detrital material and fallen trees has resulted in the island system now found within the Gearagh. As a consequence of the island and river channel formation, water flow through the Gearagh is not uniform. Heavy rainfall can result in flooding episodes which may cause the inundation of the system. River flow can undercut the islands which can result in their overturning as the tree roots which stabilize them are undermined: however, none of the islands sampled was affected by undercutting or destabilization during the current study.

B. Sampling Protocol

Within the braided river system, 16 islands were sampled for ground-dwelling terrestrial invertebrates. The islands were selected to minimize variation in size (13.85 ± 7.28 m^2 SD) and were similar in their vegetation composition. These islands were sampled bimonthly from February 2005 to 2006, inclusive. Heavy rainfall and site flooding in October and December 2005 limited the sampling of vegetation (October and December) and invertebrates (October).

Invertebrate communities on each island were sampled using six pitfall traps (8 cm height, 6.5 cm diameter), which were positioned randomly in the ground on each island. Ethylene glycol and water (3:1 ratio) were placed into each cup to a depth of 1 cm. The pitfall trap samples were collected from the field, bagged and immediately returned to the laboratory where they were cleaned and preserved with 70% ethanol prior to identification.

We sampled 18 quadrats (0.25 m^2–50 cm $\times$ 50 cm) for above-ground biomass of vegetation and detritus on each sampling date. All living above-ground vegetation was removed from each quadrat, identified to species and placed into individual labelled bags and subsequently transported to the laboratory. Each plant species sample was dried to constant mass at 60 °C for 48 h to obtain estimates of dry weight. The biomass of each plant species was measured and converted to biomass abundance (dry weight g/m^2). All detritus in each quadrat was also collected, bagged, labelled and brought back to the laboratory. Each detrital sample was placed into an oven at 60 °C for 48 h to obtain dry weights, which was subsequently converted to biomass abundance (dry weight g/m^2).

C. Food Web Construction

Invertebrates collected in the pitfall traps were identified to species, or the lowest taxonomic level that was feasible, using published keys (see Appendix I) and a reference collection assembled from the Gearagh. Individual body lengths ($n = 12{,}075$) were measured using a calibrated eyepiece graticule on a microscope (Nikon® SMZ645, $\times 0.5$ magnification). Once the invertebrates were identified, they were each oven-dried to a constant mass at 60 °C for 48 h to obtain individual dry masses. Individuals of the smallest taxa were pooled into bulk samples (minimum 20 individuals per size class) for the orders Collembola (springtails), Nematoda (nematode worms) and Acari (mites), in order to obtain sufficient material or accurate measurement.

The gut contents of $n > 250$ invertebrates were dissected, but in no instances were identifiable prey body parts found. In general, terrestrial invertebrates do not engulf their prey items whole, but rather ingest parts of the organism or feed suctorially (Memmott *et al.*, 2000). Also, many terrestrial invertebrates feed on soft-bodied organisms such as slugs, worms and springtails. This feeding strategy leaves no hard chitinous body parts in the gut. Consequently, gut content analysis could not be used to construct the food webs. Instead, the scholastic search engines Web of Science and Google Scholar were used to conduct an in-depth literature search of known prey items of the invertebrates found in the Gearagh, using strict criteria for all species-specific searches. The consumer species name plus the keywords 'diet' and 'predator' were first submitted to the search engines. Using the results of these searches, the search criteria were expanded to include any prey species identified from the initial search. The trophic interactions reported in the literature were only included in the Gearagh food webs if the feeding links were documented in three or more peer-reviewed papers or book chapters. We placed the further constraint that trophic interactions between predators and prey could only be included in the food web if both the predator and the prey were also empirically recorded in the Gearagh (from our pitfall trap samples) during the same sampling period. Species-specific trophic interactions were not always detailed in the literature, and in these circumstances, the taxonomic resolution was reduced from species to genus or family level. When this occurred, the higher taxonomic group, for example diptera, was included as a taxonomic node within the food web, along with predator–prey interactions to this group (see Appendix II for a full reference list of the literature used in the construction of the Gearagh food webs).

Using these literature searches, six plausible bimonthly food webs were constructed for the Gearagh, that is the six season webs were each pooled across islands. Finally, a composite annual food web was constructed, containing all links (375 links) found in the six bimonthly food webs. All food web figures were drawn using the software package Matlab (Version 1.0.0.1).

Length–weight relationships for all taxa and orders found in the Gearagh were also calculated directly, rather than being extracted from regressions extracted from the literature, to avoid any potential site-specific biases. Taxa and orders with significant regression relationships (44 taxa and 15 orders) are included in the appendix (see Appendix III: Tables S1 and S2).

D. Food Web Patterns

Univariate food web metrics were calculated to compare between the Gearagh food webs and 16 other well-resolved food webs available in the literature (see Table 2). The number of species, number of links, link density, directed connectance and complexity were calculated for each bimonthly food web and the composite food web. Link density (d) is the average number of links per species (L/S), where L is the total number of links in the food web and S is the number of species. Directed connectance (C') is the proportion of observed links to all possible links, including cannibalism and intraguild predation, in a food web (Martinez, 1991) and is defined as L/S^2. Complexity is the connectance of the web multiplied by the species richness (SC'). The relative proportion of top predators, intermediate consumers and basal species were calculated to assess the partitioning of resources (Briand and Cohen, 1984) for each of the bimonthly food webs and the composite food web. For a small subset of primary consumers, no predatory interactions (i.e. predators of those consumers) were documented in the literature. As a result, the percentage of taxa on the second trophic level without consumers was calculated, as these primary consumers would otherwise appear as top predators. The predator j to prey i species ratio (S_i/S_j) was quantified to determine whether the system was predator or prey dominated (Briand and Cohen, 1984). To quantify the relative importance of detritivory and herbivory (and hence the source of energy for secondary production) within the system, the percentage of detritivore and herbivore species was calculated. Trophic height is the average trophic position of a species in all food chains of which it is a member (Jonsson *et al.*, 2005), and mean chain length was calculated as the mean number of trophic nodes linking a basal resource and its top predator (Woodward and Hildrew, 2001). Predator diet breadth (importance of top-down processes) was determined by calculating the generality of the food web (the mean number of prey per predator). In turn, the importance of bottom-up processes was quantified by calculating the vulnerability (the mean number of predators per prey) of the prey in the food web (Schoener, 1989).

The importance of size-based predation in the food web was determined by calculating both the morphological niche width (MNW; the proportion of prey species consumed by a predator within the body mass range known to be consumed by the predator; see Leaper and Huxham, 2002; Warren, 1996) and the morphological niche range (MNR; the range of prey body sizes consumed by a predator relative to its own body size) as

$$\mathrm{MNR} = [\max(\mathrm{MNR}_i) - \min(\mathrm{MNR}_i)]/\mathrm{MNC}_i$$

where MNC_i is the body mass of consumer i, and MNR_i is the set of body masses of the resource species of consumer species i. Mean MNR and width values were calculated for each of the bimonthly webs as well as the composite food web (by averaging across the MNR/width values of all consumer species in a given food web). It should be noted that a limitation of the present study is that the feeding links were constructed using literature searches, and consequently some species may be documented more commonly than others. For certain taxa (24%), only one prey species was recorded in the literature, and therefore an MNW value could not be calculated. Here, those 25 taxa were removed from the calculation of the mean MNW value.

We calculated the rank–biomass abundance and rank–numerical abundance relationships across all species for the composite and all seasonal webs. Based on our predictions above, the Gearagh should have a pyramidal food web, in which body size increases and numerical abundance decreases with increasing trophic height, such that both biomass and numerical abundance should follow log-normal distributions. We also calculated the frequency distributions of the number of species by numerical and biomass abundance.

Body size distributions within communities are often undersampled and typically skewed to the right, due primarily to a bias in sampling larger organisms. Very rarely will an entire community be sampled with sufficient effort to be able to adequately describe the distribution of body masses. A limited number of aquatic studies (Jonsson *et al.*, 2005; Woodward *et al.*, 2005b) have, however, shown that body mass distributions of natural food webs can be approximated by log-normal distributions. Here, we tested the hypothesis that the body size distribution of ground-dwelling terrestrial invertebrates will also conform to a log-normal distribution. If the body size is allometrically related to rank, then rank plot for the body sizes should follow a log-hyperbolic pattern.

E. Statistical Analyses

The Shapiro–Wilk test was used to determine whether the body size distributions of ground-dwelling invertebrates were normally distributed (Shapiro and Wilk, 1965). The Kolmogorov–Smirnov test was used to assess whether there was a significant difference between the rank numerical and biomass abundances and a log-normal distribution.

Linear regressions were calculated for $\log_{10}$ body mass versus MNW (Leaper and Huxham, 2002) and MNR versus $\log_{10}$ body mass and $\log_{10}$ numerical abundance (White *et al.*, 2007), $\log_{10}$ numerical abundance versus trophic height and $\log_{10}$ body mass versus trophic height. Pearson's product moment correlation was calculated for predator and prey body mass, predator and prey biomass abundance, predator and prey numerical abundance, $\log_{10}$ body mass and predator generality and $\log_{10}$ body mass and prey vulnerability (Jonsson *et al.*, 2005). To determine whether a significant fraction of these data points were located in either the upper or lower triangular compartment of the plot, a proportion test was conducted (Wilson, 1927). Sorensen's similarity index was used to quantify changes in species composition and food web topology (i.e. presence/absence of interactions) between the seasonal and composite food webs. All statistical analyses were performed using R (Version 2.9.0).

III. RESULTS

A. Univariate Patterns

The composite (regional and time-integrated) Gearagh food web (Figure 2) included all observed species (for the year February 2005–2006) for which appropriate diet data were documented and contains 116 trophic units

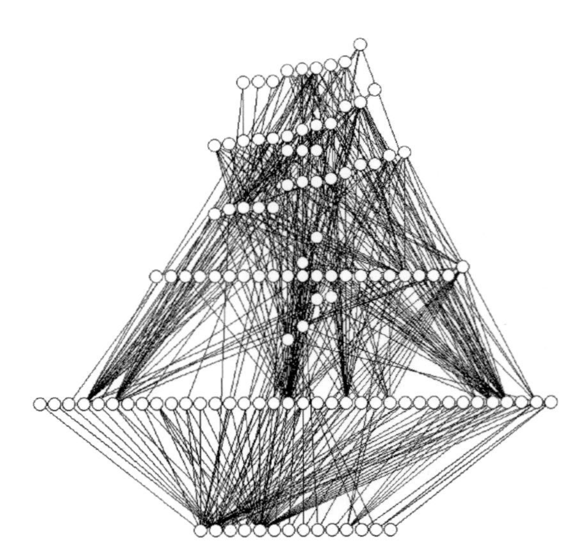

Figure 2 The Gearagh food web. This food web represents the composite food web for 2005/2006, including all links found over the six seasonal sampling sessions.

(henceforth called species) and 375 links (see Table 1 for a summary of web statistics of composite and seasonal webs and Appendix IV for primary data tables for all webs). In the laboratory, 70% of taxa were described to species level; however, species-level predator–prey interactions were not always available. Subsequently, 41% of taxa within the food webs here were detailed at species level, where the taxonomic resolution was reduced from species to genus or family level to retain the node.

There were 14 basal species in the composite web, including 10 species of plant, detritus, bacteria and microfungi. Among the consumer assemblage, there were 34 detritivore species (including seven species of slugs and snails, oligochaetes, enchytraeids, nematodes, springtails and mites), five herbivore species (including minute pirate bugs, leaf hoppers, leaf-footed bugs, water striders and lace bugs), and 63 predator species (including 17 species of carabid beetle, eight species of harvestmen, seven species of spiders and eight species of ants and parasitic wasps). Twenty-two observed taxa were not included in the webs due to unknown feeding links or being found in a non-feeding state, for example pupal stage (see Appendix V: Table S10).

The food web statistics for each of the seasonal webs revealed clear temporal changes in the structure of the networks (Table 1). Species richness and the total number of species links increased from February 2005 ($S=39$, $L=90$) to June 2005 ($S=79$, $L=256$), and subsequently decreased to February 2006 ($S=33$, $L=68$). The link density was highest in April 2005 ($d=3.32$) and lowest in the two February months (February 2005 = 2.31, February 2006 = 2.06). The number of links per species did not increase linearly with the number of species. Directed connectance demonstrated the opposite pattern to species richness, with a decrease in the spring and summer months and higher values in the winter. Mean chain length was greatest in August 2005 (2.49), and web complexity was greatest in April 2005 (6.66), both corresponding to periods when many terrestrial invertebrates emerge as adults and become more abundant. Both mean chain length and web complexity increased from February 2005 to August 2005, and subsequently decreased. Predator to prey ratios followed approximately the same pattern as connectance, with an increase in December 2005. There was no consistent pattern seen in the percentage of basal species, intermediate consumers and top predators throughout the year. In contrast, neither generality nor vulnerability appeared to be affected seasonally, and the pattern was consistent throughout the year. Similarly, there was also no clear seasonal pattern in MNW over the year (range 0.41–0.75).

The forest floor community of the Gearagh was dynamic, with marked changes in species composition and food web structure over the course of one year (February 2005–2006) (Figure 3). Of the 39 species found in February 2005, 19 (49%) were also found in all subsequent webs, creating a core community within the Gearagh to which species were added and lost during

Table 1 Food web statistics for the composite Gearagh food web and the six seasonal food webs

	February 2005	April 2005	June 2005	August 2005	December 2005	February 2006	Composite 2005/2006
S	39	74	79	70	52	33	116
L	90	246	256	207	167	68	375
C'	0.12	0.09	0.08	0.09	0.13	0.13	0.06
d	2.31	3.32	3.24	2.96	3.21	2.06	3.23
SC_{max}'	4.68	6.66	6.32	6.30	6.24	4.29	6.96
S_j:S_i	0.63	0.57	0.49	0.38	0.80	0.48	0.46
Mean chain length	1.99	2.47	2.46	2.49	2.11	2.14	2.53
Generality	2.31	3.32	3.24	2.96	3.21	2.06	2.23
Vulnerability	2.31	3.30	3.23	2.97	3.19	2.06	3.36
Morphological niche width	0.55	0.41	0.45	0.58	0.59	0.75	0.42
Basal species (%)	23	11	8	10	23	15	12
Intermediate species (%)	26	39	38	27	37	24	29
Top predators (%)	51	50	54	63	40	61	59
TL2 without predators (%)	36	47	42	53	44	65	48
TL2 detritivores (%)	88	76	74	82	89	90	80
TL2 herbivores (%)	42	38	42	50	36	35	43
Nodes to species (%)	41	46	34	30	33	22	37
Nodes to family (%)	27	28	41	44	23	48	42

S = number of species, L = number of trophic links, C' = directed connectance, d = link density (number of links per species), SC_{max}' = food web complexity, p:p = proportion of number of predators to prey, mean chain length = average number of links from a basal resource to a top predator, generality = number of prey per predator, vulnerability = number of predators per prey, morphological niche width = proportion of prey consumed within the total available prey size range known to be consumed by a predator, basal species = percentage of species which do not consume any other species in the food webs, intermediate species = percentage of species which consume basal resource and are consumed by top predators, top predators = percentage of species which consume intermediate species but are not themselves predated, TL2 without predators = percentage of species on trophic level 2 without top predators, TL2 detritivores = percentage of detritivores on trophic level 2, TL2 herbivores = percentage of herbivores on trophic level 2, nodes to species = percentage of nodes in the food webs resolved to species level, nodes to family = percentage of nodes in the food webs resolved to family level.

Table 2 Comparison of the Gearagh to 17 well-resolved food webs

Food web	*S*	*L*	*C′*	*d*
Marine				
Benguela (Yodzis, 1998)	29	191	0.23	6.59
Weddell Sea (Jacob *et al.*; see Brose *et al.*, 2005)	491	16200	0.07	32.99
Lough Hyne (O'Gorman and Emmerson, 2010)	168	1461	0.05	8.69
Caribbean (Bascompte *et al.*, 2005)	249	3313	0.05	13.31
Estuarine				
St. Mark's Seagrass (Christian and Luczkovich, 1999)	48	221	0.09	4.60
Chesapeake Bay (Baird and Ulanowicz, 1989)	31	68	0.07	2.19
Ythan Estuary (Huxham *et al.*, 1996)	134	626	0.04	4.67
Freshwater				
Skipwith Pond (Warren, 1989)	35	276	0.32	7.88
Lake Tahoe (Dunne *et al.*, 2002)	172	3885	0.13	22.59
Little Rock Lake (Martinez, 1991)	182	1973	0.12	10.84
Broadstone Stream (Woodward *et al.*, 2005a,b)	62	400	0.10	6.45
Tuesday Lake (Jonsson *et al.*, 2005)	56	269	0.08	4.80
Canton Creek (Townsend *et al.*, 1998)	102	697	0.06	6.83
Terrestrial				
Coachella Valley Polis (1991)	29	262	0.31	9.03
El Verde Rainforest (Reagan and Waide, 1996)	155	1510	0.06	9.74
The Gearagh	116	375	0.06	3.23
Scotch Broom (Memmott *et al.*, 2000)	154	403	0.03	2.62
Grassland (Dawah *et al.*, 1995)	75	119	0.03	1.59

S = number of trophic units, *L* = number of trophic links, *C′* = directed connectance, *d* = link density (number of links per species).

the year. This core was comprised mainly of detritivores (slugs and snails, oligochaetes, enchytraeids, mites and springtails) and one predator (staphylinid beetles); Vegetation and detritus accounted for 49% of all taxa found at the beginning of the study (February 2005) and 58% at the end (February 2006). The food webs for February 2005 and 2006 are dissimilar to the composite food web (0.50 and 0.44, respectively, Sorensen's similarity index, see Table 3A). A comparison of the two food webs from February 2005 and 2006 demonstrated a higher level of similarity (0.64) to one another than to any other month or to the composite web. From February 2005 to December 2005, the food webs became more species rich, and hence more similar to the composite food web. June 2005 was the most similar to the composite food web (0.81). Table 3B presents Sorensen's index for species interactions, that is how structurally similar the food webs are. February 2005 and 2006 were the most structurally dissimilar in terms of species interactions (0.39 and 0.31, respectively) to the composite food web. April 2005 and June 2005 were most structurally similar to the composite food web (0.81). Webs within the same season, for example summer, were generally

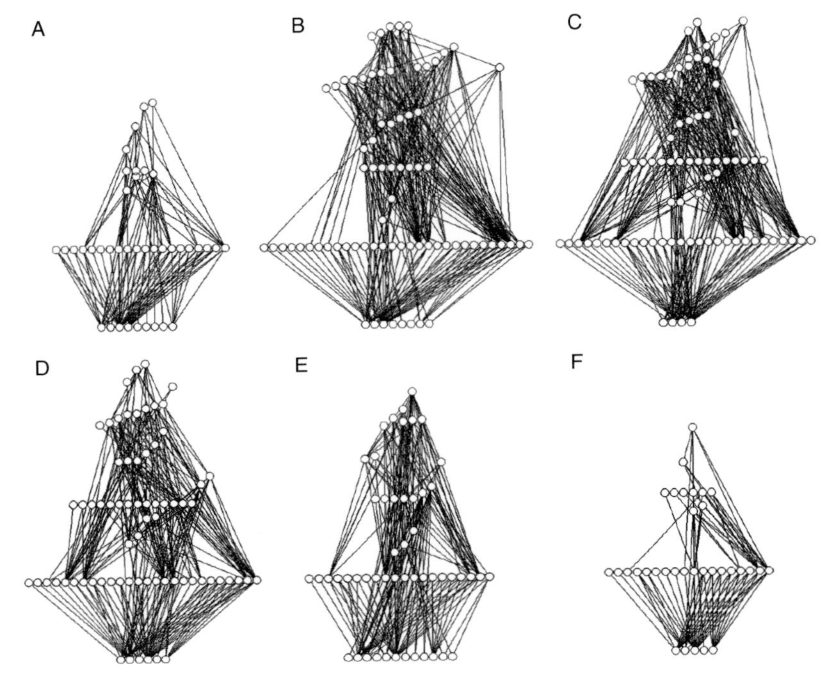

Figure 3 Illustration of the six seasonal food webs: (A) February 2005, (B) April 2005, (C) June 2005, (D) August 2005, (E) December 2005, and (F) February 2006.

more similar to one another than was the case when comparing across seasons, although December 2005 was more similar to the summer than to the winter webs.

There was no significant difference between either the rank numerical (Figure 4) or biomass (Figure 5) abundance and a log-normal distribution for the Gearagh composite food web or any of the seasonal webs (Kolmogorov–Smirnov: $p < 0.0001$). The solid lines in Figures 4 and 5 are the expected rank–abundance relationship for a log-normal distribution with the same mean and variance as that in the observed distribution. The dashed line is the log-scale frequency distribution of the number of species (top horizontal axis) by log-scale numerical abundance in Figure 4 and by (log scale) biomass abundance in Figure 5.

The distributions of $\log_{10}$ body mass of the invertebrates of the composite and all seasonal webs (Figure 6) showed non-normal distributions (Shapiro–Wilk: $p < 0.001$) (the dashed line was the expected log-normal distribution with the same mean and variance as the observed distribution) but there was no skew in either direction. There were no obvious gaps (except in February 2005) in the body size distributions, which suggested that representatives of

Table 3 Sorensen's index values corresponding to the similarities between the species (A) composition of the six seasonal food webs (B) interactions of the six seasonal food webs

	February 2005	April 2005	June 2005	August 2005	December 2005	February 2006	Composite 2005/2006
(A)							
February 2005	1	0.60	0.54	0.57	0.70	0.64	0.50
April 2005	–	1	0.75	0.63	0.64	0.50	0.78
June 2005	–	–	1	0.78	0.58	0.46	0.81
August 2005	–	–	–	1	0.62	0.52	0.75
December 2005	–	–	–	–	1	0.59	0.62
February 2005	–	–	–	–	–	1	0.44
Composite 2005/2006	–	–	–	–	–	–	1
(B)							
February 2005	1	0.46	0.44	0.52	0.58	0.65	0.39
April 2005	–	1	0.73	0.56	0.65	0.37	0.81
June 2005	–	–	1	0.77	0.57	0.35	0.81
August 2005	–	–	–	1	0.63	0.43	0.70
December 2005	–	–	–	–	1	0.47	0.63
February 2006	–	–	–	–	–	1	0.31
Composite 2005/2006	–	–	–	–	–	–	1

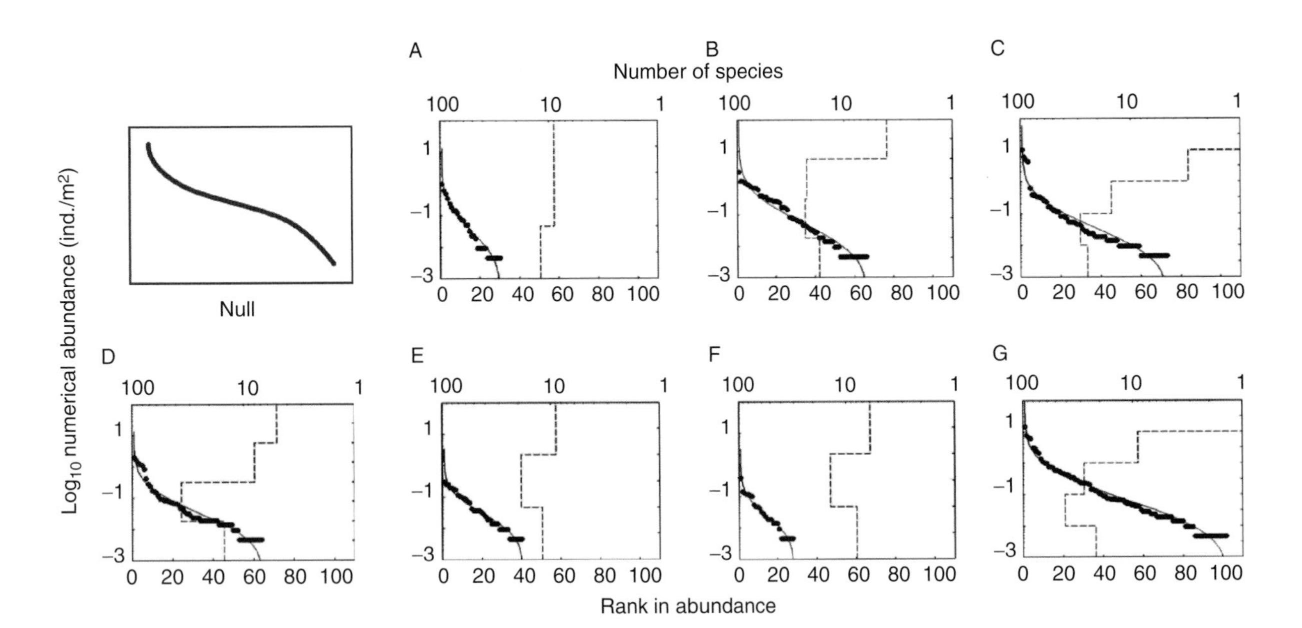

Figure 4 Abundance versus the rank in numerical abundance (ind./m^2) for all six seasonal webs and the composite food web: (A) February 2005, (B) April 2005, (C) June 2005, (D) August 2005, (E) December 2005, (F) February 2006, and (G) the composite food web. The solid line is the expected rank–numerical abundance relationship assuming that abundance is log-normally distributed. (10,000 values were drawn from a normal distribution with the same mean and variance as the observed distribution of log abundance.) The dashed line is the (log scale) frequency distribution of the number of species (top horizontal axis) by (log scale) numerical abundance.

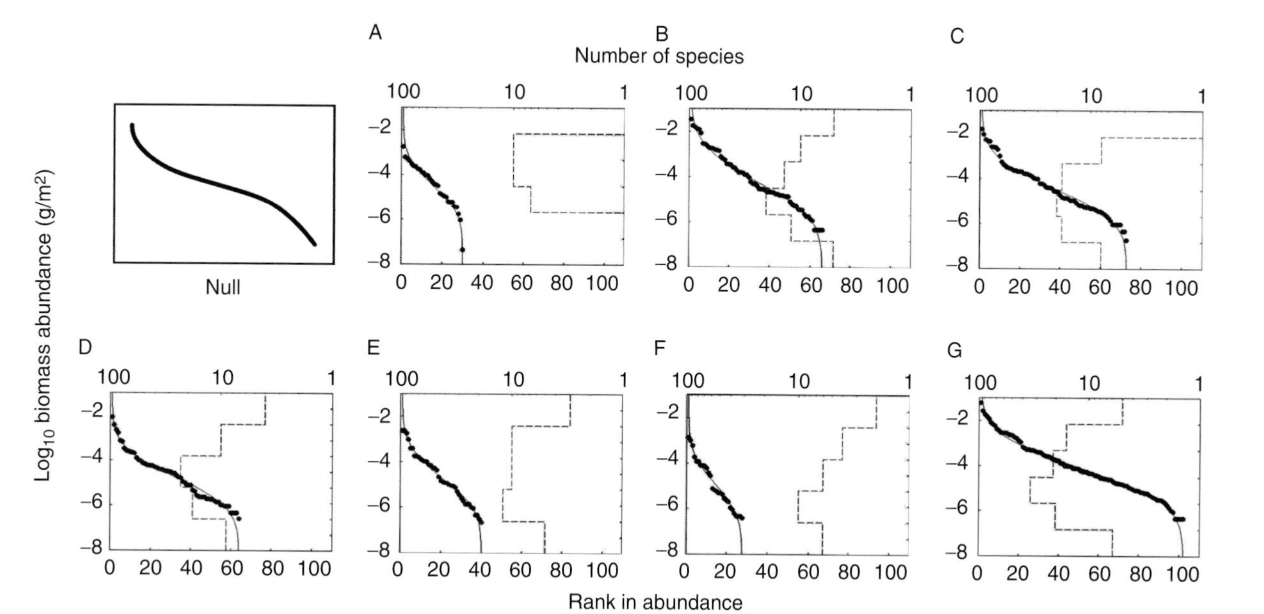

Figure 5 Abundance versus the rank in biomass abundance (g/m^2) for all six seasonal webs and the composite food web. The solid line is the expected rank–biomass abundance relationship assuming that abundance is log-normally distributed. (10,000 values were drawn from a normal distribution with the same mean and variance as the observed distribution of log abundance.) The dashed line is the (log scale) frequency distribution of the number of species (top horizontal axis) by (log scale) biomass abundance.

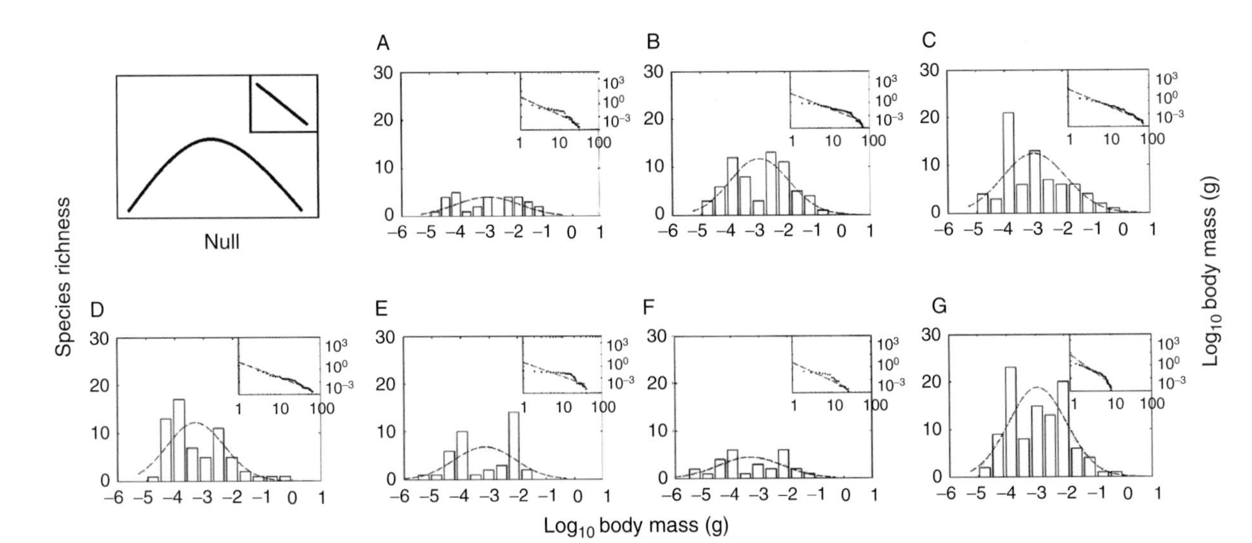

Figure 6 The frequency of species richness (number of species) versus $\log_{10}$ body mass (g). The dashed line is the log-normal distribution with the same mean and variance as the observed distribution. Sub-plot: Body mass (g) versus the rank in body mass (g). The dashed lines are the ordinary least squares regression lines, using all species.

all body sizes of invertebrates were present and sampled. The inset in each subplot of Figure 6 shows the body mass data plotted by rank. For June 2005 and August 2005, the body size–rank relationship was approximately linear, whereas there were non-linearities in the relationship for the other seasonal and composite webs. These latter webs showed a decrease in the size of species at higher ranks on these plots. Where the body size–rank relationship was close to linear, this suggested a non-log-normal distribution of body sizes. If body mass is allometrically related to rank ($BM \propto rank\ \alpha$), as predicted under our alternative hypothesis, then the frequency distribution of body mass should be more log-hyperbolic than log-normal (Jonsson *et al.*, 2005). Here, we show the rank abundances to be non-log-normal in their distributions.

B. Bivariate Patterns

There was no significant linear regression relationship between body mass (g) and MNW for the composite Gearagh food web (Figure 7 and Table 4A). There was a significant linear regression relationship between body mass (g) and MNR for the composite web (Figure 8 and Table 4B).

In many systems, predator body mass tends to be greater than prey body mass and so predator body size increases as trophic height increases (Figure 9, null), and in all the Gearagh webs except February 2006, significant positive relationships between body size and trophic height were evident (Figure 9 and Table 5).

As the body mass of an organism increases, its numerical abundance is expected to decrease (null hypothesis, Figure 10, null), and there was a significant negative linear relationship between $\log_{10}$ consumer body mass and $\log_{10}$ numerical abundance for the composite Gearagh food web (Figure 10G and Table 6). The slope of the regression line for the composite food web (-0.18) was significantly different from slopes of -0.75 and -1 ($p<0.001$ and $p<0.001$, respectively) but not from -0.25 ($p=0.18$), similar to the values reported for other local food webs. The slopes for the regression line for all bimonthly webs were significantly different from -1 (Figure 10A–G: $p<0.001$), and all but December 2005 ($p=0.14$) were significantly different from -0.75 (Figure 10A–G, excluding E: $p<0.001$). The slopes of the seasonal regression equations were significantly different from -0.25 ($p<0.001$) for June 2005 and December 2005, but not for the other seasonal webs.

Similarly, as trophic height increases, $\log_{10}$ predator numerical abundance should decrease (null hypothesis, Figure 11, null), and there was a significant negative linear relationship for the composite Gearagh food web, as well as all seasonal webs (Figure 11A–F and Table 7).

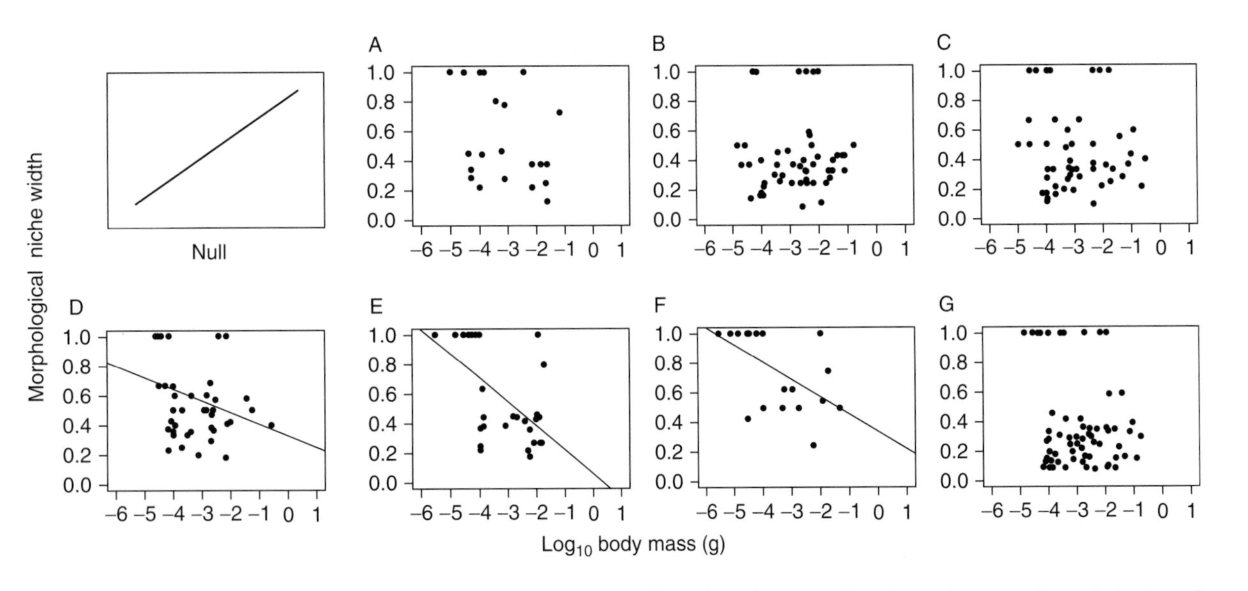

Figure 7 $\log_{10}$ consumer body mass (g) versus morphological niche width for all seasonal webs and composite web (web codes as in Figure 4). Regression equations are included for relationships with significant linear relationships.

Table 4 (A) Morphological niche width; (B) Morphological niche range and consumer body mass (g)

Web	*a*	*b*	df	r^2	*p*
(A)					
February 2005	0.35	−0.06	21	0.06	0.25
April 2005	0.48	0.02	53	0.01	0.53
June 2005	0.42	−0.01	54	0.001	0.75
August 2005	0.33	−0.08	45	0.09	<0.05
December 2005	0.05	−0.16	28	0.34	<0.001
February 2006	0.34	−0.12	17	0.30	<0.05
Composite web	0.27	−0.05	75	0.02	0.19
(B)					
February 2005	−5.03	−4.26	4	0.62	0.06
April 2005	−28.77	−17.70	24	0.25	<0.01
June 2005	−119.99	−66.28	28	0.38	<0.001
August 2005	−17.79	−8.33	22	0.11	0.12
December 2005	−87.75	−39.64	9	0.43	<0.05
February 2006	−363.51	−118.00	4	0.43	0.16
Composite web	−115.53	−55.17	42	0.36	<0.001

df is the degrees of freedom, r^2 is the coefficient of determination, p is the significance level (null hypothesis: no linear relationship). In the regression equation $Y=a+bX$, a is the intercept and b is the slope.

In all webs, the consumer was larger (typically three orders of magnitude) than the resource in a majority of the links and, for all but the two February webs, the proportion of data points located in the upper left triangle was significantly different from what would be expected if links were distributed at random (see Figure 12). A significant fraction of links described predators feeding on larger prey for the composite web: 33% of the links were of this type and between 22% and 38% for the bimonthly webs. In all webs, all parasitoid–host links were located in the lower right triangle of the plot, but these accounted for only 3% of total data points in the composite web (1–8% in the seasonal webs). Harvestmen and small carabid beetles dominated as predators in the remainder of species interactions (30%) in which predators fed on prey larger than themselves. For the composite Gearagh food web, there was a significant positive correlation between predator body mass and prey body mass (Figure 12G: $r=0.15$, $p<0.001$). For the six seasonal food webs, only April 2005 (Figure 12B) and August 2005 (Figure 12D) showed significant positive correlations between predator and prey body mass ($r=0.24$, $p<0.01$; $r=0.25$, $p<0.01$, respectively) (Table 8).

In all the food webs, the proportion of data points located in the upper left triangle of the plots was significantly different from a random distribution of trophic links (with respect to numerical abundance) (see Figure 13A–G). For the Gearagh, there was no significant positive correlation between predator

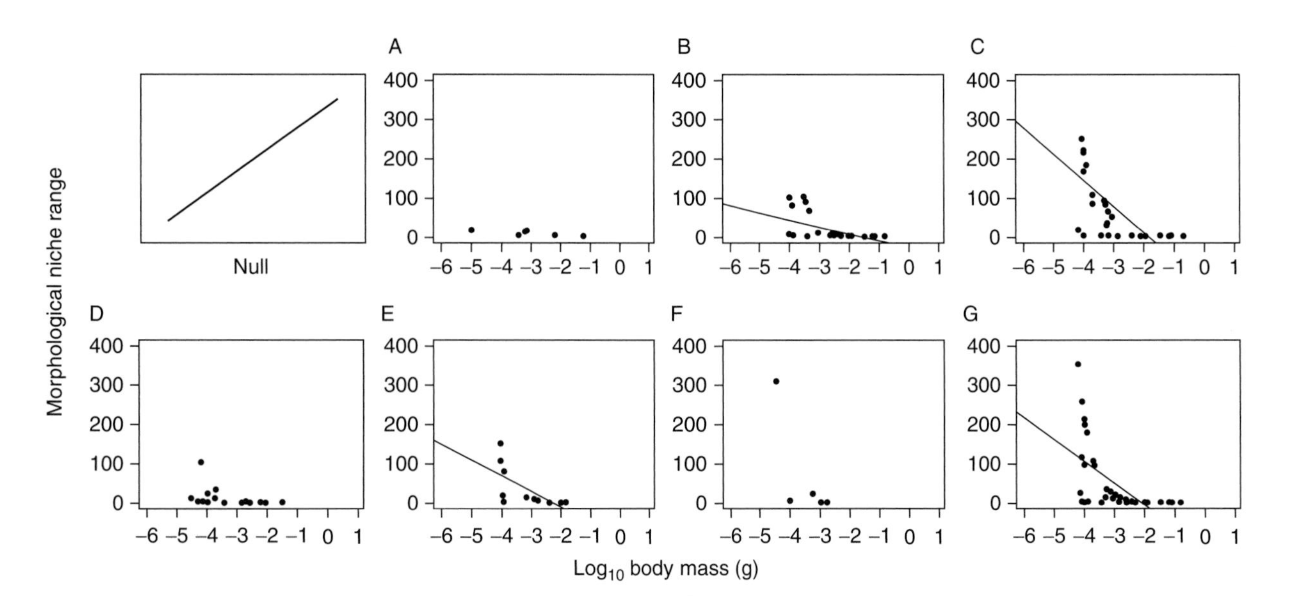

Figure 8 $\log_{10}$ consumer body mass (g) versus morphological niche range for the seven seasonal food webs (web codes as in Figure 4). Regression equations are included for relationships with significant linear relationships.

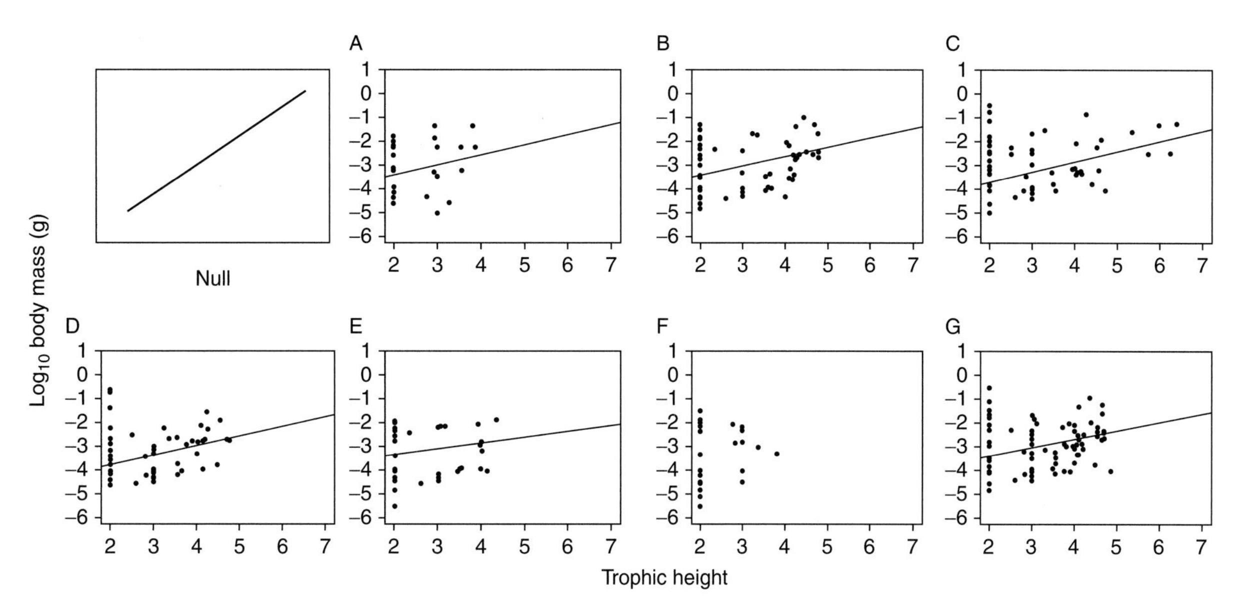

Figure 9 Consumer trophic height versus $\log_{10}$ consumer body mass (g) for the seven seasonal food webs (web codes as in Figure 4). Regression equations are included for relationships with significant linear relationships.

Table 5 Linear regression analysis for consumer trophic height and $\log_{10}$ consumer body mass (g)

Web	a	b	df	r^2	p
February 2005	−4.11	0.43	90	0.06	<0.05
April 2005	−4.12	0.41	257	0.15	<0.001
June 2005	−4.49	0.44	257	0.25	<0.001
August 2005	−4.49	0.41	205	0.15	<0.001
December 2005	−3.79	0.26	172	0.04	<0.01
February 2006	−4.55	0.43	68	0.05	0.06
Composite web	−3.99	0.36	381	0.11	<0.01

df is the degrees of freedom, r^2 is the coefficient of determination, p is the significance level (null hypothesis: no linear relationship). In the regression equation $Y=a+bX$, a is the intercept and b is the slope.

and prey numerical abundance in neither the composite food web ($r=0.003$, $p=0.94$) nor in the six seasonal food webs (Table 9).

The freshwater community of Tuesday Lake predator biomass abundance (g/m^2) increased as prey biomass abundance increased (Jonsson *et al.*, 2005). For the composite food web of April 2005, June 2005 and August 2005, the proportion of the data points located in the lower triangle of the plots was also significantly different from a random distribution of trophic links (see Figure 14). Consequently, there was a significant positive correlation between predator and prey biomass abundance (Table 10) for the composite web of the Gearagh and two of the six seasonal webs April 2005 and August 2005.

There were no significant correlations between predator body size and predator generality or vulnerability for any of the seasonal webs or the composite food webs (Tables 11 and 12, respectively).

C. Trivariate Patterns

There were clear trivariate relationships between body size, numerical abundance, biomass abundance and trophic height for the composite and seasonal webs (Figures 15 and 17). Small, abundant invertebrates were typically found at lower trophic levels in the food web, with larger, less abundant invertebrates found higher in the food web (with the exception of parasitoids). Trophic height explained more of the variability in numerical abundance (23%) than did body mass (0.11%) in the composite web.

Body mass, biomass abundance and numerical abundance varied over 5, 5 and 3 orders of magnitude, respectively, which is much lower than both Tuesday Lake (Jonsson *et al.*, 2005) and Broadstone Stream (Woodward *et al.*, 2005b). Figures 16 and 17 illustrate the trivariate patterns for the seasonal

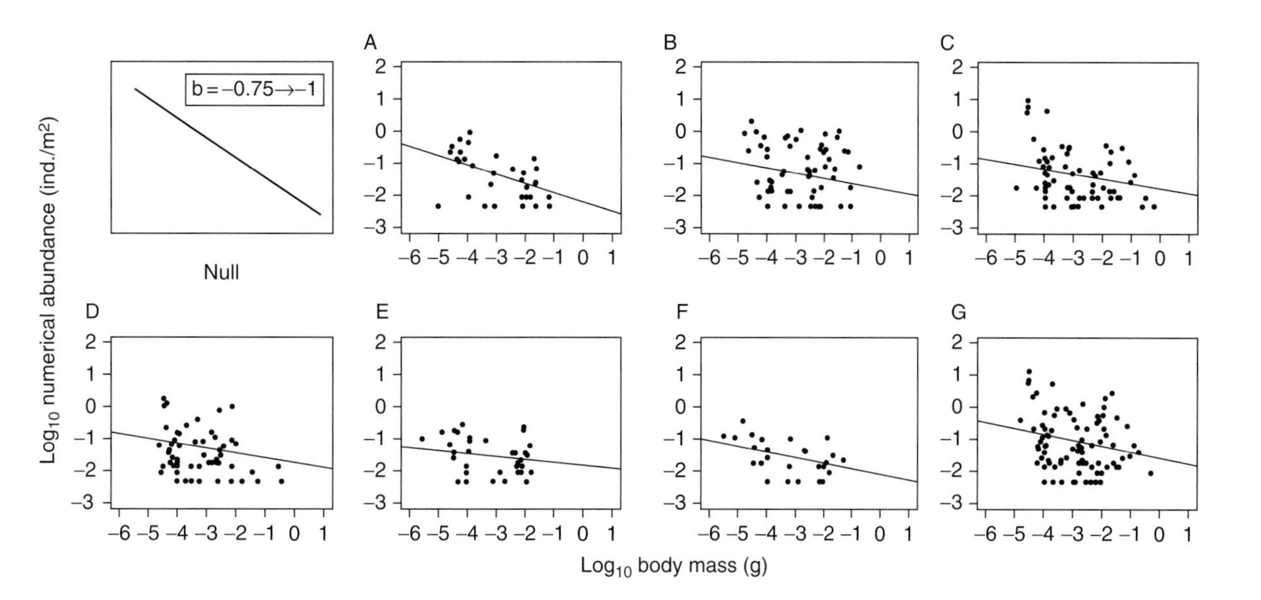

Figure 10 $\log_{10}$ consumer body mass (g) versus $\log_{10}$ consumer numerical abundance (ind./m^2) for the seven food webs (web codes as in Figure 4). Regression equations are included for relationships with significant linear relationships.

Table 6 Linear regressions for $\log_{10}$ consumer body mass (g) and $\log_{10}$ consumer numerical abundance (ind./m^2)

Web	*a*	*b*	df	r^2	*p*
February 2005	−2.20	−0.29	90	0.21	<0.001
April 2005	−1.77	−0.16	257	0.03	<0.01
June 2005	−1.78	−0.16	257	0.05	<0.001
August 2005	−1.77	−0.16	205	0.04	<0.01
December 2005	−1.81	−0.09	172	0.02	<0.05
February 2006	−2.07	−0.17	68	0.13	<0.01
Composite web	−1.55	−0.18	381	0.04	<0.001

df is the degrees of freedom, r^2 is the coefficient of determination, *p* is the significance level (null hypothesis: no linear relationship). In the regression equation $Y=a+bX$, *a* is the intercept and *b* is the slope.

webs with respect to numerical and biomass abundance, respectively, all of which follow the same patterns observed in the composite web above.

D. Temporal Changes in Biomass Abundances

In February 2005 (Figure 18A), where there were no predators on the 4th, 5th or 6th trophic levels, with the biomass of consumers on the 3rd trophic levels higher than that on the 2nd trophic level. In April 2005 (Figure 18B) a 4th trophic level developed with the seasonal emergence of the adult top predator, the carabid beetle, *Agonum muelleri*. There was a small decline in the biomass of the consumers in the 3rd trophic level, and in turn the 2nd trophic level, but to a lesser degree. With the seasonal emergence in June 2005 of two new top predators in the upper trophic levels, from the families Pompilidae and Sphecidae (Order Hymenoptera), there were decreases in the biomass abundances in the 2nd, 3rd and 4th trophic levels in June 2005, with the most noticeable decreases in the 4th and 2nd levels. In August 2005, although there was a recovery in predator biomass abundance in the 4th trophic level relative to the preceding sampling occasion, two top predators were lost (*A muelleri* and Family Pompilidae). There was an increase in the biomass abundance in both the 2nd and 3rd trophic levels relative to the preceding seasonal sampling occasion. There was a complete loss of all top invertebrate predators in December 2005 (5th and 6th trophic levels). The biomass abundance on the 4th trophic level increased in December 2005 with respect to that on August 2005, with a subsequent decrease in the 3rd trophic level, but a recovery of biomass abundance in the 2nd trophic level. In February 2006, all consumers on the 4th, 5th and 6th trophic level were lost. A further increase in the 3rd trophic level was seen, but a minor decrease in the 2nd.

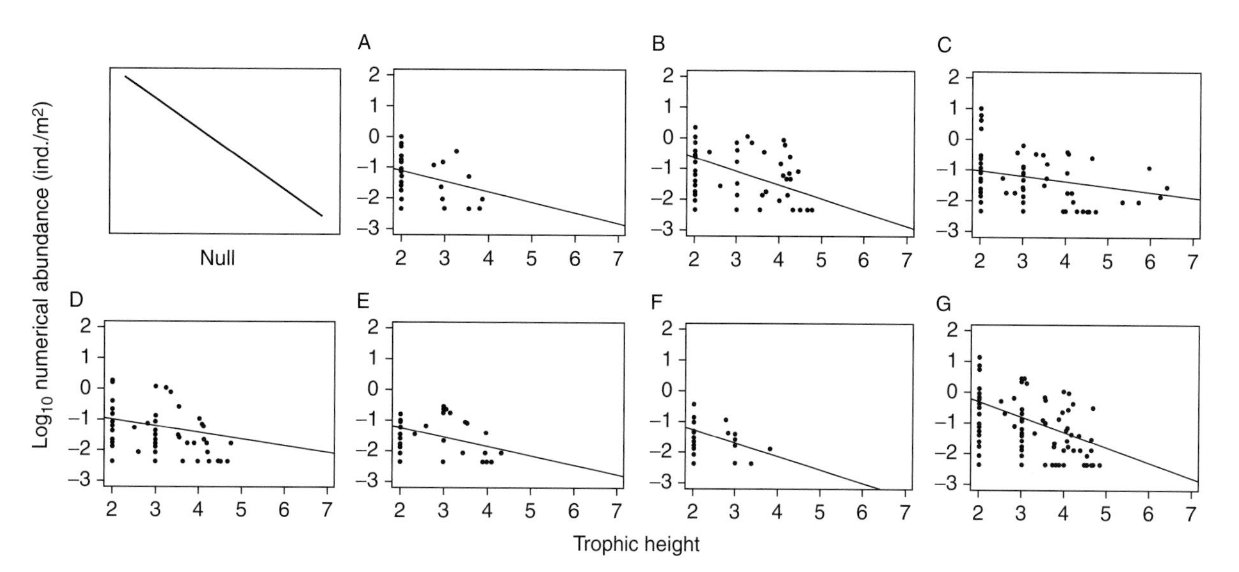

Figure 11 $\log_{10}$ consumer numerical abundance (ind./m^2) versus trophic height for the seven food webs (web codes as in Figure 4). Regression equations are included for relationships with significant linear relationships.

Table 7 Linear regressions for consumer trophic height and $\log_{10}$ consumer numerical abundance (ind./m^2)

Web	a	b	df	r^2	p
February 2005	0.45	−0.49	90	0.09	<0.001
April 2005	−0.25	−0.45	257	0.25	<0.001
June 2005	−0.70	−0.17	257	0.08	<0.001
August 2005	−0.61	−0.20	205	0.06	<0.001
December 2005	−0.58	−0.30	172	0.17	<0.001
February 2006	−0.42	−0.42	68	0.21	<0.001
Composite web	0.69	−0.49	381	0.23	<0.001

df is the degrees of freedom, r^2 is the coefficient of determination, p is the significance level (null hypothesis: no linear relationship). In the regression equation $Y=a+bX$, a is the intercept and b is the slope.

For clarity, changes in the biomass of plants and detritus (basal trophic levels) are shown separately to those within the consumer assemblage (Figure 19). Overall there was no significant variation in the biomass abundance of vegetation (one-way ANOVA, $F_{1,39}=0.305$, $p=0.58$) found in the Gearagh throughout the year. In contrast, the detrital biomass abundance did fluctuate over the year although no significant temporal pattern was evident ($F_{1,89}=0.448$; $p=0.51$).

In summary, no significant relationships were found between MNW/MNR and predator body mass. A positive linear relationship was evident for predator body mass versus trophic height. Negative linear relationships were observed for both predator body mass versus numerical abundance and for trophic height versus numerical abundance. The predator–prey biomass abundance was lower triangular in its relationship, whereas both predator–prey body mass and predator–prey numerical abundance exhibited upper triangularity.

IV. DISCUSSION

A. Food Web Patterns

In order to assess the generality of food web patterns reported in the literature, it is important to examine them in as wide a variety of habitats and seasons as possible. It has often been argued that there are fundamental differences between terrestrial and aquatic food webs (Chase, 2000; Nowlin *et al.*, 2008; Shurin *et al.*, 2006). However, many of the food webs that have formed the basis of such suggestions have often not been directly

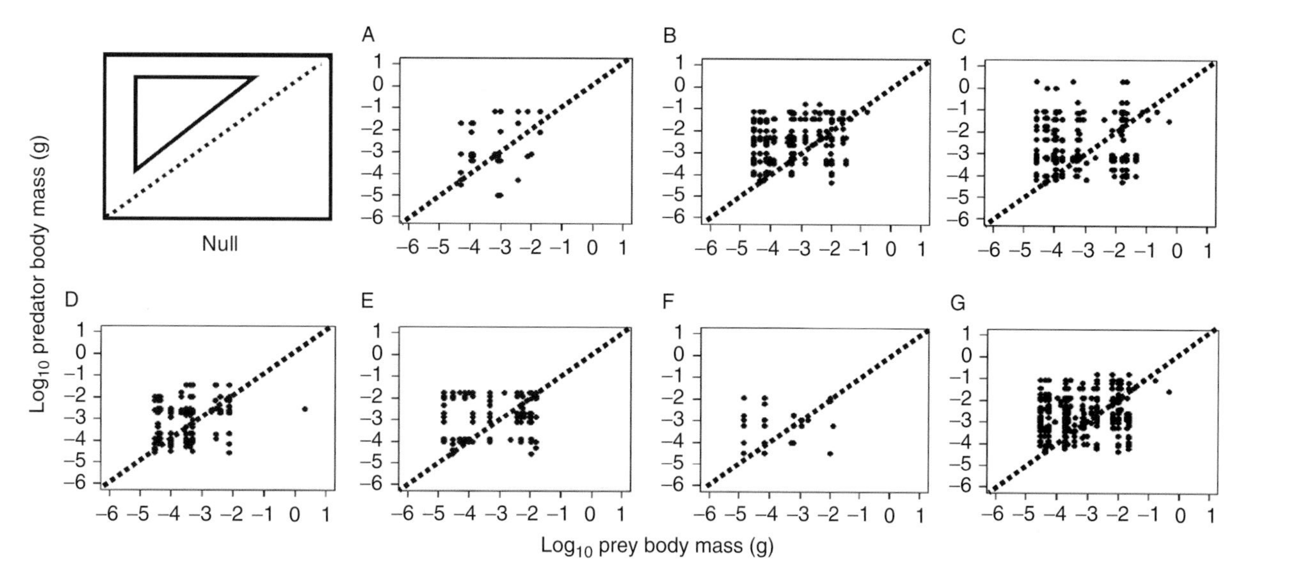

Figure 12 $\log_{10}$ predator body mass (g) versus $\log_{10}$ prey body mass (g) for the seven food webs (web codes as in Figure 4). April 2005, June 2005, August 2005, December 2005, and the composite web showed a significant proportion of data points were located in the upper left triangle (12B: prop=0.66, $p<0.001$; 12C: prop=0.74, $p<0.001$; 12D: prop=0.73, $p<0.001$; 12E: prop=0.69, $p<0.001$; 12G: prop=0.60, $p<0.001$).

Table 8 Pearson correlations for $\log_{10}$ predator body mass (g) and $\log_{10}$ prey body mass (g)

Web	t	df	r	p
February 2005	1.75	32	0.30	0.09
April 2005	3.32	175	0.24	<0.01
June 2005	1.33	183	0.10	0.18
August 2005	3.00	133	0.25	<0.01
December 2005	1.05	94	0.11	0.30
February 2006	0.48	24	0.10	0.63
Composite web	2.60	275	0.16	<0.01

t is the t-statistic, df is the degrees of freedom, r is the correlation coefficient, p is the significance level (null hypothesis: no correlation).

comparable, due to a lack of standardization across studies and systems (Olesen *et al.*, 2010; Riede *et al.*, 2010; Woodward *et al.*, 2010). Many of the best characterized food webs have been described in aquatic habitats, making such cross-habitat comparisons difficult to validate (Jonsson *et al.*, 2005; Woodward *et al.*, 2005b). Table 2 lists 17 well-resolved published food webs from both aquatic and terrestrial habitats, and includes both predator–prey and parasitoid–host webs. The Gearagh contained similar univariate patterns to the other terrestrial webs, but was less well connected than the aquatic webs. It is worth noting that the webs documented here are dominated by terrestrial invertebrates and, therefore, birds, mammals and amphibians as well as parasitoids are under-represented: their inclusion could alter the connectivity of the resulting food webs. Nonetheless, the low values of MNW in the webs reflected the low levels of connectivity in these webs relative to aquatic webs. In the Gearagh food webs, most consumers utilized less than 40% of prey species within the range of prey species body sizes available. One potential explanation for this observation is that the diets of the invertebrate consumers from the Gearagh were less well documented in the literature and, therefore, the possible complexity is not represented fully. In addition to the typical binary connectance webs most often encountered in the literature, the Gearagh webs also contained information on the body mass and abundance of all species which enabled us to make comparisons between with other similarly well-resolved food webs (see Table 2).

Body size has become an increasingly commonly measured trait that can be used to understand the patterning of trophic interactions within a community (Cohen *et al.*, 1993; Layman *et al.*, 2005; Leaper and Huxham, 2002; Warren and Lawton, 1987; Williams and Martinez, 2000; Woodward *et al.*, 2010). For instance, larger predators tend to occupy the higher trophic levels in the food web and they also often have broader diets than smaller species (Cohen *et al.*, 2003; Jennings and Mackinson, 2003; Woodward *et al.*, 2005a).

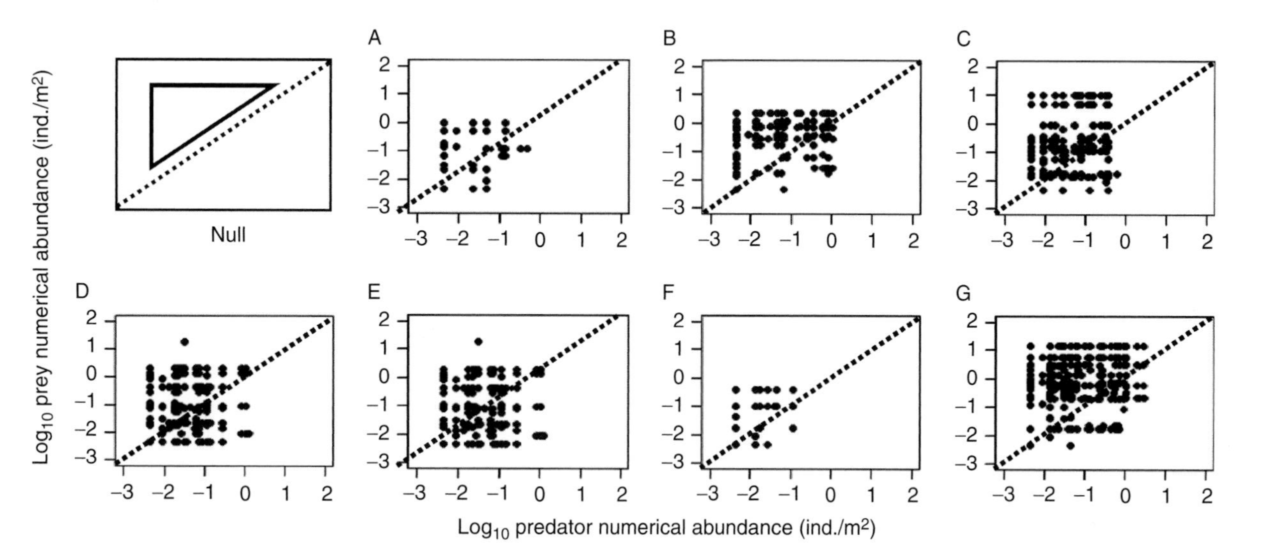

Figure 13 $\log_{10}$ predator numerical abundance (ind./m^2) versus $\log_{10}$ prey numerical abundance (ind./m^2) for the seven seasonal food webs (web codes as in Figure 4). All the food webs showed that a significant proportion of data points were located in the upper triangle of the plots (13A: prop=0.74, $p<0.001$; 13B: prop=0.86, $p<0.001$; 13C: prop=0.83, $p<0.001$; 13D: prop=0.70, $p<0.001$; 13E: prop=0.82, $p<0.001$; 13F: prop=0.81, $p<0.001$; 13G: prop=0.86, $p<0.001$).

Table 9 Pearson correlations for $\log_{10}$ predator numerical abundance (ind./m^2) and $\log_{10}$ prey numerical abundance (ind./m^2)

Web	t	df	r	p
February 2005	0.46	32	0.08	0.65
April 2005	0.73	175	0.06	0.47
June 2005	1.76	183	0.13	0.08
August 2005	0.45	133	0.04	0.66
December 2005	0.64	94	0.07	0.53
February 2006	0.77	24	0.16	0.45
Composite web	0.06	274	0.004	0.95

t is the t-statistic, df is the degrees of freedom, r is the correlation coefficient, p is the significance level (null hypothesis: no correlation).

The relationship between body mass and trophic level is, however, likely to vary among ecosystem types: for instance, aquatic communities are often strongly size-structured (Jennings and Mackinson, 2003; Jonsson *et al.*, 2005) and trophic levels often form a continuum with body mass (Jennings *et al.*, 2001, 2002; Reuman and Cohen, 2005; but see Layman *et al.*, 2005). Within terrestrial systems, however, the relationship between trophic levels and body mass often appears to be not so clear (Eggers and Jones, 2000; Ings *et al.*, 2009). The relationship between body mass and trophic height in the Gearagh had a positive slope that was significantly different from 0 for all the bimonthly and composite webs (excluding February 2006), suggesting that as trophic height increased, so too did the body mass of consumers. Importantly, this pattern held over a large part of the year, and it seems likely that the lack of this measure of size structure in February 2006 ($p = 0.06$) resulted from this month being less species rich (33 species) and, therefore, patterns of changing body size with trophic height were less pronounced.

The body mass–abundance relationship is possibly the most studied relationship in community ecology (Brown and Gillooly, 2003; Cohen *et al.*, 2003; Damuth, 1981; Enquist and Niklas, 2001; Jennings and Mackinson, 2003; Kerr and Dickie, 2001; White *et al.*, 2007). Globally, body mass often scales with numerical abundance with an exponent -0.75 (White *et al.*, 2007). This scaling exponent has been revealed when size–density relationships are examined that span 10 orders of magnitude (White *et al.*, 2007), but when abundance–body mass relationships are examined at a local scale (5–7 orders of magnitude), the relationship is sometimes claimed to be polygonal in shape and shallower, with a much lower exponent (average -0.25 in the studies reviewed by Blackburn and Gaston, 1997). Other studies, such as the rocky intertidal community investigated by Marquet *et al.* (1990), the aquatic communities reported by Cyr *et al.* (1997) and Jonsson *et al.* (2005) and

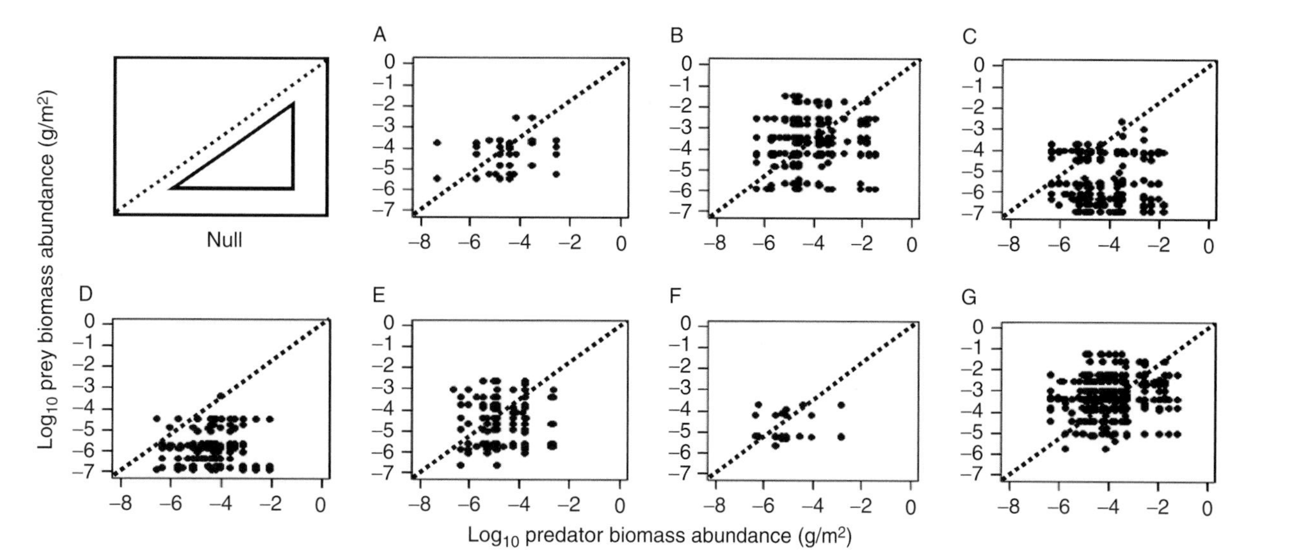

Figure 14 $\log_{10}$ predator biomass abundance (g/m^2) versus $\log_{10}$ prey biomass abundance (g/m^2) for the seven food webs (web codes as in Figure 4). April 2005, June 2005, August 2005, and the composite food web showed a significant proportion of the data points were located in the lower triangle of the plots (14C: prop=0.69, $p<0.001$; 14D: prop=0.58, $p<0.001$; 14E: prop=0.63, $p<0.001$; 14G: prop=0.58, $p<0.001$).

Table 10 Pearson correlations for $\log_{10}$ predator biomass abundance (g/m^2) and $\log_{10}$ prey biomass abundance (g/m^2)

Web	*t*	df	*r*	*p*
February 2005	1.23	32	0.21	0.22
April 2005	1.26	175	0.10	0.21
June 2005	−0.36	183	−0.03	0.72
August 2005	0.84	133	0.07	0.40
December 2005	0.28	93	0.03	0.77
February 2006	0.56	24	0.11	0.58
Composite web	1.56	275	0.09	0.12

t is the *t*-statistic, df is the degrees of freedom, *r* is the correlation coefficient, *p* is the significance level (null hypothesis: no correlation).

Table 11 Linear regression analysis for $\log_{10}$ predator body mass (g) and predator generality

Web	*t*	df	*r*	*p*
February 2005	−0.31	28	−0.06	0.76
April 2005	1.55	64	0.19	0.13
June 2005	0.54	71	0.06	0.60
August 2005	1.42	62	0.18	0.16
December 2005	0.36	38	0.06	0.72
February 2006	−1.57	26	−0.29	0.12
Composite web	1.95	100	0.19	0.05

t is the *t*-statistic, df is the degrees of freedom, *r* is the correlation coefficient, *p* is the significance level (null hypothesis: no correlation).

Table 12 Linear regression analysis for $\log_{10}$ predator body mass (g) and predator vulnerability

Web	*t*	df	*r*	*p*
February 2005	−0.06	28	−0.01	0.96
April 2005	−0.77	64	−0.10	0.45
June 2005	0.85	71	0.10	0.40
August 2005	−0.33	62	−0.04	0.74
December 2005	0.07	38	0.01	0.94
February 2006	−0.26	26	−0.05	0.80
Composite web	−0.26	100	−0.03	0.79

t is the *t*-statistic, df is the degrees of freedom, *r* is the correlation coefficient, *p* is the significance level (null hypothesis: no correlation).

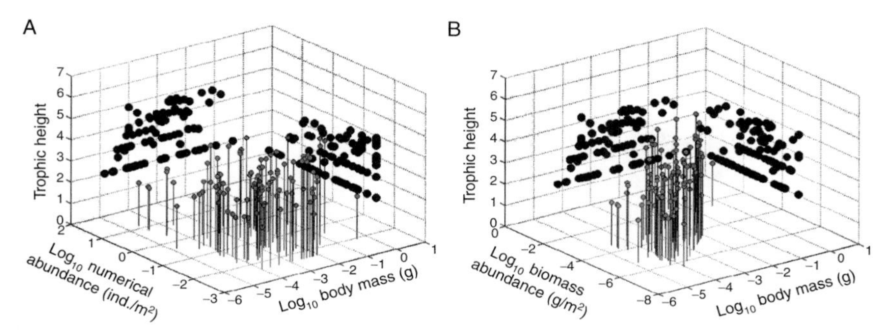

Figure 15 (A) Trivariate patterns for the Gearagh, comprising $\log_{10}$ predator body mass (g), trophic height, and $\log_{10}$ numerical abundance (g/m^2). (B) Trivariate patterns for the Gearagh, comprising of $\log_{10}$ predator body mass (g), trophic height and $\log_{10}$ biomass abundance (g/m^2).

Reuman *et al.* (2009) contradict the −0.25 scaling of local communities. For instance, in Tuesday Lake the slope was not significantly different from −0.75 across all species (from phytoplankton to fish) but was significantly shallower within phytoplankton and zooplankton, whilst Reuman *et al.* (2009) recently reported more negative scaling than −0.75 in a set of 166 food webs, including both terrestrial and aquatic systems. The abundance–body mass relationship for the composite food web of the Gearagh (Figure 10G) had an exponent of −0.18, which is significantly lower than −0.75, but not significantly different from −0.25. The absence of −0.75 scaling may be due to the limited range of body masses quantified in the present study. The body mass of the invertebrates quantified here spanned five orders of magnitude, much less than has been examined in other food webs (Jonsson *et al.*, 2005; Woodward *et al.*, 2005b). All bimonthly food webs showed significant body mass–abundance relationships, but again the exponents were much closer to −0.25 than −0.75 (see Table 6). This result differs from Tuesday Lake, even though the relationship was also examined at a local scale (Jonsson *et al.*, 2005). This steeper relationship may be due to the presence of fish in the community, which increases the range of body sizes (Tuesday Lake spanned 12 orders of magnitude), whereas the Gearagh web did not contain any large vertebrates, for example birds, and body masses spanned only five orders of magnitude. When the abundance–body mass relationship is examined at the global scale, body mass can explain approximately 80% or more of the variability in abundance; when this relationship is investigated at the local scale, body mass typically has a greatly reduced explanatory power (Layer *et al.*, 2010a,b; White *et al.*, 2007). A possible explanation for the reduced slopes at local scales might simply be that shallow scaling between body mass and abundance is obtained when only a

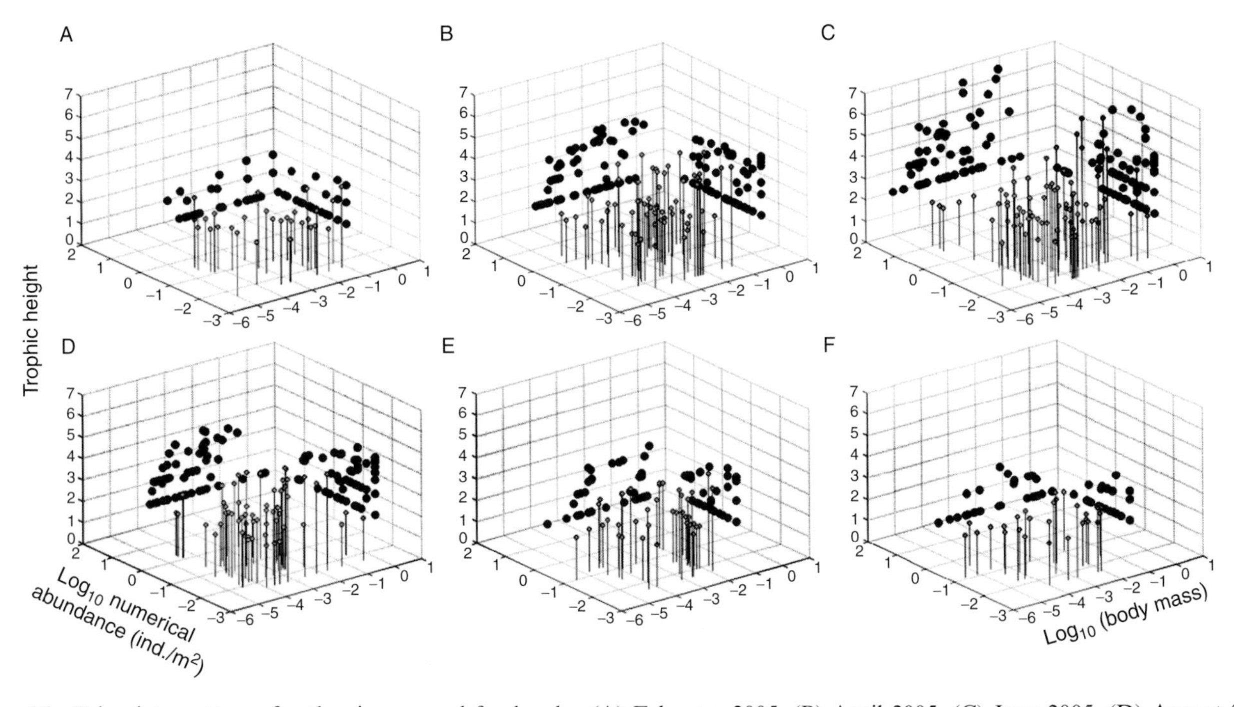

Figure 16 Trivariate patterns for the six seasonal food webs: (A) February 2005, (B) April 2005, (C) June 2005, (D) August 2005, (E) December 2005, (F) February 2006 showing $\log_{10}$ body mass (g), trophic height, $\log_{10}$ numerical abundance (ind./m^2).

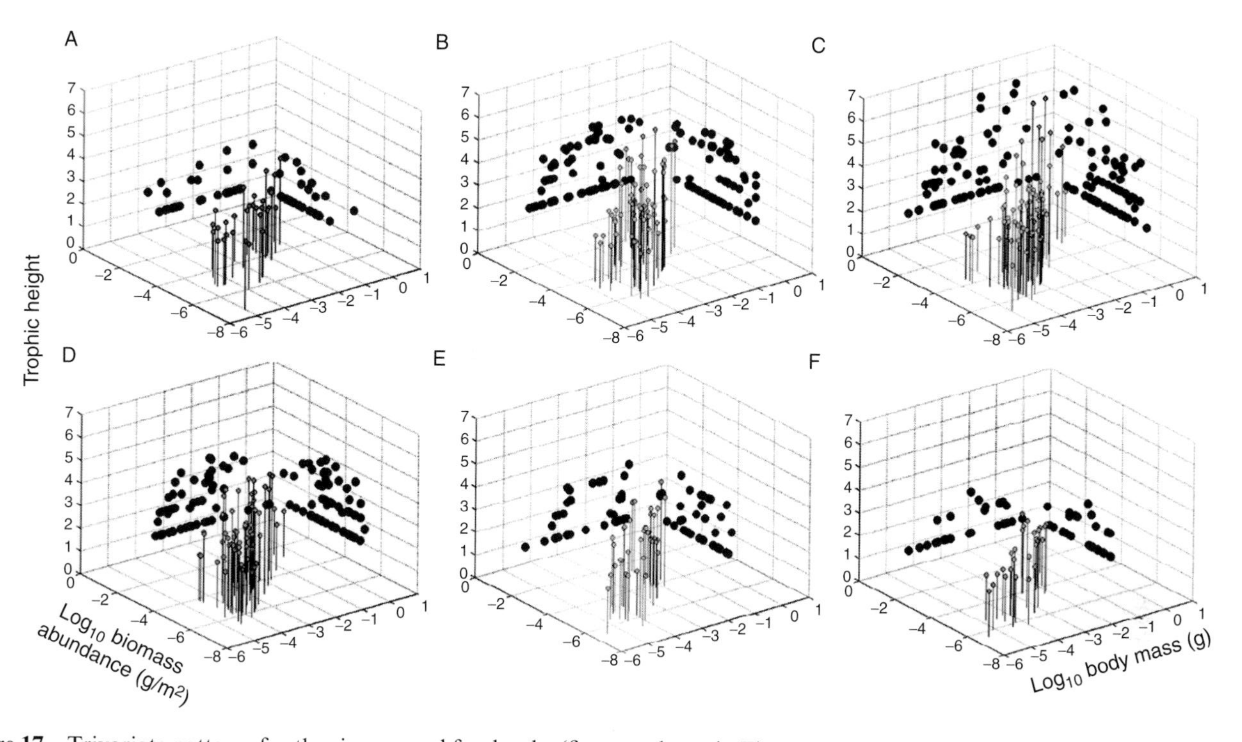

Figure 17 Trivariate patterns for the six seasonal food webs (figure codes as in Figure 16) showing $\log_{10}$ body mass (g), trophic height, $\log_{10}$ biomass abundance (g/m^2).

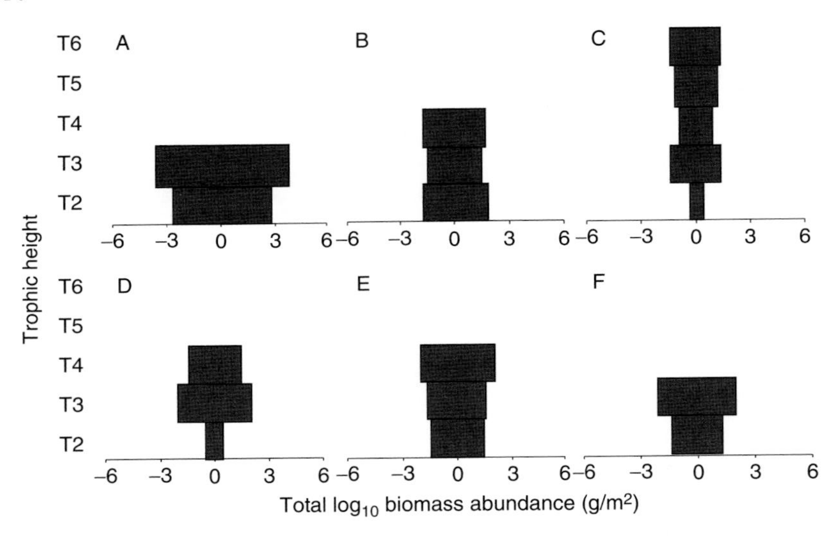

Figure 18 $\log_{10}$ biomass abundance pyramids (figure codes as in Figure 16) for bi-monthly samples (mean ± 1SE).

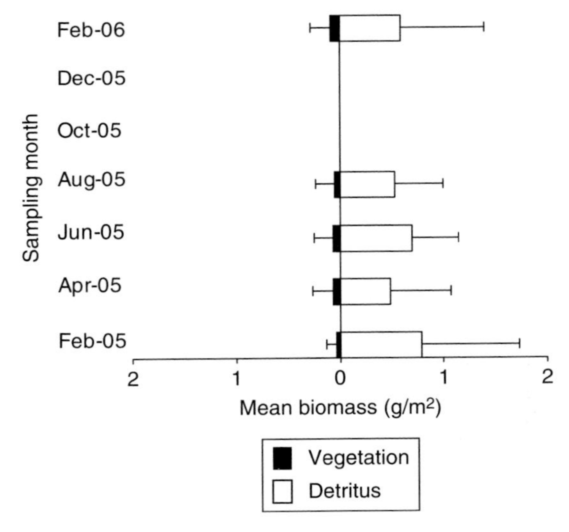

Figure 19 $\log_{10}$ biomass abundance of vegetation and detritus for bi-monthly samples (mean ± 1SE).

limited range of body masses are included in such relationships (e.g. five orders of magnitude). For global datasets, where species body masses and abundances have been synthesized from the literature, up to 14 orders of magnitude can be included in these relationships, resulting in steeper scaling of body mass and abundance (White *et al.*, 2007). The ecological significance of the body mass–abundance slopes (range -0.29 to -0.09) obtained in this study relates to the use of energy by species in different trophic levels. It is well known that metabolism E scales with body mass M to the power of 0.75 (Peters, 1983). If population density N scales with body mass to -0.75, then there is equivalent energy use by different species populations within the community (If $E = M^{0.75}$ and $N = -0.75$, then $EN = M^{0.75-0.75} \sim M^0$). If, however, the scaling of population density with body mass has a less negative slope, for example $b \sim -0.25$, then population level energy use EN scales positively with body mass, indicating that larger body-sized organisms require relatively more energy. A greater energy demand would equate with increased consumption rates and, therefore, increased interaction strengths, leading to a clear pattern of stronger interactions at higher trophic levels within food webs. Previous studies have shown that, for ground-dwelling predators such as beetles and spiders, metabolism and consumption are directly linked (Brose *et al.*, 2006a,b; Rall *et al.*, 2009), but how such patterns of interaction strength contribute to the persistence and stability of such food webs remains to be explored experimentally.

Excluding parasitoid–host relationships or situations when predators hunt in groups, predators are typically larger than their prey (Brose *et al.*, 2006a,b; Cohen *et al.*, 1993; Memmott *et al.*, 2000; Warren and Lawton, 1987; Williams and Martinez, 2000). Although predator and prey body mass may be correlated, the relationship between the two is not necessarily linear (Jonsson *et al.*, 2005). When the predators and their prey are ranked in increasing order of body size, the resulting feeding matrix often contains the majority of data points in the upper left triangle of the plot. For this study, the composite web and four of the bimonthly webs had a significant proportion of the data points in the upper left triangle of the plot (February 2005 and 2006 were non-significant: see Section III): that is 67% of the feeding links were between predators feeding on prey smaller than themselves (see also O'Gorman and Emmerson, 2010). This percentage is lower than the 80% previously reported by Cohen *et al.* (1993). One possible reason for the difference is the highly omnivorous nature of some of the predators: 24% of the predators in the composite web fed on prey that were (on average) both larger and smaller than themselves. Studies supporting this observation include the Coachella (Polis and Holt, 1992) and Broadstone Stream (Woodward *et al.*, 2005a) food webs, where both showed high levels of omnivory in their taxa (83% and 89%, respectively) (Petchey *et al.*, 2008). Both webs also had a much lower percentage of feeding links between large predators and prey than reported by Cohen *et al.* (1993) (4% and 0%, respectively), although it has been suggested that some of this discrepancy is due to underestimation of the predator–prey mass ratio due to artefacts resulting from species-averaging and that predators may only be feeding on a (smaller) subset of the prey population (Ings *et al.*, 2009; Woodward and Warren, 2007). Here, this pattern was seen primarily among the carabid beetles and harvestmen (Order Opiliones): it would be instructive for further studies to determine whether this is a particular form of feeding strategy and/or whether a bias might be present due to methodological artefacts (e.g. predators exploiting smaller sizes of prey than the average of those in the environment).

As prey numerical abundance increases, predator numerical abundance should increase (Cohen *et al.*, 2003). As with predator–prey body mass, this relationship is not expected to be linear, but should result in the data points being located in the upper left triangle of the sorted predation matrix. Although there was no significant correlation between predator–prey numerical abundance for the composite food web, a significant proportion of these same data points were found in the upper triangle. All six bimonthly food webs also showed statistically significant proportions of the data points in the upper left triangle of the plot (see Table 9). Overall, predators were numerically less abundant than their corresponding prey (86% of the data points were in the upper triangle). In cases where data points were located in the lower triangle, these links typically involved Lepidoptera larvae (butterflies and moths) and Hemiptera (true bugs). As the islands were sampled using pitfall traps, it is likely that true bugs and larvae of certain species, for example Lepidoptera, were undersampled, reflecting the way in which these taxa are more intimately associated with vegetation, either as a food source or habitat. Such differences may therefore reflect resource partitioning by these different taxa in this system (Lewinsohn *et al.*, 2005).

For the composite food web, there was a significant positive correlation between predator biomass abundance and prey biomass abundance, which corresponds with the pattern seen in Tuesday Lake (Cohen *et al.*, 2003; Jonsson *et al.*, 2005). When the predator species are ranked by biomass in the present study, a high proportion of the data points is located in the lower triangle of the predation matrix, indicating that predator biomass abundance must be greater than its corresponding prey biomass abundance. The lower right triangularity was significant in the composite Gearagh food web in April 2005 and August 2005. As with the predator–prey body mass relationship, there were data points that were located in the upper triangle. These data points were comprised of the parasitoid–host relationships, and the predators that fed on both large and small prey items. These patterns are suggestive of a high level of productivity and high turnover of biomass at lower trophic levels, which is consumed rapidly by a relatively constant biomass of predators. A snapshot of such a system could reveal low biomass at low trophic levels and high biomass of consumers: such inverted patterns of biomass have been documented extensively in aquatic systems characterized by trophic cascades, where predatory fish can have standing stocks that are large compared to their invertebrate prey (Casini *et al.*, 2008).

As a predator grows, often a greater range of prey body sizes becomes available and, hence, a predator is more likely to consume any particular prey item. Based on the same line of reasoning, as prey size decreases, the possibility of being consumed increases (e.g. Woodward and Hildrew, 2002a). This does not seem to be the case in the Gearagh, however, and there were no significant relationships between predator generality or vulnerability and body size for the composite web or any seasonal webs. However, a predator will not necessarily consume all the prey species within the potentially available size range, and Warren (1996) enumerated the MNW to describe this phenomenon. For the Gearagh, both MNW and MNR tended to decrease with increasing consumer body size, rather than the suggested increase (Warren, 1996), indicating that larger body-sized organisms at higher trophic levels feed over a relatively small range of body sizes and consumed prey that were not so dissimilar in size to themselves, that is the predator--prey body mass ratios (M_j/M_i) would be small. Such a pattern might have consequences for the arrangement of interaction strengths within these food webs, as previous studies have shown that interaction strengths scale as a function of body mass ratios (Emmerson and Raffaelli, 2004). Again, the implications for such patterns need to be explored further, using experiments and theoretical simulations.

Although species-level tritrophic interactions have previously been investigated, primarily within fisheries and biological control of pest species (Bascompte *et al.*, 2005; Huffaker and Messenger, 1976), these types of interactions have rarely been examined at a food web level (Cohen *et al.*, 2009). Trivariate patterns have, as far as we are currently aware, been described for 25 empirical food webs to date: Tuesday Lake (Cohen *et al.*, 2003; Jonsson *et al.*, 2005), Ythan Estuary (Cohen *et al.*, 2009; Woodward *et al.*, 2005a), Lochnagar (Layer *et al.*, 2010a), Broadstone Stream (Woodward *et al.*, 2005b), Lough Hyne (O'Gorman and Emmerson, 2010) and 20 stream food webs (Layer *et al.*, 2010b). The trivariate approach provides a useful technique for illustrating how various ecological patterns can be integrated to provide new information about a community. The body size range that the Gearagh web spans (five orders of magnitude) is much lower than that for Tuesday Lake and Broadstone Stream. Despite the smaller size range, the trivariate relationships seen in the Gearagh are remarkably similar to that of the aquatic webs. Different bivariate relationships are often examined depending on which system is under study; body mass spectra are typically studied in aquatic systems (Jennings *et al.*, 2001; Kerr and Dickie, 2001), whereas rank–abundance and predator–prey mass relationships are often studied in terrestrial systems (Cohen *et al.*, 2003). Terrestrial systems are typically described as being non-size-structured (Cyr *et al.*, 1997; Pace *et al.*, 1999; Shurin *et al.*, 2006). Here, however, using the trivariate approach, the Gearagh (terrestrial ecosystem) displayed clear size-structuredness indicating that the trivariate pattern approach is capable of revealing emergent properties of food webs not previously documented with bivariate pattern analyses and that there might be more of a size structure within terrestrial systems than previously envisaged (Yvon-Durocher *et al.*, 2010).

B. Temporal Variability in Food Web Structure

Many food webs are constructed from aggregated structures based on repeated sampling over an extended period, whereas a few documented webs represent a snapshot of the whole community. Both these aggregates or snapshots tend to miss seasonal changes in both species composition and, as a result, changes in food web structure (Olesen *et al.*, 2010; Woodward and Hildrew, 2002b). The Gearagh food web described here is unique in that it presents, for the first time to our knowledge, seasonal snapshots of *MN* webs as well as an aggregated picture. This means that temporal variability in a large number of food web relationships can be analysed, and in addition the data provide an alternative method to determine how well the aggregated food web represents the seasonal patterns found. Table 3A and B shows the Sorensen's similarity index for species composition and species interactions. Both indices show that February 2005 and 2006 are the most dissimilar to the composite food web, with the summer months, in particular June 2005, being the most similar. From February 2005–2006, the bimonthly food webs become more similar to the composite food web. It is also evident that February 2005 and 2006 are not identical (with approximately 36% turnover in species in the space of one year: Sorensen's index = 0.64). This illustrates the importance of temporal turnover in species composition and food web structure, which would not have been detected if the bimonthly food webs had not been examined. Despite this turnover of species, both the bivariate and trivariate patterns for the Gearagh are remarkably similar to those seen for Broadstone Stream (Woodward *et al.*, 2005b), Ythan Estuary (Cohen *et al.*, 2009; Woodward *et al.*, 2005a) and Tuesday Lake (Jonsson *et al.*, 2005), suggesting a possible functional redundancy of the species that are lost during the year (Fonseca and Ganade, 2001).

Within the entire community, 19 species were found in all seasonal webs. This core community was comprised mainly of detritivores, for example springtails, mites, nematodes, oligochaetes, enchytraeids and gastropods. Within the composite food web, 52% of feeding links involved the core community species. The core community comprised 49% of taxa in February 2005 and 58% of all taxa in February 2006. February 2006 contained more predators (spiders) than February 2005, but fewer primary consumers (millipedes and slugs), which explains much of the dissimilarity. The majority of new species which emerged seasonally entered the food web on the 3rd trophic level (e.g. from February 2005 to April 2005, 12 and 23 new species enter the 2nd and 3rd trophic levels, respectively). The addition and loss of species, however, did not significantly alter the slopes in the relationship between numerical abundance and body mass over the year. One possibility for the lack of variation in the slopes is that the newly emerging species replace similar sized and/or functioning species which have been lost through either emigration or local extinction via mortality. Alternatively, the emerging species could enter into a completely vacant section of the body mass–numerical abundance relationship: for example the emerging species are higher trophic level, larger taxa, which are also rarer than the already present smaller species. The emergence of these new species would not affect the body mass–numerical abundance relationship.

C. Morphological Niche

Warren (1996) hypothesized that, as predator body mass increased, the range of prey masses consumed would also increase, as would the percentage of prey it can utilize within that range (see also similar ideas developed by Williams and Martinez, 2000). Warren (1996) termed this metric the morphological niche. Here, a related niche metric was developed: the MNR. Warren (1996) found that the average value of the MNW for the Skipwith pond food web was 0.5. Leaper and Huxham (2002) also examined the MNW using the Ythan Estuary food web, but included parasites. For six versions of the Ythan Estuary food web, they found MNW values to be lower than at Skipwith (0.28–0.45). The Gearagh food web, also including parasitoids, had a morphological niche value of 0.42 for the composite web, which was lower (but not significantly so) than Warren's (1996) and within the range detailed by Leaper and Huxham (2002). When the bimonthly webs were examined, the mean values for the MNW was at the higher end of the range detailed by Leaper and Huxham (2002) (0.41–0.75). The inclusion of certain carabid beetles, such as *Pterostichus melanarius* and *Nebria brevicollis*, played an important role in driving the increased value. Both beetles are voracious omnivores had MNW values of above 0.50. Leaper and Huxham (2002) examined the role of body size on MNW and found no significant relationship in any of their six webs. Using the Gearagh dataset presented here, there was no significant relationship for the composite web; however, significant negative relationships were found for August 2005, December 2005 and February 2006.

We have developed a new metric termed MNR. The MNW metric as defined by Warren (1996) describes how much of the predator's niche is 'filled' by measuring the fraction of prey within a consumer's niche range that is actually eaten by the consumer. With the MNR, we were interested in how wide this niche range was in terms of prey body sizes relative to consumer body size, and how completely it was occupied. Negative relationships between the $\log_{10}$ body mass of the consumers and MNW were found, in contrast to the MNW expectations of Warren (1996). There is a significant negative relationship between MNR and log consumer body mass in the composite web (Figure 7). The higher the MNR value, the wider the range of body sizes that the consumer will feed upon. This suggests that, in relation to their body size, the consumers higher in the web do not exploit as wide a range of prey (or prey sizes) as the smaller consumers (in particular small carabid beetles and parasitoid wasps). This confirmed the observations using the MNW metric, and further supports our suggestion that larger predators might not perceive (or be able to handle) very small prey and may feed preferentially on prey of similar body sizes to themselves. Brose *et al.* (2006a,b) showed that there are some invertebrates that have optimized both their morphology and physiology in order to handle prey of a similar and larger size to the predator, but not smaller. Brose *et al.* (2006a,b) suggested that prey that are much smaller than the predator might not be energetically worthwhile with respect to prey capture, handling time and consumption, as they may contain too little energy, for example filter feeders such as baleen whales. Woodward and Hildrew (2002c) in a study of the Broadstone stream showed that whilst the predator *C. boltonii* could prey on all other macro-invertebrates in the system, the vulnerability of these prey were highly variable in response to both prey and predator sizes. They further showed that there was a strong ontogenic shift in the diet of the dragonfly predator larger prey being added as the predator grew to maturity. In the Gearagh, the predators were in their adult stage, yet fed mainly on prey of similar body sizes suggesting that the morphological differences in the predators between ecosystem types could result in variable MNW and MNR.

D. Trophic Cascades and Energetics

The presence of trophic cascades across different ecosystem types has generated much debate in the literature (Borer *et al.*, 2005; Halaj and Wise, 2001; Pace *et al.*, 1999; Polis *et al.*, 2000; Shurin *et al.*, 2006; Strong, 1992), and considerable evidence exists to suggest that they are strongest in marine and freshwater systems and weakest in terrestrial systems (Borer *et al.*, 2005). One suggested reason for this is that terrestrial habitats are often much more complex than aquatic habitats (at least in the case of pelagic systems), and that species interactions are more diffuse. Despite these predictions, trophic cascades have been detected in some structurally complex terrestrial systems (Dyer and Letourneau, 1999; Elmhagen and Rushton, 2007; Snyder *et al.*, 2006). To assess the possibility of a cascading effect of the seasonal emergence of adult predators within the summer months, the patterning of biomass abundance for trophic levels 2–6 by season within the Gearagh ecosystem was examined (Figure 18). The seasonal emergence of adult top predators (carabid beetles and wasps) into the 4th, 5th and 6th trophic levels on average was accompanied by a decrease in the biomass abundance of the consumers on the adjacent lower trophic level. The alteration in the biomass abundance of consumers on the 3rd trophic level between February 2005 and April 2005 was most likely due to the emergence of predators in the 4th trophic level. The predation pressure of the 3rd level consumers was most likely reduced by the top 4th trophic level predators, which, although it appears to have alleviated some predation pressure on the 2nd trophic level, did not suppress it entirely. June 2005 showed the greatest species diversity, with six trophic levels. Here, the presence of the top consumers on the 5th and 6th trophic levels reduced the biomass abundance of the intermediate consumers on the 4th trophic level. This alleviation of predation pressure on the consumers could possibly have allowed greater consumption of the herbivores and detritivores, resulting in the greatly reduced biomass abundance on the 2nd trophic level, suggesting a possible trophic cascade of the predatory interactions through the community. Between August 2005 and December 2005, the consumers on the 4th trophic level increase their predation on the 3rd trophic level, which allows the biomass abundance of herbivores and detritivores on the 2nd trophic level to recover. In February 2006, with the loss of the 4th trophic level there was an increase in the biomass abundance in the intermediate consumers of the 3rd level, with only a slight decrease in the 2nd trophic level. Between February 2005 and 2006, there are distinct differences in the biomass abundances of both the 2nd and 3rd trophic levels. A possible explanation for the difference in the biomass abundances between the same months one year apart is the shift in species composition (36% species turnover) across the year. The differences in the species interactions could result in the alteration seen here in the biomass abundances. These empirical patterns and the suggestion of possible cascading interactions merit further study, via both *in silico* modelling approaches and experimental manipulations.

E. Food Web Construction

The construction of any natural food web is logistically challenging, and is often based on only a snapshot of the nodes and links that make up the trophic network. There are several different approaches taken in the construction of food webs, which are often determined largely by whether the food web is aquatic or terrestrial.

Gut content analysis is the most typically used method, especially in systems dominated by engulfing predators, such as many aquatic food webs (Arim *et al.*, 2010; Christian and Luczkovich, 1999; Layman *et al.*, 2005; O'Gorman and Emmerson, 2010; Woodward *et al.*, 2005b). This method is useful for determining feeding preferences of a specific organism (i.e. preference for a particular prey item will be dependent on the quantity of that prey item in the gut), which typically reflect the numerical or biomass abundance of organisms in the environment. Often consumers will eat many more organisms than they would typically feed upon via active discrimination and selection of food items (e.g. earthworms are typically known as detritivores but may have bacteria in their guts through passive consumption, Mülder *et al.*, 2006). For many species in terrestrial systems, gut content analysis is often logistically too challenging, or in some cases simply not possible. Insects can be both highly selective and general in their consumption of prey items, and it is easy to miss feeding links for many terrestrial invertebrates. It is also often quite rare for detectable remains of prey items to be found in predator guts (Memmott *et al.*, 2000), with many carabid beetle predators feeding on soft-bodied invertebrates such as springtails and slugs (Agusti *et al.*, 2003; Harper *et al.*, 2005). DNA barcoding techniques (Greenstone *et al.*, 2005; Jurado-Rivera *et al.*, 2009) or identifying gut contents of terrestrial predators could be a useful alternative in the future, but such techniques are not fully developed or widely available at the moment, and they too are also biassed since they can only detect species in the gut for which a suitable set of markers are available.

The food webs presented in this paper were constructed using an in-depth literature search (see Section II above for details). By examining the literature for specific feeding links, it was possible to utilize peer-reviewed research employing an extensive variety of techniques, including direct observations, gut content, stable isotope and molecular analyses. Although issues relating to the use of some of these techniques have been discussed, these techniques are the most commonly used to determine predator–prey feeding links and subsequently used to determine the links in the Gearagh food web. By using the literature information from a wide range of sources, no bias by the authors was placed on any one particular method; however, the different techniques used in other studies would not necessarily have been used equally.

V. CONCLUSION

Many studies have shown that a food web can be much more than just the feeding interactions among the species (Cohen *et al.*, 2009; Jonsson *et al.*, 2005; O'Gorman and Emmerson, 2010; Woodward *et al.*, 2005b). It is a flow chart for energy and nutrient cycling that, if combined with the distribution of body sizes and metabolic theory (Brown *et al.*, 2004; Peters, 1983), can help to explain the distribution of abundances within a community. This in turn may provide a key for predicting the consequences of perturbations, such as biodiversity change, on ecosystems. Further studies are essential for the advancement of food web ecology, through the incorporation of the distributions of body mass and abundance and the elemental composition and stable isotope ratios of species. This will allow us to describe food webs as not only wiring diagrams that connect species but also as useful maps of fluxes that, together with *MN* data, predict the pattern of matter and energy flow in ecosystems. Having achieved this, we will move closer to a more complete understanding of the functioning of ecological systems. More specifically, the patterns in the distribution of, and relationships amongst, species-level characteristics such as body mass as well as numerical and biomass abundance are partly a function of food web structure and constraints on species caused by predator–prey interactions. Establishing the expected relationships among body mass as well as numerical and biomass abundance and food web topology in a range of undisturbed as well as perturbed systems is a vital step towards obtaining a better understanding of the functional consequences of biodiversity loss. Here, the Gearagh adds to an increasing list of well-resolved trivariate food webs and the bivariate patterns observed within the food web, (e.g. body mass–abundance, predator–prey body mass, body mass–trophic height) show that, regardless of ecosystem type, these relationships appear to exhibit consistent patterns in both direction and slope.

ACKNOWLEDGEMENTS

The authors would like to thank Nora Buttimer, Mick Mackay and Jonathan Percy for their assistance in the field and Eoin O'Gorman for valuable discussion and comments. This study was supported by a research studentship from the Department of Zoology, Ecology and Plant Sciences, University College Cork and the Donegal County Council.

APPENDIX I. LIST OF DICHOTOMOUS KEYS USED IN THE IDENTIFICATION OF THE TERRESTRIAL INVERTEBRATES IN THE SEASONAL AND COMPOSITE FOOD WEBS

Dichotomous Keys

1. Askew, R. R. (1968) *Handbook for the Identification of British Insects: Hymenoptera, Chalcidoidea, Section (b)*. Royal Entomological Society, London.

2. Blower, J. G. (1985) *British Millipedes (Diplopoda) with Keys to Species. Synopses of the British Fauna (New Series). No 35*. The Linnaean Society of London, Backhuys Publishers, The Netherlands.
3. Bolton, B. & Collingwood, C.A. (1975) *Handbook for the Identification of British Insects: Hymenoptera, Formicidae*. Royal Entomological Society, London.
4. Cameron, R. A. D. (2003) *Land Snails in the British Isles*. Field Studies Council, Shrewsbury.
5. Cameron, R. A. D., Jackson, N. & and Eversham, B. (1983) *A Field Guide to the Slugs of the British Isles*. Backhuys Publishers, The Netherlands.
6. Day, M. C. (1988) *Handbook for the Identification of British Insects: Spider Wasps. Hymenoptera: Pompilidae*. Royal Entomological Society, London.
7. Dindal, D. L. (Ed.) (1990) *Soil Biology Guide*. John Wiley & Sons, New York.
8. Eady, R. D. & Quinlan, J. (1963) *Handbook for the Identification of British Insects: Hymenoptera, Cynipoidea. Key to Families and Subfamilies and Cynipinae (Including Galls)*. Royal Entomological Society, London.
9. Eason, E. H. (1964) *Centipedes of the British Isles*. Warne, London.
10. Fitton, M. G., Shaw, M. R. & Gauld, I. D. (1988) *Handbook for the Identification of British Insects: Pimpline Ichneumon-Flies. Hymenoptera, Ichneumonidae (Pimplinae)*. Royal Entomological Society, London.
11. Forsythe, T. G. (1987) *Common Ground Beetles*. Richmond Publishing Company Ltd., Slough.
12. Hodge, P. J. & Jones, R. A. (1995) *New British Beetles (Species not in Joy's Practical Handbook)*. British Entomological and Natural History Society, UK.
13. Hopkin, S. (1991) *A Key to the Woodlice of Britain and Ireland*. Backhuys Publishers, The Netherlands.
14. Hopkin, S. (2007) *A Key to the Springtails of Britain and Ireland*. Field Studies Council, Shrewsbury.
15. Jones-Walters, L.M. (1989) *Keys to the Families of British Spiders*. Field Studies Council, Shrewsbury.
16. Joy, N. H. (1932) *A Practical Handbook of British Beetles. Volumes 1 and 2*. H. F. & G. Witherby, London.
17. Krantz, G. W. (1978) *A Manual of Acarology*. Oregon State University Bookstores, Oregon.
18. Le Quesne, W. J. (1960) *Handbook for the Identification of British Insects: Hemiptera: Fulgoromorpha*. Royal Entomological Society, London.
19. Le Quesne, W. J. (1965) *Handbook for the Identification of British insects: Hemiptera: Cicadomorpha (Excluding Deltocephalinae and Typhlocybinae)*. Royal Entomological Society, London.

20. Le Quesne, W. E. & Payne, K. R. (1981) *Handbook for the Identification of British Insects: Cicadellidae (Typhlocybinae) with a Checklist of the British Auchenorhyncha (Hemiptera, Homoptera)*. Royal Entomological Society, London.
21. Lindroth, C. H. (1974) *Handbooks for the Identification of British Insects. Coleoptera Carabidae*. Royal Entomological Society, London.
22. Morris, M. G. (1990) *Handbook for the Identification of British Insects: Orthocerous weevils. Coleoptera: Curculionoidea (Nemonychidae, Anthribidae, Urodontidae, Attelabidae and Apionidae)*. Royal Entomological Society, London.
23. Morris, M. G. (1997) *Handbooks for the Identification of British Insects: Broad-nosed Weevils. Coleoptera: Curculionidae (Entiminae)*. Royal Entomological Society, London.
24. Morris, M. G. (2002) *Handbook for the Identification of British Insects: True Weevils (Part 1). Coleoptera: Curculionidae (Subfamilies Raymondionyminae to Smicronychinae)*. Royal Entomological Society, London.
25. Nixon, G. E. J. (1980) *Handbook for the Identification of British Insects: Hymenoptera, Proctotrupoidea, Diapriidae Subfamily Belytinae*. Royal Entomological Society, London.
26. Perkins, J. F. (1960) *Handbook for the Identification of British Insects: Hymenoptera, Ichneumonoidea. Ichneumonidae, Subfamilies Ichneumoninae II, Alomyinae, Agriotypinae and Lycorininae*. Royal Entomological Society, London.
27. Perkins, J. F. (1976) *Handbook for the Identification of British Insects: Hymenoptera, Bethyloidea (Excluding Chrysididae)*. Royal Entomological Society, London.
28. Quinlan, J. & Gauld, I. D. (1981) *Handbook for the Identification of British insects: Symphyta (Except Tenthredinidae), Hymenoptera*. Royal Entomological Society, London.
29. Richards, O. W. (1980) *Handbook for the Identification of British Insects: Scolioidea, Vespoidea and Sphecoidea. Hymenoptera, Aculeata*. Royal Entomological Society, London.
30. Roberts, M. J. (1996) *The Spiders of Great Britain and Ireland, Vol. 1*. Harley Books, UK.
31. Sankey, J. H. P. & Savory, T. H. (1974) *British Harvestmen (Arachnida: Opiliones): Key and Notes for the Identification of the Species*. Academic Press Ltd., New York.
32. Savage, A. A. (1989) *Adults of the British Aquatic Hemiptera Heteroptera*. Freshwater Biological Association, Ambleside.
33. Shaw, M. R. & Huddleston, T. (1991) *Handbook for the Identification of British Insects: Classification and Biology of Braconid Wasps. Hymenoptera, Braconidae*. Royal Entomological Society, London.

34. Sims, R. W. & Gerard, B. M. (1985) *Earthworms*. Backhuys Publishers, The Netherlands.
35. Skinner, G. J. & Allen, G. W. (1996) *Ants*. Richmond Publishing Company, Slough.
36. Stroyan, H. L. G. (1977) *Handbook for the Identification of British Insects: Homoptera: Aphidoidea, Chaitophoridae and Callaphididae*. Royal Entomological Society, London.
37. Tilling, S. M. (1987) *A Key to the Major Groups of British Terrestrial Invertebrates*. Field Studies Council, Shrewsbury.
38. Unwin, D. M. (1988) *A Key to the Families of British Beetles*. Field Studies Council, Shrewsbury.
39. Unwin, D. M. (2001) *A Key to the Families of British Bugs (Insecta, Hemiptera)*. Field Studies Council, Shrewsbury.
40. Wilmer, P. (1985) *Bees, Ants and Wasps: A Key to Genera of the British Aculeates*. Field Studies Council, Shrewsbury.

APPENDIX II. LITERATURE USED IN THE CONSTRUCTION OF THE SEASONAL AND COMPOSITE FOOD WEBS

References

1. Aberle, N., *et al.* (2005) Selectivity and competitive interactions between two benthic invertebrate grazers (*Asellus aquaticus* and *Potamopyrgus antipodarum*): An experimental study using 13C- and 15N-labelled diatoms. *Freshwater Biology*, 50, 369–379.
2. Adams, J. (1984) The habitat and feeding ecology of woodland harvestmen (Opiliones) in England. *Oikos*, 42, 361–370.
3. Agustí, N. & Symondson, W. O. C. (2001) Molecular diagnosis of predation. *Antenna*, 25, 250–253.
4. Albert, A. M. (1983) Characteristics of two populations of Lithobiidae (Chilopoda) determined in the laboratory and their relevance with regard to their ecological role as predators. *Zoologischer Anzeiger*, 211, 214–226.
5. Allen, A. A. (1953) Predatory behaviour of an Amara (Col., Carabidae). *Entomogists' Monthly Magazine*, 89, 236–236.
6. Altieri, M.A., *et al.* (1982) Biological control of *Limax maximus* and *Helix aspersa* by indigenous predators in a daisy field in central coastal California. *Acta Oecologia/Oecologia Applicata*, 3, 387–390.
7. Amalin D.M., *et al.* (2001a) Comparison of the survival of three species of sac spiders on natural and artificial diets. *Journal of Arachnology*, 29, 253–262.

8. Amalin, D.M., *et al.* (2001b) Predatory behavior of three species of sac spiders attacking citrus leaf miner. *Journal of Arachnology*, 29, 72–81.
9. Andrén, O. & Schnürer, J. (1985) Barley straw decomposition with varied levels of microbial grazing by *Folsomia fimetaria* (L.) (Collembola, Isotomidae). *Oecologia*, 68, 57–62.
10. Armer C. A., Wiedenmann, R. N. & Bush, D. R. (1998) Plant feeding site selection on soybean by the facultatively phytophagous predator *Orius insidiosus*. *Entomologia Experimentalis et Applicata*, 86, 109–118.
11. Ayre, K. (2001) Effect of predator size and temperature on the predation of *Deroceras reticulatum* (Müller) (Mollusca) by carabid beetles. *Journal of Applied Entomology*, 125, 389–395.
12. Baines, D., Stewart, R. & Boivin, G. (1990) Consumption of carrot weevil (Coleoptera: Curculionidae) by five species of carabids (Coleoptera: Carabidae) abundant in carrot fields in southwestern Quebec. *Ecological Entomology*, 19, 1146–1149.
13. Bakker, K., *et al.* (1967) Host discrimination in *Pseudeucoila bochei* (Hymenoptera–Cynipidae). *Entomologia Experimentalis et Applicata*, 10, 295–311.
14. Bardgett, R. D., Whittaker, J. B. & Frankland, J. C. (1993a) The effect of collembolan grazing on fungal activity in differently managed upland pastures—A microcosm study. *Biology and Fertility of Soils*, 16, 255–262.
15. Bardgett R. D., Whittaker, J. B. & Frankland, J. C. (1993b) The diet and food preferences of *Onychiurus procampatus* (collembola) from upland grassland soils. *Biology and Fertility of Soils*, 16, 296–298.
16. Barr, T. C. (1974) The eyeless beetles of the genus *Arianops Brendel* (Coleoptera, Pselaphidae). *Bulletin of the American Museum of Natural History*, 154, 51.
17. Barta, M. & Cagáň, L. (2003) Entomophthoralean fungi associated with the common nettle aphid (*Microlophium carnosum* Buckton) and the potential role of nettle patches as reservoirs for the pathogens in landscape. *Anzeiger für Schädlingskunde/Journal of Pest Science*, 76, 1–13.
18. Bauer L. S., Meerschaert, J. & Forrester, O. T. (1990) An artificial diet for cottonwood and imported willow leaf beetles (Coleoptera: Chrysomelidae) and comparative performance on poplar foliage. *Journal of Entomological Science*, 19, 428–431.
19. Bay, E. C. (1974) Predator–prey relationships among aquatic insects. *Annual Review of Entomology*, 19, 441–453.
20. Bequaert, J. (1925) The arthropod enemies of mollusks, with description of a new dipterous parasite from Brazil. *Journal of Parasitology*, 11, 201–212.
21. Beyer, W. M. & Saari, D. M. (1978) Activity and ecological distribution of the slug, *Arion subfuscus* (Draparnaud) (Stylommatophora, Arionidae). *American Midland Naturalist*, 100, 359–367.

22. Bilde, T. & Toft, S. (1994) Prey preference and egg production of the carabid beetle *Agonum dorsale*. *Entomologia Experimentalis et Applicata*, 73, 151–156.
23. Bohan, D.A., *et al.* (2000) Spatial dynamics of predation by carabid beetles on slugs: A response to Mair *et al. Journal of Animal Ecology*, 69, 367–379.
24. Bristowe, W. S. (1949) The distribution of harvestmen (Phalangida) in Great Britain and Ireland, with notes on their names, enemies and food. *Journal of Animal Ecology*, 18, 100–114.
25. Burgess, R. S. L. & Ennos, R. A. (1987) Selective grazing of acyanogenic white clover. Variation in behaviour among populations of the slug *Deroceras reticulatum*. *Oecologia*, 73, 423–435.
26. Butaye, L. & Degheele, D. (1995) Benzoylphenyl uras effect on growth and development of *Eulophus pennicornis* (Hymenoptera: Eulophidae), a larval ectoparasite of the cabbage moth (Lepidoptera: Noctuidae). *Journal of Economic Entomology*, 88, 600–605.
27. Caesar, R. M. (2004) Population structure of *Acrotrichis xanthocera* (Matthews) (Coleoptera: Ptiliidae) in the Klamath Ecoregion of northwestern California, inferred from mitochondrial DNA sequence variation. Master's thesis, Texas A&M University.
28. Cates, R. G. & Orians, G. H. (1975) Successional status and the palatability of plants to generalized herbivores. *Ecology*, 56, 410–418.
29. Chandler, D. S. (1990) Insecta: Coleoptera Scarabaeidae (larvae), In *Soil Biology Guide* (Ed. Dindal, D. L.). John Wiley & Sons, New York, 1349 pp.
30. Chapman, P. A. (1994) Control of leatherjackets by natural enemies: The potential role of the ground beetle *Pterostichus melanarius*. *Proceedings of the Brighton Conference—Pests and diseases*, pp. 933–944.
31. Cheshire, J.M., *et al.* (1987) Seed treatment with rubidium for monitoring wireworm (Coleoptera, Elateridae) feeding on corn. *Environmental Entomology*, 16, 475–480.
32. Coaker, T. H. & Williams, D. A. (1963) The importance of some Carabidae and Staphylinidae as predators of the cabbage root fly, *Erioischia brassicae* (Bouché). *Entomologia Experimentalis et Applicata*, 6, 156–164.
33. Coleman, D. C., Crossley, Jr., D. A. & Hendrix, P. F. (2004) *Fundamentals of Soil Ecology*. Academic Press, Oxford, 408 pp.
34. Cornelius, M. L. & Bernays, E. A. (1995) The effect of plant chemistry on the acceptability of caterpillar prey to the Argentine ant *Iridomyrmex humilils* (Hymenoptera, Formicidae). *Journal of Insect Behaviour*, 8, 579–593.
35. Damman, H. (1987) Leaf quality and enemy avoidance by the larvae of a pyralid moth. *Ecology*, 68, 88–97.

36. Davies, M. J. (1953) Crop contents of some British carabids. *Entomologists' Monthly Magazine*, 89, 18–23.
37. Dawson, N. (1965) A comparative study of the ecology of 8 species of fenland Carabidae (Coleoptera). *Journal of Animal Ecology*, 34, 299–314.
38. de Ruiter, P. C., Neutel, A.-M. & Moore, J. C. (1995) Energetics, patterns of interaction strengths, and stability in real ecosystems. *Science*, 269, 1257–1260.
39. Dennison, D. F. & Hodkinson, I. D. (1983) Structure of the predatory beetle community in a woodland soil ecosystem. 1. Prey selection. *Pedobiologia*, 25, 109–115.
40. Dinter, A. (1998) Intraguild predation between erigonid spiders, lacewing larvae, and carabids. *Journal of Applied Entomology*, 122, 163–167.
41. Dirzo, R. & Harper, J. L. (1982a) Experimental studies on slug–plant interactions: III. Differences in the acceptability of individual plants of *Trifolium repens* to slugs and snails. *Journal of Ecology*, 70, 101–117.
42. Dirzo, R. & Harper, J. L. (1982b) Experimental studies on slug–plant interactions: IV. The performance of cyanogenic and acyanogenic morphs of *Trifolium repens* in the field. *Journal of Ecology*, 70, 119–138.
43. Dixon, P. L. & McKinlay, R. G. (1989) Aphid predation by harvestmen in potato fields in Scotland. *Journal of Arachnology*, 17, 253–255.
44. Dixon, P. L. & McKinlay, R. G. (1992) Pitfall trap catches of and aphid predation by *Pterostichus melanarius* and *Pterostichus madidus* in insecticide treated and untreated potatoes. *Entomologia Experimentalis et Applicata*, 64, 63–72.
45. Dybas, H. S. & Dybas, L. (1990) Insecta: Coleoptera Ptiliidae. In *Soil Biology Guide* (Ed. Dindal, D. L.). John Wiley & Sons, New York, 1349 pp.
46. Eisenbeis, G. & Wichard, W. (1987) *Atlas on the Biology of Soil Arthropods*. Springer-Verlag, Berlin, 437 pp.
47. El Moursy, A. A. (1961) A tentative classification of and a key to the North American genera of the Family Byrrhidae (New Sense) and Family Syncalyptidae (New Status) (Coleoptera, Polyphaga, Byrrhoidea). *The Coleopterists Bulletin*, 15, 9–15.
48. Engelmann, M. D. (1956) The role of soil arthropods in the energetics of an old field community. *Ecological Monographs*, 31, 221–238.
49. Erlandsson, A. (1992) Asymmetric interactions in semiaquatic insects. *Oecologia*, 90, 153–157.
50. Essig, E. O. (1958) *Insects and Mites of Western North America.* New York, Macmillan, 1050 pp.
51. Evan, H. E. (1984) A revision of spider wasps of the genus Turneromyia (Hymenoptera: Pompilidae). *Australian Journal of Zoology Supplementary Series*, 32, 1–59.

52. Fenner, M., Hanley, M. E. & Lawrence, R. (1999) Comparison of seedling and adult palatability in annual and perennial plants. *Functional Ecology*, 13, 546–551.
53. Flinn, P. W. (1991) Temperature-dependent functional response of the parasitoid *Cephalonomia waterstoni* (Gahan) (Hymenoptera: Bethylidae) attacking rusty grain beetle larvae (Coleoptera: Cucujidae). *Environmental Entomology*, 20, 872–876.
54. Foltan, P. (2004) Influence of slug defence mechanisms on the prey preferences of the carabid predator *Pterostichus melanarius* (Coleoptera: Carabidae). *European Journal of Entomology*, 101, 359–364.
55. Forsythe, T. G. (1983) Mouthparts and feeding of certain ground beetles (Coleoptera: Carabidae). *Zoological Journal of the Linnean Society*, 79, 319–376.
56. Fox, C. J. S. & MacLellan, C. R. (1956) Some Carabidae and Staphylinidae shown to feed on a wire worm, *Agriotes sputator* (L.) by the precipitin test. *Canadian Entomologist*, 88, 228–231.
57. Freckman, D. W. & Baldwin, J. G. (1990) Terrestrial Gastropoda, In *Soil Biology Guide* (Ed. Dindal, D. L.). John Wiley & Sons, New York, 1349 pp.
58. Getz, L. L. (1959) Notes on the ecology of slugs: *Arion circumscriptus*, *Derocerus reticulatum*, and *D. Laeve*. *American Midland Naturalist*, 61, 485–498.
59. Goldwasser, L. & Roughgarden, J. (1993) Construction and analysis of a large Caribbean food web. *Ecology*, 74, 1216–1233.
60. Gordh, G. & Hawkins, G. (1981) *Goniozus emigratus* (Rohwer), a primary external parasite of *Paramyelois transitella* (Walker), and comments on bethylids attacking Lepidoptera (Hymenoptera: Bethylidae; Lepidoptera: Pyralidae). *Journal of the Kansas Entomological Society*, 54, 787–803.
61. Graça, M. A. S., Maltby, L. & Calow, P. (1993) Importance of fungi in the diet of *Gammarus pulex* and *Asellus aquaticus*, II. Effects on growth, reproduction and physiology. *Oecologia*, 96, 304–309.
62. Greig-Smith, P. (1948) Biological flora of the British Isles: *Urtica* L. *Journal of Ecology*, 36, 339–355.
63. Hall W. E. (2001) Ptiliidae. In *American Beetles: Archostemata, Myxophaga, Adephaga, Polyphaga: Staphyliniformia Vol. 1.* (Eds. Arnett, R. H. and M. C. Thomas). CRC Press LLC, 464 pp.
64. Hansen, M. (1997) Observations on the immature stages of Georissidae (Coleoptera: Hydrophiloidea), with remarks on the evolution of the hydrophiloid egg cocoon. *Invertebrate Taxonomy*, 14, 907–916.
65. Harper, G. L., *et al.* (2005) Rapid screening of invertebrate predators for multiple prey DNA targets. *Molecular Ecology*, 14, 819–827.

66. Harwood, J. D. & Obrycki, J. J. (2005) Quantifying aphid predation rates of generalist predators in the field. *European Journal of Entomology*, 102, 335–350.
67. Hölldobler, B. & Wilson, E. O. (1994) *Journey to the Ants: A Story of Scientific Exploration*. Harvard University Press, 240 pp.
68. Horgan, F. G. & Myers, J. H. (2004) Interactions between predatory ground beetles, the winter moth and an introduced parasitoid on the Lower Mainland of British Columbia. *Pedobiologia*, 48, 23–35.
69. Hunt, G. (1992) Life cycles of a gall-forming gall midge (Diptera: Cecidomyiidae) and associated parasitoids, on Putaputaweta (*Carpodetus serratus*). *New Zealand Entomologist,* 15, 14–21.
70. Ienista, M. A. (1978) Hydradephaga und Palpicornia. In *Limnofauna Europaea* (Ed. Illies, J.). G. Fischer, 532 pp.
71. Iglesias, J. & Castillejo, J. (1999) Field observations on feeding of the land snail *Helix aspersa* Müller. *Journal of Molluscan Studies*, 65, 411–423.
72. Infante, F., Mumford, J. & Baker, P. (2005) Life history studies of *Prorops nasuta*, a parasitoid of the coffee berry borer. *Biocontrol*, 50, 259–270.
73. Isenhour, D. J., Wiseman, B. R. & Layton, R. C. (1989) Enhanced predation by *Orius insidiosus* (Hemiptera: Anthocoridae) on larvae of *Heliothis zea* and *Spodoptera frugiperda* (Lepidoptera: Noctuidae) caused by prey feeding on resistant corn genotypes. *Environmental Entomology*, 18, 418–422.
74. Jablonski, P. G. (1996) Intruder pressure affects territory size and foraging success in asymmetric contests in the water strider *Gerris lacustris*. *Ethology*, 102, 22–31.
75. Jackson, R. R. & Walls, E. I. (1998) Predatory and scavenging behaviour of *Microvelia macgregori* (Hemiptera: Veliidae), a water-surface bug from New Zealand. *New Zealand Journal of Zoology*, 25, 23–28.
76. Jaffé, K., *et al.* (1993) Chemical ecology of the palm weevil *Rhynchophorus palmarum* (L.) (Coleoptera: Curculionidae): Attraction to host plants and to a male-produced aggregation pheromone. *Journal of Chemical Ecology*, 19, 1703–1720.
77. Jennings, T. J., and Barkham, J. P. (1979) Litter decomposition by slugs in mixed deciduous woodland. *Ecography*, 2, 21–29.
78. Johnson, R. K. (1987) Seasonal variation in diet of *Chironomus plumosus* (L.) and *C. anthracinus* Zett. (Diptera: Chironomidae) in mesotrophic Lake Erken. *Freshwater Biology*, 17, 525–532.
79. Johnston, R. A. (2000) Seed-harvester ants (Hymenoptera: Formicidae) of North America: An overview of ecology and biogeography. *Sociobiology*, 36, 89–122.
80. Kalaskar, A. & Evans, E. W. (2000) Larval responses of aphidophagous lady beetles (Coleoptera: Coccinellidae) to weevil larvae versus aphids as prey. *Annals of the Entomological Society of America*, 94, 76–81.

81. Kautz, G., Zimmer, M. & Werner, T. (2002a) Leaf litter-colonizing microbiota: Supplementary food source or indicator of food quality for Porcellio scaber (Isopoda: Oniscidea)? *European Journal of Soil Biology*, 39, 209–216.
82. Kautz, G., Zimmer, M. & Topp, W. (2002b) Does *Porcellio scaber* (Isopoda: Oniscidea) gain from coprophagy? *Soil Biology and Biochemistry*, 34, 1253–1259.
83. Kfir, R. (1990) Parasites of the spotted stalk borer, Chilo partellus [Lepidoptera: Pyralidae] in South Africa. *Biocontrol*, 35, 403–410.
84. Klausnitzer, B. (1994) *Die Larven der Käfer Mitteleuropas. 2.* Band, Myxophaga/Polyphaga. Goecke & Evers, Krefeld.
85. Klimaszewski, J. & Watt, J.C. (1998) Coleoptera: family-group review and keys to identification. *Fauna of New Zealand*, 37. Manaaki Whenua Press.
86. Kline, J. A. & Beckage, N. E. (1990) Comparative suitability of *Trogoderma variabile* and *T. glabrum* (Coleoptera: Dermistidae) as hosts for the ectoparasite *Laelius pedatus* (Hymenoptera: Bethylidae). *Annals of the Entomological Society of America*, 83, 809–816.
87. Komonen, A. (2001) Structure of insect communities inhabiting old-growth forest specialist bracket fungi. *Ecological Entomology*, 26, 63–75.
88. Larochelle, A. (1990) The food of carabid beetles (Coleoptera: Carabidae, including Cicindelinae). *Fabreries Supplement*, 5, 1–132.
89. Lawrence, J. F. & Britton, E. B. (1994) *Australian Beetles*. Melbourne University Press, Melbourne, 192 pp.
90. Lord, J. C. (2001) Response of the wasp Cephalonomia tarsalis (Hymenoptera: Bethylidae) to *Beauveria bassiana* (Hyphomycetes: Moniliales) as free conidia or infection in its host, the Sawtoothed Grain Beetle, *Oryzaephilus surinamensis* (Coleoptera: Silvanidae). *Biocontrol*, 21, 300–304.
91. Loreau, M. (1984) Population density and biomass of carabidae (Coleoptera) in a forest community. *Pedobiologia*, 27, 269–278.
92. Lukasiewicz, J. (1996) Predation by the beetle *Carabus granulatus* L. (Coleoptera, Carabidae) on soil macrofauna in grassland on drained peats. *Pedobiologia*, 40, 364–376.
93. Mair, J. & Port, G. R. (2001) Predation by the carabid beetles *Pterostichus madidus* and *Nebria brevicollis* is affected by size and condition of the prey slug *Deroceras reticulatum*. *Agricultural and Forest Entomology*, 3, 99–106.
94. Mair, J., *et al.* (2001) Spatial dynamics of predation by carabid beetles on slugs. *Journal of Animal Ecology*, 70, 875–876.
95. Majka, C. G., Noronha, C. & Smith, M. (2006) Adventive and native Byrrhidae (Coleoptera) newly recorded from Prince Edward Island, Canada. *Zootaxa*, 1168, 21–30.

96. Martin, M. M., *et al.* (1980) The digestion of protein and carbohydrate by the stream detritivore, *Tipula abdominalis* (Diptera, Tipulidae). *Oecologia*, 46, 360–364.
97. Mason, C. F. (1970) Snail populations, beech litter production, and the role of snails in litter decomposition. *Oecologia*, 5, 215–239.
98. McKemey, A. R., Symondson, W. O. C. & Glen, D. M. (2003) Predation and prey size choice by the carabid beetle *Pterostichus melanarius* (Coleoptera: Carabidae): The dangers of extrapolating from laboratory to field. *Bulletin of Entomological Research*, 93, 227–234.
99. Medler, J. T. (1967) Biology of trypoxylon in trap nests in Wisconsin (Hymenoptera: Sphecidae). *American Midland Naturalist*, 78, 344–358.
100. Messer, C., *et al.* (2000) Chemical deterrents in podurid Collembola. *Pedobiologia*, 44, 210–220.
101. Mitchell, B. (1963) Ecology of two carabid beetles, *Bembidion lampros* (Herbst) and *Trechus quadristriatus* (Schrank). *Journal of Animal Ecology*, 32, 377–392.
102. Moore, J. W. (1975) The role of algae in the diet of *Asellus aquaticus* L. and *Gammarus pulex* L. *Journal of Animal Ecology*, 44, 719–730.
103. Mundy, C. A., *et al.* (2000) Prey selection and foraging behaviour by *Pterostichus cupreus* L. (Col., Carabidae) under laboratory conditions. *Journal of Applied Entomology*, 124, 349–358.
104. Naeem, S. (1990) Resource heterogeneity and community structure: A case study in *Heliconia imbricata* phytotelmata. *Oecologia*, 84, 29–38.
105. Nakasuji, F. & Dyck, V. A. (1984) Evaluation of the role of *Microvelia douglasi atrolineata* (Bergroth) (Heteroptera: Veliidae) as predator of the brown planthopper *Nilaparvata lugens* (Stäl) (Homoptera: Delphacidae). *Researches on Population Ecology*, 26, 134–149.
106. Newton, Jr., A. F. (1990) Insecta: Coleoptera Pselaphidae, In *Soil Biology Guide* (Ed. Dindal, D. L.). John Wiley & Sons, New York, 1349 pp.
107. Norton, R. A. (1990) Terrestrial Isopoda, In *Soil Biology Guide* (Ed. Dindal, D. L.). John Wiley & Sons, New York, 1349 pp.
108. Nyffeler, M. (1999) Prey selection of spiders in the field. *Journal of Arachnology*, 27, 317–324.
109. Nyffeler, M., Moor, H., and Foelix, R. F. (2001) Spiders feeding on earthworms. *Journal of Arachnology*, 29, 119–124.
110. Ohmart, C. P., Stewart, L. G. & Thomas, J. R. (1985) Effects of food quality, particularly nitrogen concentrations, of *Eucalyptus blakelyi* foliage on the growth of *Paropsis atomaria* larvae (Coleoptera: Chrysomelidae). *Oecologia*, 65, 543–549.
111. Pallant, D. (1969) The food of the grey field slug (*Agriolimax reticulatus* (Müller)) in woodland. *Journal of Animal Ecology*, 38, 391–397.

112. Pallant, D. (1972) The food of the grey field slug, *Agriolimax reticulatus* (Müller), on grassland. *Journal of Animal Ecology*, 41, 761–769.
113. Palmer, J. L., Donnelly, M. J. & Corbet, S. A. (1998) *Hydrometra stagnorum* (L.) (Hem., Hydrometridae) feeding on mosquitoes. *Entomologists' Monthly Magazine*, 133, 65–68.
114. Parkyn, S. M. & Winterbourn, M. J. (1997) Leaf breakdown and colonisation by invertebrates in a headwater stream: Comparisons of native and introduced tree species. *New Zealand Journal of Marine and Freshwater Research*, 31, 301—312.
115. Patel, K. J. & Schuster, D. J. (1991) Temperature-dependent fecundity, longevity, and host-killing activity of *Diglyphus intermedius* (Hymenoptera: Eulophidae) on third instars of *Liriomyza trifolii* (Burgess) (Diptera: Agromyzidae). *Environmental Entomology*, 20, 1195–1199.
116. Patrick, R. (1984) Some thoughts concerning the importance of pattern in diverse systems. *Proceedings of the American Philosophical Society*, 128, 48–78.
117. Penney, M. M. (1966) Diapause and reproduction in *Nebria brevicollis* (F.) (Coleoptera: Carabidae). *Journal of Animal Ecology*, 38, 219–233.
118. Nienstedt, K. M. & Poehling, M.-H. (2004) Invertebrate predation of 15N-marked prey in semi-field wheat enclosures. *Entomologia Experimentalis et Applicata*, 112, 191–200.
119. Nienstedt, K. M. & Poehling, M.-H. (2005) Prey to predator transfer of enriched 15N-contents: Basic laboratory data for predation studies using 15N as marker. *Entomologia Experimentalis et Applicata*, 112, 183–190.
120. Pollard, A. J. (1972) The importance of deterrence: Responses of grazing animals to plant variation. In *Plant Resistance to Herbivores and Pathogens: Ecology, Evolution and Genetics* (Eds. Fritz, R. S. and Simms, E. L.), Chicago University Press, 600 pp.
121. Pollard, A. J., & Briggs, D. (1984) Genecological studies of *Urtica dioica* L. III. Stinging hairs and plant–herbivore interactions. *New Phytologist*, 97, 507–522.
122. Pollard, E. (1968) Hedges. III. The effect of removal of the bottom flora of a hawthorn hedgerow on the carabidae of the hedge bottom. *Journal of Applied Ecology*, 5, 125–139.
123. Ponsard, S. & Arditi, R. (2000) What can stable isotopes (δ15N and δ13C) tell about the food web of soil macro-invertebrates? *Ecology*, 81, 852–864.
124. Rajeswaran J., Duraimurugan, P. & Shanmugam, P. S. (2005) Role of spiders in agriculture and horticulture ecosystem. *Journal of Food, Agriculture and Environment*, 3, 147–152.
125. Rathcke, B. (1985) Slugs as generalist herbivores: Tests of three hypotheses on plant choices. *Ecology*, 66, 828–836.

126. Reddy, M. V. & Das, P. K. (1983) Microfungal food preference of soil microarthropods in a pine plantation ecosystem. *Journal of Soil Biology and Ecology*, 3, 1–6.
127. Reissig, W. H., Heinrichs, E. A. & Valencia, S. L. (1982) Effects of insecticides on *Nilaparvata lugens* and its predators: Spiders, *Microvelia atrolineata* and *Cyrtorhinus lividipennis*. *Environmental Entomology*, 11, 193–199.
128. Riihimaki, J., *et al.* (2005) Testing the enemies hypothesis in forest stands: The important role of tree species composition. *Oecologia*, 142, 90–97.
129. Ryan, M. F. (1973) The natural mortality of wheat-bulb fly larvae. *Journal of Applied Ecology*, 10, 875–879.
130. Santos, F. H. & Gnaspini, P. (2002) Notes on the foraging behavior of the Brazilian cave harvestman *Goniosoma spelaeum* (Opiliones, Gonyleptidae). *Journal of Arachnology*, 30, 177–180.
131. Seagle, Jr., H. H. (1982) Comparison of the food habits of three species of riffle beetles, *Stenelmis crenata*, *Stenelmis mera*, and *Optioservus trivittatus* (Coleoptera: Dryopoidea: Elmidae). *Freshwater Invertebrate Biology*, 1, 33–38.
132. Sigsgaard, L., Toft, S. & Villareal, S. (2001a) Diet-dependent survival, development and fecundity of the spider *Atypena formosana* (Oi) (Araneae: Linyphiidae). Implications for biological control in rice. *Biocontrol Science and Technology*, 11, 233–244.
133. Sigsgaard, L., Toft S. & Villareal, S. (2001b) Diet-dependent fecundity of the spiders *Atypena formosana* and *Pardosa pseudoannulata*, predators in irrigated rice. *Agricultural and Forest Entomology*, 3, 285–295.
134. Skinner, G. J. (1980) The feeding habits of the wood-ant, *Formica rufa* (Hymenoptera: Formicidae), in limestone woodland in north-west England. *Journal of Animal Ecology*, 49, 417–433.
135. Spangler, P. J. (1991) Hydraenidae (Staphylinoidea) (=Limnebiidae). In *Immature Insects, Volume 2* (Ed. Stehr, F. W.). Kendall/Hunt, 975 pp
136. Srinivasa Murthy, K., Jalali, S. K. & Venkatesan, T. (2005) Performance of *Goniozus nephantidis* (Bethylidae: Hymenoptera) multiplied on artificial diet-reared *Opisina arenosella* at variable high temperature. *Indian Journal of Agricultural Science*, 75, 529–531.
137. Steinbauer, M. J. (1999) The population ecology of Amorbus Dallas (Hemiptera: Coreidae) species in Australia. *Entomologia Experimentalis et Applicata*, 91, 175–182.
138. Stephenson, J. W. (1965) Slug parasites and predators. *Rothamsted Experimental Station Report*, 188 pp.
139. Stille, B. (1984) The effect of host plant and parasitoids on the reproductive success of the parthenogenetic gall wasp *Diplolepis rosae* (Hymenoptera, Cynipidae). *Oecologia*, 63, 364–369.

140. Sunderland, K. D. (1975) The diet of some predatory arthropods in cereal crops. *Journal of Applied Ecology*, 12, 507–515.
141. Sunderland, K. D. & Vickerman, G. P. (1980) Aphid feeding by some polyphagous predators in relation to aphid density in cereal fields. *Journal of Applied Ecology*, 17, 389–396.
142. Sunderland, K. D., *et al.* (1987) A study of feeding by polyphagous predators on cereal aphids using ELISA and gut dissection. *Journal of Applied Ecology*, 24, 907–933.
143. Sunderland, K. D., Fraser, A. M. & Dixon, A. F. G. (1986) Field and laboratory studies on money spiders (Linyphiidae) as predators of cereal aphids. *Journal of Applied Ecology*, 23, 433–447.
144. Šustr, V. & Frouz, J. (2002) Activity of carbohydrases in the gut of Bibionidae (Diptera) larvae. *European Journal of Soil Biology*, 38, 75–77.
145. Symondson, W. O. C. (1989). Biological control of slugs by carabids. Slugs and Snails in World Agriculture (Ed. Henderson, I. F.). *BCPC Monograph* No. 41. BCPC Publications, pp. 295–300.
146. Symondson, W. O. C. (2002) Molecular identification of prey in predator diets. *Molecular Ecology*, 11, 627–641.
147. Symondson, W. O. C., Erickson, M. L. & Liddell, J. E. (1999) Development of a monoclonal antibody for the detection and quantification of predation on slugs within the *Arion hortensis* agg. (Mollusca: Pulmonata). *Biological Control*, 16, 274–282.
148. Symondson, W. O. C., *et al.* (2000) Do earthworms help to sustain the slug predator *Pterostichus melanarius* (Coleoptera: Carabidae) within crops? Investigations using monoclonal antibodies. *Molecular Ecology*, 9, 1279–1292.
149. Tahvanainen, J. O. & Root, R. B. (1972) The influence of vegetational diversity on the population ecology of a specialized herbivore, *Phyllotreta cruciferae* (Coleoptera: Chrysomelidae). *Oecologia*, 10, 321–346.
150. Tashiro, H. (1990) Insecta: Diptera (adults), In *Soil Biology Guide* (Ed. Dindal, D. L.). John Wiley & Sons, New York, 1349 pp.
151. Tavares, A. F. & Williams, D. D. (1990) Life histories, diet, and niche overlap of three sympatric species of Elmidae (Coleoptera) in a temperate stream. *Canadian Entomologist*, 122, 563–577.
152. Taylor, S. J. & Wood, D. L. (2000) Rearing *Hydrometra martini* (Heteroptera: Hydrometridae): Food and substrate effects. *Florida Entomologist*, 83, 17–25.
153. Telang, A., *et al.* (1999) Feeding damage by *Diuraphis noxia* results in a nutritionally enhanced phloem diet. *Entomologia Experimentalis et Applicata*, 91, 403–412.
154. Teskey, H. J. (1990) Insecta: Hymenoptera Formicidae, In *Soil Biology Guide* (Ed. Dindal, D. L.). John Wiley & Sons, New York, 1349 pp.

155. Toft, S. (2005) The quality of aphids as food for generalist predators: Implications for natural control of aphids. *European Journal of Entomology*, 102, 371–383.
156. Toft, T., and Bilde, S. (1997) Consumption by carabid beetles of three cereal aphid species relative to other prey types. *Biocontrol*, 42, 21–32.
157. Toft, S. & Bilde, T. (2002) Carabid diets and food value. In *The Agroecology of Carabid Beetles* (Ed. Holland, J. M.), Intercept Ltd., Andover, UK, 356 pp.
158. Turner, B. D. (1974) The population dynamics of tropical arboreal Psocoptera (Insecta) on two species of conifers in the Blue Mountains, Jamaica. *Journal of Animal Ecology*, 43, 323–337.
159. Valentine, E. W. (1975) Additions and corrections to Hymenoptera hyperparasitic on aphids in New Zealand. *New Zealand Entomologist*, 6, 59–61.
160. van Amelsvoort, P. A. M. & Usher, M. B. (1989) Egg production related to food quality in Folsomia candida (Collembola: Isotomidae): Effects on life history strategies. *Pedobiologia*, 33, 61–66.
161. Vickerman, G. P. & Sunderland, K. D. (1975) Arthropods in cereal crops: Nocturnal activity, vertical distribution and aphid predation. *Journal of Applied Ecology*, 12, 755–766.
162. Von Bless, R. (1977) Untersuchungen zur Frage des Räuber-Beute-Verhältnisses von Carabiden und Gastropoden. *Anz. Schädingskd. Pflanzenschutz, Umweltschutz*, 50, 55–57.
163. Wallace, S. K. (2004) Molecular gut analysis of carabids. (Coleoptera: Carabidae) using aphid primers. Masters thesis, Montana State University, Bozeman.
164. Wallwork, J. A. (1970) *Ecology of Soil Animals*. McGraw-Hill, New York, 280 pp.
165. Wegener, C. (1998) Predation on the grass bug species *Notostira elongata* (Heteroptera: Miridae) by Nabidae (Heteroptera) and selected non-web building spiders (Araneae). *Entomologia Generalis*, 22, 295–304.
166. Winterbourn, M. J. (1982) The invertebrate fauna of a forest stream and its association with fine particulate matter. *New Zealand Journal of Marine and Freshwater Research*, 16, 271–281.
167. Yano, K., Tsuchiya, K. & Hamasaki, S. (1984) Aphid-feeding by adult elaterid beetles (Coleoptera, Elateridae). *Japanese Journal of Entomology*, 52, 441–444
168. Young, F. N. (1958) Notes on the Care and Rearing of Tropisternus in the Laboratory (Coleoptera: Hydrophilidae). *Ecology*, 39, 166–167.
169. Zimmer, M. & Bartholme, S. (2003) Bacterial endosymbionts in *Asellus aquaticus* (Isopoda) and *Gammarus pulex* (Amphipoda) and their contribution to digestion. *Limnology and Oceanography*, 48, 2208–2213.

APPENDIX III. LENGTH–WEIGHT RELATIONSHIPS FOR ALL TAXA (TABLE S1) AND ORDERS (TABLE S2) FOUND IN THE SEASONAL AND COMPOSITE FOOD WEBS

Table S1

Taxa	Order	Abundance	Regression equation	r^2	p-Value
Abax parallelepipedus	Order Coleoptera	9	$y = 3.36x - 5.49$	0.55	0.05
Arion ater	Order Gastropoda	27	$y = 2.18x - 4.08$	0.69	0.01
Arion distinctus	Order Gastropoda	48	$y = 3.24x - 5.46$	0.82	0.01
Arion hortensis	Order Gastropoda	17	$y = 1.91x - 3.95$	0.40	0.01
Arion subfuscus	Order Gastropoda	161	$y = 2.77x - 5.05$	0.79	0.01
Bembidion lampros	Order Coleoptera	10	$y = 4.06x - 5.60$	0.44	0.05
Derocerus reticulatum	Order Gastropoda	577	$y = 2.33x - 4.76$	0.72	0.01
Family Aphididae	Order Hemiptera	89	$y = 0.99x - 4.37$	0.24	0.01
Family Clubionidae	Order Aranae	30	$y = 1.33x - 4.34$	0.35	0.01
Family Curculionidae	Order Coleoptera	70	$y = 1.63x - 4.01$	0.31	0.01
Family Cynipidae	Order Hymenoptera	40	$y = 1.60x - 4.55$	0.42	0.01
Family Delphacidae	Order Hemiptera	13	$y = 1.31x - 4.29$	0.65	0.01
Family Diapriidae	Order Hymenoptera	12	$y = 1.85x - 4.50$	0.67	0.01
Family Dictynidae	Order Aranae	47	$y = 1.77x - 4.52$	0.47	0.01
Family Elateridae	Order Coleoptera	20	$y = 2.75x - 4.78$	0.66	0.01
Family Gerridae	Order Hemiptera	18	$y = 6.00x - 8.98$	0.83	0.01
Family Hydrometridae	Order Hemiptera	16	$y = 1.73x - 4.53$	0.81	0.01
Family Isotomidae	Order Collembola	1863	$y = 1.42x - 4.93$	0.57	0.01
Family Linyphiidae	Order Aranae	29	$y = 1.16x - 4.38$	0.18	0.05
Family Neanuridae	Order Collembola	393	$y = 1.27x - 5.15$	0.11	0.01
Family Philodromidae	Order Aranae	4	$y = 2.36x - 4.33$	0.96	0.05
Family Pselaphidae	Order Coleoptera	39	$y = 2.69x - 4.84$	0.26	0.01
Family Scymaenidae	Order Coleoptera	3	$y = 2.29x - 4.59$	1.00	0.01

Family Sminthuridae	Order Collembola	2062	$y=2.35x-4.78$	0.73	0.01
Family Staphylinidae	Order Coleoptera	328	$y=2.09x-4.99$	0.51	0.01
Family Tenthredinidae	Order Hymenoptera	29	$y=1.20x-4.30$	0.47	0.01
Genus Microvelia	Order Hemiptera	71	$y=1.24x-4.36$	0.46	0.01
Gerris lacustris	Order Hemiptera	7	$y=1.54x-4.28$	0.88	0.01
Lacinius ephippiatus	Order Opiliones	38	$y=1.83x-4.06$	0.38	0.01
Leiobunum blackwalli	Order Opiliones	11	$y=2.26x-4.35$	0.89	0.01
Nebria brevicollis	Order Coleoptera	12	$y=4.64x-6.84$	0.47	0.05
Non-Oribatidae	Order Acari	469	$y=3.28x-4.24$	0.77	0.01
Odiellus spinosus	Order Opiliones	12	$y=2.07x-4.10$	0.90	0.01
Oniscus asellus	Order Isopoda	82	$y=2.44x-4.54$	0.69	0.01
Ophyiulus pilosus	Order Diplopoda	25	$y=2.99x-6.06$	0.77	0.01
Oribatidae	Order Acari	1290	$y=1.7x-4.29$	0.70	0.01
Polydesmus angustus	Order Diplopoda	91	$y=2.74x-5.49$	0.82	0.01
Porcellio scaber	Order Isopoda	22	$y=3.30x-5.42$	0.83	0.01
Porcellionides cingendus	Order Isopoda	12	$y=3.62x-5.79$	0.61	0.01
Pterostichus crenatus	Order Coleoptera	6	$y=3.19x-5.14$	0.93	0.01
Pterostichus nigrita	Order Coleoptera	32	$y=2.87x-4.93$	0.48	0.01
Pterostichus strenuus	Order Coleoptera	16	$y=2.13x-4.28$	0.51	0.01
Rilaena triangularis	Order Opiliones	4	$y=2.95x-4.19$	1.00	0.01

Table S2

Taxa	Abundance	Regression equation	r^2	p-Value
Family Carabidae	524	$y=2.82x-4.92$	0.88	0.01
Order Acari	1759	$y=2.33x-4.23$	0.73	0.01
Order Aranae	68	$y=1.52x-4.39$	0.41	0.01
Order Coleoptera	1083	$y=2.91x-5.13$	0.81	0.01
Order Coleoptera larvae	186	$y=1.33x-4.43$	0.43	0.01
Order Collembola	4318	$y=1.75x-4.92$	0.62	0.01
Order Diptera	2089	$y=2.01x-4.79$	0.76	0.01
Order Diptera larvae	365	$y=2.22x-5.28$	0.68	0.01
Order Enchytraeidae	211	$y=0.91x-4.83$	0.23	0.01
Order Gastropoda	831	$y=2.49x-4.86$	0.73	0.01
Order Hemiptera	240	$y=1.77x-4.39$	0.72	0.01
Order Hymenoptera	156	$y=0.99x-4.30$	0.28	0.01
Order Isopoda	132	$y=2.91x-5.07$	0.69	0.01
Order Oligochaeta	172	$y=1.66x-4.69$	0.50	0.01
Order Opiliones	84	$y=1.74x-3.95$	0.56	0.01
Phylum Myriapoda	120	$y=2.65x-5.43$	0.77	0.01

APPENDIX IV. PRIMARY DATA TABLES FOR ALL SEASONAL AND COMPOSITE WEBS

Table S3 February 2005

ID#	Taxa name	Prey ID#	BM	BA	NA	TH
1	*Arion ater*	33, 37, 38	2.25×10^{-2}	0.0005	2.26×10^{-2}	2
2	*Arion distinctus*	38	2.42×10^{-2}	0.0017	7.22×10^{-2}	2
3	*Arion hortensis*	33, 38	2.16×10^{-2}	1×10^{-4}	4.51×10^{-3}	2
4	*Arion subfuscus*	33, 34, 38	1.48×10^{-2}	0.0001	9.02×10^{-3}	2
5	*Carabus granulatus*	6, 21, 23, 24, 27, 20, 22, 26, 27, 31, 33, 35	6.74×10^{-2}	0.0003	4.51×10^{-3}	3.24
6	*Deroceras reticulatum*	36, 37, 38, 39	6.74×10^{-2}	0.0006	9.02×10^{-3}	2.41
7	Family Aphididae	37, 38	8.00×10^{-4}	4×10^{-6}	4.51×10^{-3}	2
8	Family Clubionidae	7, 23	1.00×10^{-5}	5×10^{-8}	4.51×10^{-3}	3
9	Family Isotomidae	32, 33, 34	5.35×10^{-5}	3×10^{-5}	6.00×10^{-1}	2
10	Family Linyphiidae	7, 20, 22, 23	3.90×10^{-4}	2×10^{-6}	4.51×10^{-3}	3
11	Family Neanuridae	32, 33, 34	2.65×10^{-5}	6×10^{-6}	2.35×10^{-1}	2
12	Family Ptiliidae	33, 34	1.05×10^{-4}	9×10^{-7}	9.02×10^{-3}	2
13	Family Sminthuridae	32, 33, 34, 38	4.29×10^{-5}	6×10^{-6}	1.40×10^{-1}	2
14	Family Staphylinidae	7, 19, 20, 21, 22, 23, 26	7.71×10^{-4}	4×10^{-5}	4.96×10^{-2}	3.19
15	*Gerris lacustris*	23	8.05×10^{-3}	4×10^{-5}	4.51×10^{-3}	3
16	Non-Oribatidae	34	7.99×10^{-5}	1×10^{-5}	1.40×10^{-1}	2
17	*Odellius spinosus*	33	8.50×10^{-4}	4×10^{-6}	4.51×10^{-3}	2
18	*Oniscus ascellus*	33	8.52×10^{-3}	0.0004	4.96×10^{-2}	2
19	*Ophyiulus pilosus*	33	1.04×10^{-2}	9×10^{-5}	9.02×10^{-3}	2
20	Order Acari	26, 32, 33, 34	1.09×10^{-4}	5×10^{-5}	4.78×10^{-1}	2
21	Order Coleoptera larvae	7, 20, 21, 22, 24, 38	6.46×10^{-4}	1×10^{-5}	2.26×10^{-2}	2.80

(*continued*)

Table S3 (*continued*)

ID#	Taxa name	Prey ID#	BM	BA	NA	TH
22	Order Collembola	32, 33, 34, 38	1.23×10^{-4}	0.0001	9.75×10^{-1}	2
23	Order Diptera	33	9.69×10^{-4}	0.0002	1.76×10^{-1}	2
24	Order Diptera larvae	33, 34, 38	7.33×10^{-3}	0.0002	3.16×10^{-2}	2
25	Order Enchytraeidae	32, 33, 34	1.49×10^{-4}	1×10^{-5}	8.57×10^{-2}	2
26	Order Nematoda	26, 27, 34, 38	5.04×10^{-5}	6×10^{-6}	1.17×10^{-1}	2.50
27	Order Oligochaeta	32, 33, 34	3.59×10^{-3}	0.0002	6.77×10^{-2}	2
28	Oribatidae	26, 32, 33, 34	2.91×10^{-5}	1×10^{-5}	3.38×10^{-1}	2.38
29	*Polydesmus angustus*	33	1.04×10^{-2}	0.0002	1.80×10^{-2}	2
30	*Pterostichus strenuus*	6, 20	7.52×10^{-3}	7×10^{-5}	9.02×10^{-3}	3.20
31	*Anemone nemorosa*	–	–	–	–	1
32	Bacteria	–	–	–	–	1
33	Detritus	–	–	–	–	1
34	Microfungi	–	–	–	–	1
35	*Osmunda regalis*	–	–	–	–	1
36	*Ranunculus ficaria*	–	–	–	–	1
37	*Urtica dioica*	–	–	–	–	1
38	Vegetation	–	–	–	–	1
39	*Veronica montana*	–	–	–	–	1

Table S4 April 2005

ID#	Taxa name	Prey ID#	BM	BA	NA	TH
1	*Abax parallelepipedus*	10, 11, 12, 16, 44, 45, 46, 48, 52, 53, 56, 61, 71	7.68×10^{-2}	0.0003	4.51×10^{-3}	3.26
2	*Agonum muelleri*	42, 43	4.00×10^{-3}	2×10^{-5}	4.51×10^{-3}	4.33
3	*Amara plebeja*	46, 47, 48	5.80×10^{-3}	3×10^{-5}	4.51×10^{-3}	3
4	*Arion ater*	69, 74	7.79×10^{-2}	0.0014	1.80×10^{-2}	2
5	*Arion distinctus*	74	1.82×10^{-2}	0.0007	3.61×10^{-2}	2
6	*Arion hortensis*	69, 74	2.12×10^{-2}	0.0015	7.22×10^{-2}	2
7	*Arion subfuscus*	69, 71, 74	4.65×10^{-2}	0.0122	2.62×10^{-1}	2
8	*Asellus aquaticus*	69, 71	2.38×10^{-3}	1×10^{-4}	4.06×10^{-2}	2
9	*Bembidion lampros*	9, 41, 43, 44, 46, 51, 53, 55	2.28×10^{-3}	0.0002	6.77×10^{-2}	3.45
10	*Carabus granulatus*	12, 43, 46, 47, 53, 61	6.28×10^{-2}	0.0156	2.48×10^{-1}	3.33
11	*Carabus nemoralis*	47, 61	1.59×10^{-1}	0.0129	8.12×10^{-2}	3.67
12	*Deroceras reticulatum*	41, 44, 52, 53, 67, 69, 73, 74	2.54×10^{-2}	0.0178	6.99×10^{-1}	2.67
13	Family Byhrridae	72, 74	1.00×10^{-4}	5×10^{-7}	4.51×10^{-3}	2
14	Family Chrysomelidae	74	1.20×10^{-3}	5×10^{-6}	4.51×10^{-3}	2
15	Family Clubionidae	46	1.00×10^{-4}	5×10^{-7}	4.51×10^{-3}	3
16	Family Curculionidae	74	2.63×10^{-3}	0.0002	5.87×10^{-2}	2
17	Family Elateridae	74	2.34×10^{-3}	0.0004	1.58×10^{-1}	2
18	Family Elmidae	69	5.00×10^{-5}	5×10^{-7}	9.02×10^{-3}	2
19	Family Formicidae	44, 46	1.22×10^{-4}	4×10^{-6}	3.16×10^{-2}	3
20	Family Hydraenidae	69	1.17×10^{-4}	4×10^{-6}	3.16×10^{-2}	2
21	Family Hydrophilidae	69, 74	3.35×10^{-3}	3×10^{-5}	9.02×10^{-3}	2
22	Family Isotomidae	68, 69, 71	4.03×10^{-5}	4×10^{-5}	1.01	2
23	Family Linyphiidae	41, 44, 46	1.33×10^{-4}	2×10^{-6}	1.35×10^{-2}	3.11
24	Family Neanuridae	68, 69, 71	2.12×10^{-5}	6×10^{-6}	2.62×10^{-1}	2

(*continued*)

Table S4 *(continued)*

ID#	Taxa name	Prey ID#	BM	BA	NA	TH
25	Family Pselaphidae	41, 44, 52, 53	1.00×10^{-4}	5×10^{-7}	4.51×10^{-3}	3.17
26	Family Ptiliidae	69, 71	1.33×10^{-4}	4×10^{-6}	2.71×10^{-2}	2
27	Family Scarabaeidae	69, 71	6.50×10^{-3}	3×10^{-5}	4.51×10^{-3}	2
28	Family Scydmaenidae	56	5.00×10^{-5}	5×10^{-7}	9.02×10^{-3}	3.33
29	Family Sminthuridae	68, 69, 71, 74	1.56×10^{-5}	1×10^{-5}	9.25×10^{-1}	2
30	Family Staphylinidae	40, 41, 43, 44, 46, 52	9.09×10^{-4}	0.0005	6.00×10^{-1}	3.33
31	Genus Microvelia	44, 46	9.30×10^{-5}	2×10^{-5}	1.67×10^{-1}	3
32	*Glomeris marginata*	69, 70	6.98×10^{-3}	0.0001	1.80×10^{-2}	2
33	*Lacinius ephippiatus*	42, 44, 45, 46, 49, 50, 51, 53	3.00×10^{-4}	1×10^{-5}	4.51×10^{-2}	3.41
34	*Loricera pilicornis*	12, 41, 42, 43, 44, 46, 47, 52	3.80×10^{-3}	2×10^{-5}	4.51×10^{-3}	3.51
35	*Nebria brevicollis*	12, 26, 41, 42, 43, 44, 46, 47, 48, 49, 50, 51, 52, 53, 54, 71	4.90×10^{-3}	2×10^{-5}	4.51×10^{-3}	3.38
36	*Nemastoma bimaculatum*	42, 44, 46, 49, 50, 51, 53	4.67×10^{-4}	6×10^{-6}	1.35×10^{-2}	3.46
37	Non-Oribatidae	71	9.75×10^{-5}	3×10^{-5}	2.93×10^{-1}	2
38	*Odiellus spinosus*	69	1.00×10^{-4}	1×10^{-6}	1.35×10^{-2}	2
39	*Oniscus asellus*	69	8.30×10^{-3}	0.0019	2.30×10^{-1}	2
40	*Ophyiulus pilosus*	69	9.68×10^{-3}	0.0008	8.12×10^{-2}	2
41	Order Acari	52, 68, 69, 71	7.39×10^{-5}	5×10^{-5}	6.81×10^{-1}	2.33
42	Order Aranae	41, 44, 46, 50, 53	1.25×10^{-4}	2×10^{-6}	1.80×10^{-2}	3.39
43	Order Coleoptera larvae	41, 43, 44, 47, 50, 74	5.49×10^{-4}	0.0002	3.56×10^{-1}	3.33
44	Order Collembola	68, 69, 71, 74	2.74×10^{-5}	6×10^{-5}	2.20	2

45	Order Diplopoda	69, 70	7.03×10^{-3}	0.0028	3.93×10^{-1}	2
46	Order Diptera	69	4.61×10^{-4}	0.0004	7.62×10^{-1}	2
47	Order Diptera larvae	69, 71, 74	1.33×10^{-3}	0.0015	1.12	2
48	Order Enchytraeidae	68, 69, 71	3.87×10^{-4}	0.0003	6.63×10^{-1}	2
47	Order Diptera larvae	69, 71, 74	1.33×10^{-3}	0.0015	1.12	2
48	Order Enchytraeidae	68, 69, 71	3.87×10^{-4}	0.0003	6.63×10^{-1}	2
49	Order Gastropoda	41, 44, 52, 53, 67, 69, 71, 73, 74	3.09×10^{-2}	0.0336	1.09	2.52
50	Order Hemiptera	44, 46	9.30×10^{-5}	2×10^{-5}	1.67×10^{-1}	3
51	Order Isopoda	61, 69, 71	6.70×10^{-3}	0.0024	3.52×10^{-1}	2.33
52	Order Nematoda	52, 53, 71, 74	4.37×10^{-5}	1×10^{-6}	2.71×10^{-2}	2.33
53	Order Oligochaeta	68, 69, 71	1.00×10^{-2}	0.0029	2.84×10^{-1}	2
54	Order Opiliones	42, 44, 45, 46, 47, 49, 50, 51, 53, 69	3.50×10^{-4}	2×10^{-5}	5.87×10^{-2}	3.14
55	Order Pseudoscorpionidae	41, 44, 48, 52, 55	4.00×10^{-4}	2×10^{-6}	4.51×10^{-3}	3.17
56	Oribatidae	52, 68, 69, 71	5.66×10^{-5}	2×10^{-5}	3.88×10^{-1}	2.33
57	*Platynus assimile*	41, 42, 44, 46, 47, 51, 52, 61	9.65×10^{-3}	0.0083	8.62×10^{-1}	3.30
58	*Platynus dorsale*	12, 42, 43, 44, 46, 47, 53	3.70×10^{-3}	5×10^{-5}	1.35×10^{-2}	3.33
59	*Platynus obscurus*	41, 42, 44, 46, 47, 54	3.00×10^{-3}	1×10^{-5}	4.51×10^{-3}	3.48
60	*Polydesmus angustus*	69	6.29×10^{-3}	0.0018	2.93×10^{-1}	2
61	*Porcellio scaber*	61, 69	3.93×10^{-3}	0.0002	6.32×10^{-2}	2
62	*Pteroshicus melanarius*	5, 6, 7, 9, 12, 16, 23, 30, 42, 43, 44, 46, 47, 53, 61, 71,	3.32×10^{-2}	0.0001	4.51×10^{-3}	3.37
63	*Pterostichus diligens*	41, 42, 44, 46, 47, 54	4.80×10^{-3}	2×10^{-5}	4.51×10^{-3}	3.47
64	*Pterostichus nigrita*	12, 44, 53	1.25×10^{-2}	0.0018	1.40×10^{-1}	3.22
65	*Pterostichus strenuus*	12, 41	2.70×10^{-3}	0.0001	4.51×10^{-2}	3.50

(*continued*)

Table S4 (*continued*)

ID#	Taxa name	Prey ID#	BM	BA	NA	TH
66	*Trechus obtusus*	46	6.00×10^{-4}	8×10^{-6}	1.35×10^{-2}	3
67	*Anemone nemorosa*	–	–	–	–	1
68	Bacteria	–	–	–	–	1
69	Detritus	–	–	–	–	1
70	*Fraxinus excelsior*	–	–	–	–	1
71	Microfungi	–	–	–	–	1
72	Phylum Bryophyta	–	–	–	–	1
73	*Ranunculus ficaria*	–	–	–	–	1
74	Vegetation	–	–	–	–	1

Table S5 June 2005

ID#	Taxa name	Prey ID#	BM	BA	NA	TH
1	*Abax parallelepipedus*	8, 9, 10, 18, 54, 55, 56, 58, 63, 64, 66, 70, 77	8.71×10^{-2}	2.36×10^{-3}	2.71×10^{-2}	3.2365
2	*Agonum muelleri*	52, 53	8.10×10^{-3}	3.65×10^{-5}	4.51×10^{-3}	4.0201
3	*Agonum viduum*	56	5.95×10^{-3}	5.37×10^{-5}	9.02×10^{-3}	3
4	*Arion ater*	76, 79	2.76×10^{-1}	2.49×10^{-3}	9.02×10^{-3}	2
5	*Arion distinctus*	79	4.22×10^{-3}	9.52×10^{-5}	2.26×10^{-2}	2
6	*Arion subfuscus*	76, 77, 79	1.13×10^{-1}	5.10×10^{-3}	4.51×10^{-2}	2
7	*Bembidion lampros*	7, 11, 51, 53, 54, 56, 61, 64	6.00×10^{-4}	1.08×10^{-5}	1.80×10^{-2}	3.2455
8	*Carabus granulatus*	10, 53, 56, 57, 62, 64, 70	7.26×10^{-2}	8.84×10^{-3}	1.22×10^{-1}	3.2497
9	*Carabus nemoralis*	57, 62, 70	2.17×10^{-1}	9.80×10^{-4}	4.51×10^{-3}	3.2951
10	*Deroceras reticulatum*	51, 54, 63, 64, 74, 76, 79	1.76×10^{-2}	4.36×10^{-3}	2.48×10^{-1}	2.8626
11	Family Aphididae	79	1.80×10^{-4}	2.85×10^{-5}	1.58×10^{-1}	2
12	Family Bethylidae	53, 62	2.00×10^{-4}	9.02×10^{-7}	4.51×10^{-3}	3.4427
13	Family Byrrihdae	78, 79	2.00×10^{-4}	9.02×10^{-7}	4.51×10^{-3}	2
14	Family Ceraphronidae	11, 62	1.00×10^{-4}	1.35×10^{-6}	1.35×10^{-2}	3
15	Family Chrysomelidae	79	1.40×10^{-4}	3.16×10^{-6}	2.26×10^{-2}	3
16	Family Clubionidae	11, 56, 62	1.00×10^{-4}	2.71×10^{-6}	2.71×10^{-2}	3
17	Family Coreidae	79	1.20×10^{-3}	5.41×10^{-6}	4.51×10^{-3}	2
18	Family Curculionidae	79	8.55×10^{-4}	4.24×10^{-5}	4.96×10^{-2}	2
19	Family Cynipidae	11, 57, 79	1.00×10^{-4}	1.80×10^{-6}	1.80×10^{-2}	2.6667
20	Family Delphacidae	79	1.10×10^{-3}	4.96×10^{-6}	4.51×10^{-3}	2
21	Family Dictynidae	11, 56, 62	8.75×10^{-5}	7.11×10^{-6}	8.12×10^{-2}	3
22	Family Elateridae	11, 79	4.17×10^{-3}	2.26×10^{-4}	5.41×10^{-2}	2.5
23	Family Eulophidae	57, 62	1.00×10^{-4}	4.51×10^{-7}	4.51×10^{-3}	3

(continued)

Table S5 *(continued)*

ID#	Taxa name	Prey ID#	BM	BA	NA	TH
24	Family Formicidae	11, 54, 56, 62	1.21×10^{-4}	1.37×10^{-5}	1.13×10^{-1}	3
25	Family Gnaphosidae	62	1.22×10^{-4}	5.51×10^{-6}	4.51×10^{-2}	3
26	Family Gyrinidae	57	1.50×10^{-3}	1.35×10^{-5}	9.02×10^{-3}	3
27	Family Helidae	76	5.32×10^{-1}	2.40×10^{-3}	4.51×10^{-3}	2
28	Family Hydraenidae	76	1.00×10^{-4}	2.26×10^{-6}	2.26×10^{-2}	2
29	Family Hydrometridae	54, 56	1.48×10^{-3}	9.98×10^{-5}	6.77×10^{-2}	3
30	Family Hydrophilidae	76, 79	8.00×10^{-4}	3.61×10^{-6}	4.51×10^{-3}	2
31	Family Isotomidae	75, 76, 77	2.66×10^{-5}	1.59×10^{-4}	5.95	2
32	Family Neanuridae	75, 76, 77	1.05×10^{-5}	1.90×10^{-7}	1.80×10^{-2}	2
33	Family Platygastridae	57	1.00×10^{-4}	9.02×10^{-7}	9.02×10^{-3}	3
34	Family Pompilidae	16	1.00×10^{-3}	4.51×10^{-6}	4.51×10^{-3}	4
35	Family Pselaphidae	51, 54, 63, 64	1.00×10^{-4}	5.41×10^{-6}	5.41×10^{-2}	3.1667
36	Family Ptiliidae	76, 77	1.00×10^{-4}	9.02×10^{-7}	9.02×10^{-3}	2
37	Family Silphidae	27	3.08×10^{-2}	2.77×10^{-4}	9.02×10^{-3}	3
38	Family Sminthuridae	75, 76, 77, 79	2.50×10^{-5}	1.05×10^{-4}	4.19	2
39	Family Sphecidae	52	8.00×10^{-4}	3.61×10^{-6}	4.51×10^{-3}	4.1548
40	Family Staphylinidae	11, 50, 51, 53, 54, 56, 63	6.80×10^{-4}	2.27×10^{-4}	3.34×10^{-1}	3.2217
41	Genus Microvelia	20, 54, 56	6.76×10^{-5}	8.84×10^{-6}	1.31×10^{-1}	3
42	*Lacinius epphipiatus*	52, 54, 55, 56, 59, 60, 61, 64	7.00×10^{-4}	1.26×10^{-5}	1.80×10^{-2}	3.3735
43	*Leiobunum blackwalli*	55, 56, 61, 64, 65	2.00×10^{-4}	6.32×10^{-6}	3.16×10^{-2}	3.25
44	*Loricera pilicornis*	10, 11, 51, 52, 53, 54, 56, 57, 63	4.35×10^{-3}	3.93×10^{-5}	9.02×10^{-3}	3.3966
45	*Nemastoma bimaculatum*	52, 54, 56, 59, 60, 61, 64	5.50×10^{-4}	4.96×10^{-6}	9.02×10^{-3}	3.4269
46	Non-Oribatidae	77	9.03×10^{-5}	2.61×10^{-5}	2.89×10^{-1}	2

47	*Odiellus spinosus*	76	8.00×10^{-4}	1.08×10^{-5}	1.35×10^{-2}	2
48	*Oligolophus agrestis*	54, 57, 59, 61, 62	9.00×10^{-4}	4.06×10^{-6}	4.51×10^{-3}	3.2167
49	*Oniscus asellus*	76	5.13×10^{-3}	2.08×10^{-4}	4.06×10^{-2}	2
50	*Ophyiulus pilosus*	76	2.05×10^{-3}	1.85×10^{-5}	9.02×10^{-3}	2
51	Order Acari	63, 75, 76, 77	4.37×10^{-3}	5.91×10^{-5}	1.35×10^{-2}	2.3333
52	Order Aranae	11, 51, 54, 56, 60, 62, 64	1.00×10^{-4}	1.53×10^{-5}	1.53×10^{-1}	3.1548
53	Order Coleoptera larvae	11, 51, 53, 54, 57, 60, 62, 79	6.33×10^{-4}	1.97×10^{-4}	3.11×10^{-1}	2.8854
54	Order Collembola	75, 76, 77, 79	2.60×10^{-5}	2.64×10^{-4}	1.02×10	2
55	Order Diplopoda	76	9.20×10^{-3}	1.66×10^{-4}	1.80×10^{-2}	2
56	Order Diptera	76	1.11×10^{-4}	5.30×10^{-4}	4.76	2
57	Order Diptera larvae	76, 77, 79	5.55×10^{-4}	1.25×10^{-4}	2.26×10^{-1}	2
58	Order Enchytraeidae	75, 76, 77	1.40×10^{-4}	1.08×10^{-5}	7.67×10^{-2}	2
59	Order Gastropoda	51, 54, 63, 64, 74, 76, 77, 79	4.46×10^{-2}	1.47×10^{-2}	3.29×10^{-1}	2.5833
60	Order Hemiptera	20, 54, 56, 79	3.89×10^{-4}	1.42×10^{-4}	3.65×10^{-1}	2.75
61	Order Isopoda	70, 76	7.60×10^{-3}	4.11×10^{-4}	5.41×10^{-2}	2.5
62	Order Lepidoptera larvae	79	2.21×10^{-2}	2.99×10^{-4}	1.35×10^{-2}	2
63	Order Nematoda	63, 64, 77, 79	5.00×10^{-5}	9.02×10^{-7}	1.80×10^{-2}	2.3333
64	Order Oligochaeta	75, 76, 77	1.68×10^{-2}	1.90×10^{-3}	1.13×10^{-1}	2
65	Order Opiliones	11, 52, 54, 55, 56, 57, 59, 60, 61, 62, 64, 65, 76	4.88×10^{-4}	3.96×10^{-5}	8.12×10^{-2}	3.1657
66	Oribatidae	63, 75, 76, 77	4.37×10^{-5}	2.72×10^{-5}	6.23×10^{-1}	2.3333
67	*Phalangium opilio*	11, 52, 55, 59, 62, 64, 65	1.00×10^{-4}	4.51×10^{-7}	4.51×10^{-3}	3.4148
68	*Platynus assimile*	51, 52, 54, 56, 57, 61, 63, 70	1.20×10^{-2}	4.59×10^{-3}	3.84×10^{-1}	3.2902
69	*Polydesmus angustus*	76	1.74×10^{-2}	1.57×10^{-4}	9.02×10^{-3}	2

(*continued*)

Table S5 (*continued*)

ID#	Taxa name	Prey ID#	BM	BA	NA	TH
70	*Porcellio scaber*	70, 76	1.50×10^{-2}	2.03×10^{-4}	1.35×10^{-2}	2
71	*Pterostichus melanarius*	5, 6, 7, 10, 11, 18, 40, 52, 53, 54, 56, 57, 62, 64, 70, 77	3.72×10^{-2}	3.36×10^{-4}	9.02×10^{-3}	3.2338
72	*Pterostichus strenuus*	10, 51	4.37×10^{-3}	5.91×10^{-5}	1.35×10^{-2}	3.5979
73	*Rilaena triangularis*	11, 56	1.40×10^{-3}	6.32×10^{-6}	4.51×10^{-3}	3
74	*Anemone nemorosa*	–	–	–	–	1
75	Bacteria	–	–	–	–	1
76	Detritus	–	–	–	–	1
77	Microfungi	–	–	–	–	1
78	Phylum Bryophyta	–	–	–	–	1
79	Vegetation	–	–	–	–	1

Table S6 August 2005

ID#	Taxa name	Prey ID#	BM	BA	NA	TH
1	*Arion ater*	66, 69	2.69×10^{-1}	3.64×10^{-3}	1.35×10^{-2}	2
2	*Arion distinctus*	69	7.67×10^{-4}	6.23×10^{-5}	8.12×10^{-2}	2
3	*Arion subfuscus*	66, 67, 69	1.50×10^{-3}	2.51×10^{-4}	1.67×10^{-1}	2
4	*Carabus granulatus*	5, 44, 47, 48, 53, 55, 61	3.68×10^{-2}	3.32×10^{-4}	9.02×10^{-3}	3.2391
5	*Deroceras reticulatum*	42, 45, 54, 55, 66, 68, 69, 70	2.68×10^{-3}	2.21×10^{-3}	8.26×10^{-1}	2.6667
6	Discus rotundatus	66, 69	5.72×10^{-2}	2.58×10^{-4}	4.51×10^{-3}	2
7	Family Anthocoridae	51, 53	1.00×10^{-4}	4.51×10^{-7}	4.51×10^{-3}	3.2381
8	Family Aphididae	68, 69	1.00×10^{-4}	2.26×10^{-6}	2.26×10^{-2}	2
9	Family Bethylidae	44, 53	2.00×10^{-4}	9.02×10^{-7}	4.51×10^{-3}	3.5034
10	Family Ceraphronidae	8, 53	5.21×10^{-5}	1.88×10^{-6}	3.61×10^{-2}	3
11	Family Clubionidae	8, 47, 53	8.33×10^{-5}	7.52×10^{-6}	9.02×10^{-2}	3
12	Family Cynipidae	8, 48, 69	6.63×10^{-5}	4.49×10^{-6}	6.77×10^{-2}	2.6667
13	Family Delphacidae	69	3.00×10^{-4}	4.06×10^{-6}	1.35×10^{-2}	2
14	Family Diapriidae	48	5.83×10^{-5}	1.05×10^{-6}	1.80×10^{-2}	3
15	Family Dictynidae	8, 47, 53	1.21×10^{-4}	1.74×10^{-5}	1.44×10^{-1}	3
16	Family Elateridae	8, 69	1.20×10^{-3}	2.17×10^{-5}	1.80×10^{-2}	3
17	Family Elmidae	66	1.00×10^{-4}	1.35×10^{-6}	1.35×10^{-2}	2
18	Family Eulophidae	48, 53	1.00×10^{-4}	9.02×10^{-7}	9.02×10^{-3}	3
19	Family Formicidae	8, 45, 47, 53	1.00×10^{-4}	4.51×10^{-7}	4.51×10^{-3}	3
20	Family Gerridae	47	8.33×10^{-4}	2.63×10^{-5}	3.16×10^{-2}	3
21	Family Gnaphosidae	53	1.00×10^{-4}	2.26×10^{-6}	2.26×10^{-2}	3
22	Family Helidae	66	3.61×10^{-1}	1.63×10^{-3}	4.51×10^{-3}	2
23	Family Hydraenidae	66	1.00×10^{-4}	2.26×10^{-6}	2.26×10^{-2}	2
24	Family Hydrophilidae	66, 69	2.00×10^{-4}	9.02×10^{-7}	4.51×10^{-3}	2
25	Family Isotomidae	65, 66, 67	4.39×10^{-5}	1.01×10^{-5}	2.30×10^{-1}	2
26	Family Neanuridae	65, 66, 67	2.42×10^{-5}	1.53×10^{-6}	6.32×10^{-2}	2
27	Family Philodromidae	51, 53	1.53×10^{-3}	2.75×10^{-5}	1.80×10^{-2}	3.3571

(continued)

Table S6 *(continued)*

1D#	Taxa name	Prey ID#	BM	BA	NA	TH
28	Family Platygastridae	48	5.30×10^{-5}	2.39×10^{-6}	4.51×10^{-2}	3
29	Family Pselaphidae	42, 45, 54, 55	7.00×10^{-5}	1.89×10^{-6}	2.71×10^{-2}	3.1667
30	Family Scymaenidae	57	6.00×10^{-4}	2.71×10^{-6}	4.51×10^{-3}	3.3333
31	Family Sminthuridae	65, 66, 67, 69	3.56×10^{-5}	5.94×10^{-5}	1.67	2
32	Family Sphecidae	43	1.67×10^{-2}	7.53×10^{-5}	4.51×10^{-3}	4.131
33	Family Staphylinidae	8, 42, 44, 45, 47, 54	1.20×10^{-4}	7.59×10^{-6}	6.32×10^{-2}	3.2789
34	Family Tingidae	69	1.00×10^{-4}	4.51×10^{-7}	4.51×10^{-3}	2
35	*Lacinius epphipiatus*	43, 45, 46, 47, 50, 51, 52, 55	2.22×10^{-3}	5.01×10^{-5}	2.26×10^{-2}	3.358
36	*Leiobunum blackwalli*	46, 47, 52, 55, 56	2.13×10^{-3}	3.84×10^{-5}	1.80×10^{-2}	3.3949
37	*Leiobunum rotundum*	43, 46, 47, 50, 51, 52, 55	2.60×10^{-3}	4.69×10^{-5}	1.80×10^{-2}	3.4091
38	Non-Oribatidae	67	1.03×10^{-4}	1.68×10^{-5}	1.62×10^{-1}	2
39	*Odiellus spinosus*	66	1.90×10^{-3}	3.43×10^{-5}	1.80×10^{-2}	2
40	*Oligolophus agrestis*	45, 48, 50, 52, 53	2.20×10^{-3}	3.97×10^{-5}	1.80×10^{-2}	3.2037
41	*Oniscus asellus*	66	2.77×10^{-3}	1.25×10^{-4}	4.51×10^{-2}	2
42	Order Acari	54, 65, 66, 67	4.42×10^{-5}	5.82×10^{-5}	1.32	2.3333
43	Order Aranae	8, 13, 42, 45, 47, 51, 53, 55	2.11×10^{-4}	5.81×10^{-5}	2.75×10^{-1}	3.131
44	Order Coleoptera larvae	8, 42, 44, 45, 48, 51, 53, 69	2.92×10^{-3}	9.22×10^{-5}	3.16×10^{-2}	3.0068
45	Order Collembola	65, 66, 67, 69	3.62×10^{-5}	7.11×10^{-5}	1.96	2
46	Order Diplopoda	66	1.00×10^{-4}	1.80×10^{-6}	1.80×10^{-2}	2
47	Order Diptera	66	4.88×10^{-4}	2.02×10^{-4}	4.15×10^{-1}	2
48	Order Diptera larvae	66, 67, 69	4.12×10^{-4}	3.35×10^{-5}	8.12×10^{-2}	2
49	Order Enchytraeidae	65, 66, 67	7.33×10^{-5}	9.93×10^{-7}	1.35×10^{-2}	2

50	Order Gastropoda	42, 45, 54, 55, 66, 67, 68, 69, 70	7.34×10^{-3}	8.05×10^{-3}	1.10	2.5185
51	Order Hemiptera	45, 47, 53, 68, 69	4.20×10^{-4}	3.22×10^{-5}	7.67×10^{-2}	2.7143
52	Order Isopoda	61, 66	3.86×10^{-3}	2.26×10^{-4}	5.87×10^{-2}	2.5
53	Order Lepidoptera larvae	69	3.00×10^{-4}	1.35×10^{-6}	4.51×10^{-3}	2
54	Order Nematoda	54, 55, 67, 69	2.86×10^{-5}	2.58×10^{-7}	9.02×10^{-3}	2.3333
55	Order Oligochaeta	65, 66, 67	7.34×10^{-3}	6.62×10^{-4}	9.02×10^{-2}	2
56	Order Opiliones	8, 43, 45, 46, 47, 48, 50, 51, 52, 53, 55, 56	1.96×10^{-3}	2.21×10^{-4}	1.13×10^{-1}	3.1553
57	Oribatidae	54, 65, 66, 67	3.63×10^{-5}	4.20×10^{-5}	1.16	2.3333
58	*Phalangium opilio*	8, 43, 46, 50, 53, 55, 56	2.50×10^{-3}	1.13×10^{-5}	4.51×10^{-3}	3.4007
59	*Platynus assimile*	42, 43, 45, 47, 48, 52, 54, 61	1.02×10^{-2}	7.37×10^{-4}	7.22×10^{-2}	3.2872
60	*Polydesmus angustus*	66	1.00×10^{-4}	1.80×10^{-6}	1.80×10^{-2}	2
61	*Porcellio scaber*	61, 66	7.50×10^{-3}	1.02×10^{-4}	1.35×10^{-2}	2
62	*Pterostichus strenuus*	5, 42	6.90×10^{-3}	3.11×10^{-5}	4.51×10^{-3}	3.5
63	*Rilaena triangularis*	8, 47	3.33×10^{-5}	4.51×10^{-7}	1.35×10^{-2}	3
64	*Trechus obtusus*	47	5.67×10^{-4}	7.67×10^{-6}	1.35×10^{-2}	3
65	Bacteria	–	–	–	–	1
66	Detritus	–	–	–	–	1
67	Microfungi	–	–	–	–	1
68	*Urtica dioica*	–	–	–	–	1
69	Vegetation	–	–	–	–	1
70	*Veronica montana*	–	–	–	–	1

Table S7 December 2005

ID#	Taxa name	Prey ID#	BM	BA	NA	TH
1	*Arion ater*	44, 50, 51	7.80×10^{-3}	7.04×10^{-5}	9.02×10^{-3}	2
2	*Arion distinctus*	51	5.65×10^{-3}	1.02×10^{-4}	1.80×10^{-2}	2
3	*Arion subfuscus*	44, 45, 51	1.23×10^{-2}	3.87×10^{-4}	3.16×10^{-2}	2
4	*Asellus aquaticus*	44, 45	1.00×10^{-4}	1.35×10^{-6}	1.35×10^{-2}	2
5	*Deroceras reticulatum*	21, 24, 33, 34, 42, 44, 46, 48, 49, 50, 51, 52	9.52×10^{-3}	1.80×10^{-3}	1.89×10^{-1}	2.42
6	Family Aphididae	50, 51	1.00×10^{-4}	1.35×10^{-6}	1.35×10^{-2}	2
7	Family Gerridae	26	7.82×10^{-3}	1.76×10^{-4}	2.26×10^{-2}	3
8	Family Isotomidae	43, 44, 45	3.83×10^{-5}	1.38×10^{-6}	3.61×10^{-2}	2
9	Family Linyphiidae	6, 21, 24, 26	1.33×10^{-4}	1.14×10^{-5}	8.57×10^{-2}	3.08
10	Family Neanuridae	43, 44, 45	3.05×10^{-6}	3.16×10^{-7}	1.04×10^{-1}	2
11	Family Platygastridae	27	5.00×10^{-5}	2.26×10^{-7}	4.51×10^{-3}	3
12	Family Sminthuridae	43, 44, 45, 51	3.78×10^{-5}	1.02×10^{-6}	2.71×10^{-2}	2
13	Family Staphylinidae	6, 20, 21, 23, 24, 26, 33	1.22×10^{-4}	4.96×10^{-6}	4.06×10^{-2}	3.26
14	Family Thomisidae	6, 22, 26, 30, 31, 34	1.00×10^{-4}	4.51×10^{-7}	4.51×10^{-3}	3.43
15	*Lacinius ephipiatus*	22, 24, 25, 26, 29, 30, 32, 34	2.00×10^{-3}	9.02×10^{-6}	4.51×10^{-3}	3.27
16	*Nebria brevicollis*	5, 6, 21, 22, 23, 24, 26, 27, 28, 29, 30, 32, 33, 34, 35, 45	1.79×10^{-2}	1.61×10^{-4}	9.02×10^{-3}	3.28
17	*Nemastoma bimaculatum*	22, 24, 26, 29, 30, 32, 34	8.00×10^{-4}	3.61×10^{-6}	4.51×10^{-3}	3.30
18	Non-Oribatidae	45	1.29×10^{-4}	1.46×10^{-5}	1.13×10^{-1}	2
19	*Oniscus asellus*	44	6.15×10^{-3}	5.55×10^{-5}	9.02×10^{-3}	2
20	*Ophyiulus pilosus*	44	2.20×10^{-3}	9.93×10^{-6}	4.51×10^{-3}	2
21	Order Acari	33, 43, 44, 45	7.39×10^{-5}	2.20×10^{-5}	2.98×10^{-1}	2.33
22	Order Aranae	6, 21, 24, 26, 30, 31, 34	1.32×10^{-4}	1.19×10^{-5}	9.02×10^{-2}	3.24
23	Order Coleoptera larvae	6, 21, 23, 24, 27, 30, 51	1.00×10^{-4}	9.02×10^{-7}	9.02×10^{-3}	3.13
24	Order Collembola	43, 44, 45, 51	1.54×10^{-5}	2.57×10^{-6}	1.67×10^{-1}	2
25	Order Diplopoda	44	6.40×10^{-3}	1.16×10^{-4}	1.80×10^{-2}	2

26	Order Diptera	44	4.63×10^{-4}	4.39×10^{-5}	9.47×10^{-2}	2
27	Order Diptera larvae	44, 45, 51	1.52×10^{-2}	9.62×10^{-4}	6.32×10^{-2}	2
28	Order Enchytraeidae	43, 44, 45	5.05×10^{-5}	8.65×10^{-6}	1.71×10^{-1}	2
29	Order Gastropoda	21, 24, 33, 34, 42, 44, 45, 46, 48, 49, 50, 51, 52	9.52×10^{-3}	2.36×10^{-3}	2.48×10^{-1}	2.36
30	Order Hemiptera	24, 26, 50, 51	4.93×10^{-3}	1.78×10^{-4}	3.61×10^{-2}	2.33
31	Order Hymenoptera	27	9.52×10^{-3}	2.36×10^{-3}	2.48×10^{-1}	3
32	Order Isopoda	40, 41, 44, 45, 47	3.66×10^{-3}	1.32×10^{-4}	3.61×10^{-2}	2.20
33	Order Nematoda	33, 34, 45, 51	2.86×10^{-5}	1.93×10^{-6}	6.77×10^{-2}	2.33
34	Order Oligochaeta	43, 44, 45	1.08×10^{-2}	3.88×10^{-4}	3.61×10^{-2}	2
35	Order Opiliones	22, 24, 25, 26, 27, 29, 30, 32, 34, 44	1.40×10^{-3}	1.26×10^{-5}	9.02×10^{-3}	3.13
36	Order Psocoptera	45	1.00×10^{-4}	4.51×10^{-7}	4.51×10^{-3}	2
37	Oribatidae	33, 43, 44, 45	3.94×10^{-5}	7.29×10^{-6}	1.85×10^{-1}	2.33
38	*Platynus assimile*	21, 22, 24, 26, 27, 32, 33, 40	1.21×10^{-2}	5.46×10^{-5}	4.51×10^{-3}	3.26
39	*Polydesmus angustus*	44	7.80×10^{-3}	1.06×10^{-4}	1.35×10^{-2}	2
40	*Porcellio scaber*	40, 41, 44, 47	5.57×10^{-3}	7.53×10^{-5}	1.35×10^{-2}	2
41	*Alnus glutinosa*	–	–	–	–	1
42	*Anemone nemorosa*	–	–	–	–	1
43	Bacteria	–	–	–	–	1
44	Detritus	–	–	–	–	1
45	Microfungi	–	–	–	–	1
46	*Osmunda regalis*	–	–	–	–	1
47	*Quercus robur*	–	–	–	–	1
48	*Ranunculus ficaria*	–	–	–	–	1
49	*Ranunculus repens*	–	–	–	–	1
50	*Urtica dioica*	–	–	–	–	1
51	Vegetation	–	–	–	–	1
52	*Veronica montana*	–	–	–	–	1

Table S8 February 2006

ID#	Taxa name	Prey ID#	BM	BA	NA	TH
1	*Arion distinctus*	33	5.67×10^{-3}	7.67×10^{-5}	1.35×10^{-2}	2
2	*Arion subfuscus*	31, 32, 33	1.88×10^{-2}	5.95×10^{-4}	3.16×10^{-2}	2
3	*Deroceras reticulatum*	21, 23, 27, 29, 31, 33	1.16×10^{-2}	1.36×10^{-3}	1.17×10^{-1}	2.50
4	*Discus rotundatus*	31, 33	4.45×10^{-2}	1.00×10^{-3}	2.26×10^{-2}	2
5	Family Amaurobiidae	27	8.50×10^{-3}	3.84×10^{-5}	4.51×10^{-3}	3
6	Family Delphacidae	33	5.00×10^{-4}	2.26×10^{-6}	4.51×10^{-3}	2
7	Family Elmidae	31	1.00×10^{-4}	4.51×10^{-7}	4.51×10^{-3}	2
8	Family Gerridae	24	1.94×10^{-3}	7.90×10^{-5}	4.06×10^{-2}	3
9	Family Hydraenidae	31	1.00×10^{-4}	4.51×10^{-7}	4.51×10^{-3}	2
10	Family Isotomidae	30, 31, 32	2.96×10^{-5}	4.13×10^{-6}	1.40×10^{-1}	2
11	Family Linyphiidae	6, 21, 23, 24	1.00×10^{-4}	2.71×10^{-6}	2.71×10^{-2}	3
12	Family Lithobiidae	21	6.20×10^{-3}	2.80×10^{-5}	4.51×10^{-3}	3
13	Family Neanuridae	30, 31, 32	3.03×10^{-6}	3.83×10^{-7}	1.26×10^{-1}	2
14	Family Pselaphidae	21, 23, 27	3.33×10^{-5}	6.02×10^{-7}	1.80×10^{-2}	3
15	Family Sminthuridae	30, 31, 32, 33	7.74×10^{-6}	9.08×10^{-7}	1.17×10^{-1}	2
16	Family Staphylinidae	20, 21, 22, 23, 24	5.67×10^{-4}	7.67×10^{-6}	1.35×10^{-2}	3.24
17	Family Theriidae	21, 24	1.00×10^{-4}	4.51×10^{-7}	4.51×10^{-3}	3
18	Non-Oribatidae	32	1.04×10^{-4}	4.69×10^{-6}	4.51×10^{-2}	2
19	*Oniscus asellus*	31	8.85×10^{-3}	1.20×10^{-4}	1.35×10^{-2}	2
20	*Ophyiulus pilosus*	31	1.33×10^{-2}	1.20×10^{-4}	9.02×10^{-3}	2
21	Order Acari	30, 31, 32	6.72×10^{-5}	6.67×10^{-6}	9.93×10^{-2}	2
22	Order Coleoptera larvae	21, 22, 23, 26, 33	1.10×10^{-3}	4.96×10^{-6}	4.51×10^{-3}	3.22
23	Order Collembola	30, 31, 32, 33	1.46×10^{-5}	5.60×10^{-6}	3.84×10^{-1}	2
24	Order Diptera	31	6.02×10^{-4}	6.25×10^{-5}	1.04×10^{-1}	2
25	Order Enchytraeidae	30, 31, 32	6.06×10^{-5}	1.09×10^{-6}	1.80×10^{-2}	2
26	Order Hemiptera	23, 24, 33	1.80×10^{-3}	8.12×10^{-5}	4.51×10^{-2}	2.67
27	Order Oligochaeta	30, 31, 32	1.04×10^{-2}	1.87×10^{-4}	1.80×10^{-2}	2
28	Oribatidae	30, 31, 32	3.65×10^{-5}	1.98×10^{-6}	5.41×10^{-2}	2
29	*Anemone nemorosa*		–	–	–	1
30	Bacteria		–	–	–	1
31	Detritus		–	–	–	1
32	Microfungi		–	–	–	1
33	Vegetation		–	–	–	1

Table S9 Composite 2005–2006

ID#	Taxa name	Prey ID#	BM	BA	NA	TH
1	*Abax parallelepipedus*	11, 12, 13, 24, 75, 76, 77, 79, 85, 86, 90, 96, 108	8.56×10^{-2}	2.70×10^{-3}	3.16×10^{-2}	3.20
2	*Agonum muelleri*	73, 74	6.05×10^{-3}	5.46×10^{-5}	9.02×10^{-3}	4.07
3	*Agonum viduum*	77	5.95×10^{-3}	5.37×10^{-5}	9.02×10^{-3}	3
4	*Amara plebja*	77, 78, 79	5.80×10^{-3}	2.62×10^{-5}	4.51×10^{-3}	3
5	*Arion ater*	106, 115	1.16×10^{-1}	7.85×10^{-3}	6.77×10^{-2}	2
6	*Arion distinctus*	115	1.10×10^{-2}	2.38×10^{-3}	2.17×10^{-1}	2
7	*Arion hortensis*	106, 115	2.20×10^{-2}	1.59×10^{-3}	7.22×10^{-2}	2
8	*Arion subfuscus*	106, 108, 115	3.11×10^{-2}	1.51×10^{-2}	4.87×10^{-1}	2
9	*Asellus aquaticus*	106, 108	1.81×10^{-3}	9.79×10^{-5}	5.41×10^{-2}	2
10	*Bembidion lampros*	10, 17, 72, 74, 75, 77, 83, 86, 88	1.61×10^{-3}	7.26×10^{-5}	4.51×10^{-2}	3.34
11	*Carabus granulatus*	13, 74, 77, 78, 84, 86, 96	6.64×10^{-2}	1.86×10^{-2}	2.80×10^{-1}	3.20
12	*Carabus nemoralis*	78, 84, 96	1.65×10^{-1}	6.69×10^{-3}	4.06×10^{-2}	3.34
13	*Deroceras reticulatum*	72, 75, 85, 86, 104, 106, 109, 112, 113, 114, 115, 116	1.24×10^{-2}	2.57×10^{-2}	2.07	2.39
14	*Discus rotundatus*	106, 115	4.66×10^{-2}	1.26×10^{-3}	2.71×10^{-2}	2
15	Family Amaurobiidae	86	8.50×10^{-3}	3.84×10^{-5}	4.51×10^{-3}	3.50
16	Family Anthocoridae	81, 84	1.00×10^{-4}	4.51×10^{-7}	4.51×10^{-3}	3.36
17	Family Aphididae	114, 115	1.80×10^{-4}	3.65×10^{-5}	2.03×10^{-1}	2
18	Family Bethylidae	74, 84	2.00×10^{-4}	9.02×10^{-7}	4.51×10^{-3}	3.50
19	Family Byrrhidae	110, 115	1.50×10^{-4}	2.71×10^{-6}	1.80×10^{-2}	2
20	Family Ceraphronidae	17, 84	6.17×10^{-5}	3.06×10^{-6}	4.96×10^{-2}	3
21	Family Chrysomelidae	115	1.40×10^{-4}	3.16×10^{-6}	2.26×10^{-2}	2
22	Family Clubionidae	17, 77, 84	8.44×10^{-5}	1.03×10^{-5}	1.22×10^{-1}	3
23	Family Coreidae	81	1.20×10^{-3}	5.41×10^{-6}	4.51×10^{-3}	3.71

(continued)

Table S9 (*continued*)

ID#	Taxa name	Prey ID#	BM	BA	NA	TH
24	Family Curculionidae	115	1.95×10^{-3}	1.93×10^{-4}	9.93×10^{-2}	2
25	Family Cynipidae	17, 78, 115	7.26×10^{-5}	6.23×10^{-6}	8.57×10^{-2}	2.67
26	Family Delphacidae	115	3.67×10^{-4}	8.27×10^{-6}	2.26×10^{-2}	2
27	Family Diapriidae	78	5.83×10^{-5}	1.05×10^{-6}	1.80×10^{-2}	3
28	Family Dictynidae	17, 77, 84	1.09×10^{-4}	2.46×10^{-5}	2.26×10^{-1}	3
29	Family Elateridae	17, 115	2.71×10^{-3}	5.99×10^{-4}	2.21×10^{-1}	3
30	Family Elmidae	105	8.33×10^{-5}	2.26×10^{-6}	2.71×10^{-2}	2
31	Family Eulophidae	78, 84	1.00×10^{-4}	1.35×10^{-6}	1.35×10^{-2}	3
32	Family Formicidae	17, 75, 77, 84	1.22×10^{-4}	1.70×10^{-5}	1.40×10^{-1}	3
33	Family Gerridae	77	4.04×10^{-3}	2.55×10^{-4}	6.32×10^{-2}	3
34	Family Gnaphosidae	84	1.17×10^{-4}	7.90×10^{-6}	6.77×10^{-2}	3
35	Family Gyrinidae	78	1.50×10^{-3}	1.35×10^{-5}	9.02×10^{-3}	3
36	Family Helidae	106	4.46×10^{-1}	4.03×10^{-3}	9.02×10^{-3}	2
37	Family Hydraenidae	106	1.08×10^{-4}	6.84×10^{-6}	6.32×10^{-2}	2
38	Family Hydrometridae	75, 77	1.48×10^{-3}	9.98×10^{-5}	6.77×10^{-2}	3
39	Family Hydrophilidae	106, 115	1.93×10^{-3}	3.47×10^{-5}	1.80×10^{-2}	2
40	Family Isotomidae	105, 106, 108	3.05×10^{-5}	2.31×10^{-4}	7.59	2
41	Family Linyphiidae	17, 26, 72, 75, 77	1.35×10^{-4}	1.77×10^{-5}	1.31×10^{-1}	3.07
42	Family Lithobiidae	72	6.20×10^{-3}	2.80×10^{-5}	4.51×10^{-3}	3.33
43	Family Neanuridae	105, 106, 108	1.47×10^{-5}	6.29×10^{-6}	4.29×10^{-1}	2
44	Family Philodromidae	81, 84	1.53×10^{-3}	2.75×10^{-5}	1.80×10^{-2}	3.36
45	Family Platygastridae	78	5.94×10^{-5}	3.76×10^{-6}	6.32×10^{-2}	3
46	Family Pompilidae	22	1.00×10^{-3}	4.51×10^{-6}	4.51×10^{-3}	4
47	Family Pselaphidae	72, 75, 85, 86	8.25×10^{-5}	8.56×10^{-6}	1.04×10^{-1}	3.17
48	Family Ptiliidae	106, 108	1.01×10^{-4}	4.11×10^{-6}	4.06×10^{-2}	2
49	Family Scarabaeidae	106, 108	6.50×10^{-3}	2.93×10^{-5}	4.51×10^{-3}	2

50	Family Scydmaenidae	90	2.33×10^{-4}	3.16×10^{-6}	1.35×10^{-2}	3.33
51	Family Silphidae	36	2.79×10^{-2}	3.78×10^{-4}	1.35×10^{-2}	3
52	Family Sminthuridae	105, 106, 108, 115	2.76×10^{-5}	1.74×10^{-4}	6.28	2
53	Family Sphecidae	73	8.75×10^{-3}	7.90×10^{-5}	9.02×10^{-3}	4.13
54	Family Staphylinidae	17, 71, 72, 74, 75, 77, 85	5.23×10^{-4}	5.14×10^{-4}	9.84×10^{-1}	3.24
55	Family Theriidae	72, 77	1.00×10^{-4}	4.51×10^{-7}	4.51×10^{-3}	3.17
56	Family Thomisidae	17, 73, 77, 81, 82, 86	1.00×10^{-4}	4.51×10^{-7}	4.51×10^{-3}	3.52
57	Family Tingidae	115	1.00×10^{-4}	4.51×10^{-7}	4.51×10^{-3}	2
58	Genus Microvelia	26, 75, 77	8.25×10^{-5}	2.42×10^{-5}	2.93×10^{-1}	3
59	*Gerris lacustris*	77	1.86×10^{-3}	7.57×10^{-5}	4.06×10^{-2}	3
60	*Glomeris marginata*	106, 107	6.98×10^{-3}	1.26×10^{-4}	1.80×10^{-2}	2
61	*Lacinius epphipiatus*	73, 75, 76, 77, 80, 81, 83, 86	1.55×10^{-3}	9.79×10^{-5}	6.32×10^{-2}	3.30
62	*Leiobunum blackwalli*	76, 77, 83, 86, 87	9.00×10^{-4}	4.47×10^{-5}	4.96×10^{-2}	3.05
63	*Leiobunum rotundum*	73, 76, 77, 80, 81, 83, 86	2.60×10^{-3}	4.69×10^{-5}	1.80×10^{-2}	3.34
64	*Loricera pilicornis*	13, 17, 72, 73, 74, 75, 77, 78, 85	4.17×10^{-3}	5.64×10^{-5}	1.35×10^{-2}	3.35
65	*Nebria brevicollis*	13, 17, 48, 72, 73, 74, 75, 77, 78, 79, 80, 81, 83, 85, 86, 87, 108	1.35×10^{-2}	1.83×10^{-4}	1.35×10^{-2}	3.27
66	*Nemastoma bimaculatum*	73, 75, 77, 80, 81, 83, 86	5.50×10^{-4}	1.49×10^{-5}	2.71×10^{-2}	3.34
67	Non-Oribatida	108	1.03×10^{-4}	7.58×10^{-5}	7.36×10^{-1}	2
68	*Odiellus spinosus*	106	1.34×10^{-3}	6.03×10^{-5}	4.51×10^{-2}	2
69	*Oligolophus agrestis*	75, 78, 80, 83, 84	1.94×10^{-3}	4.38×10^{-5}	2.26×10^{-2}	3.11
70	*Oniscus asellus*	106	7.26×10^{-3}	2.82×10^{-3}	3.88×10^{-1}	2
71	*Ophyiulus pilosus*	106	7.75×10^{-3}	6.65×10^{-4}	8.57×10^{-2}	2
72	Order Acari	85, 105, 106, 108	5.43×10^{-5}	1.66×10^{-4}	3.06	2.33
73	Order Aranae	17, 26, 72, 75, 77, 81, 82, 84, 86	2.31×10^{-4}	1.33×10^{-4}	5.78×10^{-1}	3.13

(*continued*)

Table S9 (*continued*)

ID#	Taxa name	Prey ID#	BM	BA	NA	TH
74	Order Coleoptera larvae	17, 72, 74, 75, 78, 81, 84, 115	6.65×10^{-4}	4.86×10^{-4}	7.31×10^{-1}	3.01
75	Order Collembola	105, 106, 108, 115	2.88×10^{-5}	4.11×10^{-4}	1.43×10	2
76	Order Diplopoda	106, 107	6.82×10^{-3}	2.89×10^{-3}	4.24×10^{-1}	2
77	Order Diptera	106	1.85×10^{-4}	1.08×10^{-3}	5.86	2
78	Order Diptera larvae	106, 108, 115	2.04×10^{-3}	2.78×10^{-3}	1.37	2
79	Order Enchytraeidae	105, 106, 108	2.70×10^{-4}	2.62×10^{-4}	9.70×10^{-1}	2
80	Order Gastropoda	72, 75, 85, 86, 104, 106, 108, 109, 112, 113, 114, 115, 116	1.97×10^{-2}	5.81×10^{-2}	2.95	2.36
81	Order Hemiptera	26, 75, 77, 84, 114, 115	7.39×10^{-4}	5.17×10^{-4}	6.99×10^{-1}	2.71
82	Order Hymenoptera	17, 22, 73, 74, 75, 77, 78, 81, 84, 115	9.73×10^{-4}	4.48×10^{-4}	4.60×10^{-1}	3.29
83	Order Isopoda	96, 103, 106, 108, 111	6.48×10^{-3}	3.54×10^{-3}	5.46×10^{-1}	2.20
84	Order Lepidoptera larvae	115	2.21×10^{-2}	3.98×10^{-4}	1.80×10^{-2}	2
85	Order Nematoda	85, 86, 108, 115	4.29×10^{-5}	9.49×10^{-6}	2.21×10^{-1}	2.33
86	Order Oligochaeta	105, 106, 108	9.95×10^{-3}	5.79×10^{-3}	5.82×10^{-1}	2
87	Order Opiliones	17, 73, 75, 76, 77, 78, 80, 81, 83, 84, 86, 87, 106	1.29×10^{-3}	3.21×10^{-4}	2.48×10^{-1}	3.12
88	Order Pseudoscorpionidae	72, 75, 79, 85, 88	4.00×10^{-4}	1.80×10^{-6}	4.51×10^{-3}	3.17
89	Order Psocoptera	108	1.00×10^{-4}	4.51×10^{-7}	4.51×10^{-3}	2
92	*Platynus assimile*	72, 73, 75, 77, 78, 83, 85, 96	1.05×10^{-2}	1.15×10^{-2}	1.09	3.25
93	*Platynus dorsale*	13, 73, 74, 75, 77, 78, 86	3.70×10^{-3}	1.67×10^{-5}	4.51×10^{-3}	4.13
94	*Platynus obscurus*	72, 73, 75, 77, 78, 87	3.00×10^{-3}	1.35×10^{-5}	4.51×10^{-3}	3.43

95	*Polydesmus angustus*	106	6.57×10^{-3}	2.22×10^{-3}	3.38×10^{-1}	2
96	*Porcellio scaber*	96, 103, 106, 111	6.15×10^{-3}	2.08×10^{-3}	3.38×10^{-1}	2
97	*Pterostichus diligens*	17, 72, 73, 75, 77, 78, 87	4.80×10^{-3}	2.17×10^{-5}	4.51×10^{-3}	3.37
98	*Pterostichus melanarius*	6, 8, 10, 13, 17, 24, 54, 73, 74, 75, 77, 78, 84, 86, 96, 108	3.59×10^{-2}	4.86×10^{-4}	1.35×10^{-2}	3.26
99	*Pterostichus nigrita.*	13, 75, 86	1.25×10^{-2}	1.75×10^{-3}	1.40×10^{-1}	3.13
100	*Pterostichus strenuus*	13, 72	3.88×10^{-3}	2.80×10^{-4}	7.22×10^{-2}	3.36
101	*Rilaena triangularis*	17, 77	3.75×10^{-4}	6.77×10^{-6}	1.80×10^{-2}	3
102	*Trechus obtusus*	77	5.80×10^{-4}	1.57×10^{-5}	2.71×10^{-2}	3
103	*Alnus glutinosa*	–	–	–	–	1
104	*Anemone nemorosa*	–	–	–	–	1
105	Bacteria	–	–	–	–	1
106	Detritus	–	–	–	–	1
107	*Fraxinus excelsior*	–	–	–	–	1
108	Microfungi	–	–	–	–	1
109	*Osmunda regalis*	–	–	–	–	1
110	Phylum Bryophyta	–	–	–	–	1
111	*Quercus robur*	–	–	–	–	1
112	*Ranunculus ficaria*	–	–	–	–	1
113	*Ranunculus repens*	–	–	–	–	1
114	*Urtica dioica*	–	–	–	–	1
115	Vegetation	–	–	–	–	1
116	*Veronica montana*	–	–	–	–	1

APPENDIX V. LIST OF TAXA OMITTED FROM BOTH COMPOSITE AND SEASONAL FOOD WEBS

Table S10

	Taxa	Order
1.	*Aegopinella nitidula*	Order Gastropoda
2.	*Agonum gracile*	Order Coleoptera
3.	*Agonum piceum*	Order Coleoptera
4.	*Agonum thoreyi*	Order Coleoptera
5.	*Bembidion bruxellense*	Order Coleoptera
6.	*Bembidion dentellum*	Order Coleoptera
7.	*Bembidion femoratum*	Order Coleoptera
8.	Family Geotrupidae	Order Coleoptera
9.	Family Scelionidae	Order Hymenoptera
10.	*Mitostoma chrysomelas*	Order Opiliones
11.	*Odiellus palpinalis*	Order Opiliones
12.	*Opilio saxatalis*	Order Opiliones
13.	Order Diptera pupae	Order Diptera
14.	Order Ephemeroptera adult	Order Ephemeroptera
15.	Order Odonata	Order Odonata
16.	Order Plectoptera adult	Order Plectoptera
17.	*Platynus albipes*	Order Coleoptera
18.	*Porcellionides cingendus*	Order Isopoda
19.	*Pterostichus crenatus*	Order Coleoptera
20.	*Strigamia crassipes*	Order Geophilomorpha
21.	*Trechus rubens*	Order Coleoptera
22.	*Velia caprai*	Order Hemiptera

REFERENCES

Agusti, N., Shayler, S.P., Harwood, J.D., Vaughan, I.P., Sunderland, K.D., and Symondson, W.O.C. (2003). Collembola as alternative prey sustaining spiders in arable ecosystems: Prey detection within predators using molecular markers. *Mol. Ecol.* **12**, 3467–3475.

Amundsen, P.-A., Lafferty, K.D., Knudsen, R., Primicerio, R., Klemetsen, A., and Kuris, A.M. (2009). Food web topology and parasites in the pelagic zone of a subarctic lake. *J. Anim. Ecol.* **78**, 563–572.

Arim, M., Abades, S.R., Laufer, G., Loureiro, M., and Marquet, P. (2010). Food web structure and body size: Trophic position and resource acquisition. *Oikos* **119**, 147–153.

Baird, D., and Ulanowicz, R.E. (1989). The seasonal dynamics of the Chesapeake Bay ecosystem. *Ecol. Monogr.* **59**, 329–364.

Bascompte, J., Melián, C.J., and Sala, E. (2005). Interaction strength combinations and the overfishing of a marine food web. *Proc. Natl. Acad. Sci. USA* **102**, 5443–5447.

Blackburn, T.M., and Gaston, K.J. (1997). A critical assessment of the form of the interspecific relationship between abundance and body size in animals. *J. Anim. Ecol.* **66**, 233–249.
Borer, E.T., Seabloom, E.W., Shurin, J.B., Anderson, K.E., Blanchette, C.A., Broitman, B., Cooper, S.D., and Halpern, B.S. (2005). What determines the strength of a trophic cascade? *Ecology* **86**, 528–537.
Briand, F., and Cohen, J.E. (1984). Community food webs have scale-invariant structure. *Nature* **307**, 264–267.
Brose, U., Berlow, E.L., and Martinez, N.D. (2005). Scaling up keystone effects from simple to complex ecological networks. *Ecol. Lett.* **8**, 1317–1325.
Brose, U., Jonsson, T., Berlow, E.L., Warren, P., Banasek-Richter, C., Bersier, L.F., Blanchard, J.L., Brey, T., Carpenter, S.R., Blandenier, M.F.C., Cushing, L., Dawah, H.A., *et al.* (2006a). Consumer–resource body-size relationships in natural food webs. *Ecology* **87**, 2411–2417.
Brose, U., Jonsson, T., Berlow, E.L., Warren, P., Banaasek-Richter, C., Bersier, L.-F., Blanchard, J.L., Brey, T., Carpenter, S.R., Cattin Blandenier, M.-F., Cushin, L., Dawah, H.A., *et al.* (2006b). Consumer–resource body-size relationships in natural food webs. *Ecology* **87**, 2411–2417.
Brown, A.G. (1997). Biogeomorphology and diversity in multiple-channel river systems. *Glob. Ecol. Biogeogr. Lett.* **6**, 179–185.
Brown, J.H., and Gillooly, J.F. (2003). Ecological food webs: High-quality data facilitate theoretical unification. *Proc. Natl. Acad. Sci. USA* **100**, 1467–1468.
Brown, J.H., Gillooly, J.F., Allen, A.P., Savage, V.M., and West, G.B. (2004). Toward a metabolic theory of ecology. *Ecology* **85**, 1771–1789.
Cardinale, B.J., Srivastava, D.S., Duffy, J.E., Wright, J.P., Downing, A.L., Sankaran, M., and Jouseau, C. (2006). Effects of biodiversity on the functioning of trophic groups and ecosystems. *Nature* **443**, 989–992.
Casini, M., Lövgren, J., Hjelm, J., Cardinale, M., Molinero, J.-C., and Kornilovs, G. (2008). Multi-level trophic cascades in a heavily exploited open marine ecosystem. *Proc. Biol. Sci.* **275**, 1793–1801.
Cattin, M.-F., Bersier, L.-F., Banašek-Richter, C., Baltensperger, R., and Gabriel, J.-P. (2004). Phylogenetic constraints and adaptation explain food-web structure. *Nature* **427**, 835–839.
Chase, J.M. (1999). Food web effects of prey size refugia: variable interactions and alternative stable equilibria. *Am. Nat.* **154**, 559–570.
Chase, J.M. (2000). Are there real differences among aquatic and terrestrial food webs. *Trends Ecol. Evol.* **15**, 408–412.
Christian, R.R., and Luczkovich, J.J. (1999). Organizing and understanding a winter's seagrass foodweb network through effective trophic levels. *Ecol. Modell.* **117**, 99–124.
Cohen, J.E., and Newman, C.M. (1985). A stochastic-theory of community food webs. 1. Models and aggregated data. *Proc. R. Soc. Lond. B Biol. Sci.* **224**, 421–448.
Cohen, J.E., Pimm, S.L., Yodzis, P., and Saldana, J. (1993). Body sizes of animal predators and animal prey in food webs. *J. Anim. Ecol.* **62**, 67–78.
Cohen, J.E., Jonsson, T., and Carpenter, S.R. (2003). Ecological community description using the food web, species abundance, and body size. *Proc. Natl. Acad. Sci. USA* **100**, 1781–1786.
Cohen, J.E., Schillter, D.N., Raffaelli, D.G., and Reuman, D.C. (2009). Food webs are more than the sum of their tritrophic parts. *Proc. Natl. Acad. Sci. USA* **52**, 22335–22340.

Cross, J.R., and Kelly, D.L. (2003). Wetland woods. In: *Wetlands of Ireland, Distribution, Ecology, Uses and Economic Value* (Ed. by M.L. Otte). University College Dublin Press, Dublin.

Cyr, H., Peters, R.H., and Downing, J.A. (1997). Population density and community size structure: Comparison of aquatic and terrestrial systems. *Oikos* **80**, 139–149.

Damuth, J. (1981). Population-density and body size in mammals. *Nature* **290**, 699–700.

Dawah, H.A., Hawkins, B.A., and Claridge, M.F. (1995). Structure of the parasitoid communities of grass-feeding chalcid wasps. *J. Anim. Ecol.* **64**, 708–720.

Dyer, L.A., and Letourneau, D.K. (1999). Trophic cascades in a complex terrestrial community. *Proc. Natl. Acad. Sci. USA* **96**, 5072–5076.

Dunne, J.A., Williams, R.J., and Martinez, N.D. (2002). Network structure and biodiversity loss in food webs: robustness increases with connectance. *Ecol. Lett.* **5**, 558–567.

Eggers, T., and Jones, T.H. (2000). You are what you eat . . . or are you? *Trends Ecol. Evol.* **15**, 265–266.

Elmhagen, B., and Rushton, S.P. (2007). Trophic control of mesopredators in terrestrial ecosystems: Top-down or bottom-up? *Ecol. Lett.* **10**, 197–206.

Elton, C. (1927). *Animal Ecology*. Sidgwick and Jackson, London, UK.

Emmerson, M.C., and Raffaelli, D. (2004). Predator–prey body size, interaction strength and the stability of a real food web. *J. Anim. Ecol.* **73**, 399–409.

Emmerson, M., Montoya, J.M., and Woodward, G. (2005). Body size, interaction strength and food web dynamics. In: *Dynamic Food Webs: Multispecies Assemblages, Ecosystem Development and Environmental Change* (Ed. by P.C. deRuiter, V. Wolters and J.C. Moore). Academic Press.

Enquist, B.J., and Niklas, K.J. (2001). Invariant scaling relations across tree-dominated communities. *Nature* **410**, 655–660.

Fonseca, C.R., and Ganade, G. (2001). Species functional redundancy, random extinctions and the stability of ecosystems. *J. Anim. Ecol.* **89**, 118–125.

Greenstone, M.H., Rowley, D.L., Heimbach, U., Lundgren, J.G., Pfannenstiel, R.S., and Rehner, S.A. (2005). Barcoding generalist predators by polymerase chain reaction: carabids and spiders. *Mol. Ecol.* **14**, 3247–3266.

Halaj, J., and Wise, D.H. (2001). Terrestrial trophic cascades: How much do they trickle? *Am. Nat.* **157**, 262–281.

Harper, G.L., King, R.A., Dodd, C.S., Harwood, J.D., Glen, D.M., Bruford, M.W., and Symondson, W.O.C. (2005). Rapid screening of invertebrate predators for multiple prey DNA targets. *Mol. Ecol.* **14**, 819–827.

Hector, A., Schmid, B., Beierkuhnlein, C., Caldeira, M.C., Diemer, M., Dimitrakopoulos, P.G., Finn, J.A., Freitas, H., Giller, P.S., Good, J., Harris, R., Hogberg, P., *et al.* (1999). Plant diversity and productivity experiments in European grasslands. *Science* **286**, 1123–1127.

Hooper, D.U., Chapin, F.S., Ewel, J.J., Hector, A., Inchausti, P., Lavorel, S., Lawton, J.H., Lodge, D.M., Loreau, M., Naeem, S., Schmid, B., Setala, H., *et al.* (2005). Effects of biodiversity on ecosystem functioning: A consensus of current knowledge. *Ecol. Monogr.* **75**, 3–35.

Huffaker, C.B., and Messenger, P.S. (1976). *Theory and Practice of Biological Control*. Academic Press, New York.

Huxham, M., Beaney, S., and Raffaelli, D. (1996). Do parasites reduce the chances of triangulation in a real food web? *Oikos* **76**, 284–300.

Ings, T.C., Montoya, J.M., Bascompte, J., Blüthgen, N., Brown, L., Dormann, C.F., Edwards, F., Figueroa, D., Jacob, U., Jones, J.I., Lauridsen, R., Ledger, M.E., *et al.* (2009). Ecological networks—Beyond food webs. *J. Anim. Ecol.* **78**, 253–269.

Jennings, S., and Mackinson, S. (2003). Abundance–body mass relationships in size-structured food webs. *Ecol. Lett.* **6**, 971–974.

Jennings, S., Pinnegar, J.K., Polunin, N.V.C., and Boon, T.W. (2001). Weak cross-species relationships between body size and trophic level belie powerful size-based trophic structuring in fish communities. *J. Anim. Ecol.* **70**, 934–944.

Jennings, S., Pinnegar, J.K., Polunin, N.V.C., and Warr, K.J. (2002). Linking size-based and trophic analyses of benthic community structure. *Mar. Ecol. Prog. Ser.* **226**, 77–85.

Jonsson, T., and Ebenman, B. (1998). Effects of predator–prey body size ratios on the stability of food chains. *J. Theor. Biol.* **193**, 407–417.

Jonsson, M., and Malmqvist, B. (2000). Ecosystem process rate increases with animal species richness: Evidence from leaf-eating, aquatic insects. *Oikos* **89**, 519–523.

Jonsson, M., and Malmqvist, B. (2003). Mechanisms behind positive diversity effects on ecosystem functioning: Testing the facilitation and interference hypotheses. *Oecologia* **134**, 554–559.

Jonsson, T., Cohen, J.E., and Carpenter, S.R. (2005). Food webs, body size, and species abundance in ecological community description. *Adv. Ecol. Res.* **36**(36), 1–84.

Jurado-Rivera, J.A., Vogler, A.P., Reid, C.A.M., Petitpierre, E., and Gómez-Zurita (2009). DNA barcoding insect-host plant associations. *Proc. R. Soc. Lond. B Biol. Sci.* **276**, 639–648.

Kaiser-Bunbury, C.N., Muff, S., Memmott, J., Müller, C.B., and Caflisch, A. (2010). The robustness of pollination networks to the loss of species and interactions: A quantitative approach incorporating pollinator behaviour. *Ecol. Lett.* **13**, 442–452.

Kerr, S.R., and Dickie, L.M. (2001). *The Biomass Spectrum: A Predator–Prey Theory of Aquatic Production*. Columbia University Press, pp. 320.

Layer, K., Hildrew, A., Monteith, D., and Woodward, G. (2010a). Long-term variation in the littoral food web of an acidified mountain lake. *Glob. Change Biol.* 10.1111/j.1365-2486.2010.02195.x.

Layer, K., Riede, J.O., Hildrew, A.G., and Woodward, G. (2010b). Food web structure and stability in 20 streams across a wide pH gradient. *Adv. Ecol. Res.* **42.**

Layman, C.A., Winemiller, K.O., Arrington, D.A., and Jepsen, D.B. (2005). Body size and trophic position in a diverse tropical food web. *Ecology* **86**, 2530–2535.

Leaper, R., and Huxham, M. (2002). Size constraints in a real food web: Predator, parasite and prey body-size relationships. *Oikos* **99**, 443–456.

Leaper, R., and Raffaelli, D. (1999). Defining the abundance body-size constraint space: Data from a real food web. *Ecol. Lett.* **2**, 191–199.

Lewinsohn, T.M., Novotny, V., and Basset, Y. (2005). Insects on plants: Diversity of herbivore assemblages revisited. *Annu. Rev. Ecol. Evol. Syst.* **36**, 597–620.

Lindeman, R.L. (1942). The trophic–dynamic aspect of ecology. *Ecology* **23**, 399–417.

Loreau, M., Naeem, S., Inchausti, P., Bengtsson, J., Grime, J.P., Hector, A., Hooper, D.U., Huston, M.A., Raffaelli, D., Schmid, B., Tilman, D., and Wardle, D.A. (2001). Biodiversity and ecosystem functioning: Current knowledge and future challenges. *Science* **294**, 804–808.

Marquet, P.A., Navarrete, S.N., and Castilla, J.C. (1990). Scaling population density to body size in rocky intertidal communities. *Science* **250**, 1125–1127.

Maron, J.L., Horvitz, C.C., and Williams, J.L. (2010). Using experiments, demography and population models to estimate interaction strength based on transient and asymptotic dynamics. *J. Anim. Ecol.* **98**, 290–301.

Martinez, N.D. (1991). Artifacts or attributes—Effects of resolution on the little-rock lake food web. *Ecol. Monogr.* **61**, 367–392.

McCann, K., Hastings, A., and Huxel, G.R. (1998). Weak trophic interactions and the balance of nature. *Nature* **395**, 794–798.

Memmott, J., Martinez, N.D., and Cohen, J.E. (2000). Predators, parasitoids and pathogens: Species richness, trophic generality and body sizes in a natural food web. *J. Anim. Ecol.* **69**, 1–15.

Montoya, J.M., Woodward, G., Emmerson, M.C., and Solé, R.V. (2009). Press perturbations and indirect effects in real food webs. *Ecology* **90**, 2426–2433.

Mülder, C., den Hollander, H., Schouten, T., and Rutgers, M. (2006). Allometry, biocomplexity, and web topology of hundred agro-environments in The Netherlands. *Ecol. Complex* **3**, 219–230.

Naeem, S., and Li, S.B. (1997). Biodiversity enhances ecosystem reliability. *Nature* **390**, 507–509.

Naeem, S., Thompson, L.J., Lawler, S.P., Lawton, J.H., and Woodfin, R.M. (1994). Declining biodiversity can alter the performance of ecosystems. *Nature* **368**, 734–737.

Neutel, A.M., Heesterbeek, J.A.P., and de Ruiter, P.C. (2002). Stability in real food webs: Weak links in long loops. *Science* **296**, 1120–1123.

Nowlin, W.H., Vanni, M.J., and Yang, L.H. (2008). Comparing resource pulses in aquatic and terrestrial ecosystems. *Ecology* **89**, 647–659.

O'Gorman, E.J., and Emmerson, M.C. (2010). Manipulating interaction strengths and the consequences for trivariate patterns in a marine food web. *Adv. Ecol. Res.* **42**, 301–419.

Olesen, J.M., Dupont, Y.L., O'Gorman, E.J., Ings, T.C., Layer, K., Melián, C.J., Troejelsgaard, K., Pichler, D.E., Rasmussen, C., and Woodward, G. (2010). From Broadstone to Zackenberg: Space, time and hierarchies in ecological networks. *Adv. Ecol. Res.* **42**, 1–69.

Otto, S.B., Rall, B.C., and Brose, U. (2007). Allometric degree distributions facilitate food-web stability. *Nature* **450**, 1226–1229.

Pace, M.L., Cole, J.J., Carpenter, S.R., and Kitchell, J.F. (1999). Trophic cascades revealed in diverse ecosystems. *Trends Ecol. Evol.* **14**, 483–488.

Petchey, O.L., Beckerman, A.P., Reide, J.O., and Warren, P.H. (2008). Size, foraging, and food web structure. *Proc. Natl. Acad. Sci. USA* **105**, 4191–4196.

Petermann, J.S., Müller, C.B., Weigelt, A., Weisser, W.W., and Schmid, B. (2010). Effect of plant species loss on aphid–parasitoid communities. *J. Anim. Ecol.* **79**, 709–720.

Peters, R.H. (1983). The ecological implications of body size. *Cambridge Studies in Ecology: The Ecological Implications of Body Size*. Cambridge University Press, New York, NY, USA, pp. XII+329P.

Pimm, S.L. (1984). The complexity and stability of ecosystems. *Nature* **307**, 321–326.

Polis, G.A. (1991). Complex trophic interactions in deserts—An empirical critique of food-web theory. *Am. Nat.* **138**, 123–155.

Polis, G.A., and Holt, R.D. (1992). Intraguild predation: The dynamics of complex trophic interactions. *Trends Ecol. Evol.* **7**, 151–154.

Polis, G.A., Sears, A.L.W., Huxel, G.R., Strong, D.R., and Maron, J. (2000). When is a trophic cascade a trophic cascade? *Trends Ecol. Evol.* **15**, 473–475.

Rall, B.C., Vucic-Pestic, O., Ehnes, R.B., Emmerson, M., and Brose, U. (2009). Temperature, predator–prey interaction strength and population stability. *Glob. Change Biol.* 10.1111/j.1365-2486.2009.02124.x.

Reagan, D.P., and Waide, R.B. (1996). The food web of a tropical rain forest. University of Chicago Press, pp. 616.

Reiss, J., Bridle, J.R., Montoya, J.M., and Woodward, G. (2009). Emerging horizons in biodiversity and ecosystem functioning research. *Trends Ecol. Evol.* **24**, 505–514.

Reuman, D.C., and Cohen, J.E. (2005). Estimating relative energy fluxes using the food web, species abundance, and body size. *Adv. Ecol. Res.* **36**(36), 137–182.

Reuman, D.C., Mülder, C., Banasek-Richter, C., Cattin Blandenier, M.-F., Breure, A.M., Den Hollander, H., Kneitel, J.M., Raffaelli, D., Woodward, G., and Cohen, J.E. (2009). Allometry of body size and abundance in 166 food webs. *Adv. Ecol. Res.* **41**, 1–44.

Riede, J.O., Rall, B.C., Banasek-Richter, C., Navarrete, S.A., Wieters, E.A., and Brose, U. (2010). Scaling of food-web properties with diversity and complexity across ecosystems. *Adv. Ecol. Res.* **42**, 139–170.

Schoener, T.W. (1989). Food webs from the small to the large. *Ecology* **70**, 1559–1589.

Schmid, P.E., and Schmid-Araya, J.M. (2007). Body size and scale invariance: multifractals in invertebrate communities. In: *Body Size and the Structure and Function of Aquatic Ecosystems* (Ed. by A.G. Hildrew, D.G. Raffaelli and R.O. Edmons-Brown), pp. 140–166. Cambridge University Press.

Shapiro, S.S., and Wilk, M.B. (1965). An analysis of variance test for normality: Complete samples. *Biometrika* **52**, 591–611.

Shurin, J.B., Gruner, D.S., and Hillebrand, H. (2006). All wet or dried up? Real differences between aquatic and terrestrial food webs. *Proc. R. Soc. Lond. B Biol. Sci.* **273**, 1–9.

Snyder, W.E., Snyder, G.B., Finke, D.L., and Straub, C.S. (2006). Predator biodiversity strengthens herbivore suppression in single and multiple prey communities. *Ecol. Lett.* **9**, 789–796.

Snyder, G.B., Finke, D.L., and Snyder, W.E. (2008). Predator biodiversity strengthens aphid suppression across single- and multiple-species prey communities. *Biol. Control* **44**, 52–60.

Srivastava, D.S., and Bell, T. (2009). Reducing horizontal and vertical diversity in a foodweb triggers extinctions and impacts functions. *Ecol. Lett.* **12**, 1016–1028.

Sterner, R.W., and Elser, J.J. (2002). *Ecological Stoichiometry: The Biology of Elements from Molecules to the Biosphere*. Princeton University Press, Princeton, NJ.

Stouffer, D.B., Camacho, J., Guimera, R., Ng, C.A., and Nunes Amaral, L.A. (2005). Quantitative patterns in the structure of model and empirical food webs. *Ecology* **86**, 1301–1311.

Strong, D.R. (1992). Are trophic cascades all wet? Differentiation and donor-control in speciose ecosystems. *Ecology* **73**, 747–754.

Townsend, C.R., Thompson, R.M., McIntosh, A.R., Kilroy, C., Edwards, E., and Scarsbrook, M.R. (1998). Disturbance, resource supply, and food-web architecture in streams. *Ecol. Lett.* **1**, 200–209.

Wardle, D.A. (1999). Is "sampling effect" a problem for experiments investigating biodiversity–ecosystem function relationships? *Oikos* **87**, 403–407.

Warren, P.H. (1989). Spatial and temporal variation in the structure of a freshwater food web. *Oikos* **55**, 299–311.

Warren, P.H. (1996). Structural constraints on food web assembly. In: *Aspects of the Genesis and Maintenance of Biological Diversity* (Ed. by M.E. Hochber, J. Clobert and R. Barbault), pp. 143–161. Oxford University Press.

Warren, P.H., and Lawton, J.H. (1987). Invertebrate predator–prey body size relationships: An explanation for upper triangular food webs and patterns in food web structure. *Oecologia* **74**, 231–235.

White, J. (1985). The Gearagh woodland. *Co. Cork. Irish Nat. J.* **21**, 391–396.

White, E.P., Ernest, S.K.M., Kerkhoff, A.J., and Enquist, B.J. (2007). Relationships between body size and abundance in ecology. *Trends Ecol. Evol.* **22**, 323–330.

Williams, R.J., and Martinez, N.D. (2000). Simple rules yield complex food webs. *Nature* **404**, 180–183.

Wilson, E.B. (1927). Probable inference, the law of succession, and statistical inference. *J. Am. Stat. Assoc.* **22**, 209–212.

Woodward, G., and Hildrew, A.G. (2001). Invasion of a stream food web by a new top predator. *J. Anim. Ecol.* **70**, 273–288.

Woodward, G., and Hildrew, A.G. (2002a). Body-size determinants of niche overlap and intraguild predation within a complex food web. *J. Anim. Ecol.* **71**, 1063–1074.

Woodward, G., and Hildrew, A.G. (2002b). Food web structure in riverine landscapes. *Freshw. Biol.* **47**, 777–798.

Woodward, G., and Hildrew, A.G. (2002c). Differential vulnerability of prey to an invading top predator: Integrating field surveys and laboratory experiments. *Ecol. Entomol.* **27**, 732–744.

Woodward, G., and Warren, P. (2007). Body size and predatory interactions in freshwaters: Scaling from individuals to communities. In: *Body Size: The Structure and Function of Aquatic Ecosystems* (Ed. by A.G. Hildrew, D.G. Raffaelli and R. Edmonds-Brown). Cambridge University Press, 356 pp.

Woodward, G., Ebenman, B., Emmerson, M., Montoya, J.M., Olesen, J.M., Valido, A., and Warren, P.H. (2005a). Body size in ecological networks. *Trends Ecol. Evol.* **20**, 402–409.

Woodward, G., Speirs, D.C., and Hildrew, A.G. (2005b). Quantification and resolution of a complex, size-structured food web. *Adv. Ecol. Res.* **36**, 85–135.

Woodward, G., Benstead, J.P., Beveridge, O.S., Blanchard, J., Brey, T., Brown, L., Cross, W.F., Friberg, N., Ings, T.C., Jacob, U., Jennings, S., Ledger, M.E., *et al.* (2010). Ecological networks in a changing climate. *Adv. Ecol. Res.* **42**, 71–138.

Worm, B., Barbier, E.B., Beaumont, N., Duffy, J.E., Folke, C., Halpern, B.S., Jackson, J.B.C., Lotze, H.K., Micheli, F., Palumbi, S.R., Sala, E., Selkoe, K.A., *et al.* (2006). Impacts of biodiversity loss on ocean ecosystem services. *Science* **314**, 787–790.

Yodzis, P. (1998). Local trophodynamics and the interaction of marine mammals and fisheries in the Benguela ecosystem. *J. Anim. Ecol.* **67**, 635–658.

Yvon-Durocher, G., Reiss, J., Blanchard, J., Ebenman, B., Perkins, D.M., Reuman, D.C., Thierry, A., Woodward, G., and Petchey, O.L. (2010). Across ecosystem comparisons of size structure: Methods, approaches, and prospects. *Oikos* (in press).

Zavaleta, E.S., Pasari, J.R., Hulvey, K.B., and Tilman, G.D. (2010). Sustaining multiple ecosystem functions in grassland communities requires higher biodiversity. *Proc. Natl. Acad. Sci. USA* **107**, 1443–1446.

Food Web Structure and Stability in 20 Streams Across a Wide pH Gradient

KATRIN LAYER, JENS O. RIEDE, ALAN G. HILDREW AND GUY WOODWARD

SUMMARY

Recent attempts to include more ecological detail in connectance food webs have revealed strong relationships between food web structure, species abundance and body size. Few studies, however, have assessed these and other macroecological patterns in food webs in order to examine how network structure, dynamics and their determinants change across environmental gradients. Here, we present 20 highly resolved, standardized stream food webs along a wide pH gradient (5.0–8.4). Our main goal of this study was to assess the influence of external environmental and internal biotic influences on community structure and stability. Many structural features of the food webs changed across the gradient, with web size, linkage density and complexity all increasing with pH. Chlorophyll-*a* concentrations in epilithic biofilms, as well as the biomass of macroinvertebrates and fish, were also positively correlated with pH. Directed connectance was not correlated with pH in our study, however, although some of the smallest food webs at the lowest pH displayed the highest connectance amongst the networks in this data set. The prevalence of generalism in such food webs, with many alternative food chains passing through any given species, might serve to confer a degree of stability upon these networks, if most feeding links are weak.

ADVANCES IN ECOLOGICAL RESEARCH VOL. 42

0065-2504/10 $35.00
DOI: 10.1016/S0065-2504(10)42005-X

We found clear differences between our allometrically inferred measures of *per capita* interaction strengths within the food web, with particularly strong links existing between herbivores and their algal resources and between fish and invertebrates, both of which also became more prevalent at high pH. Predatory interactions between invertebrates and between fishes were far weaker, with the weakest links of all being the few instances of invertebrates parasitizing fishes. Dynamic modelling simulations that ran for the equivalent of 10 years revealed that fewer species were lost from the more acid food webs than those at high pH, further confirming the suggestion of a negative relationship between stability and pH. This finding, which supports earlier ideas from empirical studies that have considered other types of stability, might account for the limited biological recovery in previously acidified freshwaters that are showing evidence of chemical recovery.

I. INTRODUCTION

The study of food webs has long been central to ecology because they provide a conceptual thread that links individuals to populations, communities and ecosystems (e.g. Dunne, 2009; Ings *et al.*, 2009). Most early research was based on binary networks of trophic interactions which, although they provided no information about the relative importance of different links in the web or the dynamics within it (Benke and Wallace, 1997; Hall and Raffaelli, 1993; Paine, 1988), have nonetheless yielded valuable insights into real ecological phenomena, as pattern and process are inextricably linked in food webs. More contemporary topological-based research has explored new aspects of network structure, such as the presence of 'small world' properties, modularity and nestedness, as well as revisiting earlier ideas with the better quality data that have emerged recently (Ings *et al.*, 2009; Olesen *et al.*, 2010; Riede *et al.*, 2010).

In addition to these purely structural-based approaches, there have been increasing attempts to include more ecological detail in connectance webs to enrich their information content, and to forge stronger and more explicit links between network structure and dynamics. In particular, several recent studies have revealed strong 'trivariate' relationships between food web structure (i.e. the patterning of trophic links), species abundance and body size (e.g. Cohen *et al.*, 2003; McLaughlin *et al.*, 2010; O'Gorman and Emmerson, 2010; Woodward *et al.*, 2005a,b). In addition, a few studies have sought to produce webs that are both highly resolved and quantified (e.g. Woodward *et al.*, 2005a), based on direct observation of both the nodes and links (e.g. via gut contents analysis, GCA) that comprise the network. Unfortunately, the sample size of these webs remains small, largely due to logistic constraints. An alternative approach to these detailed studies of often unreplicated 'model' systems (e.g. Woodward *et al.*, 2005a) has been to examine macroecological patterns across a larger number of webs that have been described in a (relatively) standardized, but typically less labour-intensive, manner. Such comparative studies often necessitate using networks that have been constructed from a combination of directly observed and inferred feeding links, as relying solely on the former method is rarely feasible when dealing with a large number of complex, multispecies systems (e.g. Layer *et al.*, 2010; O'Gorman and Emmerson, 2010; Riede *et al.*, 2010). A recurrent theme that has emerged from both approaches, however, is the strong influence that body size often has on both network structure and dynamics (Beckerman *et al.*, 2006; Berlow *et al.*, 2009; Petchey *et al.*, 2008; Woodward *et al.*, 2010).

Ever since Hardy's (1924) depiction of a size-based marine food web and Elton's (1927) 'pyramid of numbers', body size has been recognized as having an important role in structuring biological communities, and ecologists are once again re-evaluating how body size affects food webs (e.g. Brown *et al.*, 2004; Cohen *et al.*, 2003; Jonsson *et al.*, 2005; Petchey *et al.*, 2008; Reuman and Cohen, 2005; Savage *et al.*, 2004; Woodward *et al.*, 2005a,b). In many food webs, and particularly in aquatic systems, larger consumers feed on smaller 'prey' (Memmott *et al.*, 2000; Mulder *et al.*, 2005, 2006; Otto *et al.*, 2007) and body size determines not only the patterning of realized links within the feeding matrix (e.g. Petchey *et al.*, 2008) but also the strength of interactions (e.g. Berlow *et al.*, 2009; Brose *et al.*, 2005a,b) and the magnitude of energy and biomass fluxes (e.g. Cohen *et al.*, 2003; Reuman and Cohen, 2004; Woodward *et al.*, 2005a).

Food web complexity and interaction strength are key determinants of stability (May, 1972, 1973; McCann, 2000) and both can influence how a community responds to environmental stress (e.g. low pH) and biotic perturbations (e.g. species invasions): if modest, perturbations may simply alter relative abundances, whereas large perturbations can lead to the loss (or gain) of species. Surprisingly, few comparative studies have explicitly and systematically examined macroecological patterns in empirical food webs (but see Petchey *et al.*, 2004; Reuman *et al.*, 2009; Riede *et al.*, 2010; Townsend *et al.*, 1998; Tylianakis *et al.*, 2007), especially in terms of investigating how network structure, dynamics and their determinants (body size, abundance) might change across environmental gradients (Woodward, 2009). One reason for this is that many published webs are summary networks collated over many sampling occasions (months to decades) and many sites, or habitat patches (e.g. Jonsson *et al.*, 2005; Martinez, 1991; Polis, 1991), and this blurring across spatio-temporal scales can make it difficult to detect responses to potential stressors (Olesen *et al.*, 2010; Woodward *et al.*, 2010). Here, we used a standardized method to construct detailed snapshots of 20 local stream food webs, from across a broad environmental pH gradient, which were sampled at comparable and ecologically relevant spatial scales (100 m reaches) that equate with the approximate territory size or home range over which the top predators (mostly brown trout, *Salmo trutta* L.) typically operate (e.g. Harcup *et al.*, 2006).

Acidification is a widespread stressor of freshwater ecosystems (e.g. Hildrew and Ormerod, 1995), as has been revealed repeatedly by its powerful effects on assemblage composition and the fact that pH is often the dominant environmental gradient in large-scale multivariate analysis of community structure (Hildrew, 2009; Rosemond *et al.*, 1992; Townsend *et al.*, 1983). The principal goals of this study were to characterize 20 stream food webs across a wide pH gradient and to assess the influence of external environmental and internal biotic influences on network topology and stability. We quantified structural properties of each food web, such as the number of species and feeding links per stream, connectance and complexity, and we also used body size data to gauge potential network stability via dynamic modelling of robustness. Robustness in this context is a measure of community stability, in terms of the species richness persisting at the end of a simulation, compared to the initial species richness and, unlike other measures of stability, it is relatively simple to assess using a bioenergetic approach (Brose *et al.*, 2005a,b, 2006; Yodzis and Innes, 1992).

We predicted, based on earlier suggestions and empirical observations (e.g. Woodward and Hildrew, 2002), that food webs from more acid sites (i.e. lower pH) would be smaller and with lower average interaction strengths than those at higher pH, and that this should make them more robust. If acid food webs are indeed relatively stable, this might at least partially account for the time lags observed between biological and chemical recovery from acidification (e.g. Layer *et al.*, 2010). This appears to be an increasingly common, but still largely unexplained, phenomenon in some parts of Europe, as previously acidified freshwaters respond chemically to the effects of large-scale and long-term reductions in acidifying emissions (e.g., Ledger and Hildrew, 2005; Monteith *et al.*, 2005; Yan *et al.*, 2003). Several possible explanations, which are not necessarily mutually exclusive, have been put forward, including the potential role of dispersal constraints (e.g. Bradley and Ormerod, 2002; Raddum and Fjellheim, 2003; Snucins and Gunn, 2003; but see Masters *et al.*, 2007), and incomplete chemical recovery and the influence of ongoing acid episodes (e.g. Kowalik *et al.*, 2007; Lepori *et al.*, 2003; Rose *et al.*, 2004). The possibility of internal inertia within acidified food webs themselves acting as a brake on recovery is particularly intriguing, as it represents a higher, network-based level of organization (Hildrew, 2009; Woodward, 2009). Despite the recent onset of chemical recovery in some areas of Europe and North America, large parts of the world that are undergoing rapid industrialization, including India and China, will face growing threats of increased acidification in the coming decades (Galloway, 1995; Seip *et al.*, 1999; Woodward, 2009), and so there is a clear and pressing need to understand the mechanisms that determine the structure and dynamics of food webs that are exposed to this environmental stressor.

II. METHODS

A. Sampling of the Biota

The biological community was sampled in 20 streams that span a pH gradient of 5.0–8.4 Table 1), in Spring 2005 (14 sites) and 2006 (6 sites in the River Duddon catchment). Spring sampling ensured the maximum duration of exposure to both recent acid episodes in winter and also to overall conditions since recruitment the previous summer for many consumer taxa. Thus, any community and food web responses to pH should be manifested most clearly at the time of sampling. Sites were: Allt a'Mharcaidh (MHA, N.E. Scotland), Allt na Coire nan Con (COI, N.W. Scotland) and Dargall Lane (DAR, S.W. Scotland); Old Lodge (OLD), Broadstone Stream (BRO) and Lone Oak (OAK; S.E. England); Duddon Pike Beck (D1), Hardknott Gill (D2), Mosedale Beck (D3), Duddon (D4), Wrynose Beck (D5), Duddon Beck a (D6) and River Etherow (ETH; N.W. England); Mill Stream (MIL), Bere Stream (BER) and Narrator Brook (NAR; S.W. England); Afon Hafren (HAF) and Afon Gwy (GWY; mid-Wales); and Beagh's Burn (BEA) and Coneyglen Burn (CON; N. Ireland). Ten of these sites (MHA, COI, HAF, GWY, NAR, ETH, OLD, DAR, BEA, CON) are part of the U.K. Acid Waters Monitoring Network (UKAWMN; http://www.ukawmn.ucl.ac.uk) and have been monitored since 1988. At each site, three replicate pH measurements were taken on the same sampling occasion using a hand-held pH metre (pH340i, Wissenschaftlich-Technische Werkstätten, Weilheim, Germany), and the mean value of these was used in this study as the measure of stream pH. No recent annual means derived from repeated sampling were available for the 10 non-UKAWMN streams, but when we compared our data with those collected at the UKAWMN sites since 1988, our spot values were well correlated with mean annual pH for the same site (correlation coefficient $r=0.77$; $p=0.015$), and even more closely with minimum annual pH (correlation coefficient $r=0.95$; $p<0.0001$).

To quantify macroinvertebrate abundance, 10 replicate Surber samples (area 0.0625 m^2; mesh aperture 330 μm) were taken from the benthos of each stream within an approximately 100 m^2 stretch and preserved in 70% industrial methylated spirit. Subsequently, all individuals were sorted from debris, identified to species wherever possible [i.e. all except Diptera (identified to family) and Annelida (identified to subclass)] and counted. For a list of identification keys used, see Layer *et al.* (2010). Body measurements were taken in the form of linear dimensions (head-capsule widths or body lengths),

Table 1 Site characteristics and food web metrics calculated for 20 food webs along a wide pH gradient

Site (code)	Lat.	Long.	pH	C	I	F	S	L	L_c	L/S	L/S^2	SC'_{max}	G	V	G_k	V_k	G_{st}	V_{st}
Old Lodge (OLD)	51.04	0.08	5.0	0.92	2.54	2.00	23	137	166	5.96	0.26	7.22	12.6	6	2.2	1.04	1.22	0.41
Lone Oak (OAK)	51.08	0.10	5.2	1.21	3	0	24	159	224	6.63	0.28	9.33	11.36	6.63	1.71	1.02	1.01	0.76
Beagh's Burn (BEA)	55.10	−6.22	5.3	1.29	0.69	2.00	30	187	215	6.23	0.21	7.17	17	6.45	2.73	1.03	1.42	0.55
Etherow (ETH)	53.49	−1.83	5.3	1.7	6.01	0	44	427	659	9.7	0.22	14.98	16.42	9.7	1.69	1	0.94	1.04
Afon Hafren (HAF)	52.47	−3.70	5.3	3.11	3.22	0.80	25	136	195	5.44	0.22	7.8	9.07	5.67	1.67	1.03	0.86	0.82
Broadstone (BROA)	51.08	0.053	5.5	1.9	2.89	0.13	25	178	245	7.12	0.28	9.8	10.41	7.38	1.47	1.04	0.9	0.66
Afon Gwy (GWY)	52.45	−3.73	5.6	3.84	3.13	0.68	24	135	189	5.63	0.23	7.88	9.64	5.87	1.71	1.04	0.92	0.81
Allt na Coire nan Con (COI)	56.76	−5.61	5.7	1.11	0.78	0.95	22	94	127	4.27	0.19	5.77	7.31	4.52	1.69	1.15	1.04	0.74
Dargall Lane (DAR)	55.08	−4.43	5.8	1.28	0.76	2.20	21	99	131	4.71	0.22	6.24	8.25	4.95	1.75	1.04	1.03	0.6
Duddon main channel (D4)	54.41	−3.16	5.8	0.59	0.71	0.50	19	71	82	3.74	0.2	4.32	7.89	3.94	2.11	1.04	1.33	0.47
Mosedale Beck (D3)	54.41	−3.14	5.9	0.87	1.47	2.20	21	108	134	5.14	0.24	6.38	9.82	5.4	1.91	1.04	1.09	0.55
Coneyglen Burn (CON)	54.74	−7.00	5.9	0.67	0.92	2.05	22	56	59	2.55	0.12	2.68	11.2	2.67	4.4	1.03	2.28	0.34
Narrator Brook (NAR)	50.50	−4.02	6.0	2.32	9.89	2.26	61	759	1345	12.44	0.2	22.05	17.88	12.52	1.45	1.03	0.81	1.18
Duddon Pike Beck (D1)	54.41	−3.17	6.1	1.01	9.82	0.60	35	285	418	8.14	0.23	11.94	13.00	8.41	1.59	1.03	0.89	0.84
Wrynose Pass Beck (D5)	54.41	−3.12	6.4	2.89	6.83	0	29	194	273	6.69	0.23	9.41	11.41	6.69	1.71	1	1.09	0.82
Allt a'Mharcaidh (MHA)	57.12	−3.85	6.5	0.18	18.52	0.55	40	335	518	8.38	0.21	12.95	13.33	8.65	1.58	1.02	0.89	0.88
Duddon Beck a (D6)	54.41	−3.15	6.5	1.32	4.57	0	20	114	147	5.7	0.29	7.35	9.58	5.75	1.67	1	1.04	0.48
Hardknott Gill (D2)	54.40	−3.17	7.0	1.1	5.23	1.40	44	384	629	8.73	0.2	14.3	14.72	9.93	1.52	1.02	0.93	1.04
Bere Stream (BER)	50.73	−2.21	7.5	4.51	39.72	17.93	66	940	1444	14.24	0.22	21.88	23.48	15.15	1.65	1.04	0.96	1
Mill Stream (MIL)	50.68	−2.18	8.4	14.16	2.59	25.25	87	1653	2489	19	0.22	28.61	30.22	19.26	1.65	1.01	1.01	0.96

Shown are site codes, decimal latitude (Lat.), decimal longitude (Long.), mean pH (pH), Chlorophyll-*a* biomass (mg/m^2) (C), invertebrate biomass (g/m^2) (I), fish biomass (g/m^2) (F), number of taxa (S), number of links (L); number of links including competitive links (shared resources; L_c); linkage density (L/S); directed connectance (L/S^2); web complexity (SC'_{max}); generality (G; number of resources per consumer); vulnerability (V; number of consumers per resource); standardized generality (G_k), vulnerability (V_k), and standard deviation of G_k and V_k (G_{st} and V_{st}, respectively; see Williams and Martinez, 2000).

and individual dry masses were determined from published length-dry mass regression equations (after Woodward *et al.*, 2005a).

Fish abundance was quantified by depletion electrofishing of the entire approximately 100 m^2 survey stretch (after Seber and LeCren, 1967). Stop-nets were installed at both ends of the stretch, and three runs were completed with a Smith-Root LR-24 backpack electrofisher, moving upstream and sweeping from one side of the stream to the other. All fishes were counted and measured (fork length and body mass) before being released back into the stream alive.

To characterize diatom assemblages and measure chlorophyll-*a*, biofilm samples were taken by scrubbing a known projected area of the upper surface of 10 permanently submerged cobbles from unshaded areas at each study site. In order to calculate the number of diatoms and chlorophyll-*a* per square metre of stone surface, the surface of scrubbed stones was traced on acetate sheets, and the sampling area subsequently calculated from the mass of the tracings (after Layer *et al.*, 2010). All algal samples were frozen within less than 2 h of collection and stored at −80 °C in the dark until further processing. In the laboratory, biofilm samples were thawed overnight in the dark (at 4 °C), thoroughly shaken and then split into two equal volumes, one of which was used for the identification and counting of diatoms, the other for determining chlorophyll-*a*.

Diatom slides were prepared after Battarbee *et al.* (2001). In brief, biofilm suspensions were placed in centrifuge tubes and centrifuged at 2000 rpm for 4 min. The supernatant was decanted, and 5 ml of hydrogen peroxide (H_2O_2; 30%) was added to each sample, which was then digested in a water bath at 80 °C for approximately 4 h. Distilled water was used to wash the cleared diatom valves, and the suspensions were plated out onto cover slips and allowed to dry overnight, prior to mounting in Naphrax (Brunel Microscopes Ltd., Chippenham, UK). On each slide, 300 valves were identified using published identification keys (see Appendix I), and the percentage abundance of each diatom species per sample was determined. In addition to allow the calculation of the absolute number of diatoms per species per unit area, the total number of diatoms was counted in three sections of known area on each slide.

As a proxy for algal biomass (Kalff, 2002), biofilm samples were also analysed for their chlorophyll-*a* content. Chlorophyll was extracted overnight using acetone and its concentration subsequently measured spectrophotometrically (after Lorenzen, 1967). Absorption of the extraction was measured at 664 (chlorophyll-*a*) and 750 nm (turbidity), and chlorophyll-*a* concentration was calculated using the following equation (after Lorenzen, 1967):

$$\mathrm{Chl}\left(\frac{\mu\mathrm{g}}{\mathrm{cm}^2}\right) = \frac{A*K*(A_{664} - A_{750})*v}{S*l} \quad (1)$$

where A is the absorption coefficient of chlorophyll-a (=11), K is a factor to equate the reduction in absorbance to initial chlorophyll concentration (1.7:0.7 or 2.43), A_{664} and A_{750} are the absorptions of the solution at 664 and 750 nm, respectively, V is the volume of acetone used for extraction (ml), S is the stone surface (in cm^2) scrubbed during biofilm sampling (taking into account that only half of each sample was being used in this analysis) and l is the path length of the light in the cuvette (1 cm).

B. Food Web Construction

Binary food webs, in which species and links are described in terms of presence/absence, were constructed for all sites (after Schmid-Araya *et al.*, 2002a,b; Woodward and Hildrew, 2001). Feeding links were established either directly, via GCA, or indirectly, by inferring links from the literature. Sampling effort was assessed by the construction of yield–effort curves (after Ings *et al.*, 2009) both for the nodes (macroinvertebrates and diatoms) and for the directly observed primary consumers' diets.

GCA was performed for primary consumers in all streams, to establish herbivore–algal links in the primary consumer assemblage. Foreguts were removed from each specimen under a dissecting microscope, the contents squeezed out and mounted in Euparal on a microscope slide, and food items identified at 400× magnification. Only limited GCA was performed for predatory macroinvertebrates and fishes, due to the small number of individuals available in these often unproductive systems (especially at low pH). Many hundreds of individuals per species would have been needed to characterize fully all the feeding links within each food web via GCA (e.g. Woodward and Hildrew, 2001), and this was impracticable and, in some cases, would have imposed excessive disturbance on the study streams. In particular, it was unlikely that asymptotes could have been achieved for the number of different items in fish diets at many of the sites, even if the entire population was removed from a given stream, judging by the yield–effort curves for brown trout in the Tadnoll Brook food web (cf. Figure 5 in Ings *et al.*, 2009). Directly observed feeding links were therefore supplemented with dietary information extracted from the literature (Brose *et al.*, 2005a,b; Layer *et al.*, 2010; Schmid-Araya *et al.*, 2002a,b; Warren, 1989; Woodward and Hildrew, 2001; Woodward *et al.*, 2008). These extra feeding links were based on the assumption that a described feeding link would be realized between two species at a given study site if the same link has been described in another system where these species co-exist (e.g. after Martinez, 1991; Woodward *et al.*, 2008). In a few instances, feeding links were assigned on the basis of taxonomic similarity, by assuming that different species within the same genus had identical links and that consumers would eat all resource species within a particular genus, if a link had been established either via direct observation or from the literature for at least one congener. The use of inferred feeding links thus enriched the directly observed data (which accounted for, on average, 14% of the links per web): this should have reduced the risk of underestimating web complexity, which is highly sensitive to sampling effort, particularly in larger networks (i.e. we aimed to remove the possible confounding effect of pH, which is also positively correlated with species richness) (e.g. Olesen *et al.*, 2010; Petchey *et al.*, 2004). Nonetheless, we have to interpret our results with some caution.

C. Calculation of Food Web Statistics

Consumer–resource matrices were constructed for each stream, and a range of food web metrics were calculated, including web size (the number of species, S), number of trophic interactions (L), number of interactions including competitive links (L_c), linkage density (L/S), directed connectance ($C=L/S^2$; Martinez, 1991) and web complexity ($SC'_{max}=S*L_c/S^2=L_c/S$). Generality ($G$; number of resource taxa per consumer) and vulnerability (V; number of consumer taxa per resource), as well as normalized generality (G_k) and vulnerability (V_k) were also calculated. For taxon k, normalized G_k and V_k are:

$$G_k = \frac{1}{L/S}\sum_{j=1}^{S} a_{jk} \tag{2}$$

$$V_k = \frac{1}{L/S}\sum_{j=1}^{S} a_{jk} \tag{3}$$

where $a_{ik}=1$ if taxon k consumes taxon i (otherwise $a_{ik}=0$), and $a_{kj}=1$ if taxon k is being consumed by taxon i (otherwise $a_{kj}=0$). Mean G_k and V_k in a food web equal 1, which allows comparisons to be made across webs of different sizes. In addition, the standard deviations of G_k and V_k (G_{st} and V_{st}, respectively; see Williams and Martinez, 2000) were calculated for all taxa in each stream, to quantify the variability within each food web. These various structural measures of the pattern and distribution of links can influence network stability: for instance, low connectance tends to be more stabilizing than high connectance at a given level of average interaction strength (e.g. Berlow *et al.*, 2004; Martinez *et al.*, 2006; May, 1972, 1973; McCann *et al.*, 1998; Neutel *et al.*, 2002).

Species abundance and mean individual body mass data were used to assess structural features of the food webs and to make inference about interaction strengths, based on allometric scaling relationships. Nodes (species) were plotted on a scatterplot of $\log_{10}$ mean individual body mass (M) versus $\log_{10}$ mean abundance (N), enabling us to depict the food webs as 'trivariate' networks (e.g. after Cohen *et al.*, 2003; McLaughlin *et al.*, 2010), whereby the feeding links were drawn between the connected nodes in the 'MN' plot (e.g. between predator and prey). We constructed MN plots for both species-averaged data (after Cohen *et al.*, 2003; Woodward *et al.*, 2005a, b) and using individual-based $\log_{10}$ size bins, irrespective of species identity (after Jennings and Brander, 2010; Yvon-Durocher *et al.*, 2008) to examine allometric scaling within the food webs from species-based and size-based perspectives (after Petchey and Belgrano, 2010). Species-averaged body mass data were also used to calculate consumer–resource body mass ratios for all trophic interactions within a stream, as a proxy measure of interaction strength (e.g. after Emmerson and Raffaelli, 2004; Emmerson *et al.*, 2005). This has been recently proposed as a relatively straightforward method derived from allometric scaling relationships of predator–prey body mass ratios (Emmerson and Raffaelli, 2004), in the form of $a_{ij} \sim (M_j/M_i)^b$, where M_j and M_i are the mean body masses of the predator and prey species, respectively, and a_{ij} is the Lotka–Volterra interaction coefficient. Theoretical studies have predicted the value of the scaling coefficient b to be close to 0.75 (Emmerson *et al.*, 2005), which is the value used in this study.

D. Modelling Robustness

We explored network robustness by simulating food web dynamics over a period of 10 years. Populations of species went extinct if their biomass fell below a critical extinction threshold ($B_i < 10^{-30}$). We defined food web robustness (R) as the fraction of initial species that persisted after species removal: $R = (S_p/S_i)$, where S_p and S_i are the number of persistent and initial species (excluding the species removed at the outset, SR), respectively. We used a bioenergetic consumer–resource model (after Yodzis and Innes, 1992) to describe the change of biomass over time (B'_i) of ith autotroph producer species (Eq. 4) and ith heterotroph consumer species (Eq. 5) in an n-species system:

$$B_i' = r_i(M_i)G_iB_i - \sum_{j=\text{consumers}}^{n} x_j(M_j)y_jB_j \tag{4}$$

$$B_i' = -x_i(M_i)B_i + \sum_{j=\text{resources}} x_i(M_i)y_iB_i - \sum_{j=\text{consumers}} x_j(M_j)y_jB_j \tag{5}$$

For each species i, B_i is its biomass, r_i is its mass-specific maximum growth rate, M_i is its average body mass, G_i is its logistic net growth ($G_i = 1 - B_i/K$)

with a carrying capacity K, x_i is its mass-specific metabolic rate and y_i is its maximum consumption rate relative to its metabolic rate.

The biological rates of production (W), metabolism (X) and maximum consumption (Y) follow negative-quarter power-law relationships with the species body masses (Brown *et al.*, 2004; Enquist *et al.*, 1999):

$$W_P = a_r M_P{}^{-0.25} \tag{6}$$

$$X_C = a_x M_C^{-0.25} \tag{7}$$

$$Y_C = a_y M_C{}^{-0.25} \tag{8}$$

where a_r, a_x and a_y are allometric constants and C and P indicate consumer and producer parameters, respectively (Yodzis and Innes, 1992). The time scale of the system was defined by normalizing the biological rates by the mass-specific growth rate of the smallest producer species P^*. Then, the maximum consumption rates, Y_C, were normalized by the metabolic rates X_C:

$$r_i = \frac{W_P}{W_{P*}} = \left(\frac{M_P}{M_{P*}}\right)^{-0.25} \tag{9}$$

$$x_i = \frac{X_C}{W_{P*}} = \frac{a_x}{a_r}\left(\frac{M_C}{M_{P*}}\right)^{-0.25} \tag{10}$$

$$y_i = \frac{Y_C}{X_C} = \frac{a_y}{a_x} \tag{11}$$

where W_{P*} is the reproduction rate of the smallest basal species. Substituting Eqs. 9–11 into 4 and 5 yields a population dynamic model with allometrically scaled and normalized parameters. We used constant values for the following model parameters: maximum ingestion rate $y_j = 12$; carrying capacity $K = 1$, half saturation density of the functional response $B_0 = 0.5$ (after Brose *et al.*, 2005a,b); y_i is the maximum ingestion rate relative to its metabolic rate (Brose *et al.*, 2006). Independent simulations of each food web started with uniformly random initial biomasses ($0.05 < B_i < 1$), and they were run for equivalent to 10 years as calculated by inserting M_{P*} in Eq. 6 and taking the inverse of W_{P*} [years^{-1}].

E. Statistical Data Analysis

All bivariate statistical analyses were performed using Minitab 15 (Minitab Inc., State College, PA, USA). Multivariate unimodal detrended correspondence analysis (DCA) was performed on $\log_{10}(x+1)$-transformed presence–absence data to assess species turnover rates across sites, using CANOCO for Windows 4.5 (ter Braak and Šmilauer, 2002), with environmental variables (i.e. pH) overlain passively on the ordination (after Woodward *et al.*, 2002).

III. RESULTS

Taxonomic resolution was high in all webs, with an average of 76.7% (±11.3 S.D.) of nodes identified to species level, and an average of 85.6% (±6.4 SD) described at least to genus. Yield–effort curves constructed for all 20 stream sites suggested that sampling effort was sufficient to characterize the nodes for both benthic macroinvertebrates (Figure 1A) and the epilithic diatom community (Figure 1B). In addition, yield–effort curves for diatoms in primary consumer guts (as shown for *Leuctra* spp. as a representative species present at most sites of the pH gradient; Figure 1C) revealed that sampling effort was sufficient to characterize feeding links at the base of the food webs, without resorting to the use of inferred links from the literature.

There was a strong gradient in community structure across the 20 food webs, with Axis I of the DCA being equivalent to 2.66 SD (i.e. almost complete turnover between the two extremes) and accounting for 20.8% of the variance in species scores. This axis was highly correlated with pH ($r=0.85$, $p<0.001$): in particular, most of the largest predators (i.e. fish species) were clustered towards the extreme right-hand side of Axis I, associated with the streams at the upper end of the pH gradient (Figure 2).

As pH increased, the food webs contained increasing numbers of both nodes and links (Figure 3). A range of key food web characteristics for each stream, as well as primary producers' and consumers' population biomasses (Table 1) were positively related to pH (Table 2), including web size (S, Figure 4A), the total number of links (L), the number of consumer links including competitive links (shared resources; L_c), linkage density (L/S), complexity (SC'_{max}; Figure 4B), consumer generality, resource vulnerability, chlorophyll-*a* density and invertebrate and fish biomass. The slopes (b) for these three measures of standing biomass, as a function of pH, increased with trophic status, from the algal basal resources ($b=2.38$) to the invertebrate assemblages ($b=4.96$) to the fish assemblages ($b=6.08$), suggesting the potential for top-down effects to strengthen with rising pH (Table 2). Standardized generality and vulnerability, as well as their standard deviations, however, did not change systematically along the pH gradient. Directed connectance (L/S^2) ranged from 0.12 (Coneyglen Burn) to 0.29 (Duddon Beck a), with a mean value of 0.22 ± 0.04 SD Connectance did not show a statistically significant correlation with either pH or web size, although the food webs with the highest connectance in the data set were also species-poor

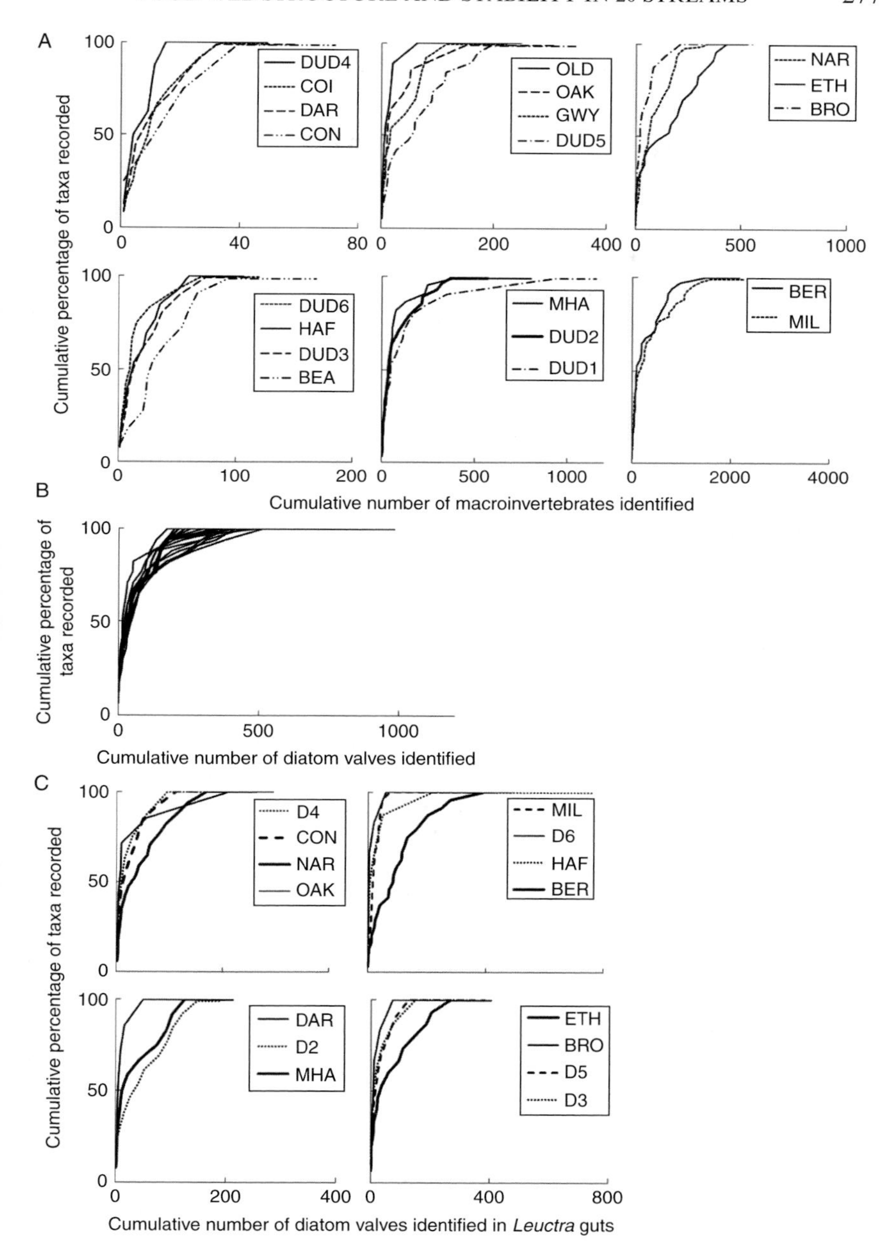

Figure 1 Yield–effort curves (A) macroinvertebrates; (B) diatoms in biofilm and (C) diatoms in the guts of a single primary consumer genus (*Leuctra* spp.).

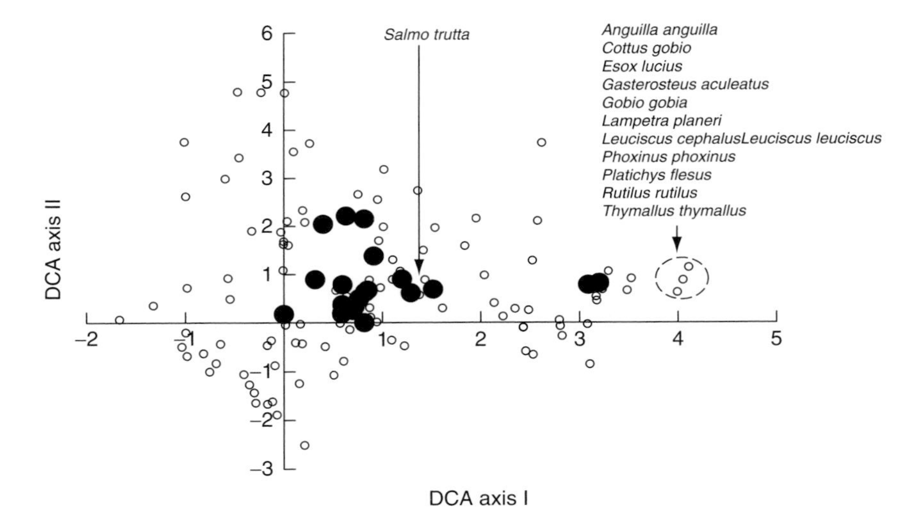

Figure 2 Detrended correspondence analysis of binary presence–absence data of nodes for consumer assemblages within the 20 stream food webs. Site scores are shown as solid circles, species scores as open circles. The positions of the fish species have been highlighted with arrows and labelled in italics.

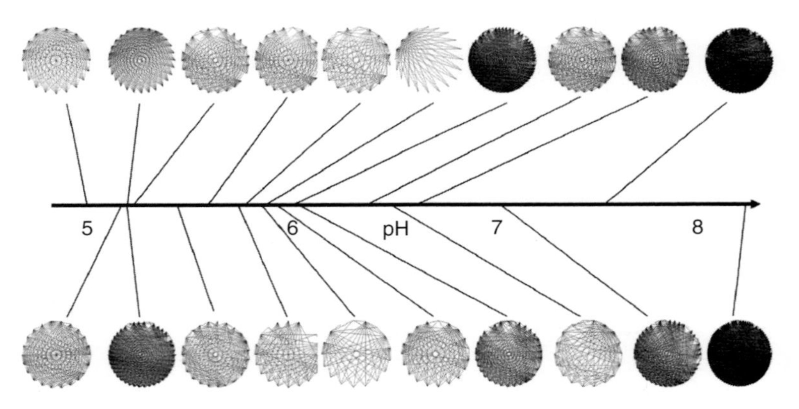

Figure 3 Binary feeding networks for 20 stream food webs across a pH gradient.

(e.g. Old Lodge, $S=23$, $C=0.26$; Broadstone Stream, $S=25$, $C=0.28$; Duddon Beck a, $S=20$, $C=0.29$).

Each of the 20 food webs exhibited a strong negative relationship between species-averaged body mass (M) and abundance (N), with the trivariate webs (Figure 5; also left-hand panels in Figure 6) revealing that the principal direction of energy flow was from the abundant, small species at the base

Table 2 OLS linear regression analysis of data from the 20 stream food webs (shown in Table 1) as a function of pH (x)

Dependent variable	Regression equation	r^2	F-ratio	p
S	$y=16.5x-65.2$	0.57	23.91	<0.001
L	$y=363x-1866$	0.62	28.72	<0.001
L_c	$y=563x-2909$	0.60	26.89	<0.001
L/S	$y=3.37x-12.8$	0.53	20.20	<0.001
SC'_{max}	$y=5.60x-22.9$	0.50	18.16	<0.001
G	$y=4.64x-14.8$	0.48	16.44	0.001
V	$y=3.55x-13.7$	0.56	22.80	<0.001
Chlorophyll-a	$y=2.38x-12.1$	0.43	13.77	0.002
Invertebrate biomass	$y=4.96x-23.7$	0.21	4.90	0.040
Fish biomass	$y=6.08x-33.6$	0.62	29.27	<0.001

Non-significant relationships (at $p>0.05$) are not shown here.

of the web to a less diverse suite of larger, rarer species higher in the food web. There was also a negative correlation ($r=-0.50$, $p=0.026$) between the MN scaling coefficients in the species-averaged webs with pH, suggesting that possible external subsidies might be more important in the more acid systems. The smallest body mass ratios and, by inference, the weakest *per capita* interaction strengths were generally located within the invertebrate–invertebrate and fish–fish subwebs, with the strongest interactions being between fish and invertebrate herbivores and algae, followed by predatory fishes feeding on invertebrates (Figures 7 and 8). There were only a few links where the (average) body mass of the consumer was smaller than that of the resource, with three of these being invertebrate fish parasites (Figure 7B). Although average inferred interaction strengths were broadly consistent among food webs across the gradient for each of the main interaction types, there was a general trend for the number of herbivorous and fish–invertebrate links to increase with pH (and with food web size, S) and at a faster rate than was the case for invertebrate–invertebrate links (Figure 9), with these relationships being best described by power laws.

Food web robustness, measured as the proportion of species still present after the simulation of 10 years of population dynamics, was negatively correlated with pH and with total fish biomass (for those streams that contained fishes), with the smaller, more acidified food webs being the most dynamically stable (Figure 10). Of the 282 extinctions that occurred in these simulations across all the food webs, 53.5% occurred within the basal resources, 27.3% among the primary consumers and 19.1% among the fish and invertebrate predators. Species that were lost were represented by a wide range of taxonomic groups and trophic levels within the food webs, and included basal resources (62.4% of all basal taxa lost), primary consumers

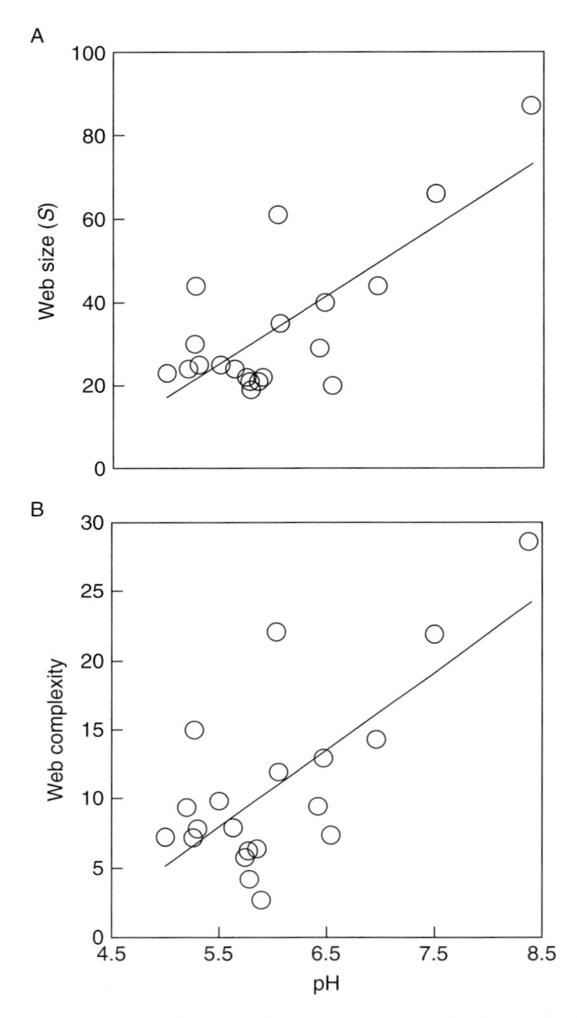

Figure 4 Web size (number of taxa, *S*) and complexity (SC'_{max}) in 20 streams as a function of stream pH (regression equations are given in Table 2).

(28.7% of all primary consumers lost) and predators (macroinvertebrates and fish: 42.9% of predatory taxa lost).

Among the predators, initial generality ($t_{d.f.115}=2.38$, $p=0.019$) and relative generality ($t_{d.f.90}=4.60$, $p<0.001$) were significantly lower for species that went extinct versus those that persisted after dynamical simulation, whereas the opposite was true for vulnerability ($t_{d.f.115}=3.42$, $p=0.001$) (Figure 11). Within the predator guild, extinctions were especially prevalent

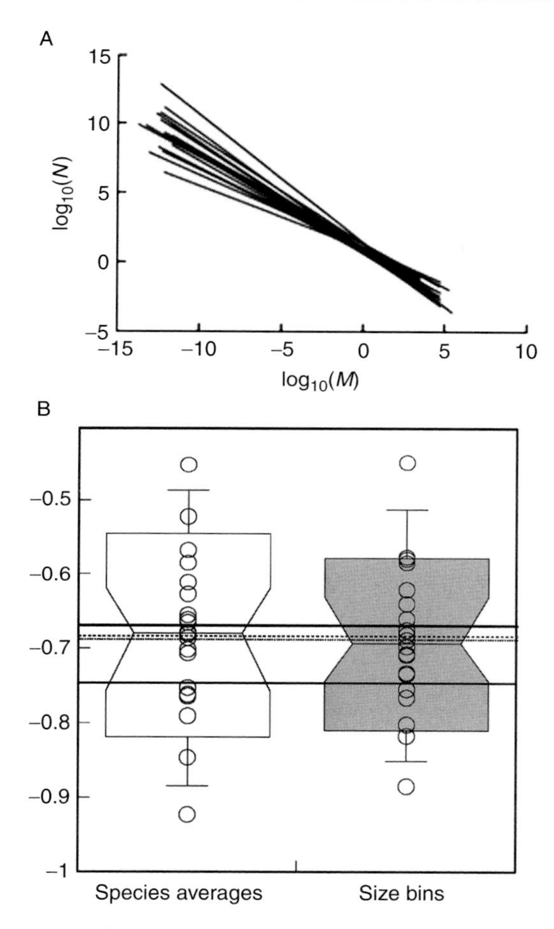

Figure 5 (A) Log-transformed body size ($\log_{10}M$; mg dry mass) versus abundance ($\log_{10}N$; individuals per square metre) regression lines for 20 stream food webs based on species-averaged data. (B) Boxplots of the distribution of allometric MN scaling coefficients derived for species-averaged and size-bin versions of the food webs (see Section II).

among some of the smaller invertebrates, such as *Siphonoperla torrentium* (lost from 12 of the 16 streams in which it occurred) and *Dicranota* sp. (lost from 7 of 8 streams), which typically occupied the lower intermediate trophic levels. Larger macroinvertebrate species with high generality were rarely lost, even if they were preyed upon by a relatively large number of predator species (i.e. high vulnerability) (e.g. *Phthersigena conspersa*, lost in only 3 of 14 streams). Fish species were also relatively robust to extinctions, and generally had low vulnerability: only 6 of the 27 species in total were lost.

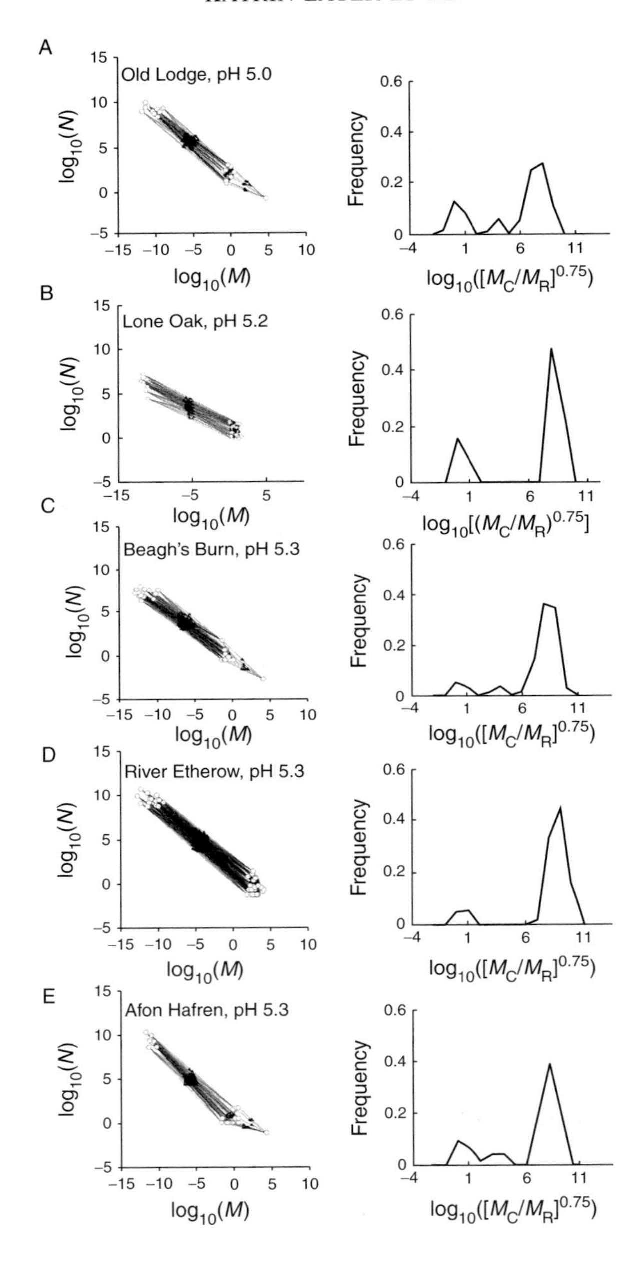
A
Old Lodge, pH 5.0
B
Lone Oak, pH 5.2
C
Beagh's Burn, pH 5.3
D
River Etherow, pH 5.3
E
Afon Hafren, pH 5.3
log10(N)
log10(M)
Frequency
log10([MC/MR]0.75)

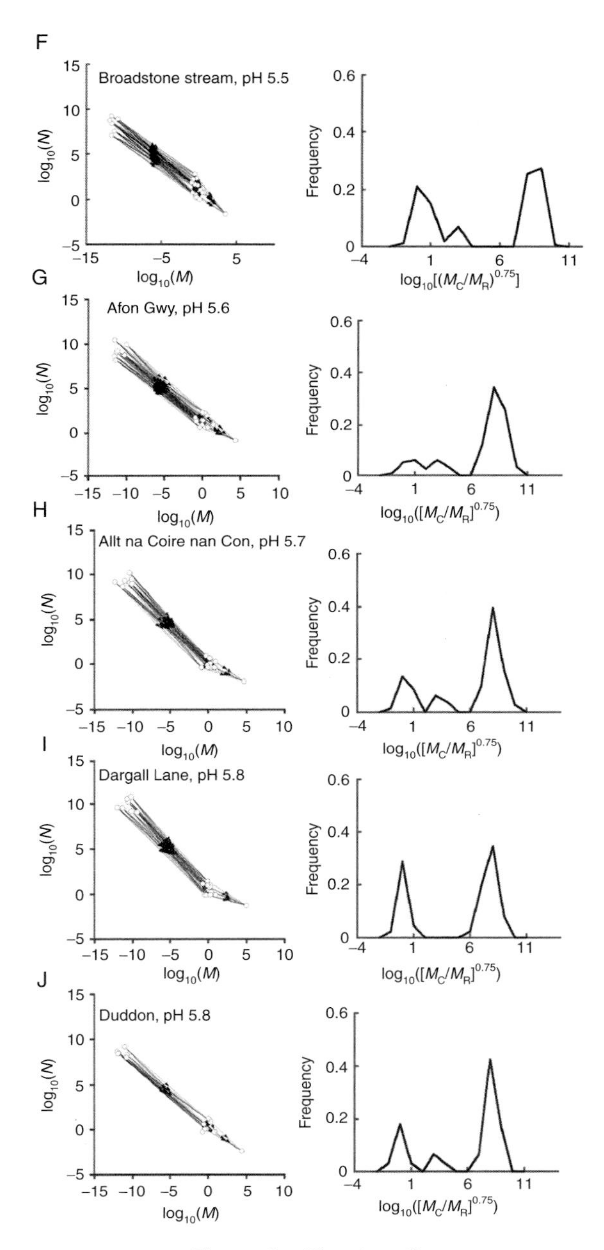

Figure 6 (Continued)

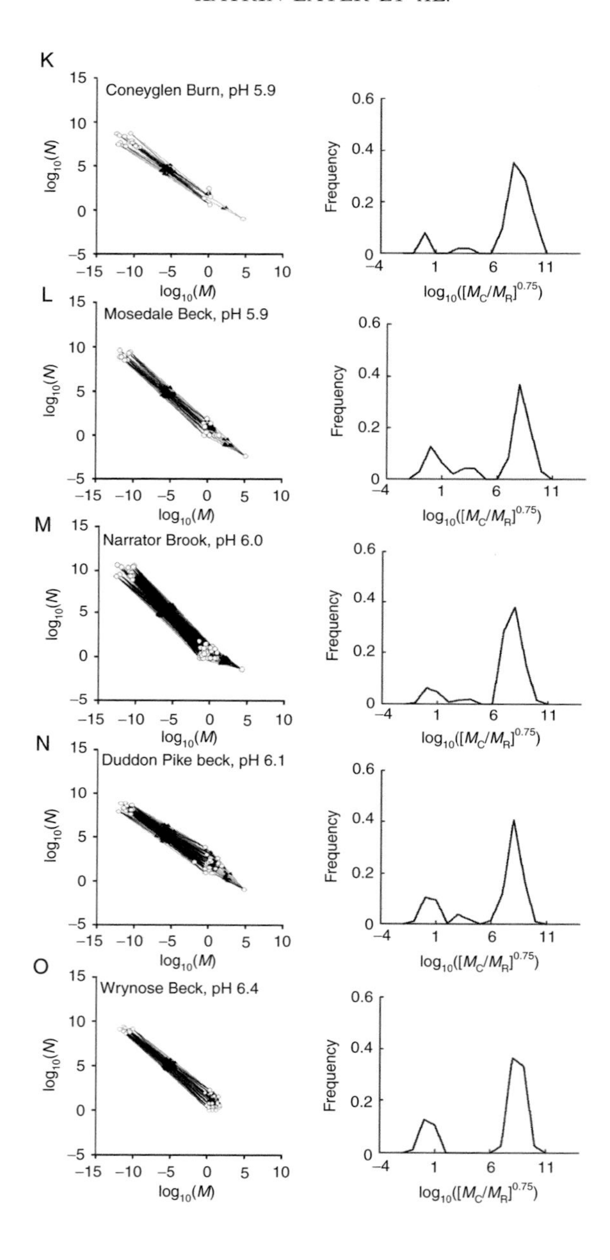
K
Coneyglen Burn, pH 5.9
L
Mosedale Beck, pH 5.9
M
Narrator Brook, pH 6.0
N
Duddon Pike beck, pH 6.1
O
Wrynose Beck, pH 6.4
$\log_{10}(N)$
$\log_{10}(M)$
Frequency
$\log_{10}([M_C/M_R]^{0.75})$

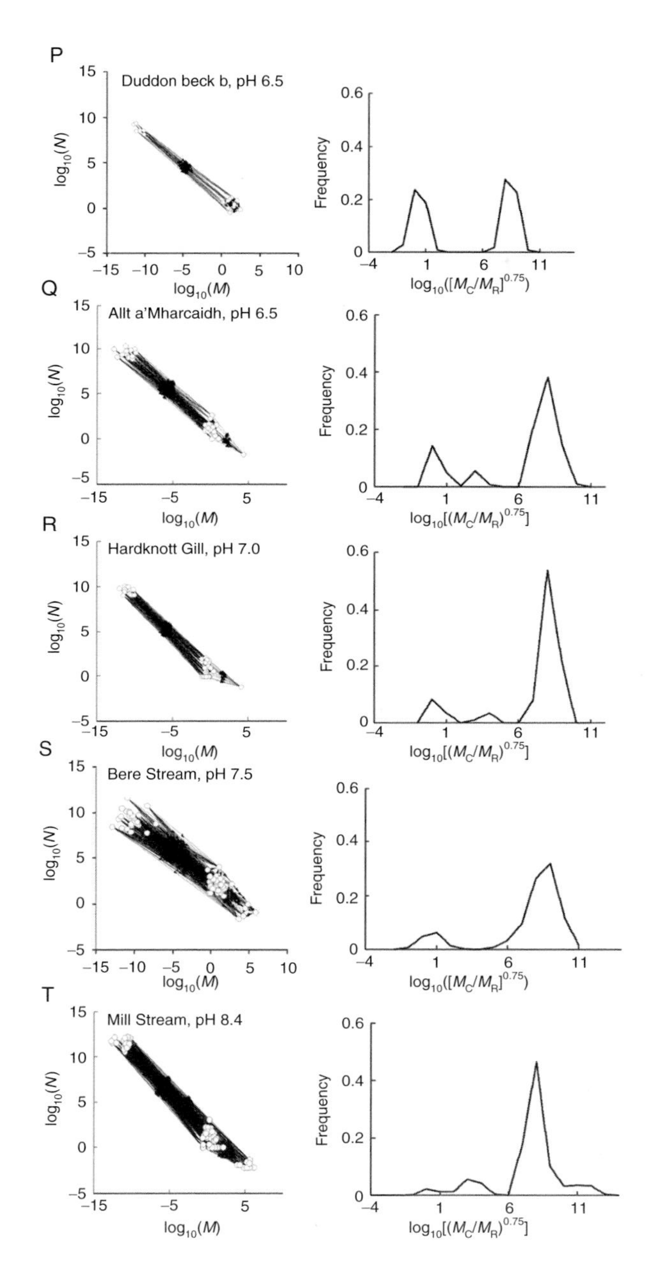

Figure 6 Food web structure in 20 streams of differing pH. Trivariate food webs are depicted for each site, in which species-averaged body mass ($\log_{10}M$; mean individual mg dry mass) is plotted against population abundance ($\log_{10}N$; individuals per square metre), with trophic links drawn in between consumer–resource pairs (left panels), and the distribution of inferred interactions strengths are shown as $\log_{10}$-transformed consumer–resource body mass ratios raised to the ¾ power (right panels).

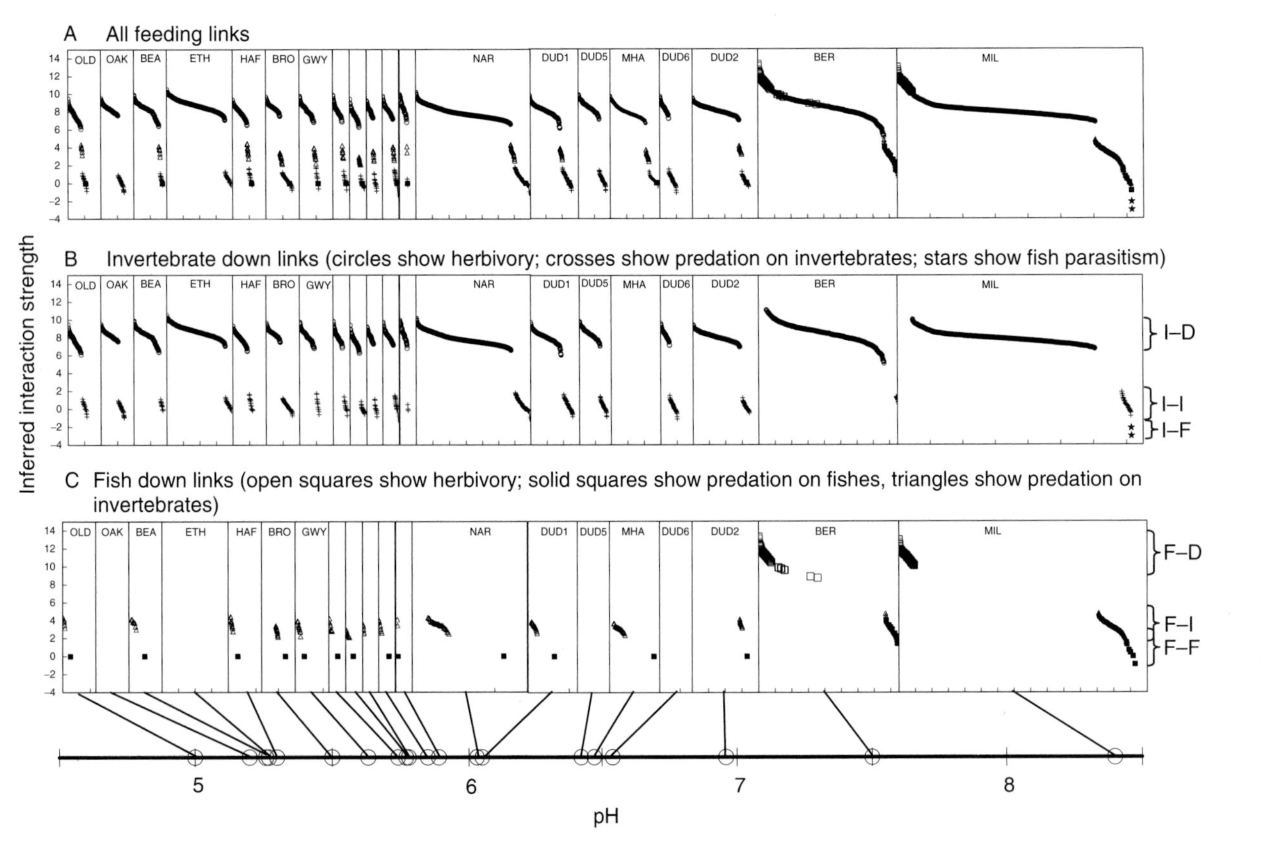

Figure 7 Ranked inferred interaction strengths (see Methods) for each of the 20 food webs across the pH gradient, for (A) the whole web; (B) invertebrate down links and (C) fish down links. (I-D = invertebrate-diatom links; I-I = invertebrate-invertebrate links; I-F = invertebrate-fish links; F-D = fish-diatom links; F-I = fish-invertebrate links; F-F = fish-fish links [the first letter denotes the consumer, the second the resource]).

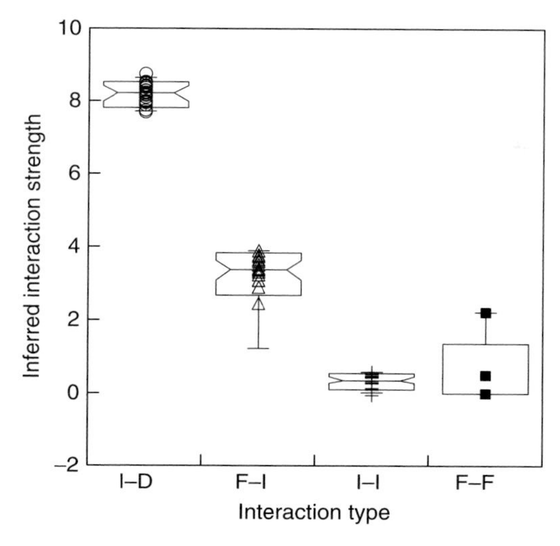

Figure 8 The distribution of mean interaction strengths across all 20 stream food webs (each network as a replicate) for the four dominant types of feeding link (invertebrate–diatom; fish–invertebrate; invertebrate–invertebrate; fish–fish). Tukey's pairwise comparisons revealed that all interactions types differed significantly from one another (ANOVA $F_{3,71} = 2{,}089$, $p < 0.001$) except I–I versus F–F ($p > 0.05$).

No clear distinction of species, based on generality or vulnerability, between those that were lost and those that persisted was evident within the primary consumer assemblage.

IV. DISCUSSION

The binary connectance food webs that have dominated much of the research in this area are increasingly being replaced by more data-rich, quantitative or semi-quantitative webs, and these new data have yielded important insights into network structure and dynamics (Ings *et al.*, 2009). For instance, remarkably consistent negative allometric scaling relationships between body mass and abundance have been described among even seemingly very different food webs that have no species in common (e.g. Jennings and Brander, 2010; Woodward *et al.*, 2005a,b) and apparently conservative patterns have also been reported from replicated experimental food webs within a single system (e.g. O'Gorman and Emmerson, 2010). The 20 highly resolved food webs described here had broadly similar ranges of mean species body mass and abundance (each ~ 12 orders of magnitude) from the base to the top of the food web, despite considerable differences in species identity, and the *MN* scaling coefficients were broadly similar, irrespective of whether species

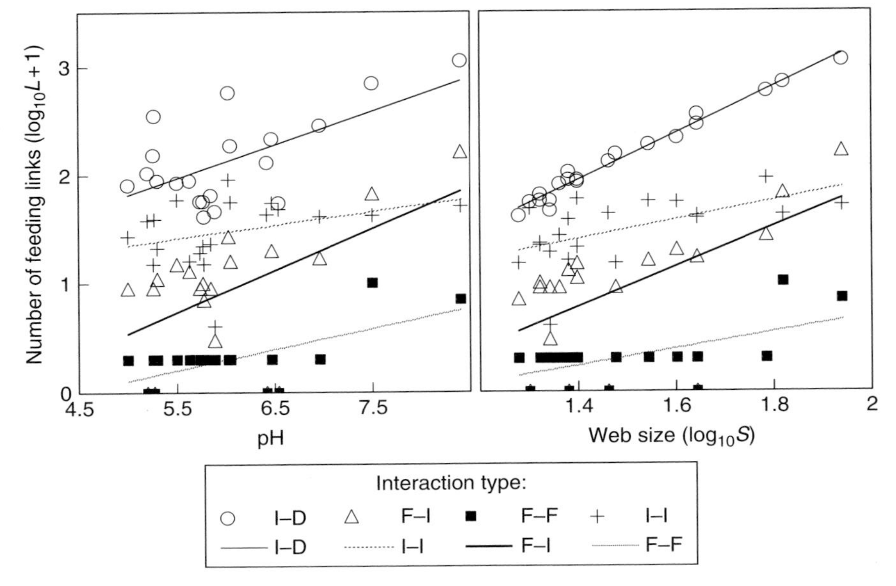

Figure 9 Link scaling relationships for the four dominant types of $\log_{10}$ feeding links, as a function of pH (left hand panel) or $\log_{10}$ food web size, S (right hand panel). OLS regression equations are: (i) I–D$=0.271+0.309$pH; $r^2=0.39$, $F_{1,19}=11.48$, $p=0.003$; (ii) F–I$=-1.39+0.386$ pH; $r^2=0.29$, $F_{1,19}=7.49$, $p=0.014$; (iii) F–F$=-0.832+0.188$ pH; $r^2=0.41$; $F_{1,19}=12.74$, $p=0.018$; (iv) I–I$=$n.s.; (v) I–D$=-1.05+2.14S$; $r^2=0.98$, $F_{1,19}=778$, $p<0.001$; (vi) F–I$=-1.82+1.85S$; $r^2=0.35$; $F_{1,19}=9.78$, $p=0.006$; (vii) F–F$=-0.819+0.754S$; $r^2=0.35$, $F_{1,19}=9.55$, $p=0.006$; (viii) I–I$=0.172+0.877S$; $r^2=0.31$, $F_{1,19}=7.91$, $p=0.012$. Note: I-D = invertebrate-diatom links; F-I = fish-invertebrate links; F-F = fish-fish links; I-I = invertebrate-invertebrate links (the first letter denotes the consumer, the second the resource).

averages or $\log_{10}$ size bins were used on the x-axis (e.g. Figure 5). Consistent trivariate patterns were evident across the 20 webs, such that the predominant direction of energy flux was from many small, abundant species at the base of the web towards fewer, larger and rarer species at the top along a similar trajectory of mass–abundance scaling (cf. Figure 6 with trivariate food webs presented in Cohen *et al.*, 2003; Layer *et al.*, 2010; Woodward *et al.*, 2005b).

All food webs studied here displayed strong size-structuring, with the log$M$–log$N$ regression lines from 20 individual food webs being constrained within a relatively small portion on the mass–abundance phase space (Figure 5), probably due to a combination of phylogenetic and metabolic constraints (Woodward *et al.*, 2005b). The mean allometric species-averaged MN scaling coefficient (mean for all webs: $b=-0.69\pm0.10$ SD) was almost identical to that found by Cohen *et al.* (2003) in their study of the Tuesday

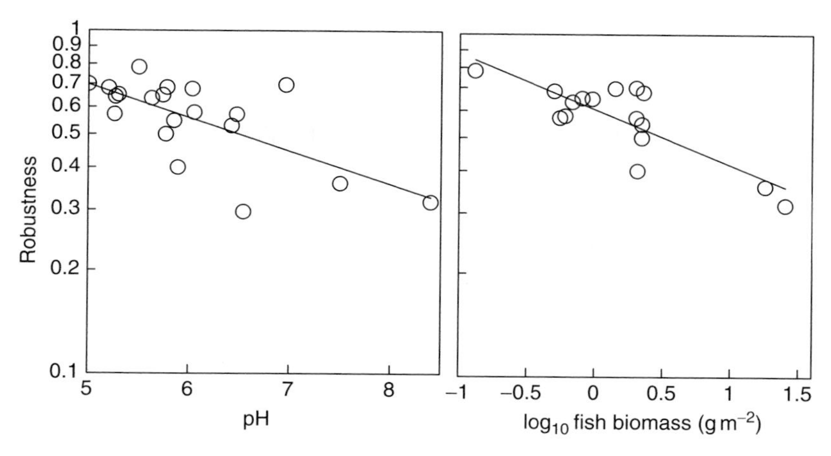

Figure 10 Robustness of food webs, measured as the $\log_{10}$ proportion of taxa remaining after simulation of 10 years of population dynamics, against mean pH (left hand panel) and empirically measured initial fish population biomass ($\log_{10}$ g/m^2) (right hand panel). OLS regression equations are: (i) $R=0.330-0.097$pH; $r^2=0.46$, $F_{1,19}=15.29$, $p=0.001$; (ii) $R=-0.218-0.162$ fish biomass; $r^2=0.67$, $F_{1,15}=28.11$, $p<0.001$.

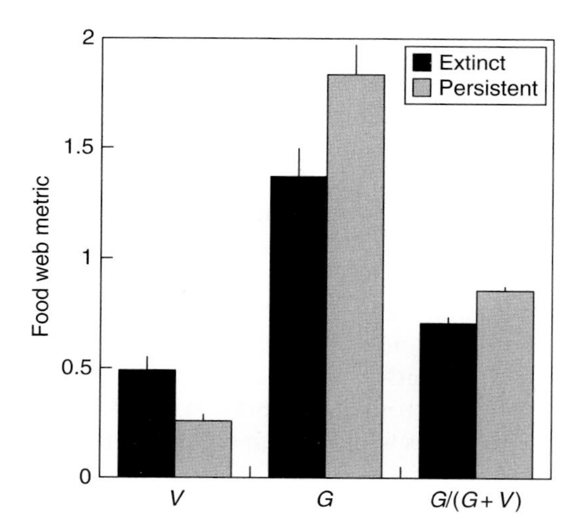

Figure 11 Mean ($\pm$SE) generality (G), vulnerability (V) and relative generality ($G/G+V$) of predators in 20 streams across a pH gradient, for taxa that went extinct (black bars) or persisted (grey bars) after simulation of 10 years of population dynamics.

Lake pelagic community ($b=-0.67$); i.e. closer to ⅔ than to a ¾ power law (cf. Savage *et al.*, 2004). The scaling coefficients for our 20 webs were shallower than the -1 predicted for energetic equivalence (i.e. where consumer–resource biomass abundance is equal). This might be at least in part due to detrital subsidies to the food webs, which cannot be captured in the analysis of log$M$–log$N$ relationships because this basal resource does not have an easily definable 'body size'. There was a significant trend for the MN scaling coefficients to become steeper (i.e. more negative) as pH increased ($r = -0.50$, $p = 0.026$), suggesting that detrital subsidies might play a more prominent role in acid systems in terms of their contribution to the food webs relative to the algal resources. This is in concordance with the widely held view that the food webs of acid freshwaters are more reliant on detrital inputs (primarily of terrestrial leaf-litter) than is the case for more circumneutral waters, which are more reliant on autochthonous algal inputs (e.g. Hildrew, 2009; Layer *et al.*, 2010; Ledger and Hildrew, 2005; Woodward *et al.*, 2005a).

Two hitherto unconnected empirical results are instructive here. First, Townsend *et al.* (1987) compared persistence in 27 stream communities across a pH gradient—including three of the acid sites covered in the present survey (Broadstone Stream, Old Lodge and Lone Oak). Communities from the small acid streams were the most persistent in Townsend *et al.*'s (1987) survey, in line with the findings of our simulations of robustness. Second, the lack of biological recovery reported for previously acidified ecosystems (e.g. Monteith *et al.*, 2005) has been attributed, at least in part, to the potential of acid food webs to be particularly stable, whereby the dynamics of the resident acid-tolerant community may prevent the re-establishment of acid-sensitive species even in the face of improving chemical conditions (Ledger and Hildrew, 2005; Woodward, 2009). The term 'stable' here refers mainly to the ability of a system to maintain its structure; for example, through the ability to recover from temporary (pulse) perturbations or the resistance to permanent (press) perturbations.

Our study provides the first formal and standardized analysis of stream food web structure and stability over a pH gradient that encompasses both acid systems and those at pH > 7. Both web size and the number of links increased rapidly with pH, and this should, in theory, make these systems less stable, if interaction strengths remain constant (McCann, 2000). Natural food webs form complex trophic networks, which, it is generally thought, consist of mainly weak links within which only a few strong ones are embedded (McCann, 2000; Woodward *et al.*, 2008). Dynamic modelling of the 20 food webs, which took into account allometrically scaled estimates of interaction strength, revealed that fewer species were lost from the more acid food webs over the course of the modelling period than those at high pH. This supported the suggestion of a negative relationship between stability and pH, and was in agreement with (and potentially accounted for) empirical observations of other forms of stability, including high persistence in acid stream communities and, suggestions of inertia in their recovery as pH ameliorates (Hildrew *et al.*, 2004; Ledger and Hildrew, 2005; Monteith *et al.*, 2005; Speirs *et al.*, 2000; Townsend *et al.*, 1987). It is intriguing that the food webs at higher pH contained a greater number of strong links, particularly between primary consumers and algae and between fish and invertebrates, than was the case for the smaller and more robust acid systems. Since the patterning and strength of interactions within a network are both important for its stability, this might explain why strong trophic cascades have been reported in systems where both fish and algae are abundant (e.g. Jones and Sayer, 2003; Power, 1990; Townsend, 2003), whereas this phenomenon seems to be far less prevalent in acid systems dominated by detrital inputs and lacking significant fish populations (Hildrew, 2009; Woodward and Hildrew, 2002). The negative relationships, we observed between network robustness versus both pH and total fish biomass, are certainly suggestive of the potential for top-down effects to strengthen as acidity ameliorates (Woodward and Hildrew, 2002). This is further supported by the progressive increase in slopes for the pH–biomass relationships with trophic level, from algal basal resources ($b=2.38$) to the assemblages of invertebrate consumers ($b=4.96$) and fishes ($b=6.08$). In addition, stable isotope data from these 20 streams (not presented here) have revealed an increasing reliance on detrital inputs both among and within primary consumer species, as pH falls (Layer, 2010). These different lines of evidence taken together offer support to earlier suggestions that the prevalence of donor control (i.e. links with an interaction strength of zero) at the base of the web, which tends to have a stabilizing effect on network dynamics, increases with acidification (Woodward and Hildrew, 2002).

Species extinctions in our simulation affected all food webs and many species over a range of trophic levels, although most of the extinctions occurred within the smallest members of the food web, the basal resources. No direct (e.g. physiological) effect of pH was included in these simulations (i.e. we made the simplifying assumption that it had no effect on growth, respiration and consumption *per se*), but rather its impact was mediated via changes in food web topology which altered the stability of the networks. Surprisingly, larger taxa were not more vulnerable to extinction than smaller species when comparing across trophic levels, in contrast to what is typically predicted (e.g. Raffaelli, 2004), and only relatively small proportions of fish and larger bodied invertebrate species were lost over the course of the simulations. However, within the predator guild there were clear differences in the relative trophic status and position within the food webs of species that were lost versus those that persisted. Essentially, higher generalism appeared to provide predators some protection from extinction when one or several of their resources were lost, or abundance became very low, whereas species that were exploited by a relatively large number of higher predators were more likely to be lost (Figure 11). Consequently, there was a tendency for intermediate consumers with relatively low generality, but often high vulnerability, to go extinct in our simulations: these might represent 'hub' species that act as important intermediate connectors within the food web (sensu Olesen *et al.*, 2010), although this suggestion requires further investigation. In general, within primary consumer assemblages, relatively specialist herbivore species, such as mayflies and snails, typically become both more diverse and abundant as pH rises, whereas in acid streams these taxa are typically replaced by generalist herbivore–detritivores, such as nemourid stoneflies, that are better able to exploit detrital subsidies of terrestrial leaf-litter (e.g., Ledger and Hildrew, 2005; Winterbourn *et al.*, 1992). Although we did not include detritus in our dynamic models, its role as a dominant basal resource at low pH could further stabilize acidified webs (e.g. Woodward and Hildrew, 2002) and damp the potentially destabilizing oscillatory effects of "fast" autochthonous-based food chains (e.g. Rooney *et al.*, 2006).

There are several caveats to our study that need to be borne in mind, including the use of (partially) inferred links within the networks. The complete documentation of links takes a considerable effort for any given system (Raffaelli, 2007; Schmid-Araya *et al.*, 2002a; Woodward *et al.*, 2005a), and many food web parameters, including connectance, are sensitive to the level of taxonomic inclusiveness, resolution and sampling effort (e.g., Martinez, 1991, 1993; Polis, 1991; Schmid-Araya *et al.*, 2002a,b). For instance, connectance in a version of the Broadstone Stream food web that included the soft-bodied meiofauna was considerably lower than for the more partial web reported here (Schmid-Araya *et al.*, 2002a; Woodward *et al.*, 2005a), although this might be due to undersampling of rare feeding links in the former (Olesen *et al.*, 2010). For logistical reasons, the soft-bodied meiofauna were not included in these food webs (as is the case the vast majority of other studies), even though such animals are numerous and diverse in riverine ecosystems and can influence the relationship between body mass and abundance, as well as estimates of connectance (Schmid-Araya *et al.*, 2002a,b; Stead *et al.*, 2005; Woodward *et al.*, 2005a). Nevertheless, the patterns revealed here are based on comparisons among food webs that were resolved and constructed in a consistent fashion.

Undersampling, especially of links, is a common problem in network analysis, particularly in large food webs where the probability of detecting links between rare predators and rare prey is often too small to be able to say with certainty if these 'missing' interactions are realized within the food web or not (Olesen *et al.*, 2010). Diets are therefore often necessarily partly or wholly inferred because of the difficulties of describing them via direct observation (e.g. Martinez, 1991; McLaughlin *et al.*, 2010; O'Gorman and Emmerson, 2010; Woodward *et al.*, 2008). Some have criticized the use of inferred feeding links on the basis that they might overestimate diet breadth for some species because of local-scale differences in consumer preferences, such that the linkages present will be a smaller subset of those that are possible (Hall and Raffaelli, 1997; Raffaelli, 2007). Clearly, there is a potential trade-off and compromise to be reached between the risk of undersampling, which increases with web size (Olesen *et al.*, 2010), and the possibility of including links that might not be realized in a given system, even though they may be common elsewhere. In our study, a combination of direct and indirect methods to establish trophic links was used, and herbivore–algal links were determined directly by gut-content analysis. Due to logistic constraints, it was not possible to document completely the diets of macroinvertebrate predators or fishes, given that several hundred guts are typically required to describe accurately the links for only one consumer species in even a single, species-poor food web (e.g. Ings *et al.*, 2009; Woodward and Hildrew, 2001). However, given the widely reported generalism of consumers in fresh waters, and that macroinvertebrate predators typically feed on virtually any prey of an appropriate size (e.g. Warren, 1989; Woodward and Hildrew, 2001), these seem reasonable justifications for the inference of feeding links in this study.

Whilst bearing in mind these potential caveats, very few well-resolved food webs have yet been constructed for a large number of systems using a standardized methodology, and our study is one of the first to address macroecological questions relating to structural and dynamical responses of food webs across a broad environmental gradient. Furthermore, our simulations of community dynamics showed that network stability and pH were correlated, with food webs in acid conditions being more robust than those at higher pH. This finding has potentially far-reaching implications in the field of applied ecology, as it could provide a possible explanation for the persistence of communities in acid waters and also the time lags often observed between chemical amelioration and the onset of biological recovery (Layer *et al.*, 2010; Ledger and Hildrew, 2005; Woodward, 2009). Given these findings, it would be instructive in future work to enrich the data further by direct characterization of predator diets (where possible) across the gradient, to examine the potential stabilizing effects of donor-controlled detrital subsidies more explicitly and also to assess the influence of the likely metabolic costs that many species will face when operating under physiological stress at low pH.

ACKNOWLEDGEMENTS

We would like to thank Mr. Paul Fletcher for help in the field. This study also relied partly on the United Kingdom Acid Waters Monitoring Network (UKAWMN), established and supported by the Department for Environment, Food and Rural Affairs (DEFRA), and now co-funded by the Scottish Government, the Welsh Assembly Government and the Natural

Environment Research Council (NERC). Additional water chemistry data for the UKAWMN were provided by the Freshwater Laboratory, Fisheries Research Services, Pitlochry, and the NERC Centre for Ecology and Hydrology. Financial support to K. L. was provided by a Queen Mary University of London postgraduate studentship, and J. R. was supported by a Research Network Programme of the European Science Foundation on body size and ecosystem dynamics (SIZEMIC). Last but not least, we would also like to thank Eoin O'Gorman and Ulrich Brose for their helpful comments on an earlier version of the manuscript.

REFERENCES

Battarbee, R.W., Jones, V.J., Flower, R.J., Cameron, N.G., Bennion, H.B., Carvalho, L., and Juggins, S. (2001). Diatoms. In: *Tracking Environmental Change Using Lake Sediments* (Ed. by J.P. Smol and H.J.B. Birks), Vol. 3, pp. 155–202. Kluwer Academic Publishers, Dordrecht, The Netherlands.

Beckerman, A.P., Petchey, O.L., and Warren, P.H. (2006). Foraging biology predicts food web complexity. *Proc. Natl. Acad. Sci. USA* **103**, 13745–13749.

Benke, A.C., and Wallace, J.B. (1997). Trophic basis of production among riverine caddisflies: Implications for food web analysis. *Ecology* **78**, 1132–1145.

Berlow, E.L., Neutel, A.-M., Cohen, J.E., De Ruiter, P.C., Ebenman, B.O., Emmerson, M., Fox, J.W., Jansen, V.A.A., Jones, J.I., Kokkoris, G.D., Logofet, D.O., McKane, A.J., *et al.* (2004). Interaction strengths in food webs: Issues and opportunities. *J. Anim. Ecol.* **73**, 585–598.

Berlow, E.L., Dunne, J.A., Martinez, N.D., Stark, P.B., Williams, R.J., and Brose, U. (2009). Simple prediction of interaction strengths in complex food webs. *Proc. Natl. Acad. Sci. USA* **106**, 187–191.

Bradley, D.C., and Ormerod, S.J. (2002). Long-term effects of catchment liming on invertebrates in upland streams. *Freshwater. Biol.* **47**, 161–171.

Brose, U., Cushing, L., Berlow, E.L., Jonsson, T., Banasek-Richter, C., Bersier, L.-F., Blanchard, J.L., Brey, T., Carpenter, S.R., Cattin Blandenier, M.-F., Cohen, J.E., Dawah, H.A., *et al.* (2005a). Body sizes of consumers and their resources. *Ecology* **86**, 2545.

Brose, U., Berlow, E.L., and Martinez, N.D. (2005b). Scaling up keystone effects from simple to complex ecological networks. *Ecol. Lett.* **8**, 1317–1325.

Brose, U., Williams, R.J., and Martinez, N.D. (2006). Allometric scaling enhances stability in complex webs. *Ecol. Lett.* **9**, 1228–1236.

Brown, J.H., Gillooly, J.F., Allen, A.P., Savage, V.M., and West, G.B. (2004). Toward a metabolic theory of ecology. *Ecology* **85**, 1771–1789.

Cohen, J.L., Jonsson, T., and Carpenter, S.R. (2003). Ecological community description using the food web, species abundance, and body size. *Proc. Natl. Acad. Sci. USA* **100**, 1781–1786.

Dunne, J.A. (2009). Food webs. In: *Encyclopedia of Complexity and Systems Science* (Ed. by R.A. Meyers), pp. 3661–3682. Springer, New York.

Elton, C.S. (1927). Animal Ecology. Sidgwick and Jackson, London, UK.

Emmerson, M.C., and Raffaelli, D.G. (2004). Predator–prey body size, interaction strength and the stability of a real food web. *J. Anim. Ecol.* **73**, 399–409.

Emmerson, M.C., Montoya, J.M., and Woodward, G. (2005). Body size, interaction strength and food web dynamics. In: *Dynamic Food Webs: Multispecies Assemblages, Ecosystem Development, and Environmental Change* (Ed. by P.C. De Ruiter, V. Wolters and J.C. Moore), pp. 167–179. Elsevier.

Enquist, B.J., West, G.B., Charnov, E.L., and Brown, J.H. (1999). Allometric scaling of production and life-history variation in vascular plants. *Nature* **401**, 907–911.

Galloway, J.N. (1995). Acid deposition: Perspectives in time and space. *Water Air Soil Pollut.* **85**, 15–24.

Hall, S.J., and Raffaelli, D.G. (1993). Food webs: Theory and reality. *Adv. Ecol. Res.* **24**, 187–239.

Hall, S.J., and Raffaelli, D.G. (1997). Food web patterns: What do we really know? In: *Multitrophic Interactions* (Ed. by A.C. Gange and V.K. Brown), pp. 395–417. Blackwells Scientific Publications, Oxford, UK.

Harcup, M.F., Williams, R., and Ellis, D.M. (2006). Movements of brown trout, *Salmo trutta* L., in the River Gwyddon, South Wales. *J. Fish Biol.* **24**, 415–426.

Hardy, A.C. (1924). The herring in relation to its animate environment. Part I. The food and feeding habits of the herring with special reference to the east coast of England. *Fish. Invest. Lond. Ser. 2* **7**(3), 1–53.

Hildrew, A.G. (2009). Sustained research on stream communities: A model system and the comparative approach. *Adv. Ecol. Res.* **41**, 175–312.

Hildrew, A.G., and Ormerod, S.J. (1995). Acidification: Causes, consequences and solutions. In: *The Ecological Basis for River Management* (Ed. by D.M. Harper and A.J.D. Ferguson), pp. 147–160. John Wiley and Sons Ltd., London.

Hildrew, A.G., Woodward, G., Winterbottom, J.H., and Orton, S. (2004). Strong density-dependence in a predatory insect: Larger scale experiments in a stream. *J. Anim. Ecol.* **73**, 448–458.

Ings, T.C., Montoya, J.M., Bascompte, J., Bluthgen, N., Brown, L., Dormann, C.F., Edwards, F., Figueroa, D., Jacob, U., Jones, J.I., Lauridsen, R.B., Ledger, M.E., *et al.* (2009). Ecological networks—Beyond food webs. *J. Anim. Ecol.* **78**, 253–269.

Jennings, S.J., and Brander (2010). Predicting the effects of climate change on marine communities and the consequences for fisheries. *J. Mar. Syst.* **79**, 418–426.

Jones, J.I., and Sayer, C.D. (2003). Does the fish–invertebrate–periphyton cascade precipitate plant loss in shallow lakes? *Ecology* **84**, 2155–2167.

Jonsson, T., Cohen, J.E., and Carpenter, S.R. (2005). Food webs, body size and species abundance in ecological community description. *Adv. Ecol. Res.* **36**, 1–84.

Kalff, J. (2002). Limnology: Inland Water Ecosystems. Prentice Hall, Upper Saddle River, NJ, USA.

Kowalik, R.A., Cooper, D.M., Evans, C.M., and Ormerod, S.J. (2007). Acid episodes retard the biological recovery of upland British streams from acidification. *Global Change Biol.* **13**, 2439–2452.

Layer, K. (2010). Responses of Freshwater Food Webs to Spatial and Temporal pH Gradients. PhD Thesis, Queen Mary & Westfield College, University of London. 236pp.

Layer, K., Hildrew, A.G., Monteith, D.T., and Woodward, G. (2010). Long-term variation in the littoral food web of an acidified mountain lake. *Global Change Biol.* 10.1111/j.1365-2486.2010.02195.x.

Ledger, M.E., and Hildrew, A.G. (2005). The ecology of acidification: Patterns in stream food webs across a pH gradient. *Environ. Pollut.* **137**, 103–118.

Lepori, F., Barbieri, A., and Ormerod, S.J. (2003). Causes of episodic acidification in Alpine streams. *Freshwater. Biol.* **48**, 175–189.

Lorenzen, C.J. (1967). Determination of chlorophyll and pheo-pigments: Spectrophotometric equations. *Limnol. Oceanogr.* **12**, 343–346.

Martinez, N.D. (1991). Artifacts or attributes? Effects of resolution on the Little Rock Lake food web. *Ecol. Monogr.* **61**, 367–392.

Martinez, N.D. (1993). Effects of resolution on food web structure. *Oikos* **66**, 403–412.

Martinez, N.D., Williams, R.J., and Dunne, J.A. (2006). Diversity, complexity, and persistence in large model ecosystems. In: *Ecological Networks: Linking Structure to Dynamics in Food Webs* (Ed. by M. Pascual and J.A. Dunne), pp. 167–185. Oxford University Press, Oxford, UK.

Masters, Z., Petersen, I., Hildrew, A.G., and Ormerod, S.J. (2007). Insect dispersal does not limit the biological recovery of streams from acidification. *Aquat. Conserv. Mar. Freshwat. Ecosyst.* **16**, 1–9.

May, R.M. (1972). Will a large complex system be stable? *Nature* **238**, 413–414.

May, R.M. (1973). Stability and Complexity in Model Ecosystems. Princeton University Press, USA.

McCann, K.S. (2000). The diversity–stability debate. *Nature* **405**, 228–233.

McCann, K.S., Hastings, A., and Huxel, G.R. (1998). Weak trophic interactions and the balance of nature. *Nature* **395**, 794–798.

McLaughlin, O.B., Jonsson, T., and Emmerson, M.C. (2010). Temporal variability in predator–prey relationships of a forest floor food web. *Adv. Ecol. Res.* **42,** 171–264.

Memmott, J., Martinez, N.D., and Cohen, J.E. (2000). Predators, parasitoids and pathogens: Species richness, trophic generality and body sizes in a natural food web. *J. Anim. Ecol.* **69**, 1–15.

Monteith, D.T., Hildrew, A.G., Flower, R.J., Raven, P.J., Beaumont, W.R.B., Collen, P., Kreiser, A.M., Shilland, E.M., and Winterbottom, J.H. (2005). Biological responses to the chemical recovery of acidified fresh waters in the UK. *Environ. Pollut.* **137**, 83–101.

Mulder, C., Cohen, J.E., Setälä, H., Bloem, J., and Breure, J.M. (2005). Bacterial traits, organism mass, and numerical abundance in the detrital soil food web of Dutch agricultural grasslands. *Ecol. Lett.* **8**, 80–90.

Mulder, C., den Hollander, H., Schouten, T., and Rutgers, M. (2006). Allometry, biocomplexity, and web topology of hundred agro-environments in The Netherlands. *Ecol. Complex.* **3**, 219–230.

Neutel, A.-M., Heesterbeek, J.A.P., and de Ruiter, P.C. (2002). Stability in real food webs: Weak links in long loops. *Science* **296**, 1120–1223.

O'Gorman, E.J., and Emmerson, M.C. (2010). Manipulating interaction strengths and the consequences for trivariate patterns in a marine food web. *Adv. Ecol. Res.* **42**, 301–419.

Olesen, J.M., Dupont, Y.L., O'Gorman, E., Ings, T.C., Layer, K., Melián, C.J., Troejelsgaard, K., Pichler, D.E., Rasmussen, C., and Woodward, G. (2010). From Broadstone to Zackenberg: Space, time and hierarchies in ecological networks. *Adv. Ecol. Res.* **42**, 1–69.

Otto, S.B., Rall, B.C., and Brose, U. (2007). Allometric degree distributions facilitate food-web stability. *Nature* **450**, 1226–1230.

Paine, R.T. (1988). Food webs—Road maps of interactions or grist for theoretical development. *Ecology* **69**, 1648–1654.

Petchey, O.L., and Belgrano, A. (2010). Body-size distributions and size-spectra: Universal indicators of ecological status? *Biol. Lett.* 10.1098/rsbl.2010.0240 (in press).

Petchey, O.L., Downing, A.L., Mittelbach, G.G., Persson, L., Steiner, C.F., Warren, P.H., and Woodward, G. (2004). Species loss and the structure and functioning of multitrophic aquatic systems. *Oikos* **104**, 467–478.

Petchey, O.L., Beckerman, A.P., Riede, J.O., and Warren, P.H. (2008). Size, foraging, and food web structure. *Proc. Natl. Acad. Sci. USA* **105**, 4191–4196.

Polis, G.A. (1991). Complex desert food webs: An empirical critique of food web theory. *Am. Nat.* **138**, 123–155.

Power, M.E. (1990). Effects of fish in river food webs. *Science* **250**, 811–814.

Raddum, G.G., and Fjellheim, A. (2003). Liming of River Audna, southern Norway: A large-scale experiment of benthic invertebrate recovery. *AMBIO* **32**, 230–234.

Raffaelli, D.G. (2004). How extinction patterns affect ecosystems. *Science* **306**, 1141–1142.

Raffaelli, D.G. (2007). Food webs, body size and the curse of the Latin binomial. In: *From Energetics to Ecosystems: The Dynamics and Structure of Ecological Systems* (Ed. by K. Rooney, K.S. McCann and D.L.G. Noakes), pp. 53–64. Springer, Dordrecht, The Netherlands.

Reuman, D.C., and Cohen, J.E. (2004). Trophic links' length and slope in the Tuesday Lake food web with species' body mass and numerical abundance. *J. Anim. Ecol.* **73**, 852–866.

Reuman, D.C., and Cohen, J.E. (2005). Estimating relative energy fluxes using the food web, species abundance, and body size. *Adv. Ecol. Res.* **36**, 137–182.

Reuman, D.C., Mulder, C., Banasek-Richter, C., Cattin Blandenier, M.F., Breure, A.M., Den Hollander, H., Knetiel, J.M., Raffaelli, D., Woodward, G., and Cohen, J.E. (2009). Allometry of body size and abundance in 166 food webs. *Adv. Ecol. Res.* **41**, 1–46.

Riede, J.O., Rall, B.C., Banasek-Richter, C., Navarrete, S.A., Wieters, E.A., and Brose, U. (2010). Scaling of food web properties with diversity and complexity across ecosystems. *Adv. Ecol. Res.* **42,** 139–170.

Rooney, N., McCann, K., Gellner, G., and Moore, J.C. (2006). Structural asymmetry and the stability of diverse food webs. *Nature* **442**, 265–269.

Rose, N., Monteith, D.T., Kettle, H., Thompson, R., Yang, H., and Muir, D. (2004). A consideration of potential confounding factors limiting chemical and biological recovery at Lochnagar, a remote mountain loch in Scotland. *J. Limnol.* **63**, 63–76.

Rosemond, A.D., Reice, S.R., Elwood, J.W., and Mulholland, P.J. (1992). The effects of stream acidity on benthic invertebrate communities in the south-eastern United States. *Freshwater. Ecol.* **27**, 193–209.

Savage, V.M., Gillooly, J.F., Brown, J.H., West, G.B., and Charnov, E.L. (2004). Effects of body size and temperature on population growth. *Am. Nat.* **163**, E429–E441.

Schmid-Araya, J.M., Hildrew, A.G., Robertson, A., Schmid, P.E., and Winterbottom, J.H. (2002a). The importance of meiofauna in food webs: Evidence from an acid stream. *Ecology* **83**, 1271–1285.

Schmid-Araya, J.M., Schmid, P.E., Robertson, A., Winterbottom, J.H., Gjerløv, C., and Hildrew, A.G. (2002b). Connectance in stream food webs. *J. Anim. Ecol.* **71**, 1056–1062.

Seber, G.A.F., and LeCren, E.D. (1967). Estimating population parameters from catches large relative to the population. *J. Anim. Ecol.* **36**, 631–643.

Seip, H.M., Aagaard, P., Angell, V., Eilertsen, O., Larssen, T., Lydersen, E., Mulder, J., Muniz, I.P., Semb, A., Dagang, T., Vogt, R.D., Jinshong, X., *et al.* (1999). Acidification in China: Assessment based on studies at forested sites from Chongqing to Guangzhou. *AMBIO* **28**, 522–528.

Snucins, E., and Gunn, H.M. (2003). Use of rehabilitation experiments to understand the recovery dynamics of acid-stressed fish populations. *AMBIO* **32**, 240–243.

Speirs, D.C., Gurney, W.S.C., Winterbottom, J.H., and Hildrew, A.G. (2000). Long-term demographic balance in the Broadstone Stream insect community. *J. Anim. Ecol.* **69**, 45–58.

Stead, T.K., Schmid-Araya, J.M., and Hildrew, A.G. (2005). Distribution of body size in a stream community: One system, many patterns. *J. Anim. Ecol.* **74**, 475–487.

ter Braak, C.J.F. and Šmilauer, P. (2002). *CANOCO Reference Manual and CanoDraw for Windows User's Guide: Software for Canonical Community Ordination* (version 4.5). Microcomputer Power. Ithaca, NY, USA.

Townsend, C.R. (2003). Individual, population, community, and ecosystem consequences of a fish invader in New Zealand streams. *Conserv. Biol.* **17**, 38–47.

Townsend, C.R., Hildrew, A.G., and Francis, J.E. (1983). Community structure in some Southern English streams: The influence of physicochemical factors. *Freshwater. Biol.* **13**, 531–544.

Townsend, C.R., Hildrew, A.G., and Schofield, K. (1987). Persistence of stream invertebrate communities in relation to environmental variability. *J. Anim. Ecol.* **56**, 597–613.

Townsend, C.R., Thompson, R.M., McIntosh, A.R., Kilroy, C., Edwards, E., and Scarsbrook, M.R. (1998). Disturbance, resource supply and food-web architecture in streams. *Ecol. Lett.* **1**, 200–209.

Tylianakis, J.M., Tscharntke, T., and Lewis, O.T. (2007). Habitat modification alters the structure of tropical host–parasitoid food webs. *Nature* **445**, 202–205.

Warren, P.H. (1989). Spatial and temporal variation in the structure of a freshwater food web. *Oikos* **55**, 299–311.

Williams, R.J., and Martinez, N.D. (2000). Simple rules yield complex food webs. *Nature* **409**, 180–183.

Winterbourn, M.J., Hildrew, A.G., and Orton, S. (1992). Nutrients, algae and grazers in some British streams of contrasting pH. *Freshwater. Biol.* **28**, 173–182.

Woodward, G. (2009). Biodiversity, ecosystem functioning and food webs in freshwaters: Assembling the jigsaw puzzle. *Freshwater. Biol.* **54**, 2171–2187.

Woodward, G., and Hildrew, A.G. (2001). Invasion of a stream food web by a new top predator. *J. Anim. Ecol.* **70**, 273–288.

Woodward, G., and Hildrew, A.G. (2002). Food web structure in riverine landscapes. *Freshwater. Biol.* **47**, 777–798.

Woodward, G., Jones, J.I., and Hildrew, A.G. (2002). Community persistence in Broadstone Stream (UK) over three decades. *Freshwat. Biol.* **47**, 1419–1435.

Woodward, G., Speirs, D.C., and Hildrew, A.G. (2005a). Quantification and temporal resolution of a complex size-structured food web. *Adv. Ecol. Res.* **36**, 85–135.

Woodward, G., Ebenman, B., Emmerson, M., Montoya, J.M., Olesen, J.M., Valido, A., and Warren, P.H. (2005b). Body size in ecological networks. *Trends Ecol. Evol.* **20**, 402–409.

Woodward, G., Papantoniou, G., Edwards, F.E., and Lauridsen, R.B. (2008). Trophic trickles and cascades in a complex food web: Impacts of a keystone predator on stream community structure and ecosystem processes. *Oikos* **117**, 683–692.

Woodward, G., Benstead, J.P., Beveridge, O.S., Blanchard, J., Brey, T., Brown, L., Cross, W.F., Friberg, N., Ings, T.C., Jacob, U., Jennings, S., Ledger, M.E., *et al.* (2010). Ecological networks in a changing climate. *Adv. Ecol. Res.* **42**, 71–138.

Yan, N.D., Leung, B., Keller, W., Arnott, S.E., Gunn, J.M., and Raddum, G.G. (2003). Developing conceptual frameworks for the recovery of aquatic biota from acidification. *AMBIO* **32**, 165–169.

Yodzis, P., and Innes, S. (1992). Body size and consumer–resource dynamics. *Am. Nat.* **139**, 1151–1175.

Yvon-Durocher, G., Montoya, J.M., Emmerson, M.C., and Woodward, G. (2008). Macroecological patterns and niche structure in a new marine food web. *Cent. Eur. J. Biol.* **3**, 91–103.

Manipulating Interaction Strengths and the Consequences for Trivariate Patterns in a Marine Food Web

EOIN J. O'GORMAN AND MARK C. EMMERSON

ADVANCES IN ECOLOGICAL RESEARCH VOL. 42

0065-2504/10 $35.00
DOI: 10.1016/S0065-2504(10)42006-1

SUMMARY

We are experiencing a global extinction crisis as a result of climate change and human-induced alteration of natural habitats, with large predators at high trophic levels in food webs being particularly vulnerable. Unfortunately, there is a scarcity of food web data that can be used to assess how species extinctions alter the structure and stability of temporally and spatially replicated networks. We established a series of large experimental mesocosms in a shallow subtidal benthic marine system and constructed food webs for each replicate. After 6 months of community assembly, we removed large predators from the core communities of 20 experimental food webs, based on the strength of their trophic interactions, and monitored the changes in the networks' structure and stability over an 8-month period. Our analyses revealed the importance of allometric relationships and size-structuring in natural communities as a means of preserving food web structure and sustainability, despite significant changes in the diversity, stability and productivity of the system.

I. INTRODUCTION

A. Why Study Food Webs?

A food web is a diagrammatic representation of energy and material fluxes among organisms via trophic interactions (Cohen *et al.*, 1993a). Connectance webs have been a central organizing concept in ecology since the late 1920s (Elton, 1927), but they are restricted to simple presence/absence data on species and feeding links. Such webs were described by Pimm (1982) as being caricatures of nature, depicting binary relationships between species and missing much of the important ecological properties of the system. For example, predators may change their diet seasonally and do not interact equally with all their prey, as temporal and energetic constraints place limitations on the interactions that can take place. Attempts have been made to address these shortcomings by incorporating a temporal analysis of food web structure (Tavares-Cromar and Williams, 1996; Thompson and Townsend, 1999; Warren, 1989; Woodward *et al.*, 2005), or by including more quantitative information about species, leading, for instance, to the development of so-called *MN* webs (Cohen *et al.*, 2003; Jonsson *et al.*, 2005; Layer *et al.*, 2010; McLaughlin *et al.*, 2010; Reuman and Cohen, 2005; Woodward *et al.*, 2005). These *MN* webs include descriptions of the body mass (*M*) and abundance (*N*) of the constituent species, often depicted as an *x*–*y* scatterplot on log–log axes with the feeding links overlain on the nodes within the network. This approach has facilitated an exploration of patterns between the trophic structure of the food web and these two fundamental attributes of species populations. Such descriptions are a step closer to bringing the caricature to life and unravelling the complex web of interactions that contribute to the functioning and stability of our natural ecosystems and understanding the consequences of the threats they face.

We are currently in the midst of a global extinction crisis (Dirzo and Raven, 2003; Pimm *et al.*, 1995, 2006; Sala *et al.*, 2000; Worm *et al.*, 2006) that has been driven largely by human transformation of the natural environment (Vitousek *et al.*, 1997). Anthropogenic impacts associated with habitat loss and fragmentation, overexploitation and introductions of exotic species have led to unprecedented levels of species loss worldwide (Fahrig, 2003; Purvis *et al.*, 2000; Seabloom *et al.*, 2006). There is also a growing body of evidence that climate change is changing the structure and composition of our ecosystems, with the potential to cause rapid declines in biodiversity on a global scale (Harmon *et al.*, 2009; Harrison, 2000; Petchey *et al.*, 1999). Top predators are especially vulnerable to human-induced disturbances on ecosystems, with environmental warming (Petchey *et al.*, 1999) and overfishing (Jackson *et al.*, 2001) proving to be particularly detrimental because of the characteristics of apex predators such as large body size, rarity, small population size, small geographical range, slow population growth and specialized ecological habits (Duffy, 2003). Analyses of the changes to *MN* food web structure that result from species loss at high trophic levels would facilitate an exploration of possible secondary extinctions (Borrvall and Ebenman, 2006) and detrimental effects on ecosystem structure and stability (Worm and Duffy, 2003).

B. The Need for Temporal and Spatial Replication of Food Webs

Due to logistical constraints, most food web studies have been restricted to single sites, often using a ‘model system’ approach. Some comparisons of food web properties have been made between different systems, but are still limited to a handful of studies that adequately describe complex natural food webs to a sufficient level of standardization and taxonomic resolution (Ings *et al.*, 2009; Olesen *et al.*, 2010). Although the body of high-quality data in the literature is growing rapidly, the best described webs are spread across a range of marine (Jacob, 2005; Opitz, 1993), freshwater (Jonsson *et al.*, 2005; Martinez, 1991; Warren, 1989; Woodward *et al.*, 2005), estuarine (Hall and Raffaelli, 1991) and terrestrial (Memmott *et al.*, 2000; Polis, 1991; Reagan and Waide, 1996) systems, often with very different study aims in mind: in many instances, particularly prior to the 1990s, the construction of a food web has been a by-product of other research activities. Combined with a lack of temporal replication, as most studies show either a single snapshot or an integrated summary network constructed over many sampling occasions (decades in some instances), this lack of standardization has prohibited an in-depth analysis of the likely consequences of realistic scenarios of extinction from these food webs (Olesen *et al.*, 2010; Woodward *et al.*, 2010). Very few food web studies have explicitly tested the impact of species loss on food web patterns (but see Jonsson *et al.*, 2005). One of the principal aims of our study was to address these shortcomings by carrying out a detailed exploration of the consequences of high trophic level extinction in benthic marine food webs via the use of an experimental setup that facilitated spatial and temporal replication of small, yet realistically complex, mesocosm food webs.

Our initial objective was to establish replicate shallow-water marine benthic communities in 24 separate mesocosms *in situ* at a sheltered sea lough. Each mesocosm consisted of a core assemblage of 10 large predators at the outset, which would be maintained or manipulated throughout the experiment, and formed the highest trophic positions in the community of benthic invertebrates that rapidly assembled through the mesh of the cages. Natural recruitment through the mesh of the cages allowed complex communities of benthic invertebrates to build up the food web around this core assemblage. Regular monitoring allowed us to construct each mesocosm food web at six different time intervals (typically 2 months apart), along with information on the average body size and abundance of each species in the webs. The removal of subsets of our core assemblage of large predators from the communities enabled us to investigate the impact of species loss on the structure of the mesocosm food webs over the remainder of the experiment. This unique experimental approach enabled us to test for common patterns and changes in our replicate food webs and link them explicitly to the targeted predator extinctions.

C. Univariate, Bivariate and Trivariate Patterns

There is a wide range of univariate food web metrics available. Connectance (the proportion of potential links that are realized) and linkage density (the average number of links per species) illustrate the degree of reticulation within the trophic network, and both have been related to the dynamic stability of a system; for example as connectance increases, communities have been shown to be more robust to secondary extinctions (Dunne *et al.*, 2002b), while mobile higher order consumers, which increase linkage density, stabilize food webs by linking them in space (McCann *et al.*, 2005). Species richness has often been described as having a positive relationship with ecosystem process rates (Loreau *et al.*, 2001), while species turnover can explain the resistance of a community to species invasions and/or extinctions (O'Gorman and Emmerson, 2009). The mean length of food chains describes the resource availability and size of a system (Kaunzinger and Morin, 1998; Post *et al.*, 2000). The fractions of top, intermediate and basal species, as well as the distribution of links between these groups, characterize the flow of energy and concentrations of direct interactions in various compartments of a food web (Havens, 1992; Schmid-Araya *et al.*, 2002). These parameters were characterized in our experimental food webs and used to quantify the effect of our removal perturbations.

Distributions of body mass within the experimental food webs will also reveal much about the changing structure of the mesocosm food webs in response to the manipulations. Individual body mass is linked to many key biological rates, including growth, respiration, reproduction and mortality (Brown and Gillooly, 2003; Kleiber, 1947; Peters, 1983), and hence is fundamental to our understanding of dynamic biological systems (Woodward *et al.*, 2010; Yvon-Durocher *et al.*, 2010). Distributions of body mass in natural communities are expected to form a pyramid of numbers, with many small and few large individuals (Elton, 1927). Changes to predator and prey body mass can also have profound implications for the strength of trophic interactions (Brose *et al.*, 2006b; Emmerson and Raffaelli, 2004; Jonsson and Ebenman, 1998), which are key determinants of the stability of food webs (McCann *et al.*, 1998; Neutel *et al.*, 2002; O'Gorman and Emmerson, 2009). The scaling of abundance with body mass is thought to be approximately inversely proportional to the scaling of metabolic rate, suggesting an energetic equivalence across species populations within a local food web (Damuth, 1981, 1987; but see Blackburn and Gaston, 1997). The biomass of a species is also often independent of its body size, leading to biomass equivalence across different trophic levels (Brown and Gillooly, 2003; Rinaldo *et al.*, 2002; but see Stork and Blackburn, 1993). Disruptions to these recurrent patterns in body mass distribution and/or abundance could have negative effects on community structure and ecosystem functioning within the food web, with skewed distributions of resources across populations leading to possible species extinctions.

By examining a number of bivariate relationships, such as the correlation between predator and prey body size, we can test the robustness of body mass and abundance patterns in our food webs to the extinction of high trophic level species. For instance, as a null model, we might expect the data points in a matrix of predator M against prey M to be distributed at random. Alternatively, if predators typically feed on smaller prey, we would expect to find a significantly higher than random proportion of data points in the upper left triangle of these plots (see Cohen *et al.*, 1993b; Warren and Lawton, 1987). If predators are larger than their prey, we would also expect body size to increase with trophic height (TH) and, based on previous studies, in a loglinear relationship (see Jennings *et al.*, 2001; Jonsson *et al.*, 2005; Woodward and Hildrew, 2002). Such a pattern might also lead to an inverted trophic pyramid of body mass in the system, with total body mass increasing with TH (see Elton, 1927). Similarly, marked, recurrent patterns are also predicted for numerical abundance (NA). Based on early observations (Elton, 1927; Hutchinson, 1957), we would expect, for instance, predators to be rarer than their prey (i.e. a significantly greater than random proportion of data points in the lower right triangle of a plot of predator vs. prey NA). We would also expect NA of a given species population to decrease with TH and total NA to decrease at each trophic level.

The bivariate relationships involving biomass abundance (BA) are less clear: for instance since BA is the product of body mass, M, and NA ($B = M \times \text{NA}$), the bivariate relationships above will depend on how body mass scales with abundance. The slope of body mass–abundance relationships in nature is often thought to lie between -0.75 and -1.1 (Brown and Gillooly, 2003; Cyr *et al.*, 1997; Damuth, 1981; but see Blackburn and Gaston, 1997). A body mass–abundance slope of -1 would imply that biomass is independent of body mass (if $\text{NA} \propto M^{-1}$, then $\text{BA} = M^{-1} \times M^{1} = M^{0}$) and we would expect no relationship between predator and prey BA or between BA and TH (and a body mass–BA relationship with a slope of 0). A slope greater than -1, that is less negative, should lead to predators feeding on a lower BA of prey and an increase in BA with TH. A slope less than -1, that is more negative, should lead to predators feeding on a higher BA of prey and a decrease in BA with TH. These relationships will have important consequences for energy flow in the food webs, determining the feeding pressure exerted by higher trophic levels on their resources.

Lastly, recent advances in food web ecology have shown the existence of so-called trivariate patterns in some aquatic food webs (Jonsson *et al.*, 2005; Layer *et al.*, 2010; Woodward *et al.*, 2005). The communities in these studies are punctuated by clear size-structuring, which is revealed through the simultaneous exploration of the relationships between body mass, abundance and trophic interactions. Large-bodied species are often rare and occupy the higher trophic levels, whereas small-bodied species are often abundant and occupy the lower trophic levels. An exploration of trivariate patterns in this study will also reveal the extent of size-structuring in our mesocosm communities. Thus, we will employ a wide range of metrics and explore many different patterns to quantify the structural properties of our mesocosm food webs and assess the impact of the proposed species removals on the integrity of this food web structure.

II. METHODS

A. Natural History of the Study Site

The chosen study site was Lough Hyne, a sheltered yet fully marine sea lough, in County Cork, southwest Ireland (51°30′ N, 9°18′ W; see Figure 1 for an overview of the Lough, with key features labelled). It was designated Europe's first marine nature reserve in 1981, and is approximately 1 km long and 0.6 km wide. It consists of a north and south basin (both ~20 m in depth), connected by a deeper Western Trough (~50 m). The Lough opens out to the Atlantic Ocean in the south via a long, narrow inlet called Barloge Creek. A narrow constriction, known as the Rapids, regulates the water flow between the Lough and Barloge Creek. This narrow constriction results in an asymmetric tide, with ebb flow lasting twice as long as flood. In addition, the tidal range of the Lough is a little over 1 m (Renouf, 1931), while it has a long flushing time of 41 days due to reduced flow rates (Johnson *et al.*, 1995). Water

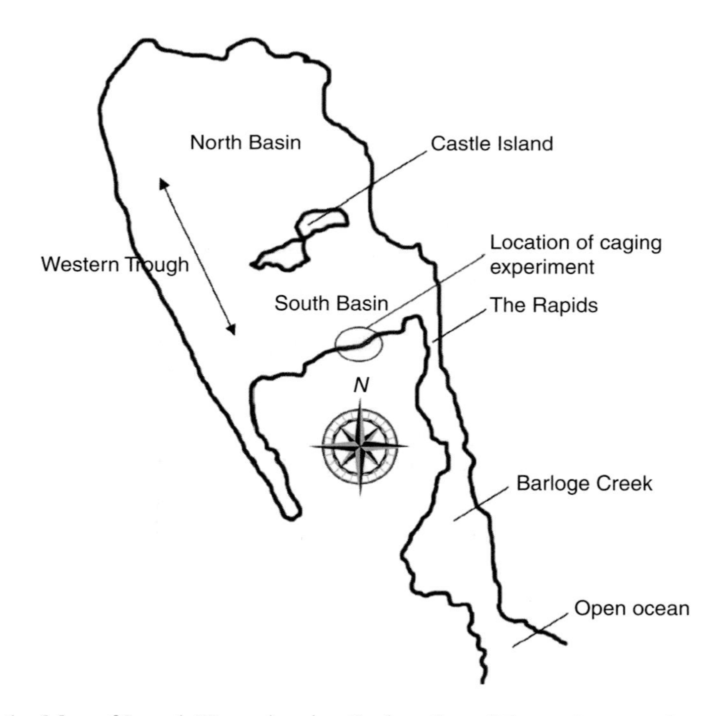

Figure 1 Map of Lough Hyne showing the location of the caging experiment on the south shoreline and other main features. The highly sheltered nature of the Lough is also clearly visible, with only a narrow body of water connecting it to the sea (Barloge Creek).

temperatures also tend to be higher inside the reserve during the summer months compared to the adjacent coastline (Rawlinson *et al.*, 2004), contributing to an oxy-thermocline in the deep Western Trough (McAllen *et al.*, 2009). Lough Hyne contains a rich biodiversity within its small area, with algal (Rees, 1935), planktonic (Holmes and O'Connor, 1990), intertidal (Little, 1990) and fish (Minchin, 1987; Rogers, 1990) communities representative of British and Irish coastlines. As such, Lough Hyne is an ideal system for ecological studies and explorations, which can be generalized to represent temperate east Atlantic communities. This is reflected in the substantial catalogue of scientific publications relating to the area (e.g. Costello, 1992; Crook *et al.*, 2000; Ebling *et al.*, 1948; Kitching *et al.*, 1959; Minchin, 1987; Muntz *et al.*, 1965; Norton *et al.*, 1977; Renouf, 1931; see Wilson, 1984 for a compilation).

B. Description of the Experimental Communities

Lough Hyne has a diverse flora and fauna and, rather than attempt to describe the entire food web, we chose to focus on one compartment of this web in a controlled, yet semi-natural experimental setting. To that end, we established a series of 24 mesocosm cages in the shallow subtidal, on the south shoreline of the Lough (see Figure 1). Each mesocosm consisted of a large cylindrical skeletal structure, made from two polypropylene rings (0.76 m in diameter), connected by six evenly spaced polypropylene struts (0.5 m tall). Polyethylene netting (5 mm mesh size) was attached to this structure to complete the exclusion cages, which had a benthic surface area of 0.45 m^2. See Figure 2 for a photograph of the mesocosms. We chose 10 abundant benthic species (comprising fish, decapods and echinoderms) that we aimed to manipulate as the largest species in the food webs. The species used were black goby (*Gobius niger*), rock goby (*Gobius paganellus*), sea scorpion (*Taurulus bubalis*), shore rockling (*Gaidropsarus mediterraneus*), goldsinny wrasse (*Ctenolabrus rupestris*), shore crab (*Carcinus maenas*), velvet swimming crab (*Necora puber*), common prawn (*Palaemon serratus*), spiny starfish (*Marthasterias glacialis*) and purple sea urchin (*Paracentrotus lividus*). All of these species are locally common in the lough, reaching densities in the shallow subtidal during summer that are approximately equal to the densities reached in our mesocosms, that is 1 individual per 0.45 m^2 (Costello, 1992; Crook *et al.*, 2000; Verling *et al.*, 2003; Yvon-Durocher *et al.*, 2008).

At the start of our experiment, we covered the bottom of each cage with a clean, stony substrate and added one individual from each of these 10 species to our exclusion cages before sealing the lids. The weight of substrate in the cages was sufficient to keep them in position in the shallow subtidal for the duration of the experiment. The 5-mm mesh aperture of the mesocosms was

Figure 2 The mesocosm cages used in the experiment. Note that they are sealed cages, with skeletal structure enclosed by mesh at all sides.

sufficiently small to contain the manipulated species whilst allowing small benthic invertebrates and fish to recruit naturally into the cages. Our aims were to: (1) maintain the 10 manipulated species in all mesocosm communities for 6 months of food web assembly, by replacing any individuals that died; (2) introduce a series of perturbations, by removing manipulated species from subsets of the communities and, again, maintain the number of individuals by replacement (if necessary) over the remaining 8 months of the experiment; (3) characterize the food web that developed in each experimental mesocosm every 2 months for the 14-month duration of the experiment.

C. Experimental Design

The experiment began on 5 October 2006. We arranged our mesocosms into four blocks of six, to facilitate the manipulation phase of the experiment after 6 months. Two of these blocks were positioned in the shallow subtidal at 1 m depth (low spring tide) and the remaining two blocks were placed at 2 m depth (low spring tide). For the first 6 months of the experiment, all 24 cages contained identical core assemblages, consisting of the 10 manipulated species detailed above. After 6 months, some of the manipulated species were removed based on the mean strength of their trophic interactions with the benthic invertebrate community typically found in the mesocosm cages. A previous study carried out at Lough Hyne ranked these 10 species according to the mean absolute strength of their interactions (see O'Gorman and Emmerson, 2009). This ranking is shown in Figure 3. We employed six

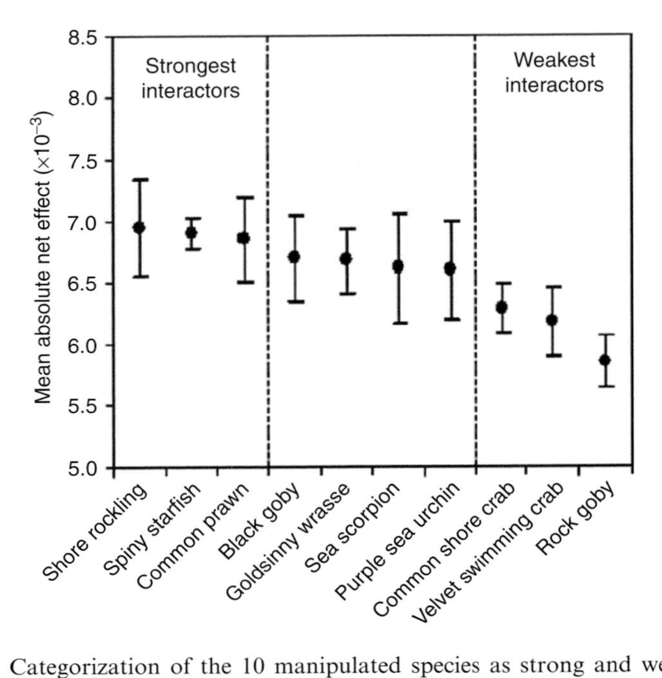

Figure 3 Categorization of the 10 manipulated species as strong and weak interactors. The manipulated species are ranked according to their mean absolute net effect (±SE), measured using the dynamic index (see O'Gorman and Emmerson, 2009 for a more detailed description of this ranking). The three strongest and weakest interactors were chosen for manipulation in the experiment.

treatments in this phase of the experiment: (1) 10 species community (W^+S^+), that is an intact community; (2) two weakest interactors removed ($W^{-2}S^+$); (3) three weakest interactors removed ($W^{-3}S^+$); (4) two strongest interactors removed (W^+S^{-2}); (5) three strongest interactors removed (W^+S^{-3}); (6) all strong and weak interactors removed, that is only intermediate interactors remaining (W^-S^-). These six treatments were randomly assigned within each of the four blocks.

D. Food Web Sampling

We sampled the sessile benthic invertebrate community in the cages using settlement panels. These panels consisted of 100 mm × 100 mm PVC squares, which are commonly used for quantifying sessile species such as sponges, bryozoans and calcareous polychaetes (Watson and Barnes, 2004). We sampled the mobile benthic invertebrate community in the cages using nylon pot scourers. The pot scourers have an approximate radius of 4 cm and represent a passive sampling device, obtaining a manageable subsample of invertebrates from the stony substrate in the cages. The network of gaps in the pot scourers simulates the form and structure of coralline algae prevalent in the study area and they are ideal for quantifying mobile species such as amphipods, isopods and gastropods (O'Gorman *et al.*, 2008; Underwood and Chapman, 2006). See Figure 4 for a photograph of these two sampling substrates. We calculated the NA of every species identified on these two substrates (species identified using Athersuch *et al.*, 1989; Graham, 1988; Hayward and Ryland, 1977, 1985, 1995; Holdich and Jones, 1983; Lincoln, 1979). From these abundances, we estimated density per square metre. We also measured the length of every individual identified ($n = 228{,}163$) and estimated its corresponding body mass using length–weight relationships defined for the study system (see Table S1 in Appendix I). While the sampling substrates detailed above allowed us to quantify most species in the cages, they overlooked two small fish species (two-spot goby, *Gobiusculus flavescens*, and painted goby, *Pomatoschistus pictus*) that were able to pass through the small mesh size of the cages. We used existing data to estimate the mean body size and abundance of these two fish species along the south shoreline

Figure 4 Sampling substrates used in the experiment: green settlement panels for sessile invertebrates; red pot scourers for mobile invertebrates; glass slides for primary production; and mesh bags to estimate decomposition of kelp. Only data from the settlement panels and mesh pads are used in the current study. Estimates of primary production were used in O'Gorman and Emmerson (2009). Kelp decomposition proved to be too rapid for the bi-monthly sampling sessions, so these data were not used in the experiments. (For interpretation of the references to colour in this figure legend, the reader is referred to the Web version of this chapter.)

of the Lough (O'Gorman and Emmerson, 2009; O'Gorman *et al.*, 2008). We also measured the body size of the manipulated species at each sampling session. Measuring the density or body mass of resources such as algae, Coarse Particulate Organic Matter (CPOM), Fine Particulate Organic Matter (FPOM) and diatoms was problematic. Therefore, we did not take any such measurements for the basal resources.

E. Food Web Construction

We used a combination of gut content analysis and intensive literature research to establish the food web structure of the benthic cages. Gut content analysis was carried out on the larger predators in the cages, that is manipulated species and the two small gobies. All animals were sacrificed for gut content analysis, with the stomach and intestines dissected out and examined under a Nikon® SMZ1000 microscope at 50× magnification. All species found (whole or body parts) were identified using a reference collection of every species found in the experiment and the identification keys outlined above. The gut content analysis was also supplemented with information on the diets of these species from studies carried out at Lough Hyne (see Appendix II). Owing to the small body size and difficulty in identifying the guts of benthic invertebrates, the remaining links were reviewed from the literature. Information from Lough Hyne was not available for the diets of many of the benthic invertebrate species identified in the study. Accordingly, we compiled data from more than 200 publications (peer-reviewed journals and books) using the approach of Martinez (1991). Here, a direct feeding link was assigned to any pair of species A and B within the benthic cages, whenever investigators from two or more publications reported that A is likely to consume B in a given year. This criterion was maintained throughout construction of the web and restricted the inclusion of prey links that may have occurred through passive consumption (see Appendix III for a list of the literature used to assign benthic invertebrate feeding links). A composite web, containing predator–prey links between all species identified in the mesocosm experiments (including the study that calculated the mean absolute strength of the manipulated species' interactions, O'Gorman and Emmerson, 2009), is shown in Figure 5. For the links in this composite web, 37% are gut content links, 3% were directly observed in previous studies from Lough Hyne and 60% are based on feeding links observed at other sites. Species names corresponding to the numbers shown in Figure 5 are supplied in Table S2 in Appendix I. All subsequent food webs in this study were drawn from this composite web.

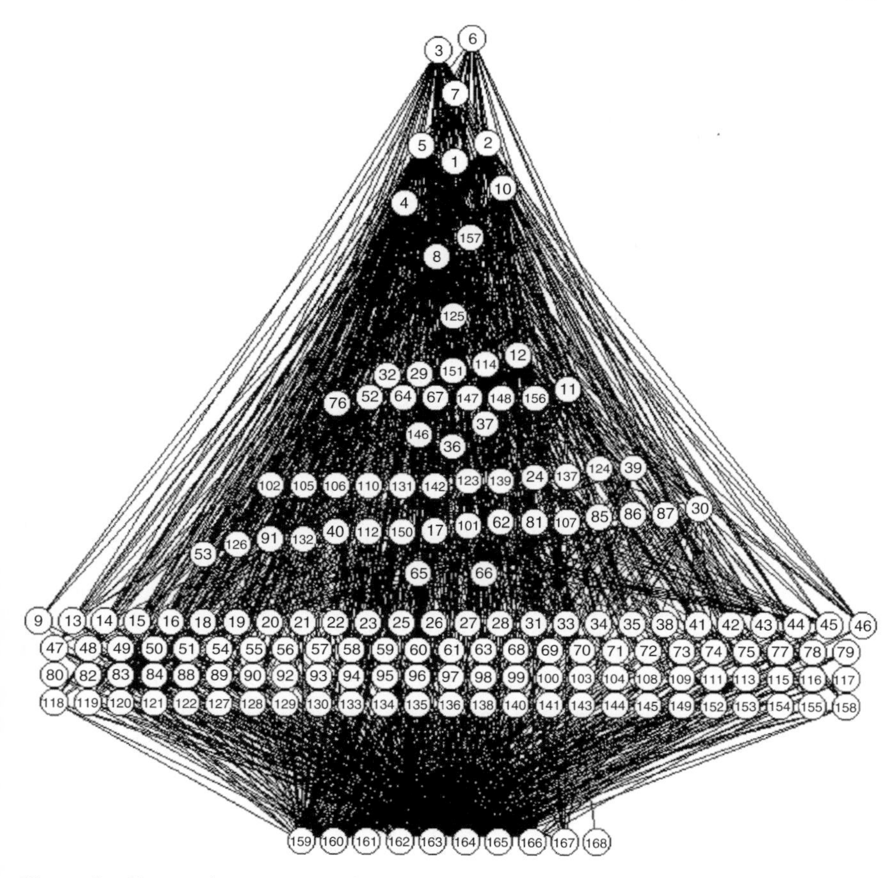

Figure 5 Composite mesocosm food web, showing all possible species interactions in the caging experiments (including the experiment that categorized the manipulated species as strong or weak interactors). A list of species names that correspond to the codes shown in the figure can be found in Table S4 of Appendix II.

F. Univariate Patterns

We quantified the food web structure for each replicate mesocosm at each of six different sampling sessions, that is we quantified 6 treatments $\times$ 4 replicates $\times$ 6 sampling sessions $=$ 144 mesocosm food webs in the study (see Appendix IV for details of the 144 webs). For every food web described, we quantified a number of food web metrics. The interactive connectance (C) of a food web was measured as $2L/(S^2 - S)$, where S is the species richness and L is the number of trophic links (TL) in the web, that is the actual number of links divided by the total number of possible links. The linkage density (LD) of a food web was measured as L/S. The Shannon index (H) was used to measure the species diversity of each food web. Whittaker's index of beta diversity (β_w) was used to measure the turnover of species in a given replicate food web between consecutive sampling sessions. β_w was measured as $(s/\alpha) - 1$, where s is the total number of species in a replicate food web over two consecutive sampling sessions and α is the average species richness of the two webs. We also calculated the mean food chain length (FCL), the proportion of top (T), intermediate (I) and basal (B) species, as well as the proportion of basal to intermediate (B–I), basal to top (B–T), intermediate to intermediate (I–I), intermediate to top (I–T) and cannibalistic (Ca) links in each food web.

Three of the sampling sessions took place before the food web manipulations (i.e. pre-manipulation) on 13 December 2006, 22 February 2007 and 18 April 2007. Three of the sampling sessions took place after the food web manipulations (i.e. post-manipulation) on 16 June 2007, 14 August 2007 and 4 December 2007. Consequently, we split our analysis of food web structure to examine responses pre- and post-manipulation. Since we were measuring the same communities through time, we employed a repeated measures design to analyse the data (Underwood, 1997). There were three repeated measurements pre-manipulation and three post-manipulation. We used a general linear model (GLM) to analyse the data, with C, LD, H, β_w and FCL as response variables. This analysis corresponded to a fully factorial two-way ANOVA for repeated measures, including the main effects and interaction terms for presence/absence of strong interactors and presence/absence of weak interactors, with the addition of a single main effect term for block. To analyse the data in a balanced statistical design, we carried out one GLM on W^+S^+, $W^{-2}S^+$, W^+S^{-2} and W^-S^- and one GLM on W^+S^+, $W^{-3}S^+$, W^+S^{-3} and W^-S^-. This approach allowed us to investigate whether effects were consistent for the removal of both two and three strong or weak interactors. We could not normalize the proportional data using an arcsine transformation. Consequently, we employed a generalized linear mixed model (GLMM) approach to analyse these data, where we specified our repeated measure (season) as a random effect and fitted a binomial distribution to the model. We analysed T, I, B, B–I, B–T, I–I, I–T and Ca in this manner, examining responses pre- and post-manipulation, as before. All analyses were carried out in R version 2.8.1.

The body mass (M) of a species was measured as the average species dry weight in milligrams. The NA of a species was measured as the average number of individuals per square metre. The BA of a species was measured as $M \times$ NA in units of mg m^{-2}. The TH was measured as the average trophic position of a species in all food chains of which it is a part. We constructed the frequency distribution of log M, log NA and log BA and also characterized the relationships between log M, log NA, log BA and the rank in log M, log NA and log BA, respectively. To simplify this analysis, we obtained the average *M*, NA, BA and TH for the composite webs shown in Figure 6 (see Appendix V for a list of species, with TH and the number of TL, for each web), that is we averaged across the 12 webs (3 sampling sessions $\times$ 4 replicates) that constituted each composite web. This reduced the comparison of the aforementioned univariate (and all subsequent) relationships to a total of 12 composite webs, rather than all 144 webs. Thus, we compared the composite webs for each of the six treatments pre-manipulation to the composite webs for each treatment post-manipulation.

G. Bivariate and Trivariate Patterns

We also investigated the bivariate and trivariate patterns of the composite food webs for each treatment, to test for differences among treatments. We characterized the predator–prey allometry of each composite web, by plotting predator *M*, NA and BA against prey *M*, NA and BA. We also examined the relationship between body size and abundance (log *M* vs. log NA and log BA) and the trophic pyramids for each composite web, that is by comparing total *M*, NA and BA at each discrete trophic level. Trivariate plots, following the approach of Jonsson *et al.* (2005), were used to show the relationship between body size, abundance (NA and BA) and TH. Such plots give a unique insight into the relationship between these three food web parameters, as well as revealing the patterns of the bivariate relationships between log *M*, log NA, log BA and TH.

III. RESULTS

A. Univariate Patterns

There was no significant difference in food web connectance between any of the treatments before the interaction strength manipulations took place (Figure 7A, B; Table 1a.i), although there was a main effect of block in the two species removal analysis ($F_{3,9}=4.091$, $p=0.044$) (note that the removals have not yet taken place in the pre-manipulation phase). This implied that connectance varied between some of the experimental blocks, but because our six treatments were randomly assigned within each block, this effect would have been consistent across all treatments. After the interaction strength manipulations had been initiated, the removal of two or three strong or weak interactors led to a significant reduction (4–32%) in the connectance of the experimental food webs (Figure 7A, B; Table 1a.ii).

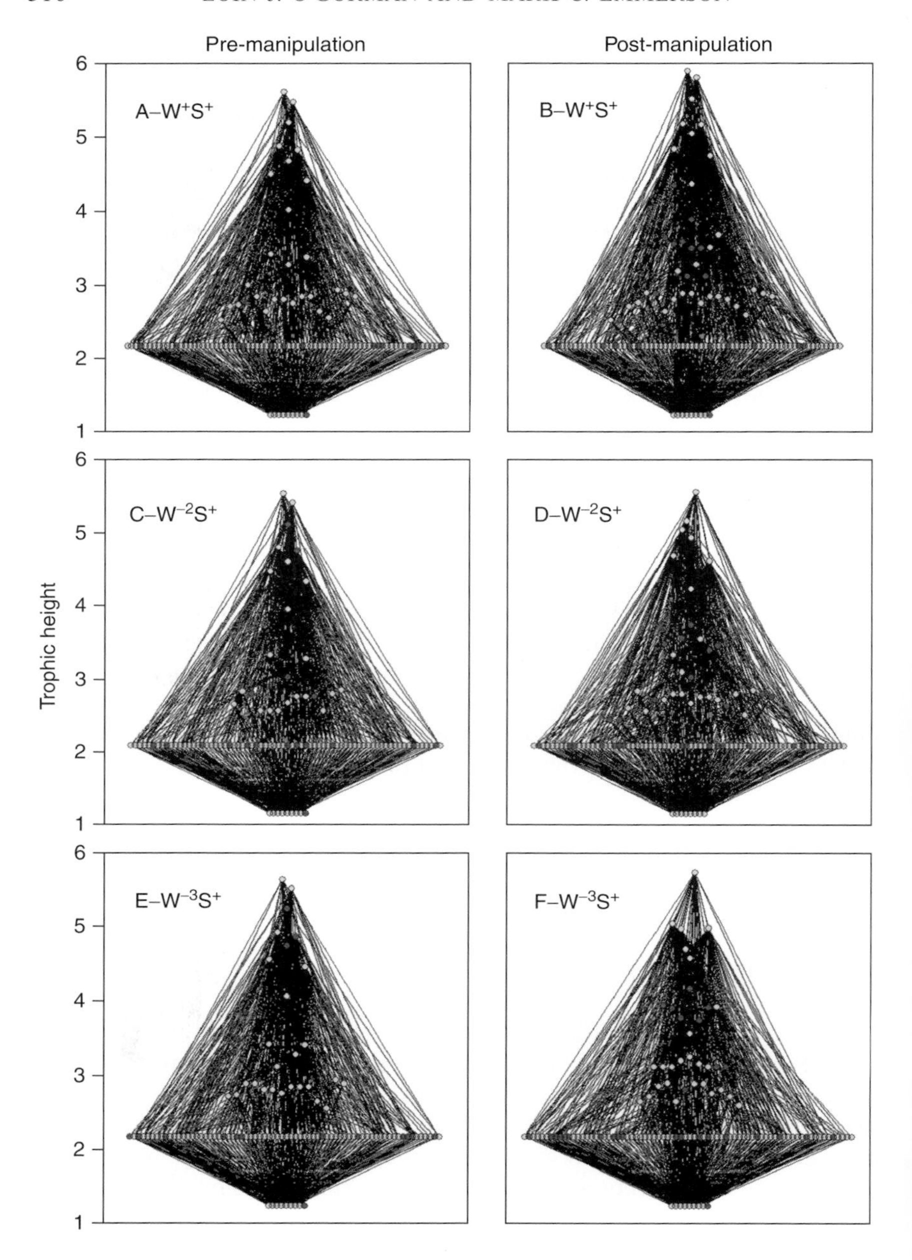

Pre-manipulation
Post-manipulation
Trophic height
A–W⁺S⁺
B–W⁺S⁺
C–W⁻²S⁺
D–W⁻²S⁺
E–W⁻³S⁺
F–W⁻³S⁺
1
2
3
4
5
6

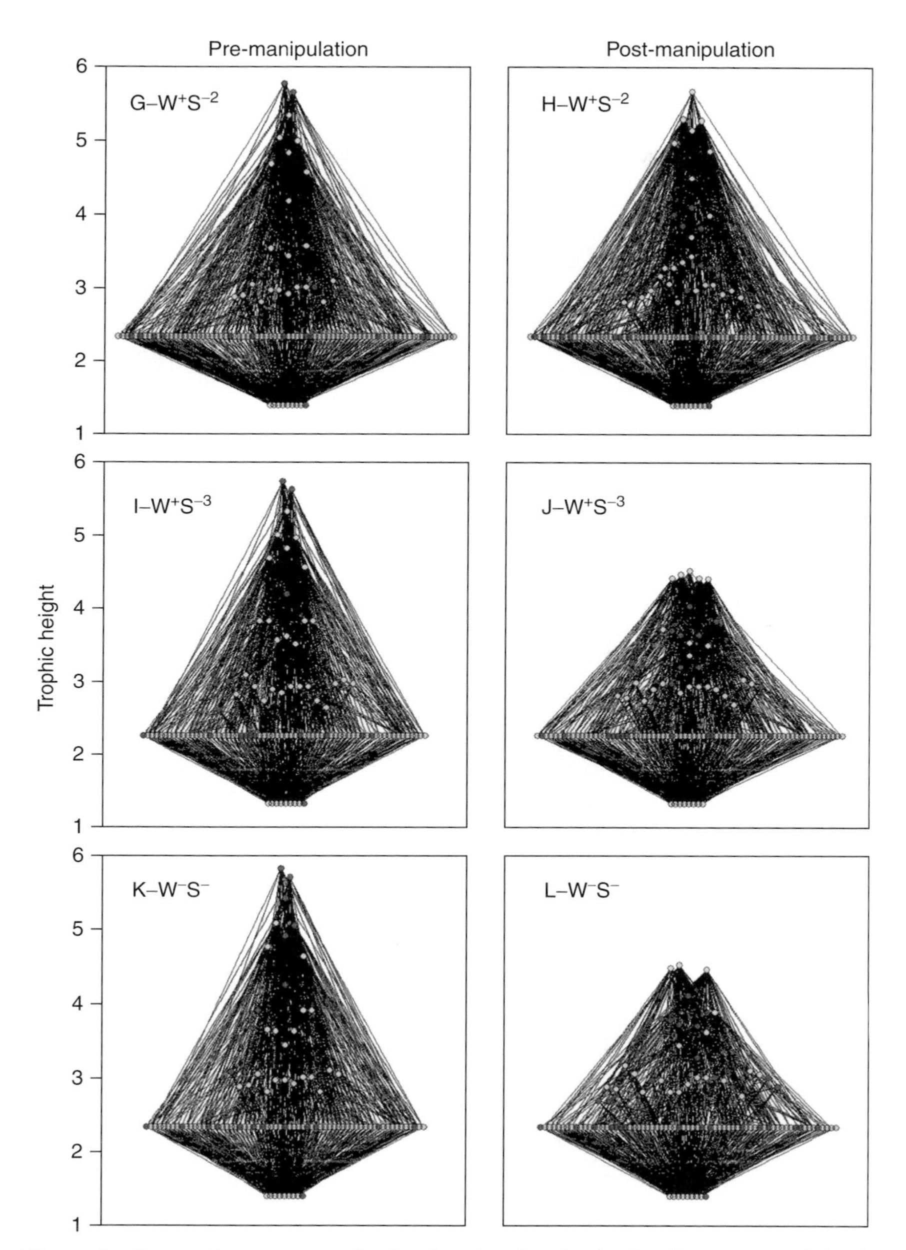

Figure 6 Composite mesocosm food webs, showing the food web structure of the six treatments, before and after the interaction strength manipulations took place. The six treatments were: (A, B) W^+S^+ = an intact community; (C, D) $W^{-2}S^+$ = two weakest interactors removed; (E, F) $W^{-3}S^+$ = three weakest interactors removed; (G, H) W^+S^{-2} = two strongest interactors removed; (I, J) W^+S^{-3} = three strongest interactors removed; and (K, L) W^-S^- = all strong and weak interactors removed. Blue nodes are species common to a treatment web, pre and post-manipulation. Red nodes indicate species which were unique to a treatment web. A list of species that were present in each composite web can be found in Appendix IV. (For interpretation of the references to colour in this figure legend, the reader is referred to the Web version of this chapter.)

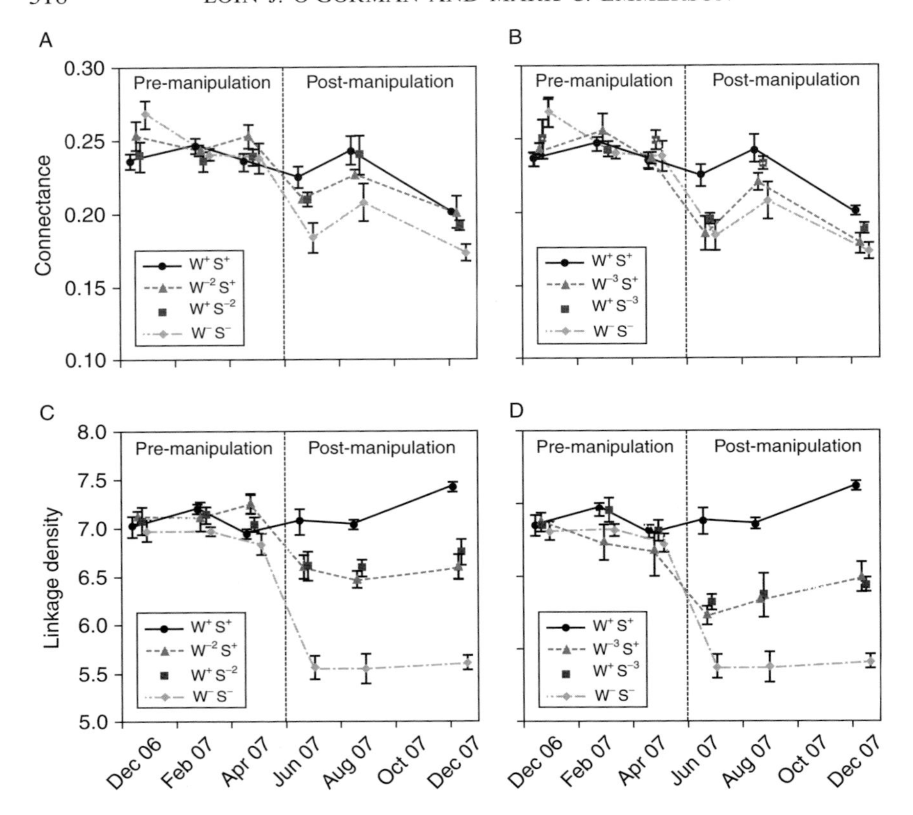

Figure 7 Effect of the interaction strength manipulations on the (A, B) connectance and (C, D) linkage density of the mesocosm food webs (±SE). Three of the food web sampling sessions took place pre-manipulation, with three more post-manipulation. In the key, W^+S^+ = an intact community; $W^{-2}S^+$ = two weakest interactors removed; $W^{-3}S^+$ = three weakest interactors removed; W^+S^{-2} = two strongest interactors removed; W^+S^{-3} = three strongest interactors removed; and W^-S^- = all strong and weak interactors removed.

There was no significant difference in linkage density between any of the treatments pre manipulation (Figure 7C, D; Table 1b.i). There were very clear effects on linkage density after the manipulations, whereby the removal of two or three strong or weak interactors led to a highly significant reduction (4–20%) in the linkage density (Figure 7C, D; Table 1b.ii). The greatest reduction in linkage density occurred when all strong and weak interactors were removed (strong × weak: $F_{1,9}=14.173$, $p=0.004$).

There was no significant difference in Shannon diversity (Figure 8A, B; Table 1c.i) or species turnover (measured as Whittaker's index of beta

Table 1 F (and t) statistics and p values for the models performed on a range of metrics, measured for each of the 144 experimental food webs

Line			Removal of strong interactors				Removal of weak interactors			
			2 removed		3 removed		2 removed		3 removed	
Reference	Food web metric	Manipulation	$F_{1,9}$	p Value	$F_{1,9}$	p Value	$F_{1,9}$	p Value	$F_{1,9}$	p Value
a.i	Connectance	Pre	3.086	0.113	0.009	0.927	2.152	0.176	0.117	0.740
a.ii		Post	13.123	0.006	6.822	0.028	14.589	0.004	24.229	0.001
b.i	Linkage density	Pre	2.910	0.122	0.124	0.732	0.046	0.836	2.595	0.142
b.ii		Post	152.479	<0.001	257.785	<0.001	189.282	<0.001	281.604	<0.001
c.i	Shannon diversity	Pre	1.421	0.264	4.190	0.071	1.848	0.207	1.153	0.311
c.ii		Post	0.956	0.354	7.166	0.025	9.417	0.013	1.411	0.265
d.i	Species turnover	Pre	0.043	0.841	1.517	0.249	0.275	0.613	0.836	0.384
d.ii		Post	5.20	0.049	1.10	0.322	0.667	0.435	5.381	0.046
e.i	Mean food chain length	Pre	6.652	0.030	0.064	0.805	1.118	0.318	1.258	0.291
e.ii		Post	3046.683	<0.001	8055.833	<0.001	772.296	<0.001	315.515	<0.001
			t Value	p Value	t Value	p Value	t Value	p Value	t Value	p Value
f.i	Proportion of top species	Pre	0.116	0.908	0.738	0.465	1.476	0.148	0.834	0.409
f.ii		Post	−14.454	<0.001	−11.223	<0.001	−4.292	<0.001	−3.240	0.002
g.i	Proportion of intermediate species	Pre	0.583	0.563	−0.315	0.755	0.784	0.437	0.202	0.841
g.ii		Post	15.869	<0.001	12.316	<0.001	5.636	<0.001	4.061	<0.001
h.i	Proportion of basal species	Pre	−0.771	0.445	−0.148	0.883	−1.799	0.079	−1.049	0.300
h.ii		Post	−1.586	0.120	−2.737	0.009	−1.538	0.132	−1.477	0.147

(continued)

Table 1 (*continued*)

			t Value	*p* Value	*t* Value	*p* Value	*t* Value	*p* Value	*t* Value	*p* Value
i.i	Proportion of basal to intermediate links	Pre	−1.450	0.155	0.029	0.977	0.350	0.728	−1.228	0.226
i.ii		Post	−12.475	<0.001	−10.079	<0.001	−14.055	<0.001	−11.482	<0.001
j.i	Proportion of basal to top links	Pre	-0.335	0.739	0.519	0.607	1.228	0.227	0.637	0.528
j.ii		Post	−12.51	<0.001	−10.39	<0.001	−10.788	<0.001	−8.642	<0.001
k.i	Proportion of intermediate to intermediate links	Pre	2.645	0.012	−0.283	0.778	0.382	0.705	2.013	0.051
k.ii		Post	46.6	<0.001	36.25	<0.001	9.487	<0.001	2.697	0.010
l.i	Proportion of intermediate to top links	Pre	-2.013	0.050	0.267	0.791	−1.684	0.100	−1.553	0.128
l.ii		Post	−59.206	<0.001	−59.134	<0.001	17.826	<0.001	20.741	<0.001
m.i	Proportion of cannibalistic links	Pre	−0.270	0.788	−0.123	0.903	0.107	0.915	0.452	0.653
m.ii		Post	−1.686	0.099	−0.135	0.894	−1.823	0.076	0.157	0.876

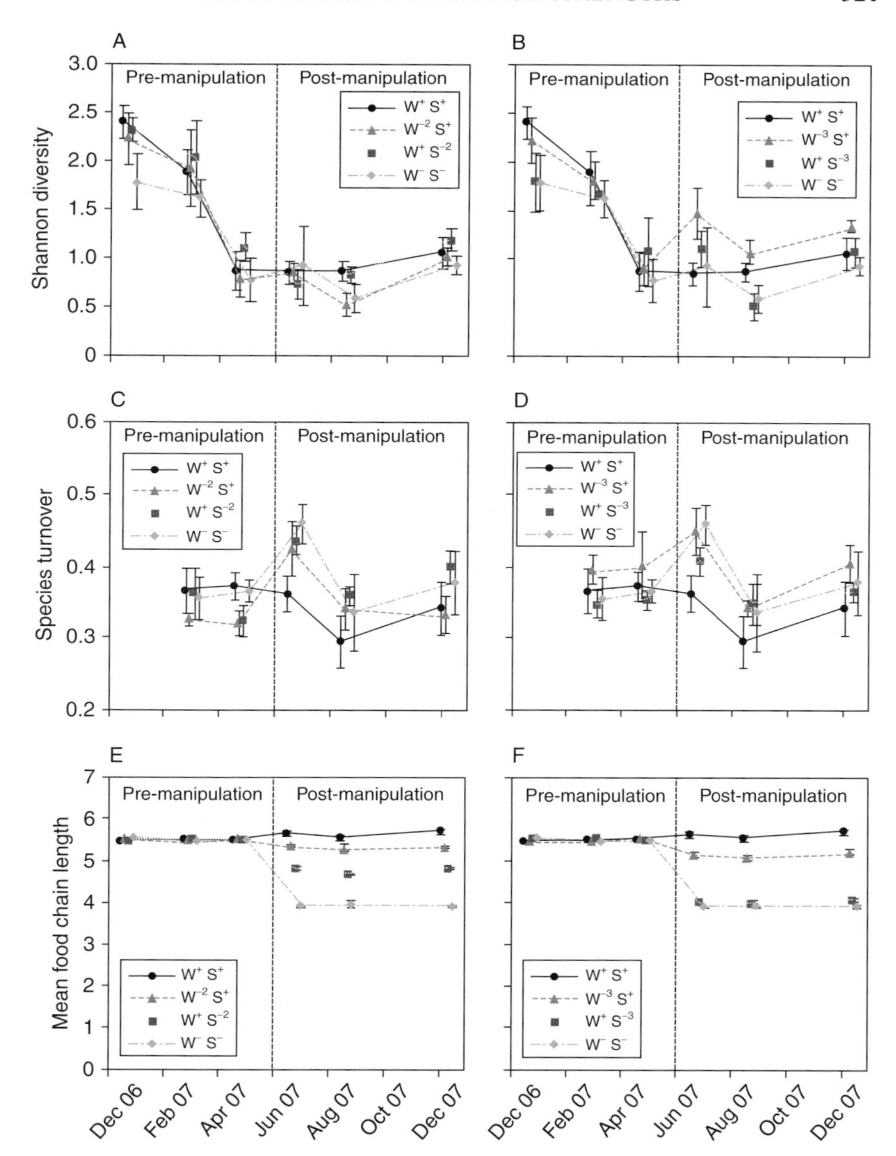

Figure 8 Effect of the interaction strength manipulations on the (A, B) Shannon diversity, (C, D) species turnover (measured as Whittaker's index of beta diversity) and (E, F) mean food chain length of the mesocosm food webs (±SE). Three of the food web sampling sessions took place pre-manipulation, with three more post-manipulation. In the key, W^+S^+ = an intact community; $W^{-2}S^+$ = two weakest interactors removed; $W^{-3}S^+$ = three weakest interactors removed; W^+S^{-2} = two strongest interactors removed; W^+S^{-3} = three strongest interactors removed; and W^-S^- = all strong and weak interactors removed.

diversity) (Figure 8C, D; Table 1d.i) between any of the treatments prior to the manipulation, and there were inconsistent effects on both measures of diversity post-manipulation. The removal of two weak interactors and three strong interactors reduced Shannon diversity (Figure 8A, B; Table 1c.ii) and the removal of two strong interactors and three weak interactors reduced species turnover (Figure 8C, D; Table 1d.ii).

There appears to be some differences in mean FCL between treatments pre-manipulation: the treatment that would later have two strong interactors removed once the interaction strength manipulations were initiated had a significantly shorter mean FCL(Figure 8E; Table 1e.i). In spite of this possible pre-manipulation effect, the dramatic reduction in the mean FCL post-manipulation is particularly clear. The removal of two or three strong or weak interactors led to a significant reduction in the mean FCL (Figure 8E, F; Table 1e.ii), with the strongest effects occurring when all strong and weak interactors were removed (strong $\times$ weak: $F_{1,9}=142.552$, $p<0.001$).

There was no significant difference in the proportion of top (Figure 9A, B; Table 1f.i), intermediate (Figure 9C, D; Table 1g.i) or basal (Figure 9E, F; Table 1h.i) species between any of the treatments pre-manipulation. There was a significant main effect of block in the three species analysis of the proportion of basal species ($t=2.664$, $p=0.011$). Due to the randomization of treatments within each block, this effect was consistent across all treatments. The removal of two or three strong or weak interactors led to a significant increase in the proportion of top species (Figure 9A, B; Table 1f. ii). The removal of two or three strong or weak interactors also led to a significant reduction in the proportion of intermediate species (Figure 9C, D; Table 1g.ii). The greatest reduction in the proportion of intermediate species occurred when both strong and weak interactors were removed (strong $\times$ weak: $t=-2.666$, $p=0.011$). The removal of three strong interactors also led to a significant increase in the proportion of basal species (Figure 9E, F; Table 1h.ii).

There was a significantly higher proportion of intermediate to intermediate links pre-manipulation in the treatment that would later have two strong interactors removed (Figure 10F; Table 1k.i), indicating the presence of possible pre-manipulation differences in the distribution of links for this treatment. All other treatments showed no significant difference in the distribution of links (Figure 10; Table 1i.i–m.i). There was a significant main effect of block pre-manipulation in the three species analysis of the proportion of intermediate to top links ($t=2.112$, $p=0.041$). There was also a significant main effect of block post-manipulation in the two ($t=2.703$, $p=0.010$) and three ($t=2.452$, $p=0.019$) species analysis of the proportion of basal to top links. Again, the use of a randomized block design made this effect consistent across all treatments. The removal of two or three strong interactors led to a significant increase in the proportion of basal to intermediate

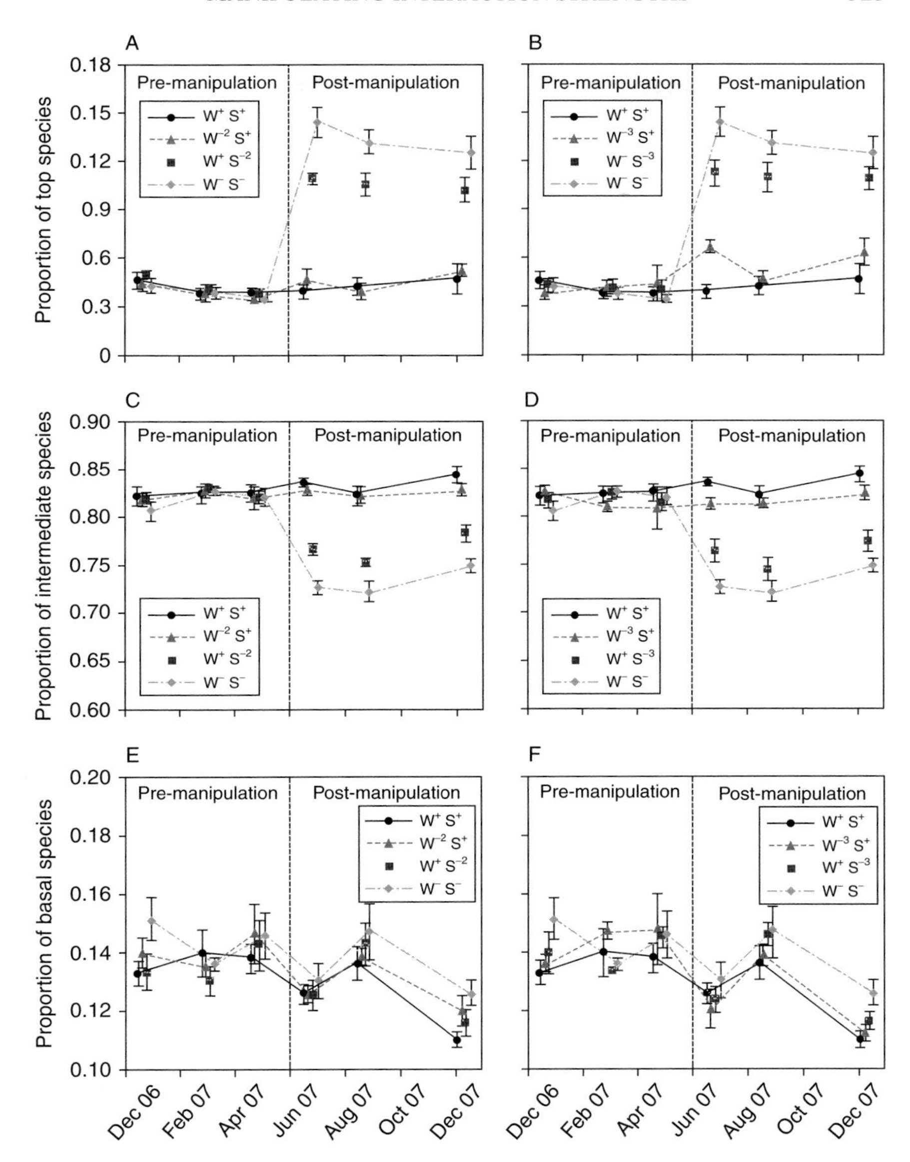

Figure 9 Effect of the interaction strength manipulations on the proportion of (A, B) top, (C, D) intermediate and (E, F) basal species in the mesocosm food webs (±SE). Three of the food web sampling sessions took place pre-manipulation, with three more post-manipulation. In the key, W^+S^+ = an intact community; $W^{-2}S^+$ = two weakest interactors removed; $W^{-3}S^+$ = three weakest interactors removed; W^+S^{-2} = two strongest interactors removed; W^+S^{-3} = three strongest interactors removed; and W^-S^- = all strong and weak interactors removed.

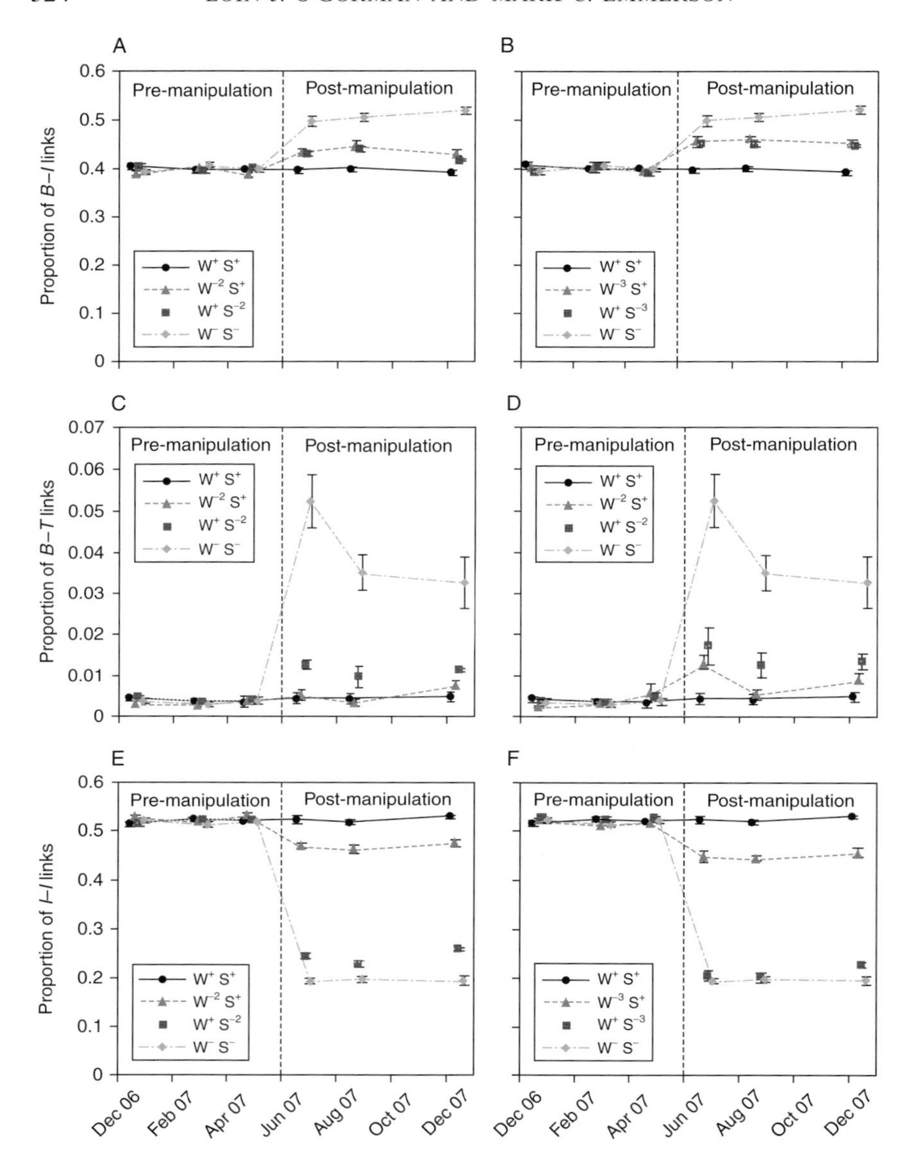
A
B
C
D
E
F
Pre-manipulation
Post-manipulation
Proportion of B–I links
Proportion of B–T links
Proportion of I–I links
W+ S+
W−2 S+
W+ S−2
W−3 S+
W+ S−3
W− S−
Dec 06
Feb 07
Apr 07
Jun 07
Aug 07
Oct 07
Dec 07

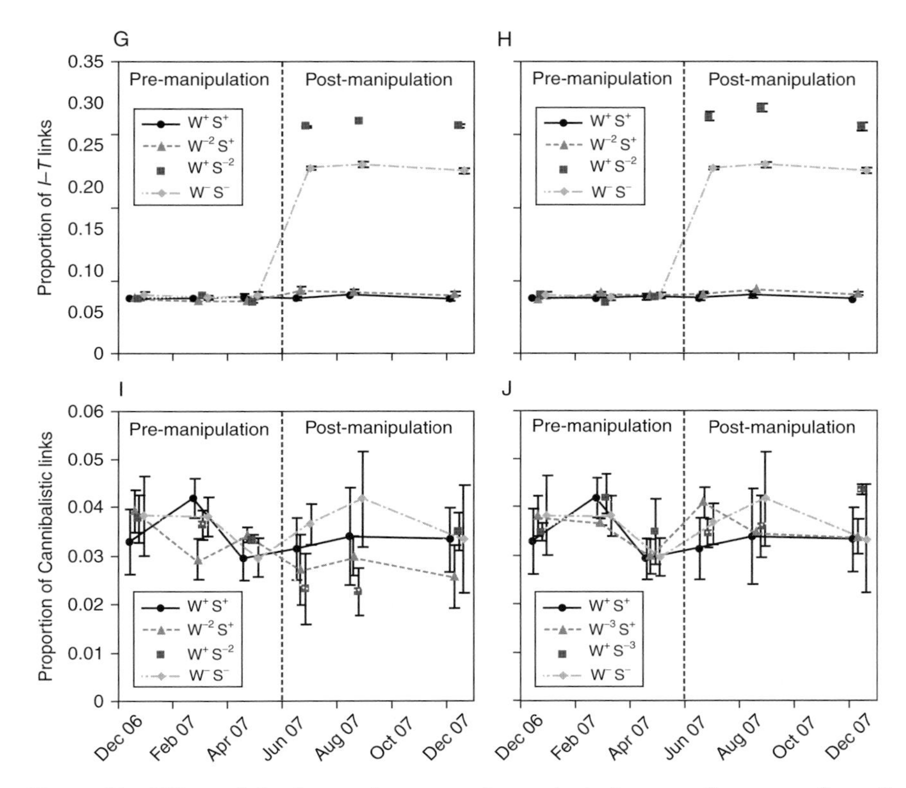

Figure 10 Effect of the interaction strength manipulations on the proportion of (A, B) basal to intermediate, (C, D) basal to top, (E, F) intermediate to intermediate, (G, H) intermediate to top and (I, J) cannibalistic links in the mesocosm food webs (±SE). Three of the food web sampling sessions took place pre manipulation, with three more post-manipulation. In the key, W^+S^+ = an intact community; $W^{-2}S^+$ = two weakest interactors removed; $W^{-3}S^+$ = three weakest interactors removed; W^+S^{-2} = two strongest interactors removed; W^+S^{-3} = three strongest interactors removed; and W^-S^- = all strong and weak interactors removed.

(Figure 10A, B; Table 1i.ii) and basal to top links (Figure 10C, D; Table 1j.ii), with the greatest increase occurring when all strong and weak interactors were removed (strong × weak: $t = 4.252$, $p < 0.001$ and $t = 4.201$, $p < 0.001$, respectively). There was also a significant increase in the proportion of intermediate to top links as a result of the interaction strength manipulations (Figure 10G, H; Table 1l.ii), with the greatest increase occurring when weak interactors were present without strong interactors (strong × weak: $t = -14.069$, $p < 0.001$). The removal of two or three strong interactors led to a significant reduction in the proportion of intermediate to intermediate links post-manipulation (Figure 10E, F; Table 1k.ii). The greatest reduction in the proportion of links occurred when all strong and weak interactors were removed (strong × weak: $t = 4.290$, $p < 0.001$). The interaction strength manipulations had no significant effects on the proportion of cannibalistic links in the mesocosm food webs (Figure 10I, J; Table 1m.ii).

The frequency distribution of body mass was bimodal for all treatments. Pre-manipulation, there was a gap in the distribution for all six treatments (panels A, C, E, G, I and K in Figure 11i). Post-manipulation, this gap in the distribution of body mass was bridged for all treatments (panels B, D, F, H, J and L in Figure 11i). The frequency distributions of body mass, NA and BA all deviated significantly from a log-normal distribution both before and after the interaction strength manipulations (Shapiro–Wilk, $p < 0.001$, $p = 0.002$ and $p = 0.014$; dashed lines in Figure 11i–iii, respectively). If the manipulated species and the two small gobies are excluded, the benthic invertebrate community still differed significantly from a log-normal distribution for all treatments pre- and post-manipulation (Shapiro–Wilk, $p = 0.015$, $p < 0.001$ and $p = 0.023$, for body mass, NA and BA, respectively) and the frequency distribution of body mass was right-skewed. There was also a deviation from a log-normal distribution (for the whole community) in the relationship between body mass/NA/BA and rank in body size/NA/BA (solid line in all panels in Figure 11i–iii). Here, a logarithmic model provided the best fit for all treatments, both pre- and post-manipulation, for body mass and BA ($r^2 > 0.94$ and $r^2 > 0.90$, respectively) and a linear model provided the best fit for all treatments, both pre- and post-manipulation, for NA ($r^2 > 0.78$).

B. Bivariate Patterns

Figure 12 reveals that the experimental food webs were strongly size-structured. If a data point lies in the upper left triangles of Figure 12i and iii, this represents a predator feeding on smaller prey, or a lower biomass of prey, respectively. If a data point lies in the lower left triangles of Figure 12ii, this represents a predator feeding on more numerically abundant prey. For body mass/NA/BA, over 96%/96%/93% of the predator–prey links in the pre-manipulation webs and 89%/91%/83% of the predator–prey links in the post-manipulation webs fell within these regions, respectively. This represented a significantly higher proportion of predators feeding on smaller/more numerically abundant/a lower biomass of prey for all treatments than would be expected if the feeding links were distributed at random (proportion test, $p < 0.001$).

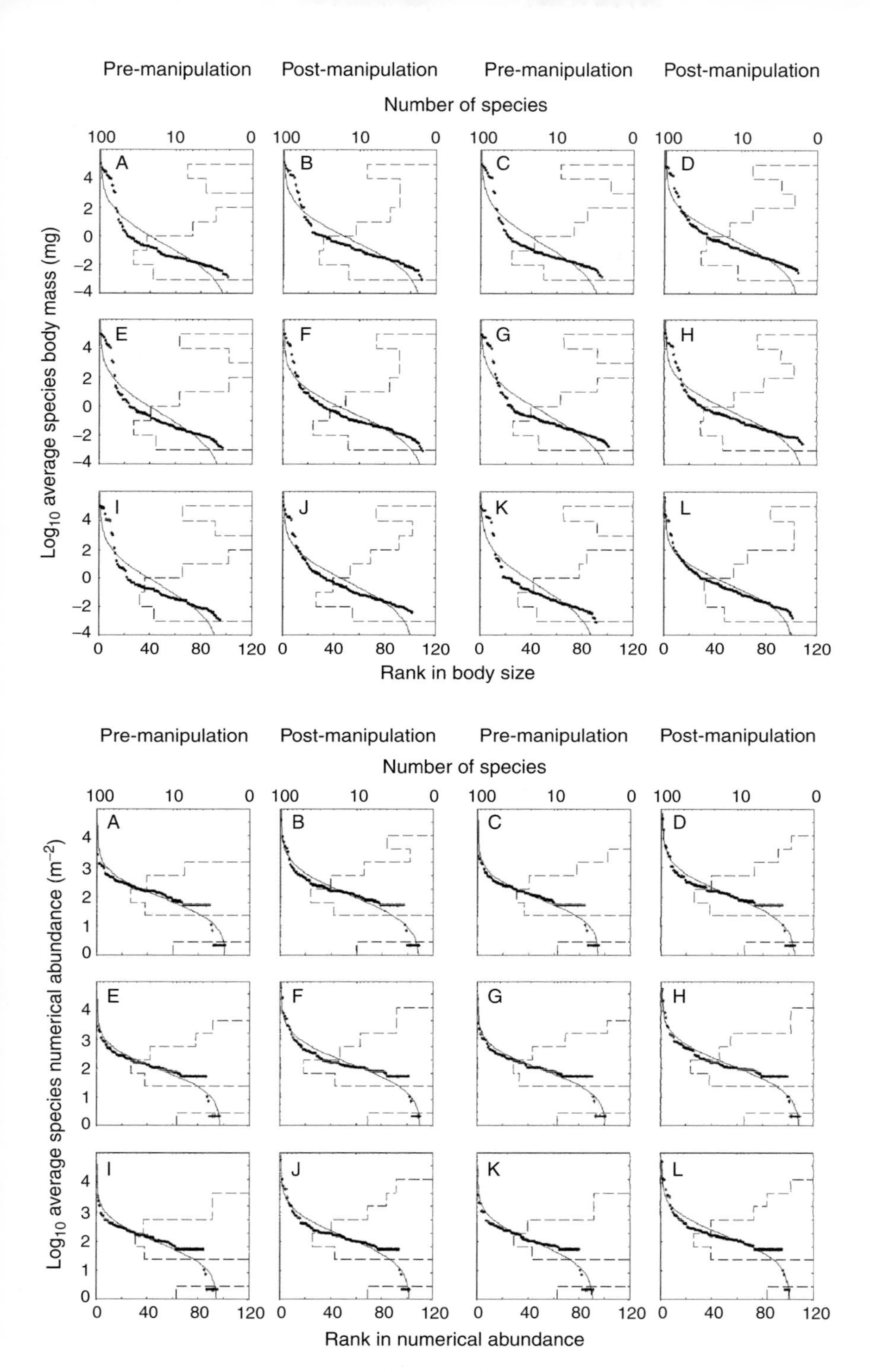

Figure 11 (Continued)

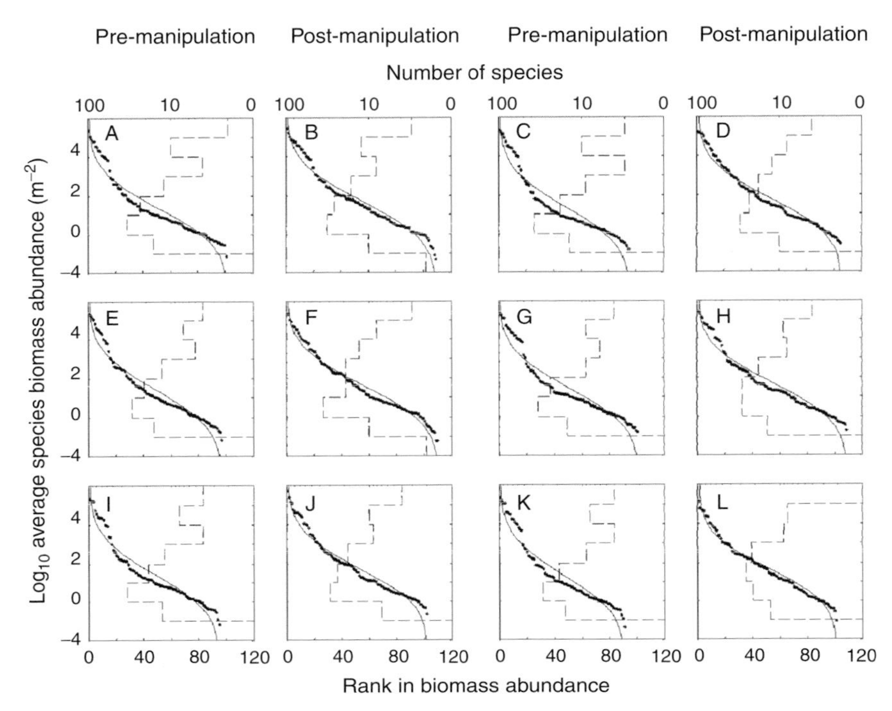

Figure 11 (i) Log body size *versus* rank in body size, (ii) log numerical abundance versus rank in numerical abundance and (iii) log biomass abundance versus rank in biomass abundance for the species in the composite treatment webs shown in Figure 4. Rank goes from greatest to smallest in each case. The solid line is the expected relationship assuming a log-normal distribution and is produced by drawing 10,000 values from a log-normal distribution with the same mean and variance as the observed distributions of body size, numerical abundance and biomass abundance. The dashed line is the frequency distribution of the number of species (top horizontal axis) by log body size/numerical abundance/biomass abundance. The panels in each figure represent: (A, B) intact community; (C, D) two weak interactors removed; (E, F) three weak interactors removed; (G, H) two strong interactors removed; (I, J) three strong interactors removed; and (K, L) all strong and weak interactors removed. Comparisons within a treatment are shown pre- and post-manipulation.

There was a weak but significant correlation between predator and prey body mass abundance in all treatments pre-manipulation, but this was not significant after the removal of three strong interactors or all strong and weak interactors (Table 2i). In a linear least-squares regression of predator and prey body mass, the intercept was significantly smaller after the interaction strength manipulations for all treatments, suggesting a seasonal effect on the relationship between predator and prey body mass. The removal of all

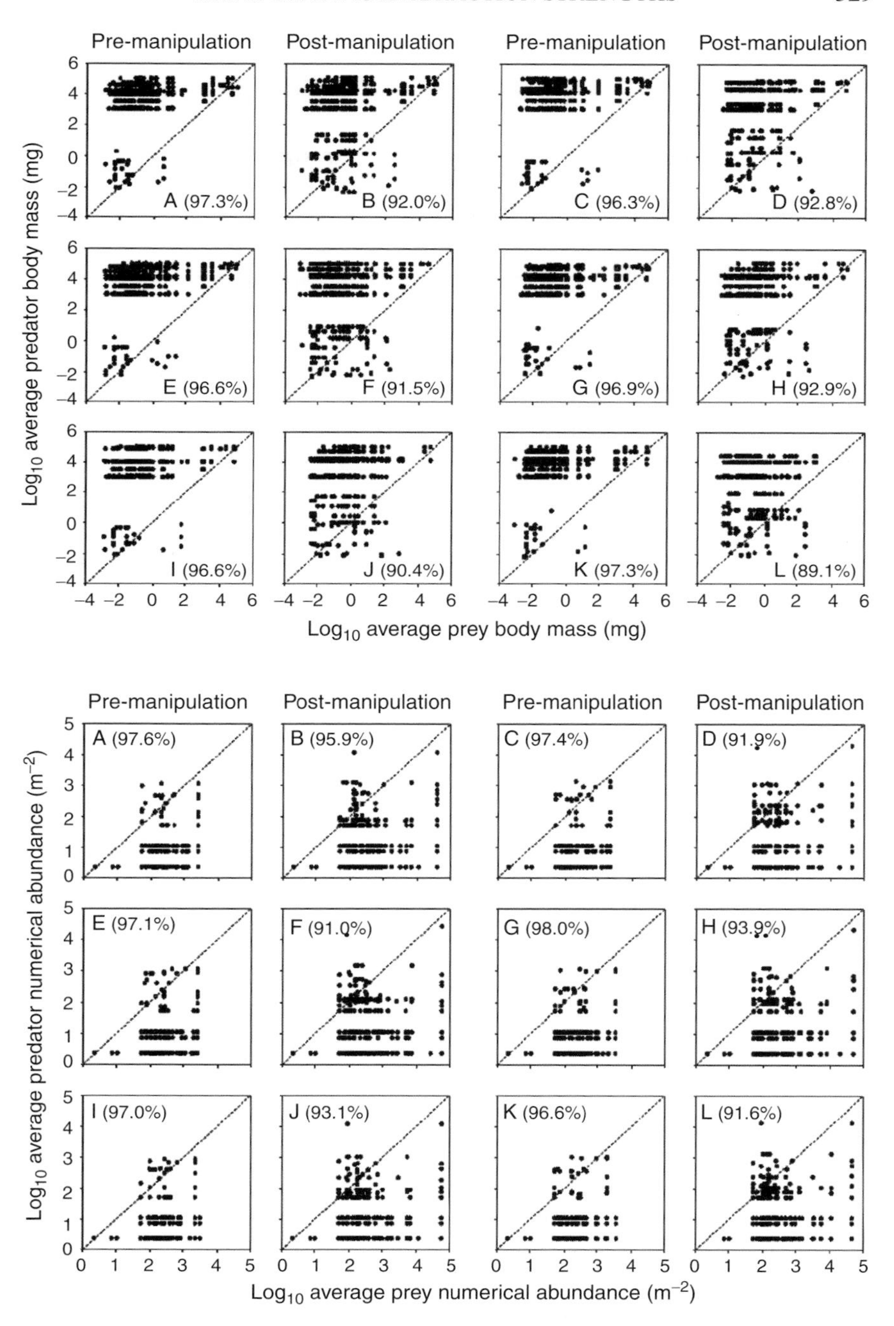

Figure 12 (Continued)

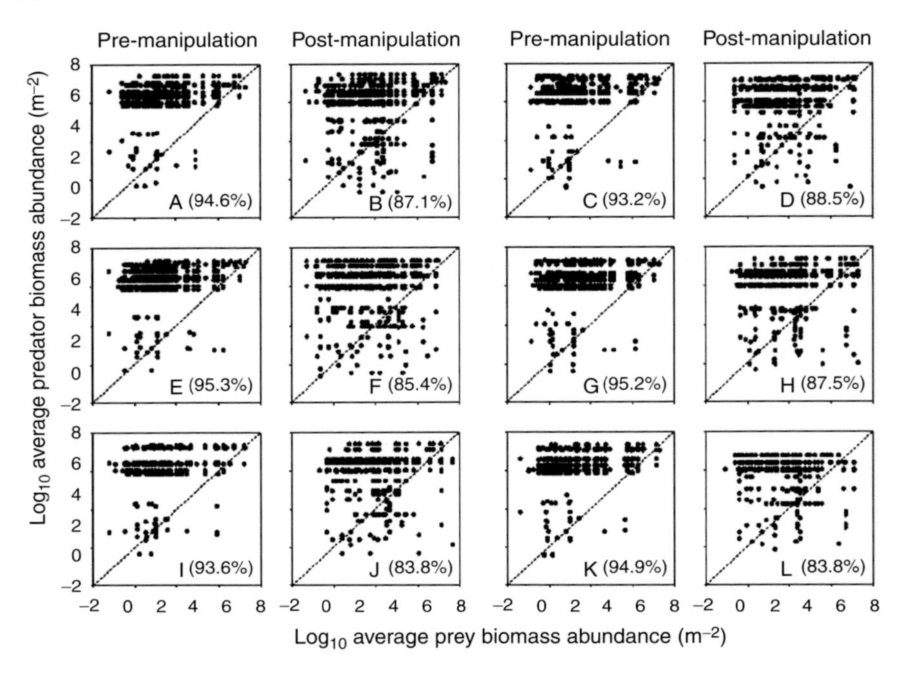

Figure 12 (i) Predator body size versus prey body size. Numbers in brackets represent the percentage of predators that consume smaller prey (data points located in the upper triangle); (ii) predator numerical abundance versus prey numerical abundance. Numbers in brackets represent the percentage of predators that consume more numerically abundant prey (data points located in the lower triangle); (iii) predator biomass abundance versus prey biomass abundance for the predator–prey. Numbers in brackets represent the percentage of predators that have a greater biomass than their prey (data points located in the upper triangle. The panels in each figure represent: (A, B) intact community; (C, D) two weak interactors removed; (E, F) three weak interactors removed; (G, H) two strong interactors removed; (I, J) three strong interactors removed; and (K, L) all strong and weak interactors removed. Comparisons within a treatment are shown pre and post-manipulation.

strong and weak interactors led to a significant reduction in the slope of this relationship (Table 2i). There was also a weak but significant correlation between predator and prey NA in all treatments pre-manipulation. This correlation was unaffected by the interaction strength manipulations (Table 2ii). In a linear least-squares regression of predator and prey NA, the removal of two or three weak interactors, three strong interactors or all strong and weak interactors led to a significant reduction in the intercept of this relationship. Again, the interaction strength manipulations had no effect on the slope of this relationship (Table 2ii). There was a significantly weak

Table 2 Regressions and correlations between (i) predator and prey body size, (ii) predator and prey numerical abundance and (iii) predator and prey biomass abundance

Treatment	Manipulation	*a*	*b*	*r*	*p*	*n*
(i)						
W^+S^+	Pre	3.768	0.233	0.248	<0.001	440
	Post	3.248	0.178	0.140	0.001	535
	p value	<0.001	0.430			
$W^{-2}S^+$	Pre	3.791	0.214	0.226	<0.001	429
	Post	3.194	0.194	0.149	0.002	442
	p value	<0.001	0.787			
$W^{-3}S^+$	Pre	3.811	0.251	0.265	<0.001	443
	Post	3.013	0.181	0.124	0.009	445
	p value	<0.001	0.389			
W^+S^{-2}	Pre	3.827	0.236	0.254	<0.001	457
	Post	3.217	0.124	0.095	0.040	462
	p value	<0.001	0.127			
W^+S^{-3}	Pre	3.783	0.222	0.241	<0.001	440
	Post	3.079	0.127	0.094	0.057	409
	p value	<0.001	0.234			
W^-S^-	pre	3.830	0.244	0.264	<0.001	416
	post	2.610	0.024	0.017	0.759	321
	p value	<0.001	0.016			
(ii)						
W^+S^+	Pre	0.222	0.170	0.193	<0.001	440
	Post	0.364	0.178	0.192	<0.001	535
	p value	0.290	0.900			
$W^{-2}S^+$	Pre	0.231	0.169	0.194	<0.001	429
	Post	0.534	0.149	0.144	0.002	442
	p value	0.043	0.756			
$W^{-3}S^+$	Pre	0.188	0.194	0.212	<0.001	443
	Post	0.596	0.144	0.135	0.004	445
	p value	0.010	0.452			
W^+S^{-2}	Pre	0.221	0.164	0.201	<0.001	457
	Post	0.498	0.135	0.130	0.005	462
	p value	0.057	0.634			
W^+S^{-3}	Pre	0.218	0.169	0.197	<0.001	440
	Post	0.547	0.141	0.136	0.006	409
	p value	0.036	0.662			
W^-S^-	Pre	0.271	0.148	0.171	<0.001	416
	Post	0.690	0.151	0.134	0.016	321
	p value	0.023	0.970			
(iii)						
W^+S^+	Pre	3.969	0.149	0.198	<0.001	440
	Post	3.837	0.055	0.058	0.179	535
	p value	0.243	0.085			
$W^{-2}S^+$	Pre	4.067	0.117	0.153	0.002	429
	Post	3.832	0.081	0.089	0.063	442
	p value	0.055	0.522			

(*continued*)

Table 2 (*continued*)

Treatment	Manipulation	*a*	*b*	*r*	*p*	*n*
$W^{-3}S^{+}$	Pre	4.021	0.152	0.205	<0.001	443
	Post	3.750	0.046	0.048	0.312	445
	p value	0.028	0.063			
$W^{+}S^{-2}$	Pre	4.046	0.137	0.186	<0.001	457
	Post	3.931	0.016	0.020	0.670	462
	p value	0.295	0.019			
$W^{+}S^{-3}$	Pre	4.008	0.142	0.192	<0.001	440
	Post	3.878	0.012	0.013	0.795	409
	p value	0.307	0.021			
$W^{-}S^{-}$	Pre	4.008	0.163	0.228	<0.001	416
	Post	3.740	-0.047	0.054	0.335	321
	p value	0.039	<0.001			

In the regression equation $y=a+bx$, a is the intercept and b is the slope. P values indicate the significance of pre- and post-manipulation comparisons of a and b. The correlation coefficient r and the significance of the correlation between predator and prey allometry p are also given, along with the number of predator–prey pairs n.

correlation between predator and prey BA in all treatments pre-manipulation. This correlation was no longer significant for any treatment post-manipulation, suggesting a seasonal effect on the relationship between predator and prey BA (Table 2iii). In a linear least-squares regression of predator and prey BA, the removal of three weak interactors or all strong and weak interactors led to a significant reduction in the intercept of this relationship. The removal of two or three strong interactors or all strong and weak interactors also led to a significant reduction in the slope of this relationship (Table 2iii).

There was a highly significant negative relationship between body mass and NA for all treatments, pre- and post-manipulation (Figure 13i). There was a highly significant positive relationship between body mass and BA for all treatments, pre- and post-manipulation (Figure 13ii). The intercept of these relationships was significantly greater after the interaction strength manipulations for all treatments, suggesting a seasonal effect on the relationship between body mass and numerical/BA. The slope of the relationship between body mass and NA varied between -0.16 and -0.25. All slopes were significantly different from -0.75 to -1 (t-test, $p<0.001$), but not from a slope of -0.25 (t-test, $p>0.057$). The interaction strength manipulations had no effect on the slope of the relationship between body mass and NA (Table 3i). The slope of the relationship between body mass and BA varied between 0.75 and 0.83. All slopes were significantly different from 0 (t-test, $p<0.001$), but not from a slope of 0.75 (t-test, $p>0.053$). The interaction strength manipulations had no effect on the slope of the relationship between body mass and BA (Table 3ii).

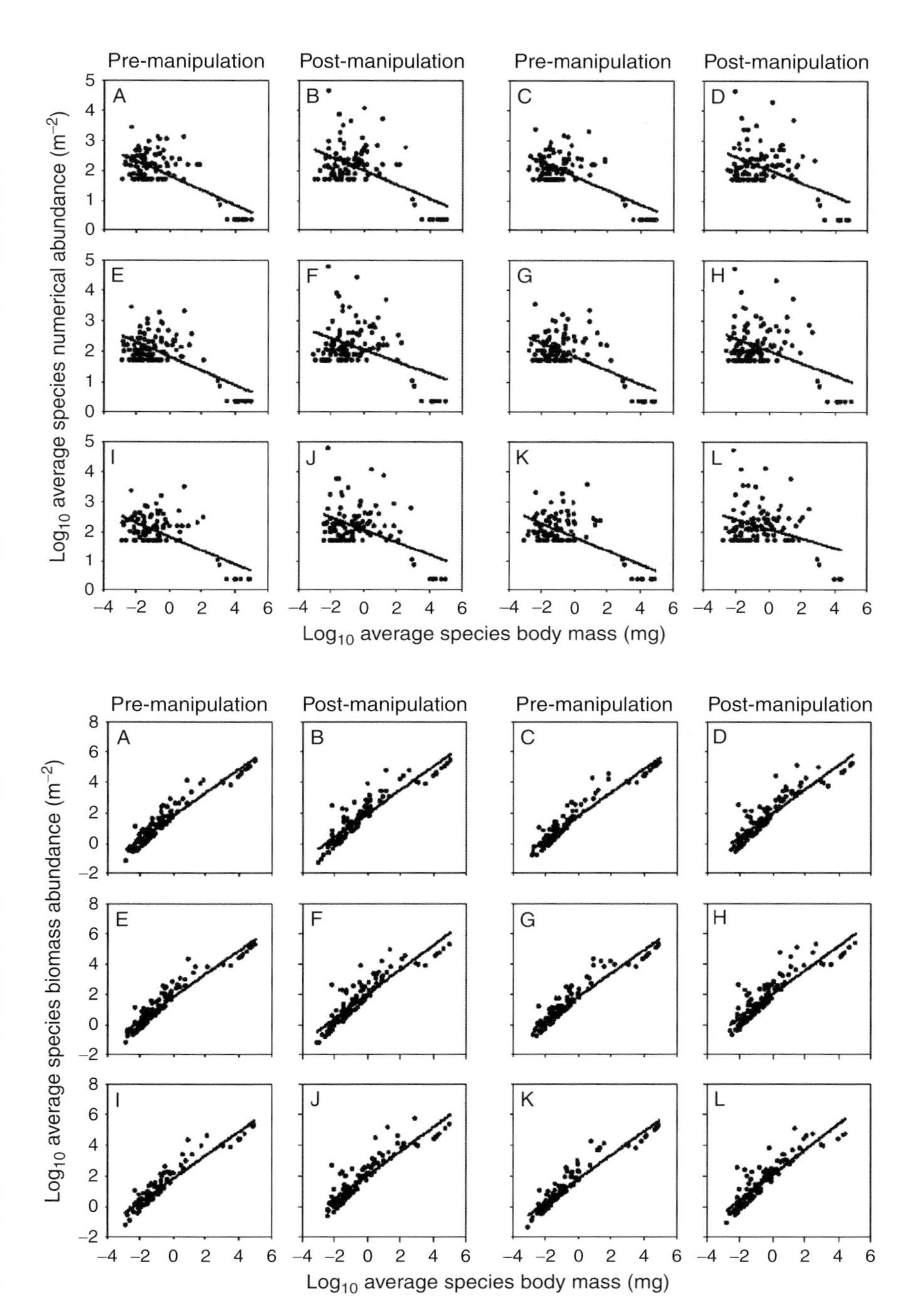

Figure 13 (i) Body mass versus numerical abundance; and (ii) body mass versus biomass abundance for the species in the composite treatment webs shown in Figure 4. The panels in each figure represent: (A, B) intact community; (C, D) two weak interactors removed; (E, F) three weak interactors removed; (G, H) two strong interactors removed; (I, J) three strong interactors removed; and (K, L) all strong and weak interactors removed. Comparisons within a treatment are shown pre- and post-manipulation.

Table 3 Regressions between (i) body size and numerical abundance and (ii) body size and biomass abundance in the experimental food webs

Treatment	Manipulation	*a*	*b*	r^2	*p*	*n*
(i)						
W^+S^+	Pre	1.829	−0.246	0.502	<0.001	101
	Post	1.994	−0.231	0.316	<0.001	109
	p value	0.044	0.713			
$W^{-2}S^+$	Pre	1.814	−0.236	0.473	<0.001	96
	Post	2.028	−0.216	0.258	<0.001	105
	p value	0.011	0.655			
$W^{-3}S^+$	Pre	1.829	−0.233	0.472	<0.001	98
	Post	2.045	−0.192	0.200	<0.001	110
	p value	0.012	0.352			
W^+S^{-2}	Pre	1.830	−0.224	0.420	<0.001	102
	Post	2.020	−0.201	0.217	<0.001	109
	p value	0.028	0.609			
W^+S^{-3}	Pre	1.839	−0.236	0.474	<0.001	97
	Post	2.071	−0.212	0.246	<0.001	102
	p value	0.008	0.597			
W^-S^-	Pre	1.819	−0.234	0.460	<0.001	93
	Post	2.103	−0.169	0.141	<0.001	101
	p value	0.002	0.192			
(ii)						
W^+S^+	Pre	1.817	0.756	0.906	<0.001	101
	Post	1.974	0.767	0.848	<0.001	109
	p value	0.049	0.780			
$W^{-2}S^+$	Pre	1.799	0.765	0.905	<0.001	96
	Post	1.996	0.780	0.833	<0.001	105
	p value	0.016	0.725			
$W^{-3}S^+$	Pre	1.828	0.766	0.906	<0.001	98
	Post	2.027	0.807	0.818	<0.001	110
	p value	0.020	0.359			
W^+S^{-2}	Pre	1.824	0.775	0.897	<0.001	102
	Post	2.015	0.792	0.815	<0.001	109
	p value	0.026	0.697			
W^+S^{-3}	Pre	1.829	0.775	0.913	<0.001	97
	Post	2.052	0.783	0.824	<0.001	102
	p value	0.009	0.858			
W^-S^-	Pre	1.810	0.776	0.904	<0.001	93
	Post	2.076	0.829	0.808	<0.001	101
	p value	0.003	0.269			

In the regression equation $y=a+bx$, a is the intercept and b is the slope. p values indicate the significance of pre- and post-manipulation comparisons of a and b. The amount of variation in log body mass explained by log numerical abundance r^2 and the significance of the regression p are also given, along with the number of species analysed n.

The distinctly size-structured nature of the experimental food webs is further highlighted by the trophic pyramids in Figure 14. Here, the TH of each species was rounded up to the nearest whole number in order to assign species to discrete trophic levels. Total body mass increased with each discrete trophic level, with the exception of trophic level 3, which had a lower total body size than trophic level 2 (Figure 14i). Total NA decreased (Figure 14ii) and BA increased with trophic level (except trophic level 3) (Figure 14iii). These patterns were consistent across all treatments, both before and after the interaction strength manipulations. However, an entire trophic level was lost in the treatments which had three strong interactors and all strong and weak interactors removed (see also Figure 6J and L).

There was a significant positive relationship between body size and TH in the experimental food webs (left rear walls in Figure 15). The removal of three strong interactors or all strong and weak interactors significantly reduced the intercept and the slope of this relationship, while the relationship was no longer significant after the removal of three strong interactors (Table 4i). There was a significantly negative relationship between body size and NA in the mesocosm food webs (right rear walls in Figure 15). The removal of three weak, three strong or all strong and weak interactors significantly reduced the intercept of this relationship. The removal of any species significantly increased the slope of the relationship (i.e. made it less negative). The relationship was no longer significant after the removal of two weak interactors or all strong and weak interactors (Table 4ii). There was a significant positive relationship between BA and TH in the experimental food webs (right rear walls in Figure 16). The removal of three strong interactors or all strong and weak interactors significantly reduced the slope of this relationship, while the relationship was no longer significant after the removal of three strong interactors (Table 4iii).

C. Trivariate Patterns

The trivariate plots in Figures 15 and 16 revealed a relationship between body size, numerical/biomass abundance and TH. For the relationship involving NA (Figure 15), many of the points lie approximately on a diagonal line from the top right-hand corner to the bottom left-hand corner of the plot. For the relationship involving BA (Figure 16), many of the points lie approximately on a diagonal line from the top centre to the bottom centre of the plot. In other words, large-bodied species tended to occur at high trophic levels and at low NAs/high BAs, whereas small-bodied species tended to occur at low trophic levels and at high NAs/low BAs. These relationships appeared consistent before and after the manipulations, although multiple linear regression (similar to that carried out for trivariate patterns by Jonsson

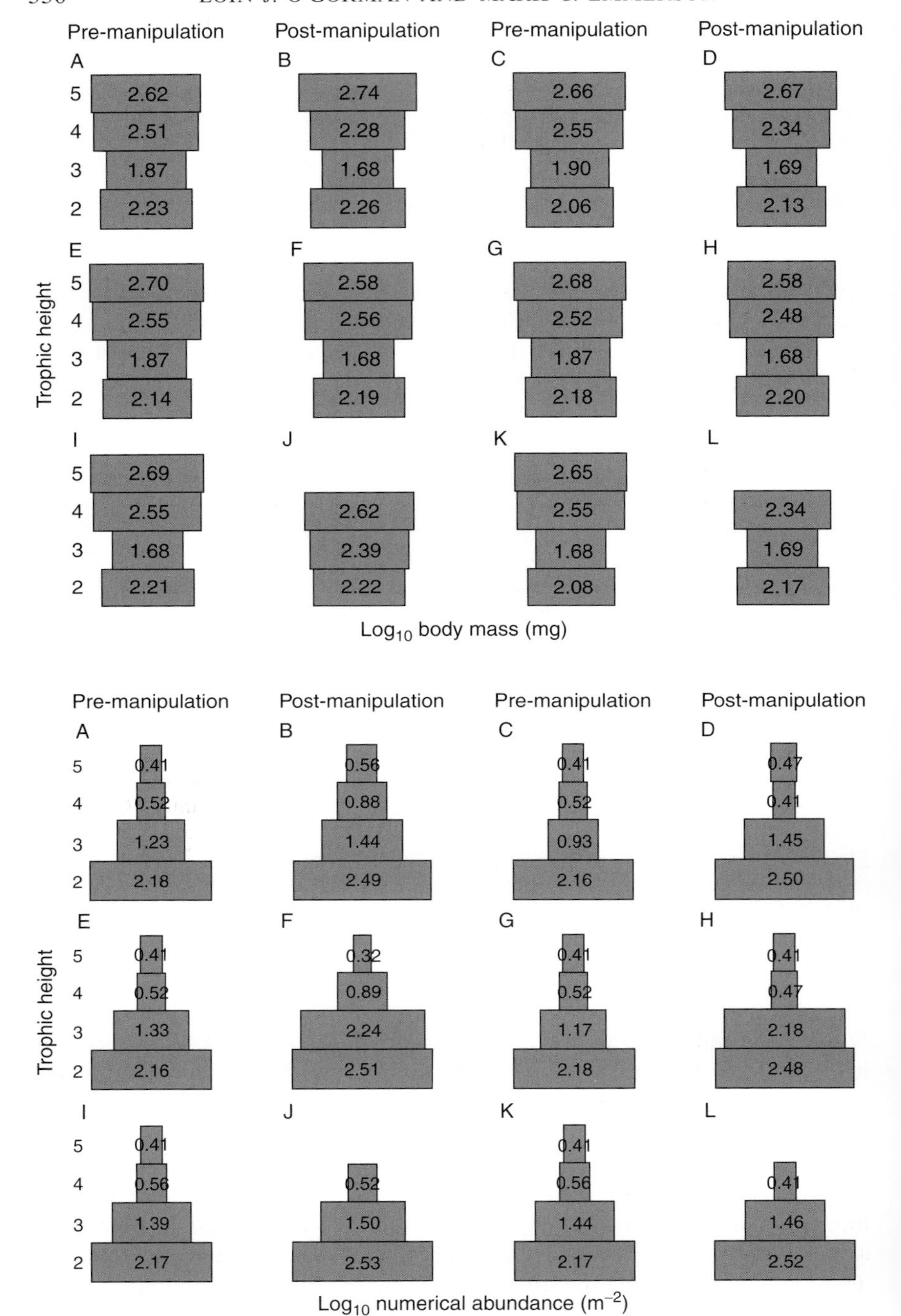

Pre-manipulation
Post-manipulation
Pre-manipulation
Post-manipulation
A
B
C
D
2.62
2.51
1.87
2.23
2.74
2.28
1.68
2.26
2.66
2.55
1.90
2.06
2.67
2.34
1.69
2.13
E
F
G
H
2.70
2.55
1.87
2.14
2.58
2.56
1.68
2.19
2.68
2.52
1.87
2.18
2.58
2.48
1.68
2.20
I
J
K
L
2.69
2.55
1.68
2.21
2.62
2.39
2.22
2.65
2.55
1.68
2.08
2.34
1.69
2.17
Trophic height
5
4
3
2
Log10 body mass (mg)
Pre-manipulation
Post-manipulation
Pre-manipulation
Post-manipulation
A
B
C
D
0.41
0.52
1.23
2.18
0.56
0.88
1.44
2.49
0.41
0.52
0.93
2.16
0.47
0.41
1.45
2.50
E
F
G
H
0.41
0.52
1.33
2.16
0.32
0.89
2.24
2.51
0.41
0.52
1.17
2.18
0.41
0.47
2.18
2.48
I
J
K
L
0.41
0.56
1.39
2.17
0.52
1.50
2.53
0.41
0.56
1.44
2.17
0.41
1.46
2.52
Trophic height
Log10 numerical abundance (m−2)

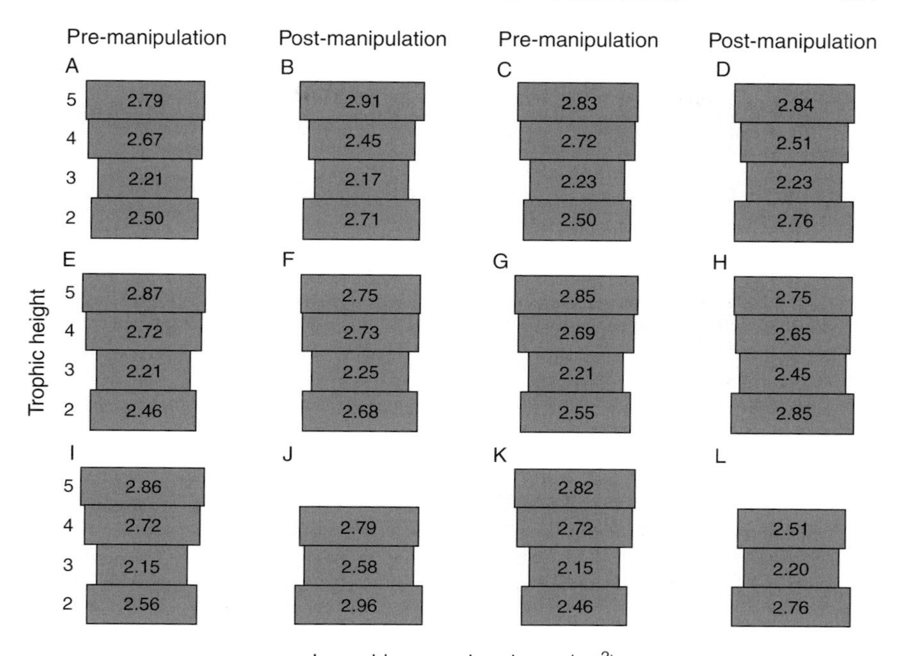

Figure 14 Trophic pyramids for (i) total body mass; (ii) total numerical abundance; and (iii) total biomass abundance at discrete trophic levels, for each of the composite treatment webs shown in Figure 4. No measurements were taken for the basal resources, so there is no description of the body mass or abundance at a trophic level of 1. The panels in each figure represent: (A, B) intact community; (C, D) two weak interactors removed; (E, F) three weak interactors removed; (G, H) two strong interactors removed; (I, J) three strong interactors removed; and (K, L) all strong and weak interactors removed. Comparisons within a treatment are shown pre- and post-manipulation.

et al., 2005) revealed that body mass and numerical/biomass abundance explained less of the variation in TH post-manipulation compared to pre manipulation (see 'r^2' in Table 5i, ii). This was particularly true in the treatments where three strong interactors were removed and where all strong and weak interactors were removed. A comparison of regressions before and after the interaction strength manipulations also showed that the slope of numerical/biomass abundance and the intercept of the relationship were significantly different for these two treatments (see '*p* values' in Table 5i). This appeared to be largely driven by the loss of a trophic level in these food webs, which is clearly visible in Figures 6 and 14.

IV. DISCUSSION

A. Univariate Patterns

Many of the univariate patterns in our experimental food webs were susceptible to our targeted extinctions, and the removal of weak interactors proved to have effects that were just as marked as those due to the removal of strong

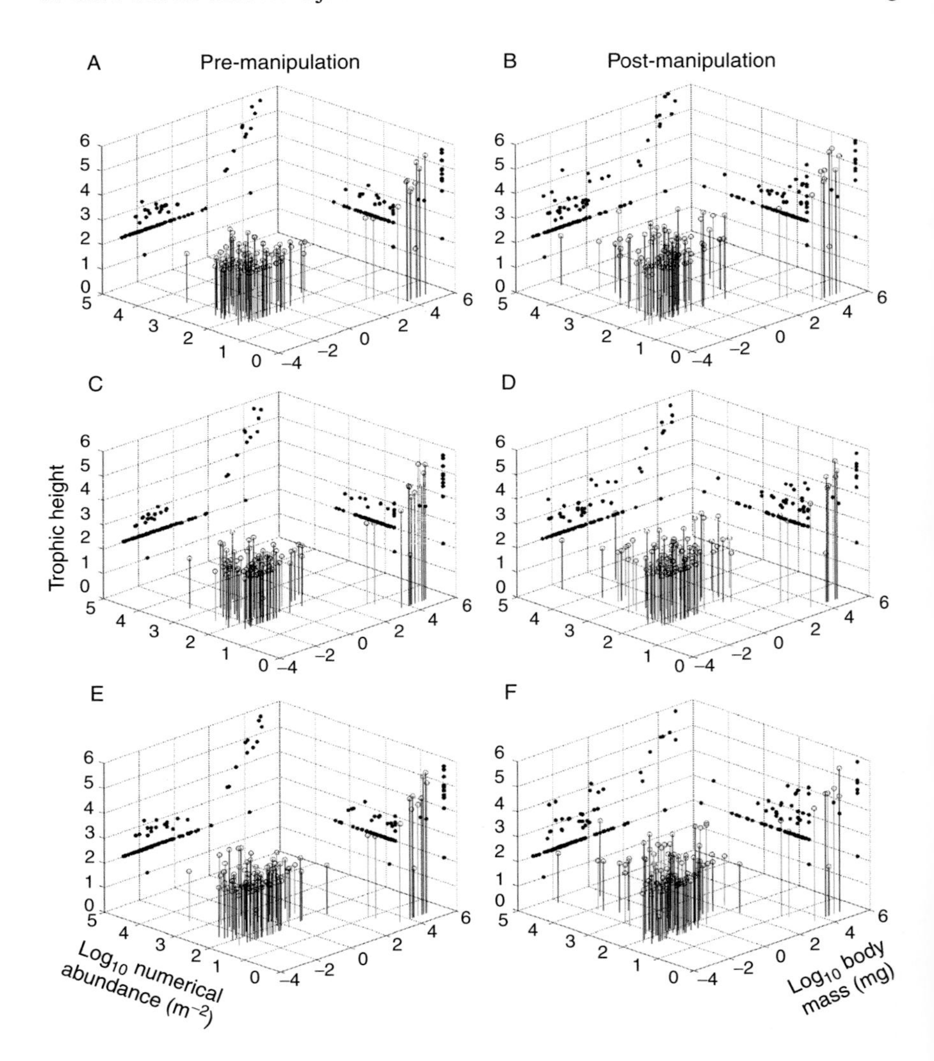

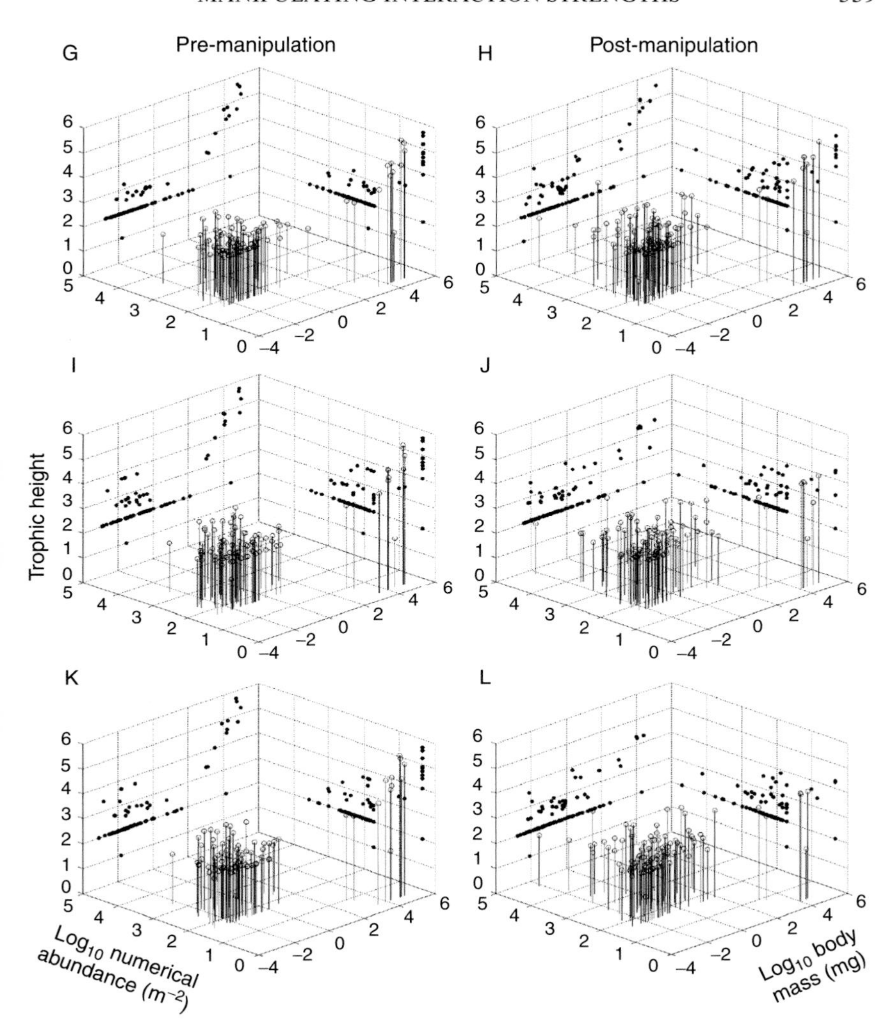

Figure 15 Trivariate relationships between log body mass, log numerical abundance and trophic height for each of the composite treatment webs shown in Figure 4. The panels in each figure represent: (A, B) intact community; (C, D) two weak interactors removed; (E, F) three weak interactors removed; (G, H) two strong interactors removed; (I, J) three strong interactors removed; and (K, L) all strong and weak interactors removed. Comparisons within a treatment are shown pre and post-manipulation.

interactors in most cases. This is most likely due to the highly connected nature of the manipulated species (strong and weak interactors), which ranged from 25 to 87 direct TL in the composite mesocosm food web

Table 4 Regressions between (i) body size and trophic height; (ii) numerical abundance and trophic height and (iii) numerical abundance and trophic height in the experimental food webs

Treatment	Manipulation	a	b	r^2	p	n
(i)						
W^+S^+	Pre	2.574	0.333	0.598	<0.001	101
	Post	2.648	0.340	0.466	<0.001	109
	p value	0.404	0.863			
$W^{-2}S^+$	Pre	2.548	0.338	0.677	<0.001	96
	Post	2.502	0.307	0.458	<0.001	105
	p value	0.575	0.470			
$W^{-3}S^+$	Pre	2.552	0.335	0.598	0.013	98
	Post	2.566	0.282	0.365	<0.001	110
	p value	0.872	0.246			
W^+S^{-2}	Pre	2.508	0.332	0.593	<0.001	102
	Post	2.549	0.306	0.442	<0.001	109
	p value	0.621	0.547			
W^+S^{-3}	Pre	2.623	0.316	0.482	<0.001	97
	Post	2.434	0.208	0.335	0.196	102
	p value	0.033	0.017			
W^-S^-	Pre	2.631	0.327	0.517	<0.001	93
	Post	2.400	0.184	0.248	<0.001	101
	p value	0.009	0.002			
(ii)						
W^+S^+	Pre	4.120	−0.882	0.585	<0.001	101
	Post	4.125	−0.768	0.468	<0.001	109
	p value	0.986	0.364			
$W^{-2}S^+$	Pre	4.140	−0.911	0.519	<0.001	96
	Post	3.681	−0.601	0.318	0.385	105
	p value	0.087	0.014			
$W^{-3}S^+$	Pre	4.138	−0.899	0.498	<0.001	98
	Post	3.514	−0.505	0.216	<0.001	110
	p value	0.028	0.003			
W^+S^{-2}	Pre	4.053	−0.877	0.496	<0.001	102
	Post	3.591	−0.555	0.279	<0.001	109
	p value	0.085	0.010			
W^+S^{-3}	Pre	4.270	−0.921	0.481	<0.001	97
	Post	3.251	−0.411	0.245	<0.001	102
	p value	<0.001	<0.001			
W^-S^-	Pre	4.228	−0.909	0.477	<0.001	93
	Post	2.967	−0.299	0.133	0.188	101
	p value	<0.001	<0.001			
W^+S^+	Pre	1.884	0.360	0.442	0.000	101
	Post	2.008	0.307	0.262	0.000	109
	p value	0.391	0.405			
$W^{-2}S^+$	Pre	1.871	0.359	0.441	0.000	96
	Post	1.932	0.275	0.268	0.000	105
	p value	0.663	0.172			

Table 4 (*continued*)

Treatment	Manipulation	*a*	*b*	r^2	*p*	*n*
$W^{-3}S^{+}$	Pre	1.872	0.356	0.442	0.000	98
	Post	2.043	0.238	0.269	0.596	110
	p value	0.211	0.052			
$W^{+}S^{-2}$	Pre	1.859	0.339	0.414	0.000	102
	Post	1.977	0.264	0.254	0.000	109
	p value	0.376	0.209			
$W^{+}S^{-3}$	Pre	1.992	0.327	0.328	0.000	97
	Post	2.049	0.178	0.184	0.000	102
	p value	0.683	0.016			
$W^{-}S^{-}$	Pre	1.996	0.331	0.356	0.593	93
	Post	2.065	0.148	0.137	0.138	101
	p value	0.607	0.003			

In the regression equation $y = a + bx$, *a* is the intercept and *b* is the slope. *p* values indicate the significance of pre- and post-manipulation comparisons of *a* and *b*. The amount of variation in log body mass explained by log numerical abundance r^2 and the significance of the regression *p* are also given, along with the number of species analysed *n*.

shown in Figure 5. Both Dunne *et al.* (2002b) and Coll *et al.* (2008) have demonstrated that targeted removal of highly connected species can cause a greater deterioration in food web structure than the loss of poorly connected species. This is clearly the case in our experimental communities, where the removal of our highly connected manipulated species altered the connectance, linkage density, mean FCL, fraction of top, intermediate and basal species and the distribution of links in the food webs.

Connectance is a fundamental property of network structure, since it describes the fraction of realized links in a food web, and as such it has been widely used in conjunction with species richness to predict key structural properties of even the most complex food webs in the literature (Williams and Martinez, 2000). Changes in connectance are also often associated with altered fractions of top, intermediate and basal species, distributions of links, FCLs and the prevalence of cannibalism and omnivory (Williams and Martinez, 2000). A reduction in connectance is also associated with decreased robustness, a measure of stability that defines the ability of a food web to resist secondary extinctions (Dunne *et al.*, 2002b). Linkage density describes the mean number of interactions per species and provides another measure of how highly connected a food web is. See Table 6 for a comparison of simple food web properties in our study (including connectance and linkage density) with some of the best described food webs in the published literature. Note that we calculated interactive connectance in this study in line with the trivariate patterns analysis of Jonsson *et al.* (2005). Many other food web studies calculate directed connectance (DC) (L/S^2), and so for consistency in Table 6 we estimated DC across all studies. The ranges of DC (0.06–0.20) and linkage density (4.0–9.7) in the experimental food webs described here overlap with many of the best resolved published food webs. Interestingly, many of the other marine food webs have much higher linkage densities than our study, although this appears to be due to the extremely large size of these webs, which often span large areas of open ocean (i.e. the Northeastern US Shelf, a Caribbean Coral Reef and the Weddell Sea in Antarctica).

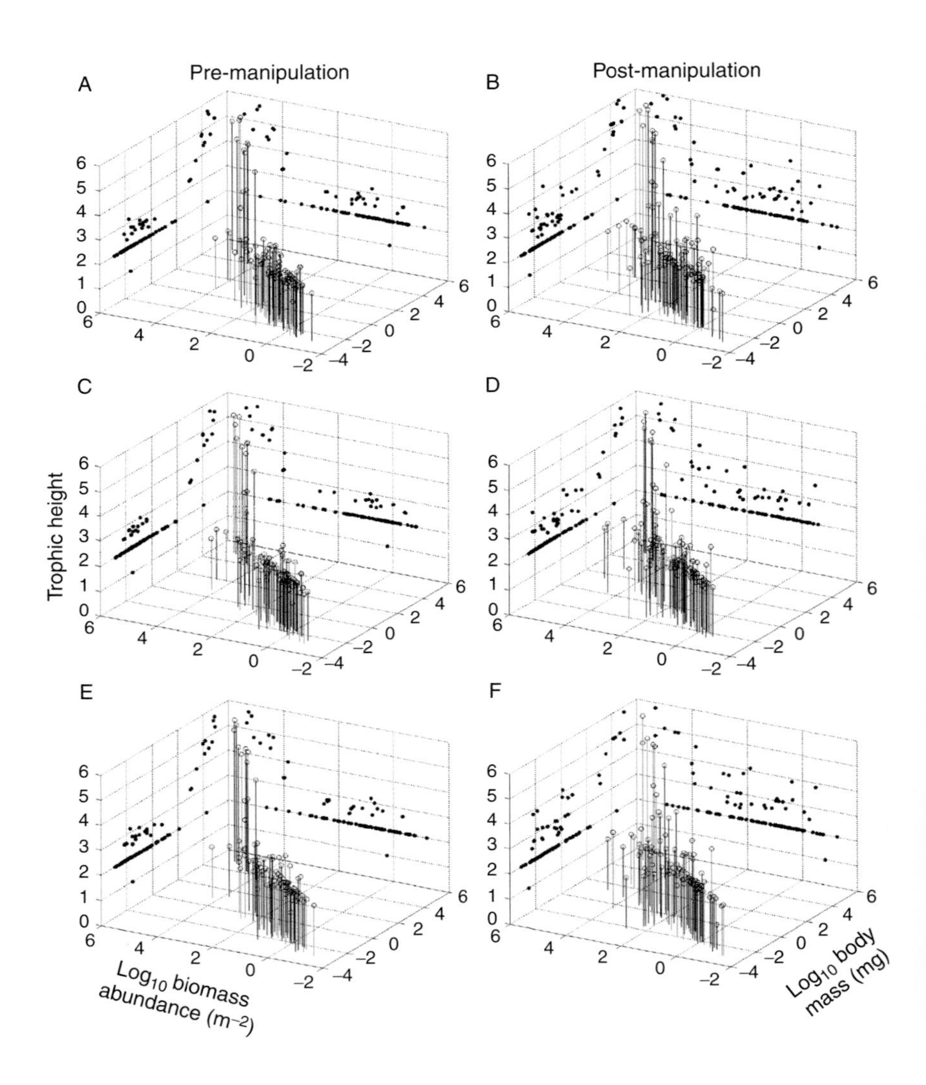

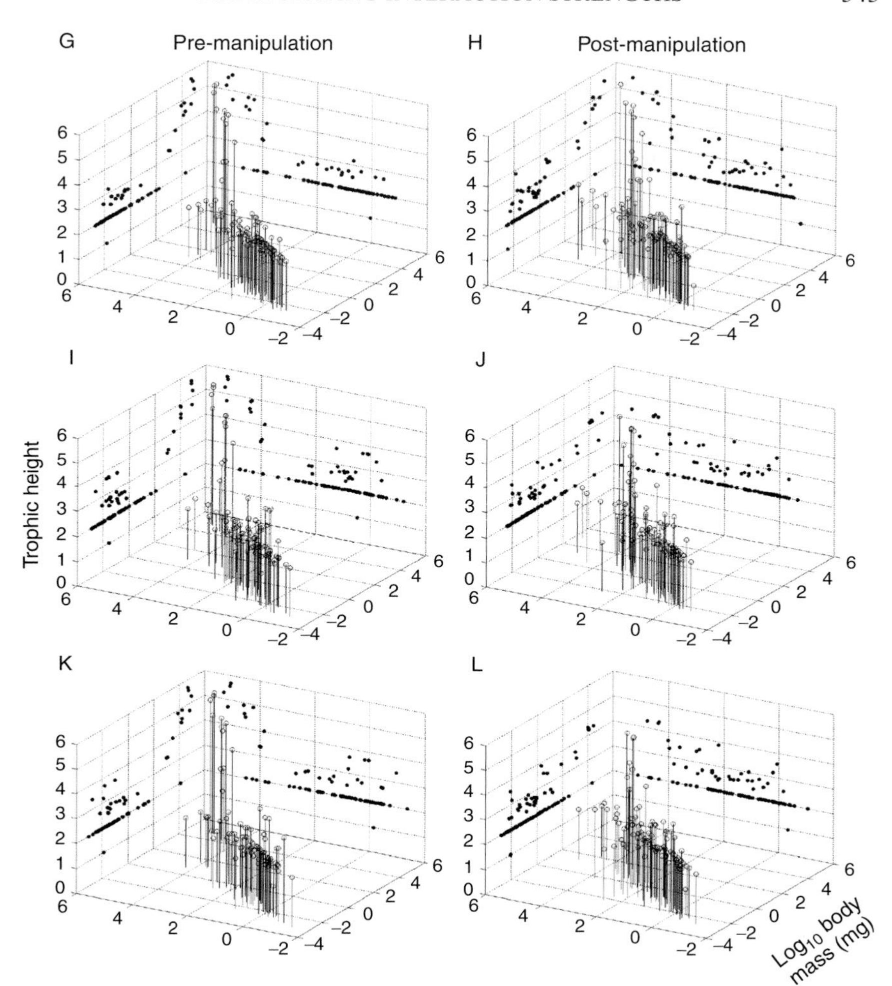

Figure 16 Trivariate relationships between log body mass, log biomass abundance and trophic height for each of the composite treatment webs shown in Figure 4. The panels in each figure represent: (A, B) intact community; (C, D) two weak interactors removed; (E, F) three weak interactors removed; (G, H) two strong interactors removed; (I, J) three strong interactors removed and (K, L) all strong and weak interactors removed. Comparisons within a treatment are shown pre- and post-manipulation.

The loss of two or more species reduced the connectance and linkage density of the mesocosm food webs by up to 32% and 20%, respectively (see also Figure 7), although there was no distinction between the effects of

Table 5 Multiple regressions for trivariate food web patterns

Treatment	Manipulation	a	b	c	r^2	p	n
(i)							
W^+S^+	Pre	0.232	−0.410	3.324	0.653	<0.001	101
	Post	0.238	−0.442	3.529	0.557	<0.001	109
	p value	0.902	0.821	0.465			
$W^{-2}S^+$	Pre	0.234	−0.442	3.351	0.672	<0.001	96
	Post	0.239	−0.316	3.143	0.523	<0.001	105
	p value	0.925	0.346	0.435			
$W^{-3}S^+$	Pre	0.237	−0.420	3.320	0.655	<0.001	98
	Post	0.231	−0.264	3.107	0.413	<0.001	110
	p value	0.916	0.264	0.447			
W^+S^{-2}	Pre	0.233	−0.441	3.315	0.665	<0.001	102
	Post	0.248	−0.287	3.128	0.499	<0.001	109
	p value	0.754	0.223	0.461			
W^+S^{-3}	Pre	0.154	−0.605	3.741	0.546	<0.001	97
	Post	0.160	−0.226	2.903	0.390	<0.001	102
	p value	0.908	0.010	0.004			
W^-S^-	Pre	0.190	−0.518	3.582	0.568	<0.001	93
	Post	0.156	−0.172	2.761	0.286	<0.001	101
	p value	0.524	0.018	0.005			
(ii)							
W^+S^+	Pre	0.642	−0.410	3.318	0.652	<0.001	101
	Post	0.693	−0.460	3.556	0.556	<0.001	109
	p value	0.666	0.729	0.407			
$W^{-2}S^+$	Pre	0.673	−0.437	3.335	0.670	<0.001	96
	Post	0.555	−0.318	3.137	0.518	<0.001	105
	p value	0.301	0.386	0.464			
$W^{-3}S^+$	Pre	0.651	−0.413	3.306	0.654	<0.001	98
	Post	0.495	−0.264	3.102	0.412	<0.001	110
	p value	0.187	0.289	0.466			
W^+S^{-2}	Pre	0.674	−0.441	3.313	0.665	<0.001	102
	Post	0.522	−0.272	3.097	0.492	<0.001	109
	p value	0.160	0.187	0.400			
W^+S^{-3}	Pre	0.753	−0.601	3.729	0.540	<0.001	97
	Post	0.387	−0.229	2.904	0.388	<0.001	102
	p value	0.004	0.014	0.006			
W^-S^-	Pre	0.713	−0.520	3.581	0.571	<0.001	93
	Post	0.319	−0.162	2.737	0.280	<0.001	101
	p value	0.002	0.014	0.004			

(i) Trophic height = a*(log$_{10}$ body mass) + b*(log$_{10}$ numerical abundance) + c.
(ii) Trophic height = a*(log$_{10}$ body mass) + b*(log$_{10}$ biomass abundance) + c. p values indicate the significance of a pre- and post-manipulation comparisons of a, b and c. The amount of variation in trophic height explained by log body mass and log numerical abundance r^2 and the significance of the regression p are also given, along with the number of species analysed n.

Table 6 A comparison of food web statistics for the current study to some of the best described published food webs

Food web	*S*	*L*	*LD*	*DC*	Reference
Marine					
Lough Hyne mesocosms	46–78	243–575	4.0–9.7	0.06–0.20	O'Gorman and Emmerson (2010)
Benguela	29	191	6.6	0.23	Yodzis (1998)
Northeast US Shelf	81	1562	19.3	0.24	Link (2002)
Caribbean	247	3288	13.3	0.05	Opitz (1993)
Weddell Sea	490	16041	32.7	0.07	Jacob (2005)
Estuarine					
Chesapeake Bay	31	68	2.2	0.07	Baird and Ulanowicz (1989)
St. Mark's Seagrass	48	221	4.6	0.10	Christian and Luczkovich (1999)
Ythan Estuary	92	409	4.4	0.05	Hall and Raffaelli (1991)
Freshwater					
River Duddon Streams	19–87	56–1653	2.6–19.0	0.12–0.29	Layer *et al.* (2010)
Broadstone Stream	24–34	109–170	4.4–5.1	0.13–0.21	Woodward and Hildrew (2001)
Duffin Creek	31–39	101–146	3.1–3.7	0.09–0.11	Tavares-Cromar and Williams (1996)
Appalachian Mountains	35–41	125–200	3.6–4.5	0.10–0.15	Hall *et al.* (2000)
New Zealand Streams	86–113	286–950	9.5–16.7	0.03–0.08	Townsend *et al.* (1998)
Skipwith Pond	35	276	7.9	0.23	Warren (1989)
Tuesday Lake	56	269	4.8	0.09	Jonsson *et al.* (2005)
Bere Stream	142	1383	9.7	0.07	Woodward *et al.* (2008)
Lake Tahoe	172	3885	22.6	0.13	Dunne *et al.* (2002a,b)
Little Rock Lake	182	1973	10.8	0.06	Martinez (1991)
Terrestrial					
The Gearagh	33–79	68–256	2.1–3.3	0.08–0.13	McLaughlin and Emmerson (this volume)
Coachella Valley	29	262	9.0	0.31	Polis (1991)
UK Grassland	75	119	1.6	0.02	Dawah *et al.* (1995)
Scotch Broom	154	403	2.6	0.02	Memmott *et al.* (2000)
El Verde Rainforest	155	1510	9.7	0.06	Reagan and Waide (1996)

S = number of taxa, L = number of links, LD = linkage density (L/S) and DC = directed connectance (L/S^2).

losing strong versus weak interactors. Rather, the number of species lost seemed to contribute to the extent of the reduction in connectance and linkage density, that is two species removed contributed to a significant reduction in these food web properties, three species removed led to even greater reductions, with the largest decrease in connectance and linkage density occurring when six species were removed (three strong and three weak interactors). It should be noted that this finding may be due to a possible relationship between connectance/number of links and number of species, although there is still much debate as to whether connectance or linkage density may remain constant regardless of the number of species (see Cohen *et al.*, 1986; Martinez, 1992; Montoya and Sole, 2003; Rejmanek and Stary, 1979; Riede *et al.*, 2010). Our results indicate that the removal of any highly connected species, whether it is a strong or a weak interactor, reduced the stability of the experimental food webs. This is in line with other studies that show the importance of weak, as well as strong, interactors for food web stability (McCann *et al.*, 1998; Neutel et al., 2002; O'Gorman and Emmerson, 2009). Species turnover provides another a measure of a system's dynamic stability (O'Gorman and Emmerson, 2009), with low rates equating to a high resistance to species invasions and/or extinctions: in our study the loss of two strong or three weak interactors led to significantly higher turnover (see Figure 8). These combined effects on connectance and species turnover highlight the importance of maintaining the natural empirical pattern of strong and weak interactors to preserve food web stability.

Long food chains are often thought to be rare in nature, due to their perceived dynamical instability (Lawler and Morin, 1993; Pimm and Lawton, 1977) and/or energetic constraints (Hutchinson, 1959; Pimm, 1982). More recently, however, some studies have suggested that these factors do not adequately explain natural patterns of FCL (Spencer and Warren, 1996; Sterner *et al.*, 1997). Experiments suggest that high productivity (Kaunzinger and Morin, 1998; Vander Zanden *et al.*, 1999) or a large ecosystem size, with high species diversity, habitat availability and habitat heterogeneity (Post *et al.*, 2000; Spencer and Warren, 1996) can sustain longer FCLs. Mean FCL in our experimental communities was relatively high, at approximately 5.5 species per food chain. This is comparable to Tuesday Lake (Jonsson *et al.*, 2005) and the Ythan Estuary (Hall and Raffaelli, 1991), but is close to the upper range for the 113 food webs analysed by Briand and Cohen (1987). The loss of two or three weak interactors led to a significant reduction in mean FCL, with the loss of two strong interactors producing even shorter food chains. While shorter FCLs may be associated with higher stability (Pimm and Lawton, 1977; but see Post, 2002), they also represent a lower productivity, diversity or habitat availability (Kaunzinger and Morin, 1998; Post *et al.*, 2000). The greatest reduction in mean FCL occurred in the treatments with three strong interactors or all strong and weak interactors removed, which reflects the simplification of food web structure in these two treatments (clearly visible through the loss of a trophic level in Figure 6J and L). There was a significant reduction in primary productivity after the removal of strong interactors from the communities (see O'Gorman and Emmerson, 2009) and a reduction in species diversity after the removal of two weak and three strong interactors (see Figure 8A and B). The shortening of mean FCL following the manipulations might, therefore, be a response to the reductions in productivity (i.e. resource availability) and/or species diversity (which is related to the ecosystem-size hypothesis; Cohen and Newman, 1991), as suggested through the findings of other experimental studies (Kaunzinger and Morin, 1998; Post, 2002; Post *et al.*, 2000; Vander Zanden *et al.*, 1999).

The fractions of top (2–16%), intermediate (69–86%) and basal (10–18%) species in the experimental communities are similar to those documented in many other aquatic systems (e.g. Link, 2002; Martinez, 1991; Warren, 1989; Woodward *et al.*, 2005). Hall and Raffaelli (1991) reported a lower percentage of basal and intermediate species and a higher percentage of top species for the Ythan Estuary food web, but this web included many birds as top species, which were not part of our experimental food webs: thus the cages effectively constrained the upper limit to body size in our study. Jonsson *et al.* (2005) report a lower percentage of intermediate species and a higher percentage of basal species for the Tuesday Lake food web compared to this study, but in contrast with the Ythan web, their study included a highly resolved description of primary producers, with many phytoplankton described to species level. The basal resources in our experimental communities suffer from a lack of resolution, which undoubtedly contributed to the observed differences with Tuesday Lake. This highlights the importance of standardizing resolution when comparing patterns both among and within food webs. Martinez (1993) demonstrated that linkage density, the fractions of top, intermediate and basal species and links between these species are very sensitive to differences in resolution, whereas the ratio of predators to prey and connectance was relatively robust. This suggests that the observed effects of species removal on the connectance of the experimental communities should be consistent regardless of the resolution of our food webs, whereas some of the other univariate patterns need to be interpreted with caution when comparing across studies (but see Bersier and Sugihara, 1997; Hall and Raffaelli, 1991; Sugihara *et al.*, 1989 for examples of studies with consistency in these properties regardless of taxonomic resolution). It would be interesting to compare the Ythan and Tuesday Lake food webs to the Lough Hyne mesocosms after removing the bird species and aggregating the primary producers, respectively. This may then reveal consistency in the fractions of top, intermediate and basal species.

The fractions of top, intermediate and basal species are intrinsically related to the distributions of links between species in these categories. For example, the increased fraction of top species resulting from the manipulations (see Figure 9A, B) led to an increase in the fraction of basal to top and intermediate to top species. This was particularly evident with the removal of two or three strong interactors, or all strong and weak interactors, with little effect of the removal of weak interactors (see Figure 10C, D and G, H). Similarly, a decrease in the fraction of intermediate species (see Figure 9C, D) led to a large reduction in the fraction of intermediate-to-intermediate links (see Figure 10E, F). However, the increase in the fraction of basal to intermediate links suggested that the removal of strong and weak interactors had little effect on species that feed on basal resources, instead altering the distribution of species and links higher in the food web (Havens, 1992; Schmid-Araya *et al.*, 2002). Given that the manipulated species themselves all occupied high trophic levels (see Figure 5; Table S2 in Appendix I), this may be evidence for a concentration of knock-on direct effects on species closest along the food chain to the manipulated species, with a dissipation of the perturbation through longer chains of indirect effects. There was little impact of the manipulations on the fraction of basal species: the removal of three strong interactors caused a significant increase in the proportion of basal species, but there was little or no change in the actual number of basal species in these webs (see Appendix IV). The apparent consistency in the number of basal species may be partly due to the poor resolution of this portion of the web, where phytoplankton, diatoms, cladocerans and algal species were aggregated during construction of the web. If these groups had been identified to changes in the fraction of basal species might have been more readily detectable.

Individual body size is closely correlated with many aspects of morphology, physiology, behaviour and ecology, through allometric relationships (Brown and Gillooly, 2003; Calder, 1984; Emmerson and Raffaelli, 2004; Peters, 1983; Schmidt-Nielsen, 1984), and as such, it is an important and easily measured community descriptor. The distribution of body masses is typically highly skewed, with many small species decreasing to few large species (Brown and Nicoletto, 1991; Holling, 1992; Hutchinson and Macarthur, 1959; Jonsson *et al.*, 2005; May, 1988), reflecting the Eltonian pyramid of sizes and numbers (Elton, 1927). We observed a similar pattern in our experimental cages (see Figure 11i), although there was a peak at the right tail of the distribution in all treatments, produced by the manipulated species. This might suggest that the number of manipulated species used in the experiment was artificially high for the size of these communities, although they were in line with summer densities along the same stretch of shoreline at Lough Hyne (Costello, 1992; Crook *et al.*, 2000; Verling *et al.*, 2003; Yvon-Durocher *et al.*, 2008). Intriguingly, there was a gap in the right tail of the distribution pre-manipulation, which disappeared post-manipulation. The consistency of this effect across all treatments suggested that it was a seasonal effect in the communities and not a result of the manipulations. The small mesh size of the cages meant that only very small species (< 5mm) could recruit into the food webs throughout the experiment, but over time the species that had settled in the cages most likely grew in size to fill the gaps in these body mass distributions (see Holling, 1992 for other possible mechanisms leading to gaps in body mass distributions).

Numerical and biomass abundance in the experimental food webs followed bimodal distributions, again most likely due to artificially raised densities of the large manipulated species. There were also gaps in all the distributions for NA, both before and after the manipulations (see Figure 11ii), which were most likely a sampling effect, that is the density of species on the settlement panels and mesh pads was limited by the area of those sampling substrates. This produced the distinctive straight line of points on the plots of log NA versus rank in NA (see Figure 11ii). These two abnormalities precluded what might otherwise have been tightly fitting normal distributions in log NA (as evidenced by the close fit of almost all the data points to the log-normal relationship in the plot of log NA vs. rank in NA in Figure 11ii). A normal distribution of log NA would be indicative of an undisturbed community at equilibrium, where competitive species interactions are abundant (May, 1975; Tokeshi, 1993). Figure 11ii indicates that such patterns were evident across all treatments pre- and post-manipulation, suggesting that the communities with strong and/or weak interactors removed might have settled at a new equilibrium, despite the observed disruption in many of the univariate patterns in these food webs (see Table 1). We cannot confirm this pattern without more detailed information on the distribution of NAs, however, and this is a major limitation of the sampling methods employed.

B. Bivariate Patterns

Many of the bivariate patterns in the experimental food webs were unaffected by the manipulations, and all the communities were strongly size-structured, as is typical of many aquatic systems (Jennings *et al.*, 2001; Jonsson *et al.*, 2005; Reuman and Cohen, 2005). A high percentage of feeding interactions (>89%) consisted of large, rare predators eating small, abundant prey (see Figure 12), and this pattern is common in many other natural systems (Brose *et al.*, 2006a; Cohen *et al.*, 1993b, 2003; Layer *et al.*, 2010; Warren and Lawton, 1987). The food webs also maintained a trophic pyramid of NA and an inverted pyramid of body size and BA, even after the manipulations (see Figure 14). The increase in total body size at trophic level 2 was due to the influence of the large herbivorous sea urchin, *Paracentrotus lividus*, which, if removed from this analysis, left a perfect inverted pyramid of body size in the cages, that is body size decreased with every discrete trophic level. Even the loss of a trophic level in the treatments where three strong interactors or all strong and weak interactors were removed did not disrupt the structure of these pyramids. A community-level trophic cascade occurs when a perturbation to a top predator (or group of predators) leads to subsequent changes in the biomass of lower trophic levels (Polis *et al.*, 2000). Trophic cascades are commonly reported phenomena (Bruno and O'Connor, 2005; Byrnes *et al.*, 2006; Carpenter *et al.*, 1985; Estes and Palmisano, 1974; Finke and Denno, 2005; O'Gorman *et al.*, 2008) and it is intriguing that the loss of an entire trophic level, through the removal of the three most strongly interacting and six of the largest predators, did not produce cascading effects on the total body size, NA or BA on adjacent trophic levels. This is especially interesting because we know a community-level cascade occurred in these treatments (O'Gorman and Emmerson, 2009), with the loss of the strongly interacting predators releasing benthic invertebrate prey from predation pressure, which in turn suppressed primary production levels in the system. Clearly, even a community-level cascade such as this was not sufficient to override the size structure maintained by these aquatic communities.

There is, however, some evidence that this pattern might have been weakened to some extent by the manipulations. Jennings *et al.* (2002) showed that body size forms a continuum with trophic level in aquatic communities, reflected in the significant positive correlation between body size and TH in all our experimental treatments pre-manipulation. The loss of three strong interactors or all strong and weak interactors significantly reduced the intercept and the slope of this relationship. This was partly driven by the aforementioned loss of a trophic level in these treatments and also by an increased prevalence of large-bodied species at lower trophic levels. Furthermore, the relationship between body size and trophic level was no longer significant in the treatment where three strong interactors were removed. The dissolution of this relationship suggested that, over time, the inverted pyramid of body size dominating our experimental communities might also eventually break down.

Neutel *et al.* (2002) found that biomass pyramids can be important for the stability of complex food webs. If a pyramid of biomass exists (i.e. biomass decreases with increasing trophic level), omnivorous predators feeding on two or more trophic levels will have a relatively small top-down effect on high trophic levels, coupled with strong top-down effects on lower trophic levels (considering per unit biomass feeding rates). This leads to an aggregation of weak links in long feeding loops, reducing the mean interaction strength of the loop and increasing the stability of the system. Our experimental food webs possessed inverted biomass pyramids and yet were seemingly stable before the interaction strength manipulations were initiated (as suggested by high connectance and low species turnover). Raffaelli (2002) speculated that the slope of the biomass pyramid might be an indicator of food web stability, with tall, thin pyramids less likely to be stable than short, relatively flat pyramids. Raffaelli (2002) also acknowledged that many systems do not have biomass pyramids, especially open water marine and freshwater systems, and so the patterns described by Neutel *et al.* (2002) are likely to be sufficiently robust to accommodate such exceptions. This appears to be the case in our food webs, which exhibited a high proportion of weak interactions, despite their inverted biomass pyramids (O'Gorman *et al.*, 2010). An investigation of loop weights in the experimental food webs might reveal interesting insights into the biological mechanisms that contribute to stability in the absence of a stabilizing pattern of biomass, as described by Neutel *et al.* (2002).

Body mass–abundance relationships in nature are often thought to exhibit slopes of -0.75 (Brown, 1995; Damuth, 1981; Enquist *et al.*, 1998; McMahon and Bonner, 1983) or -1 (Borgmann, 1987; Brown and Gillooly, 2003; Griffiths, 1992, 1998). Such relationships are typically compiled from the average body size M and abundance N of species in the literature and constitute global size density relationships (GSDR). Body mass–abundance relationships with a slope of -0.75 have been used as evidence of an energy equivalence rule (EER) in natural communities (Damuth, 1981, 1987). Since body mass is thought to scale with metabolic rate B as $B \propto M^{0.75}$ (Brown and Gillooly, 2003; Kleiber, 1947; Peters, 1983; West *et al.*, 1997), this suggests that population energy use E is approximately invariant with respect to mass ($E \propto M^{0.75} \times M^{-0.75} = M^0$). The EER proposes that a combination of physiological and ecological processes results in energetic trade-offs, such that resources are divided equally among species populations, irrespective of their body size (White *et al.*, 2007). Body mass–abundance relationships with a slope of -1 have been used as evidence of a biomass equivalence rule (BER) across species populations. Since $\text{BA} = N \times M$, and if $N \propto M^{-1}$, then $\text{BA} = M^{-1} \times M^1 = M^0$. This relationship implies that the total biomass of every species is roughly independent of body mass (Rinaldo *et al.*, 2002).

Our experimental communities exhibited markedly different relationships to those described above for global communities (see Figure 13). The slope of the body mass–NA relationship differed significantly from a slope of -0.75 and -1 for all treatments, pre- and post-manipulation. In fact, the slope of the body mass–NA relationship in these experimental communities much more closely resembled a slope of -0.25, and this was consistent across all treatments. It should be noted that the experimental communities described in this study are likely to exhibit source–sink dynamics (Holt, 1985; Pulliam, 1988) with the mesocosms being sinks fed by immigration of species from the larger Lough Hyne source pool. Local communities have been demonstrated to have body mass–abundance relationships with a slope that does not differ significantly from -0.25 (Blackburn and Gaston, 1997), similar to those in our study. This will have consequences for the EER outlined above, that is $E \propto M^{0.75} \times M^{-0.25} = M^{0.5}$. If energy use scales with body mass as $M^{0.5}$, this means that large-bodied species will consume less energy per unit biomass than small-bodied species in the experimental communities. This is perhaps confirmed by the prevalence of strong per-unit biomass trophic interactions among small-bodied species in our mesocosms (see O'Gorman *et al.*, 2010).

The body mass–abundance slope of -0.25 will also have consequences for the BER, that is $B = M^{-0.25} \times M^{1} = M^{0.75}$. If biomass scales with body mass as $M^{0.75}$, large-bodied species should have a greater population biomass than small-bodied species. Combining this pattern with the observed positive relationship between body size and TH (left rear walls in Figure 15) in turn produces the inverted trophic pyramid of BA shown in Figure 14iii, where high trophic levels tend to have higher BA than low trophic levels in the experimental communities, irrespective of the manipulations. Confirmation of the predicted scaling of BA with body size in the experimental communities can be found in the slope of the body mass–BA relationship. Here, all treatments exhibited a slope that was significantly different from 0 (predicted for the BER), but not from 0.75 (in line with the scaling predicted above).

Studies that make predictions about the strength of trophic interactions between species in a community often rely on assumptions such as the BER to estimate interaction strength from predator–prey allometry (Emmerson *et al.*, 2005; Montoya *et al.*, 2005, 2009; O'Gorman *et al.*, 2010). We have shown that this assumption does not hold for our study, and it is interesting that estimates of interaction strength in the experimental communities, predicted using the assumption of biomass equivalence across populations of species, match empirical estimates of interaction strength measured in the cages themselves (O'Gorman *et al.*, 2010). This suggests that other factors, such as the metabolic requirements and feeding preferences of the predators in these communities, are deviating from the expectations of theory to balance the inverted pyramid of population biomasses observed in the cages. It would be instructive to re-calculate interaction strength in these communities, substituting the values for body size and biomass scaling predicted by theory with those demonstrated empirically in these experiments. Such an approach could offer an insight into the additional factors that might be contributing to energy equivalence across different populations in these mesocosm communities.

C. Trivariate Patterns

There is a clear relationship between body size, abundance and TH in these experimental food webs, as revealed by the trivariate plots in Figures 15 and 16. Large-bodied species occur at high trophic levels and at low NA and high BA, while small-bodied species occur at low trophic levels and at high NA and low BA. The slope and intercept of these relationships may shift in magnitude in response to species loss in the food webs, but the relationships remained qualitatively intact against even the strongest perturbations (see Table 5). Similar trivariate relationships between body mass, abundance and TH have been described for Tuesday Lake (Jonsson *et al.*, 2005), Broadstone Stream (Woodward *et al.*, 2005), the terrestrial food web of the Gearagh in Ireland (McLaughlin *et al.*, 2010) and also in a recent study of 20 stream food webs across a broad pH gradient (Layer *et al.*, 2010). The relationships in Tuesday Lake were also robust to the removal of three highly connected fish predators. These trivariate patterns give rise to the robust body mass–abundance relationships and size-structured nature of the experimental food webs described in this study, providing an overarching structure to the communities present. While the connectance, linkage density, proportions of top, intermediate and basal species and the distribution of links in our experimental food webs often fluctuated in response to perturbations, the intricate relationship between body size, abundance and TH appeared to provide a stabilizing configuration, which might allow certain key food web properties to persist in spite of top level extinctions and high levels of species turnover.

V. CONCLUSION

Food web studies are often limited by their portrayal of species interactions at one moment in time (or as summary webs compiled over an extended period) and the reliance on correlational data (Ings *et al.*, 2009). We took an alternative approach by replicating food web structure in 24 mesocosm communities and studying the changes in structure as a result of targeted species extinctions, at a number of different points in time. We have shown that species contribute to the stability of a food web, whether they are strong or weak interactors. The removal of any of the highly connected species had profound effects on many of the univariate patterns in our experimental food webs. That these food webs persist, even after the loss of an entire trophic level and up to six of the most highly connected species in the system, seemed to depend on the preservation of key higher level (bivariate and trivariate) patterns. Pyramids of body size and abundance provided a distinctive size structuring to the communities, with the vast majority of feeding interactions taking place between large, rare species eating small, abundant species. The total body size, NA and BA at each discrete trophic level remained relatively constant, even in the face of a community-level trophic cascade (O'Gorman and Emmerson, 2009), supporting theories of niche differentiation (Hutchinson, 1957; Schoener, 1974). The relationship between body mass and abundance was extremely robust to the removal perturbations, never varying from the expected slope of local communities. Finally, the intricate relationship between body mass, abundance and TH appears to act as a consistent and recurrent property of these communities, maintaining the flow of energy and the balance of species interactions within the food web. Thus, the trivariate patterns approach to studying food web ecology (Cohen *et al.*, 2003; Jonsson *et al.*, 2005; Layer *et al.*, 2010; McLaughlin *et al.*, 2010; Woodward *et al.*, 2005) might unveil hitherto unknown stabilizing mechanisms that may be missed by considering only univariate patterns.

ACKNOWLEDGEMENTS

We thank Marion Twomey, Ute Jacob, Laura Lyons, Órla McLaughlin, Roz Anderson, Claire Passarelli, Jesus Fernandez, Maike Pohlmann, Mick Mackey and Luke Harman for field assistance. Thanks to Tomas Jonsson for providing Matlab scripts used to draw some of the figures. The work was funded by the Irish Research Council for Science Engineering and Technology.

APPENDIX I

Table S1 Length–weight (L–W) relationships used to estimate the body mass of all species in the study

Code	Species	L–W relationship	r^2
LW1	*Carcinus maenas*[a]	$y=0.2668x^{2.9545}$	0.9693
LW2	*Ctenolabrus rupestris*[b]	$y=0.0057x^{3.181}$	0.9734
LW3	*Gaidropsarus mediterraneus*[b]	$y=0.0008x^{3.3972}$	0.9847
LW4	*Gobius niger*[b]	$y=0.0074x^{3.0788}$	0.932
LW5	*Gobius paganellus*[b]	$y=0.0014x^{3.4672}$	0.9356
LW6	*Marthasterias glacialis*[c]	$y=0.3088x^{2.7417}$	0.9187
LW7	*Necora puber*[a]	$y=0.2989x^{2.9639}$	0.9204
LW8	*Palaemon serratus*[d]	$y=0.0014x^{3.3838}$	0.9201
LW9	*Paracentrotus lividus*[e]	$y=1.2774x^{2.737}$	0.9398
LW10	*Taurulus bubalis*[b]	$y=0.0032x^{3.3258}$	0.9604
LW11	*Gobiusculus flavescens*[b]	$y=0.0004x^{3.7234}$	0.9612
LW12	*Pomatoschistus pictus*[b]	$y=0.0039x^{3.1954}$	0.9733
LW13	*Alvania* spp.[f]	$y=0.1391x^{2.71}$	0.9877
LW14	*Anomia ephippium*[g]	$y=0.0304x^{2.9244}$	0.9428
LW15	*Aora gracilis*[h]	$y=0.0018x^{3.2994}$	0.9202
LW16	Aoridae[h]	$y=0.0031x^{2.8427}$	0.9596
LW17	*Ascidiella aspersa*[i]	$y=0.1159x^{2.3628}$	0.8922
LW18	*Bittium reticulatum*[f]	$y=0.1224x^{2.3117}$	0.9831
LW19	*Buccinum undatum*[f]	$y=0.0958x^{3.0601}$	0.9804
LW20	Cardiidae[g]	$y=0.1084x^{3.0951}$	0.987
LW21	*Chlamys varia*[g]	$y=0.0508x^{3.036}$	0.9893
LW22	*Clathrina coriacea*[i]	$y=0.2909x^{1.9999}$	0.9541
LW23	*Crassicorophium* spp.[h]	$y=0.0046x^{3.1972}$	0.9491
LW24	*Crisia* spp.[i]	$y=0.00004x^{2.6928}$	0.9691
LW25	Cumacea[d]	$y=0.0101x^{1.9552}$	0.8806
LW26	*Dysidea fragilis*[i]	$y=0.1435x^{1.9328}$	0.8442
LW27	*Epilepton clarkiae*[g]	$y=0.0959x^{2.8774}$	0.9805
LW28	Foraminifera[i]	$y=0.1598x^{3.2349}$	0.9801
LW29	*Galathea squamifera*[h]	$y=0.0284x^{4.3903}$	0.9353
LW30	*Hiatella arctica*[d]	$y=0.053x^{2.9161}$	0.954
LW31	*Janua pagenstecheri*[i]	$y=0.1117x^{3.0229}$	0.9314
LW32	*Lembos websteri*[h]	$y=0.0037x^{2.6724}$	0.9806
LW33	*Lysianassa ceratina*[h]	$y=0.0096x^{3.0979}$	0.9877
LW34	Melitidae[h]	$y=0.004x^{3.095}$	0.9598
LW35	*Microdeutopus anomalus*[h]	$y=0.0016x^{3.3615}$	0.9685
LW36	*Musculus discors*[g]	$y=0.0986x^{2.7968}$	0.9766
LW37	Mysidae[d]	$y=0.0006x^{3.2529}$	0.9236
LW38	Nudibranchia[i]	$y=0.0096x^{2.8116}$	0.9726
LW39	*Ophiothrix fragilis*[j]	$y=0.4875x^{2.9185}$	0.9435
LW40	*Ophiura ophiura*[j]	$y=0.2936x^{2.5329}$	0.9603
LW41	Ostracoda[i]	$y=0.1738x^{4.2678}$	0.7896
LW42	*Parvicardium exiguum*[g]	$y=0.1104x^{3.0932}$	0.9786
LW43	*Parvicardium ovale*[g]	$y=0.1018x^{3.1784}$	0.99

(*continued*)

Table S1 (*continued*)

Code	Species	L–W relationship	r^2
LW44	*Parvicardium scabrum*[g]	$y=0.1103x^{3.0607}$	0.987
LW45	Pectinidae[g]	$y=0.0698x^{2.8284}$	0.9871
LW46	*Perinereis cultrifera*[i]	$y=0.0015x^{3.0023}$	0.9733
LW47	*Pilumnus hirtellus*[a]	$y=0.1324x^{2.963}$	0.9438
LW48	*Platynereis dumerilii*[i]	$y=0.0113x^{2.2781}$	0.8051
LW49	Polychaeta[i]	$y=0.0021x^{2.395}$	0.8612
LW50	*Pomatoceros* spp.[k]	$y=0.0029x^{2.781}$	0.9688
LW51	*Rissoa* spp.[f]	$y=0.1532x^{2.3992}$	0.9691

Weight (y) is measured in milligrams. Length (x) is measured in millimetres.
[a]Widest part of the carapace.
[b]Tip of the mouth to base of the tail (straight line along the midline).
[c]Centre of the oral surface to tip of the longest arm.
[d]Tip of the rostrum to base of the telson (straight line along the top surface of the animal).
[e]Diameter of test.
[f]Straight line from the apex to the base.
[g]Straight line from the umbo to the ventral margin.
[h]Tip of the head to base of the abdomen (straight line along the top surface of the animal).
[i]Straight line along the longest dimension.
[j]Diameter of oral disc (all legs frequently damaged).
[k]Line following ridge from tip of the operculum to base of the tube.

Table S2 List of species corresponding to the codes shown in Figure 3

Code	Species	L–W	TH	Gen	Vul	TL
1	*Carcinus maenas*	LW1	5.40	33	4	37
2	*Ctenolabrus rupestris*	LW2	5.54	66	1	67
3	*Gaidropsarus mediterraneus*	LW3	6.22	46	0	46
4	*Gobius niger*	LW4	5.10	77	1	78
5	*Gobius paganellus*	LW5	5.52	86	1	87
6	*Marthasterias glacialis*	LW6	6.30	25	0	25
7	*Necora puber*	LW7	5.89	24	1	25
8	*Palaemon serratus*	LW8	4.71	44	7	51
9	*Paracentrotus lividus*	LW9	2	2	3	5
10	*Taurulus bubalis*	LW10	5.21	51	1	52
11	*Gobiusculus flavescens*	LW11	3.72	36	7	43
12	*Pomatoschistus pictus*	LW12	3.98	54	7	61
13	*Abra alba*	LW27	2	4	5	9
14	*Acanthocardia echinata*	LW20	2	4	4	8
15	*Acanthocardia tuberculata*	LW20	2	4	3	7
16	*Acanthochitona crinitus*	LW25	2	4	1	5
17	*Aequipecten opercularis*	LW45	2.68	10	5	15
18	*Alvania beani*	LW13	2	2	6	8
19	*Alvania punctura*	LW13	2	2	2	4
20	*Alvania semistriata*	LW13	2	2	3	5
21	*Ammonicera rota*	LW51	2	2	1	3

Table S2 (*continued*)

Code	Species	L–W	TH	Gen	Vul	TL
22	*Amphilochus manudens*	LW33	2	3	1	4
23	*Anomia ephippium*	LW14	2	6	9	15
24	*Aora gracilis*	LW15	3.07	8	5	13
25	*Apherusa bispinosa*	LW34	2	5	1	6
26	*Apseudes latreillei*	LW25	2	5	3	8
27	*Apseudes talpa*	LW25	2	5	1	6
28	*Ascidiella aspersa*	LW17	2	4	4	8
29	*Asterina phylactica*	LW6	3.84	11	3	14
30	*Bittium reticulatum*	LW18	2.85	14	9	23
31	*Boreotrophon truncatus*	LW19	2	4	1	5
32	*Buccinum undatum*	LW19	3.82	13	1	14
33	Calanoida	LW25	2	6	16	22
34	*Callopora lineata*	LW22	2	4	1	5
35	*Caprella acanthifera*	LW16	2	4	4	8
36	*Caprella equilibra*	LW16	3.30	8	3	11
37	*Caprella linearis*	LW16	3.47	4	4	8
38	*Ceradocus semiserratus*	LW34	2	2	1	3
39	*Cerastoderma edule*	LW20	3.14	10	2	12
40	*Cerithiopsis tubercularis*	LW18	2.67	6	8	14
41	Chironomidae spp.	LW49	2	3	5	8
42	*Chlamys varia*	LW21	2	5	5	10
43	*Circulus striatus*	LW28	2	3	3	6
44	*Clathrina coriacea*	LW22	2	4	6	10
45	*Cliona celata*	LW22	2	4	10	14
46	*Coriandria fulgida*	LW51	2	2	9	11
47	*Crassicorophium bonnellii*	LW23	2	7	6	13
48	*Crassicorophium crassicorne*	LW23	2	7	8	15
49	*Crisia denticulata*	LW24	2	3	8	11
50	*Crisia eburnea*	LW24	2	3	6	9
51	*Cumacea A*	LW25	2	4	1	5
52	*Cuthona A*	LW38	3.67	2	0	2
53	Cyclopoida	LW25	2.50	6	16	22
54	*Cythere lutea*	LW41	2	4	2	6
55	*Dexamine spinosa*	LW34	2	2	5	7
56	*Dexamine thea*	LW34	2	2	2	4
57	*Disporella hispida*	LW22	2	3	1	4
58	*Dysidea fragilis*	LW26	2	4	10	14
59	*Elasmopus rapax*	LW34	2	2	4	6
60	*Electra pilosa*	LW22	2	3	3	6
61	*Elysia viridis*	LW38	2	2	1	3
62	*Emarginula fissura*	LW28	2.75	7	1	8
63	*Epilepton clarkiae*	LW27	2	4	7	11
64	*Epitonium clathrus*	LW19	3.67	2	0	2
65	*Ericthonius brasiliensis*	LW23	2.36	8	7	15
66	*Ericthonius punctatus*	LW23	2.36	8	7	15
67	*Eubranchus farrani*	LW38	3.67	2	0	2
68	*Exogone gemmifera*	LW49	2	4	2	6
69	*Foraminifera A*	LW28	2	3	8	11

(*continued*)

Table S2 (*continued*)

Code	Species	L–W	TH	Gen	Vul	TL
70	*Foraminifera B*	LW28	2	3	8	11
71	*Foraminifera C*	LW28	2	3	4	7
72	*Foraminifera D*	LW28	2	3	3	6
73	*Foraminifera E*	LW28	2	3	4	7
74	*Foraminifera F*	LW28	2	3	7	10
75	*Foraminifera G*	LW28	2	3	6	9
76	*Galathea squamifera*	LW29	3.62	25	2	27
77	*Gammaropsis maculata*	LW16	2	3	2	5
78	*Gammarus locusta*	LW34	2	4	4	8
79	*Gammarus zaddachi*	LW34	2	4	2	6
80	*Gastropod A*	LW51	2	3	0	3
81	*Gibbula umbilicalis*	LW19	2.75	7	4	11
82	*Halacarellus basteri*	LW25	2	3	3	6
83	Harpacticoida	LW25	2	7	20	27
84	*Hiatella arctica*	LW30	2	4	9	13
85	Hydrozoa	LW24	2.79	5	12	17
86	*Idotea A*	LW25	2.80	7	3	10
87	*Idotea B*	LW25	2.80	7	4	11
88	*Iothia fulva*	LW28	2	2	0	2
89	*Janua pagenstecheri*	LW31	2	3	6	9
90	*Lasaea rubra*	LW27	2	4	4	8
91	*Lembos websteri*	LW32	2.61	10	7	17
92	*Leptocheirus tricristatus*	LW16	2	5	2	7
93	*Leptochelia savignyi*	LW25	2	4	3	7
94	*Leptocythere pellucida*	LW41	2	4	1	5
95	*Leptomysis lingvura*	LW34	2	5	2	7
96	*Loxoconcha rhomboidea*	LW41	2	5	7	12
97	*Lysianassa ceratina*	LW33	2	3	2	5
98	*Macoma balthica*	LW20	2	5	0	5
99	*Melita palmata*	LW34	2	4	5	9
100	*Microdeutopus anomalus*	LW35	2	5	6	11
101	*Microprotopus maculatus*	LW16	2.71	3	3	6
102	*Modiolula phaseolina*	LW36	3.00	8	14	22
103	*Monia patelliformis*	LW14	2	6	7	13
104	*Munna kroyeri*	LW25	2	4	4	8
105	*Musculus discors*	LW36	3.00	8	14	22
106	*Mytilus edulis*	LW36	3.00	8	14	22
107	*Nannastacus unguiculatus*	LW25	2.75	7	1	8
108	Nematoda spp.	LW49	2	4	11	15
109	*Nereis A*	LW49	2	3	8	11
110	*Odostomia plicata*	LW51	3.00	2	0	2
111	*Omalogyra atomus*	LW28	2	2	1	3
112	*Onoba semicosta*	LW51	2.67	9	1	10
113	*Ophiothrix fragilis*	LW39	2	7	7	14
114	*Ophiura ophiura*	LW40	3.91	22	6	28
115	*Ostracod A*	LW41	2	4	4	8
116	*Ostracod B*	LW41	2	4	1	5
117	*Ostracod C*	LW41	2	4	0	4

Table S2 (*continued*)

Code	Species	L–W	TH	Gen	Vul	TL
118	*Paradoxostoma variabile*	LW41	2	4	6	10
119	*Parvicardium exiguum*	LW42	2	4	10	14
120	*Parvicardium ovale*	LW43	2	4	11	15
121	*Parvicardium scabrum*	LW44	2	4	11	15
122	*Perinereis cultrifera*	LW46	2	3	7	10
123	*Phtisica marina*	LW16	3.04	7	2	9
124	*Phyllodocid A*	LW49	3.13	6	4	10
125	*Pilumnus hirtellus*	LW47	4.26	23	7	30
126	*Platynereis dumerili*	LW48	2.57	4	8	12
127	*Pomatoceros lamarcki*	LW50	2	4	10	14
128	*Pomatoceros triqueter*	LW50	2	4	10	14
129	*Pontocypris mytiloides*	LW41	2	4	3	7
130	*Pseudoparatanais batei*	LW25	2	4	3	7
131	*Retusa truncatula*	LW51	3.00	7	0	7
132	*Rissoa parva*	LW51	2.62	7	10	17
133	*Rissoa sarsi*	LW51	2	5	10	15
134	*Rissoella diaphana*	LW51	2	5	8	13
135	*Rissoella opalina*	LW51	2	5	7	12
136	*Sabella pavonina*	LW49	2	4	5	9
137	*Sagitta elegans*	LW49	3.08	7	6	13
138	*Scrupocellaria* spp.	LW24	2	3	8	11
139	*Semibalanus balanoides*	LW28	3.04	7	2	9
140	*Semicytherura nigrescens*	LW41	2	4	2	6
141	*Serpulid A*	LW50	2	4	8	12
142	*Siriella armata*	LW37	3.00	7	3	10
143	*Skenea serpuloides*	LW28	2	2	1	3
144	*Spirorbis A*	LW31	2	3	3	6
145	*Spirorbis B*	LW31	2	3	2	5
146	*Stenothoe marina*	LW16	3.39	5	2	7
147	*Syllidae A*	LW49	3.67	2	2	4
148	*Syllidae B*	LW49	3.67	2	2	4
149	*Tapes aureus*	LW27	2	3	3	6
150	*Tectura virginia*	LW28	2.67	9	0	9
151	*Tomopteris helgolandica*	LW49	3.87	3	0	3
152	*Tritaeta gibbosa*	LW34	2	2	1	3
153	*Tryphosella sarsi*	LW33	2	2	0	2
154	*Tubulipora liliacea*	LW49	2	3	1	4
155	*Turbellaria A*	LW49	2	2	3	5
156	*Typosyllis prolifera*	LW49	3.67	2	2	4
157	*Vitreolina philippi*	LW51	4.84	2	1	3
158	*Xestoleberis aurantia*	LW41	2	4	7	11
159	Algae	NA	1	0	67	67
160	Bacteria	NA	1	0	77	77
161	Cladocerans	NA	1	0	10	10
162	CPOM	NA	1	0	71	71
163	Diatoms	NA	1	0	81	81
164	FPOM	NA	1	0	122	122
165	Microphytobenthos	NA	1	0	64	64
166	Phytoplankton	NA	1	0	49	49

(*continued*)

Table S2 (*continued*)

Code	Species	L–W	TH	Gen	Vul	TL
167	Cyprid larvae	LW41	1	0	9	9
168	Hymenopteran larvae	LW25	1	0	1	1

A code for the length–weight (L–W) relationship used to estimate the body mass of each species is provided. These codes correspond to the key shown in Table 1. The trophic height (TH), generality (Gen), vulnerability (Vul) and number of trophic links (TL) of each species are provided.

APPENDIX II. GUT CONTENT ANALYSIS AND LOUGH HYNE-BASED LITERATURE USED TO CONSTRUCT THE DIET OF THE MANIPULATED SPECIES AND TWO SMALL GOBIES IN THE STUDY

Table S3 Prey links for *Gobius niger* at Lough Hyne

Prey	*N*	Prey	*N*
Abra alba	2	*Hiatella arctica*	9
Acanthocardia echinata	1	*Janua pagenstecheri*	6
Aequipecten opercularis	1	*Lasaea rubra*	1
Alvania beani	1	*Lembos websteri*	4
Anomia ephippium	18	*Leptocheirus tricristatus*	1
Aora gracilis	1	*Leptochelia savignyi*	2
Ascidiella aspersa	13	*Leptocythere pellucida*	2
Bittium reticulatum	19	*Loxoconcha rhomboidea*	1
Calanoida	1	*Microdeutopus anomalus*	2
Caprella acanthifera	2	*Modiolula phaseolina*	11
Caprella equilibra	1	*Monia patelliformis*	2
Caprella linearis	2	*Musculus discors*	12
Cerithiopsis tubercularis	1	*Mytilus edulis*	2
Chironomidae spp.	9	*Nannastacus unguiculatus*	2
Chlamys varia	6	Nematoda spp.	13
Circulus striatus	2	*Nereis A*	1
Clathrina coriacea	2	*Ostracod A*	2
Cliona celata	1	*Palaemon serratus*	10
Coriandria fulgida	8	*Paradoxostoma variabile*	1
Crassicorophium bonnellii	1	*Parvicardium exiguum*	1
Crassicorophium crassicorne	9	*Parvicardium ovale*	1
Crisia denticulata	2	*Parvicardium scabrum*	3
Crisia eburnea	1	*Pilumnus hirtellus*	1
Cumacea A	1	*Pomatoceros lamarcki*	1
Cyclopoida	8	*Pomatoceros triqueter*	8
Cythere lutea	1	*Pomatoschistus pictus*	1
Dexamine spinosa	2	*Pontocypris mytiloides*	2

Table S3 (*continued*)

Prey	N	Prey	N
Dexamine thea	1	*Rissoa parva*	9
Dysidea fragilis	2	*Rissoa sarsi*	11
Elasmopus rapax	2	*Rissoella diaphana*	1
Epilepton clarkiae	4	*Rissoella opalina*	3
Ericthonius brasiliensis	1	*Sagitta elegans*	2
Ericthonius punctatus	1	*Scrupocellaria* spp.	15
Foraminifera A	6	*Serpulid A*	1
Foraminifera B	11	*Spirorbis A*	1
Foraminifera F	8	*Tapes aureus*	1
Foraminifera G	2	*Tritaeta gibbosa*	1
Gobiusculus flavescens	4	*Xestoleberis aurantia*	2
Harpacticoida	9		

N = number of *G. niger* stomachs where a given prey item was found. A total of 68 individual *G. niger* stomachs were examined.

Table S4 Prey links for *Gobius paganellus* at Lough Hyne

Prey	N	Prey	N
Abra alba	1	*Janua pagenstecheri*	6
Acanthocardia echinata	1	*Lasaea rubra*	1
Acanthocardia tuberculata	1	*Lembos websteri*	6
Aequipecten opercularis	4	*Leptochelia savignyi*	1
Alvania beani	1	*Loxoconcha rhomboidea*	3
Alvania semistriata	1	*Melita palmata*	2
Ammonicera rota	1	*Microdeutopus anomalus*	2
Anomia ephippium	16	*Microprotopus maculatus*	2
Aora gracilis	4	*Modiolula phaseolina*	2
Apseudes latreillei	1	*Monia patelliformis*	1
Ascidiella aspersa	4	*Munna kroyeri*	1
Bittium reticulatum	11	*Musculus discors*	10
Caprella acanthifera	2	*Mytilus edulis*	1
Caprella equilibra	1	Nematoda spp.	4
Caprella linearis	2	*Nereis A*	5
Carcinus maenas	7	*Omalogyra atomus*	2
Cerithiopsis tubercularis	1	*Ophiothrix fragilis*	4
Chironomidae spp.	6	*Ostracod A*	4
Chlamys varia	9	*Ostracod B*	1
Circulus striatus	1	*Palaemon serratus*	16
Coriandria fulgida	1	*Paradoxostoma variabile*	1
Crassicorophium bonnellii	2	*Parvicardium exiguum*	1
Crassicorophium crassicorne	19	*Parvicardium ovale*	2
Crisia denticulata	1	*Parvicardium scabrum*	2
Cyclopoida	4	*Perinereis cultrifera*	1
Cyprid larvae	2	*Phtisica marina*	1
Cythere lutea	1	*Pilumnus hirtellus*	4
Dexamine spinosa	2	*Platynereis dumerilii*	4
Elasmopus rapax	2	*Pomatoceros lamarcki*	2
Epilepton clarkiae	2	*Pomatoceros triqueter*	4

(*continued*)

Table S4 (*continued*)

Prey	N	Prey	N
Ericthonius brasiliensis	8	*Pomatoschistus pictus*	2
Ericthonius punctatus	3	*Pseudoparatanais batei*	2
Foraminifera A	4	*Rissoa parva*	13
Foraminifera B	2	*Rissoa sarsi*	11
Foraminifera G	3	*Rissoella diaphana*	2
Galathea squamifera	2	*Rissoella opalina*	6
Gammaropsis maculata	1	*Sagitta elegans*	2
Gammarus locusta	2	*Scrupocellaria* spp.	7
Gobiusculus flavescens	1	*Semicytherura nigrescens*	1
Harpacticoida	13	*Serpulid A*	1
Hiatella arctica	8	*Skenea serpuloides*	1
Idotea A	4	*Spirorbis B*	1
Idotea B	1	*Xestoleberis aurantia*	1

N=number of *G. paganellus* stomachs where a given prey item was found. A total of 62 individual *G. paganellus* stomachs were examined.

Table S5 Prey links for *Taurulus bubalis* at Lough Hyne

Prey	N	Prey	N
Anomia ephippium	2	*Leptomysis lingvura*	1
Aora gracilis	1	*Microdeutopus anomalus*	2
Ascidiella aspersa	4	*Microprotopus maculatus*	1
Boreotrophon truncatus	1	*Monia patelliformis*	1
Buccinum undatum	2	Nematoda spp.	12
Callopora lineata	1	*Nereis A*	2
Coriandria fulgida	1	*Onoba semicosta*	1
Crassicorophium bonnellii	3	*Ophiothrix fragilis*	1
Crassicorophium crassicorne	11	*Ophiura ophiura*	1
Crisia denticulata	1	*Palaemon serratus*	10
Crisia eburnea	1	*Perinereis cultrifera*	1
Cyclopoida	2	*Platynereis dumerilii*	8
Cyprid larvae	6	*Pomatoschistus pictus*	2
Dexamine spinosa	2	*Rissoa parva*	1
Disporella hispida	1	*Rissoa sarsi*	4
Electra pilosa	1	*Sabella pavonina*	1
Ericthonius brasiliensis	8	*Sagitta elegans*	2
Ericthonius punctatus	4	*Scrupocellaria spp.*	2
Gammarus locusta	6	*Siriella armata*	6
Gammarus zaddachi	1	*Syllidae A*	2
Gobiusculus flavescens	1	*Syllidae B*	1
Halacarellus basteri	1	*Tubulipora liliacea*	4
Harpacticoida	2	*Turbellaria A*	2
Hymenoptera larvae	2	*Typosyllis prolifera*	1
Janua pagenstecheri	1	*Vitreolina philippi*	1
Lembos websteri	2		

N=number of *T. bubalis* stomachs where a given prey item was found. A total of 36 individual *T. bubalis* stomachs were examined.

Table S6 Prey links for *Gaidropsarus mediterraneus* at Lough Hyne

Prey	*N*	Prey	*N*
Alvania beani	2	*Gobius paganellus*	–[a]
Alvania punctura	1	*Gobiusculus flavescens*	3
Amphilochus manudens	1	Harpacticoida	7
Apherusa bispinosa	1	*Idotea A*	3
Bittium reticulatum	5	*Idotea B*	1
Caprella acanthifera	3	*Janua pagenstecheri*	2
Caprella linearis	1	*Lembos websteri*	1
Cerithiopsis tubercularis	2	*Leptomysis lingvura*	1
Crassicorophium bonnellii	1	*Lysianassa ceratina*	1
Crassicorophium crassicorne	3	*Melita palmata*	1
Ctenolabrus rupestris	–[a]	*Microdeutopus anomalus*	1
Cyclopoida	1	*Microprotopus maculatus*	1
Cyprid larvae	3	*Munna kroyeri*	1
Dexamine spinosa	1	Nematoda spp.	6
Dexamine thea	1	*Nereis A*	3
Elasmopus rapax	1	*Palaemon serratus*	12
Elysia viridis	2	*Perinereis cultrifera*	1
Epilepton clarkiae	4	*Pilumnus hirtellus*	2
Ericthonius brasiliensis	5	*Platynereis dumerilii*	8
Ericthonius punctatus	2	*Pomatoschistus pictus*	4
Gammaropsis maculata	1	*Siriella armata*	5
Gammarus locusta	2	*Taurulus bubalis*	–[a]
Gobius niger	–[a]	*Turbellaria A*	3

N=number of *G. mediterraneus* stomachs where a given prey item was found. A total of 30 individual *G. mediterraneus* stomachs were examined.
[a]Stomach content links were supplemented with personal observations of feeding links during sampling of the mesocosm experiments at Lough Hyne.

Table S7 Prey links for *Ctenolabrus rupestris* at Lough Hyne

Prey	*N*	Prey	*N*
Aequipecten opercularis	2	*Loxoconcha rhomboidea*	1
Alvania beani	1	*Melita palmata*	1
Alvania semistriata	1	*Microdeutopus anomalus*	2
Anomia ephippium	10	*Modiolula phaseolina*	3
Aora gracilis	1	*Monia patelliformis*	1
Asterina phylactica	1	*Musculus discors*	8
Bittium reticulatum	5	*Mytilus edulis*	1
Calanoida	1	*Nematoda* spp.	3
Carcinus maenas	1	*Nereis A*	1
Ceradocus semiserratus	1	*Ophiothrix fragilis*	1
Cerithiopsis tubercularis	2	*Ophiura ophiura*	1
Chlamys varia	3	*Palaemon serratus*	4
Coriandria fulgida	1	*Paradoxostoma variabile*	1

(*continued*)

Table S7 (*continued*)

Prey	*N*	Prey	*N*
Crassicorophium bonnellii	1	*Parvicardium exiguum*	1
Crassicorophium crassicorne	6	*Parvicardium ovale*	2
Crisia denticulata	2	*Parvicardium scabrum*	1
Crisia eburnea	1	*Perinereis cultrifera*	1
Cyclopoida	3	*Phyllodocid A*	2
Cyprid larvae	2	*Pilumnus hirtellus*	1
Elasmopus rapax	2	*Platynereis dumerilii*	5
Ericthonius brasiliensis	1	*Podocoryne borealis*	1
Ericthonius punctatus	1	*Pomatoceros lamarcki*	1
Foraminifera F	1	*Pomatoceros triqueter*	4
Foraminifera G	1	*Pomatoschistus pictus*	2
Galathea squamifera	1	*Rissoa parva*	2
Gammarus locusta	2	*Rissoa sarsi*	6
Gammarus zaddachi	1	*Rissoella diaphana*	1
Gobiusculus flavescens	1	*Rissoella opalina*	3
Harpacticoida	8	*Sabella pavonina*	1
Hiatella arctica	9	Scrupocellaria spp.	3
Janua pagenstecheri	2	*Serpulid A*	1
Lembos websteri	3	*Spirorbis A*	1
Leptocheirus tricristatus	1	*Xestoleberis aurantia*	2

N = number of *C. rupestris* stomachs where a given prey item was found. A total of 48 individual *C. rupestris* stomachs were examined.

Table S8 Prey links for *Gobiusculus flavescens* at Lough Hyne

Prey	*N*	Prey	*N*
Aora gracilis	2	*Modiolula phaseolina*	1
Calanoida	6	*Musculus discors*	1
Caprella acanthifera	1	*Mytilus edulis*	1
Caprella equilibra	2	*Palaemon serratus*	1
Caprella linearis	1	*Paradoxostoma variabile*	2
Chironomidae spp.	3	*Parvicardium exiguum*	1
Coriandria fulgida	1	*Parvicardium ovale*	2
Crassicorophium bonnellii	2	*Parvicardium scabrum*	2
Crassicorophium crassicorne	11	*Phtisica marina*	1
Cyclopoida	7	*Pomatoceros lamarcki*	1
Cyprid larvae	34	*Pomatoceros triqueter*	3
Epilepton clarkiae	1	*Rissoa parva*	3
Ericthonius brasiliensis	4	*Rissoa sarsi*	1
Ericthonius punctatus	2	*Rissoella diaphana*	1
Harpacticoida	25	*Rissoella opalina*	2
Lasaea rubra	1	*Serpulid A*	1
Lembos websteri	4	*Stenothoe marina*	2
Loxoconcha rhomboidea	1	*Xestoleberis aurantia*	1
Microdeutopus anomalus	1		

N = number of *G. flavescens* stomachs where a given prey item was found. A total of 38 individual *G. flavescens* stomachs were examined.

Table S9 Prey links for *Pomatoschistus pictus* at Lough Hyne

Prey	*N*	Prey	*N*
Abra alba	1	*Musculus discors*	4
Acanthochitona crinitus	1	*Mytilus edulis*	1
Anomia ephippium	2	Nematoda spp.	2
Apseudes latreillei	1	*Ophiothrix fragilis*	1
Apseudes talpa	1	*Ophiura ophiura*	1
Bittium reticulatum	5	*Ostracod A*	1
Calanoida	2	*Palaemon serratus*	2
Chironomidae spp.	1	*Paradoxostoma variabile*	1
Coriandria fulgida	1	*Parvicardium exiguum*	1
Crassicorophium crassicorne	4	*Parvicardium ovale*	1
Crisia denticulata	2	*Parvicardium scabrum*	1
Crisia eburnea	1	*Phyllodocid A*	1
Cyclopoida	4	*Platynereis dumerilii*	1
Dexamine spinosa	2	*Pomatoceros lamarcki*	1
Epilepton clarkiae	2	*Pomatoceros triqueter*	1
Foraminifera A	1	*Pontocypris mytiloides*	3
Foraminifera B	4	*Pseudoparatanais batei*	1
Foraminifera E	1	*Rissoa parva*	1
Foraminifera F	2	*Rissoa sarsi*	2
Harpacticoida	12	*Rissoella diaphana*	1
Hiatella arctica	5	*Rissoella opalina*	1
Idotea A	1	Scrupocellaria spp.	1
Idotea B	1	*Semicytherura nigrescens*	1
Lasaea rubra	1	*Stenothoe marina*	1
Loxoconcha rhomboidea	2	*Tapes aureus*	1
Lysianassa ceratina	1	*Turbellaria A*	1
Modiolula phaseolina	1	*Xestoleberis aurantia*	3
Munna kroyeri	1		

N=number of *P. pictus* stomachs where a given prey item was found. A total of 39 individual *P. pictus* stomachs were examined.

Table S10 Prey links for *Palaemon serratus* at Lough Hyne

Prey	*N*	Prey	*N*
CPOM	–[a]	*Melita palmata*	2
FPOM	–[a]	*Modiolula phaseolina*	2
Alvania beani	2	*Munna kroyeri*	2
Anomia ephippium	7	*Musculus discors*	7
Apseudes latreillei	6	*Mytilus edulis*	1
Bittium reticulatum	35	Nematoda spp.	2
Cerithiopsis tubercularis	1	*Parvicardium ovale*	2
Chironomidae spp.	19	*Parvicardium scabrum*	1
Circulus striatus	2	*Perinereis cultrifera*	2
Coriandria fulgida	12	*Pilumnus hirtellus*	3
Crassicorophium crassicorne	3	*Platynereis dumerili*	33

(*continued*)

Table S10 (*continued*)

Prey	*N*	Prey	*N*
Crisia denticulata	2	*Pomatoschistus pictus*	1
Cyclopoida	7	*Pontocypris mytiloides*	3
Exogone gemmifera	2	*Pseudoparatanais batei*	4
Foraminifera A	2	*Rissoa parva*	6
Foraminifera B	1	*Rissoa sarsi*	4
Gobiusculus flavescens	2	*Rissoella diaphana*	1
Harpacticoida	18	*Rissoella opalina*	1
Hiatella arctica	4	*Sagitta elegans*	3
Idotea B	9	Scrupocellaria spp.	2
Leptochelia savignyi	3	*Siriella armata*	2
Loxoconcha rhomboidea	2	*Xestoleberis aurantia*	4

N = number of *P. serratus* stomachs where a given prey item was found. A total of 64 individual *P. serratus* stomachs were examined.
[a]Stomach content links were supplemented with personal observations of feeding links during sampling of the mesocosm experiments at Lough Hyne.

Table S11 Prey links for *Necora puber* at Lough Hyne

Prey	*N*	Literature
Algae	24	
CPOM		O'Gorman (unpublished)
Acanthocardia echinata	1	
Aequipecten opercularis	1	
Anomia ephippium	7	
Carcinus maenas	2	
Cerastoderma edule	1	
Chlamys varia		Muntz *et al.* (1965)
Foraminifera B	2	
Gibbula umbilicalis		Muntz *et al.* (1965), Kitching and Thain (1983)
Hiatella arctica	6	
Marthasterias glacialis		Muntz *et al.* (1965)
Modiolula phaseolina	2	
Monia patelliformis	1	
Musculus discors	8	
Mytilus edulis		Ebling *et al.* (1964), Kitching *et al.* (1959)
Palaemon serratus	2	
Paracentrotus lividus	1	
Parvicardium exiguum	2	
Parvicardium ovale	3	
Parvicardium scabrum	6	
Pilumnus hirtellus	1	
Pomatoceros lamarcki	2	
Pomatoceros triqueter	4	
Serpulid A	1	

N = number of *N. puber* stomachs where a given prey item was found. A total of 36 individual *N. puber* stomachs were examined. Stomach content links were supplemented with the literature links from studies carried out at Lough Hyne. The references for these literature links are provided towards the end of this appendix.

Table S12 Prey links for *Carcinus maenas* at Lough Hyne

Prey	*N*	Literature
Algae	16	
CPOM		O'Gorman (unpublished)
Acanthocardia tuberculata	1	
Anomia ephippium	6	
Asterina phylactica	1	
Coriandria fulgida	1	
Emarginula fissura	1	
Foraminifera A	2	
Foraminifera C	1	
Foraminifera F	1	
Gibbula umbilicalis		Muntz *et al.* (1965)
Hiatella arctica	2	
Loxoconcha rhomboidea	2	
Marthasterias glacialis		Verling *et al.* (2003)
Melita palmata	3	
Modiolula phaseolina	4	
Monia patelliformis	1	
Musculus discors	5	
Mytilus edulis		Ebling *et al.* (1964); Kitching *et al.* (1959)
Ostracod A	1	
Palaemon serratus	4	
Paracentrotus lividus		Muntz *et al.* (1965)
Paradoxostoma variabile	1	
Parvicardium exiguum	2	
Parvicardium ovale	4	
Parvicardium scabrum	4	
Pomatoceros lamarcki	1	
Pomatoceros triqueter	4	
Rissoa parva	3	
Rissoa sarsi	3	
Rissoella opalina	1	
Semibalanus balanoides	2	
Serpulid A	1	
Xestoleberis aurantia	2	

N = number of *C. maenas* stomachs where a given prey item was found. A total of 39 individual *C. maenas* stomachs were examined. Stomach content links were supplemented with the literature links from studies carried out at Lough Hyne. The references for these literature links are provided towards the end of this appendix.

Table S13 Prey links for *Marthasterias glacialis* at Lough Hyne

Prey	Literature
CPOM	O'Gorman (unpublished)
Aequipecten opercularis	O'Gorman (unpublished)
Anomia ephippium	Muntz *et al.* (1965), Kitching and Thain (1983), Verling *et al.* (2003), Ebling *et al.* (1966), Verling (unpublished)
Asterina phylactica	Verling (unpublished)
Bittium reticulatum	O'Gorman (unpublished)
Carcinus maenas	Verling *et al.* (2003), Verling (unpublished)
Cerithiopsis tubercularis	Verling (unpublished)
Chlamys varia	Muntz *et al.* (1965), Kitching and Thain (1983), Verling (unpublished)
Gibbula umbilicalis	Ebling *et al.* (1966), Verling (unpublished)
Gobiusculus flavescens	O'Gorman (unpublished)
Modiolula phaseolina	O'Gorman (unpublished)
Monia patelliformis	O'Gorman (unpublished)
Musculus discors	O'Gorman (unpublished)
Mytilus edulis	Kitching *et al.* (1959), Verling (unpublished)
Necora puber	Verling *et al.* (2003), Verling (unpublished)
Ophiothrix fragilis	Verling (unpublished)
Ophiura ophiura	O'Gorman (unpublished)
Paracentrotus lividus	Kitching and Thain (1983), Verling *et al.* (2003, 2005), Verling (unpublished)
Pilumnus hirtellus	O'Gorman (unpublished)
Pomatoceros lamarcki	Verling *et al.* (2003), Verling (unpublished)
Pomatoceros triqueter	Verling *et al.* (2003), Verling (unpublished)
Pomatoschistus pictus	O'Gorman (unpublished)
Semibalanus balanoides	Verling *et al.* (2003), Verling (unpublished)
Serpulid A	Verling (unpublished)
Tapes aureus	Verling *et al.* (2003), Verling (unpublished)

The diet of *M. glacialis* is entirely constructed from the literature links from studies carried out at Lough Hyne. The references for these literature links are provided towards the end of this appendix.

Table S14 Prey links for *Paracentrotus lividus* at Lough Hyne

Prey	Literature
Algae	Kitching and Thain (1983), Kitching and Ebling (1961)
Microphytobenthos	Kitching and Thain (1983), Kitching and Ebling (1961)

The diet of *P. lividus* is entirely constructed from the literature links from studies carried out at Lough Hyne. The references for these literature links are provided towards the end of this appendix.

APPENDIX III. SOURCES OF ALL LITERATURE-BASED LINKS FOR THE FOOD WEB SHOWN IN FIGURE 3

Table S15 Source of literature-based links for the food web shown in Figure 3

Taxa	Feeding mode	References
Abra alba	Suspension feeder; deposit feeder	Amouroux *et al.* (1988, 1991), Bachelet and Cornet (1981), Rosenberg (1993), Stromgren *et al.* (1993), Dekker (1989), Dominici (2001), Rueda *et al.* (2009), Koulouri *et al.* (2006)
Acanthocardia echinata	Suspension feeder	Koulouri *et al.* (2006); Kiorboe and Mohlenberg (1981), Tillin *et al.* (2006)
Acanthocardia tuberculata	Suspension feeder	Rueda *et al.* (2009); Koulouri *et al.* (2006), Tagmouti-Talha *et al.* (2000), Vale and Sampayo (2002)
Acanthochitona crinitus	Grazer; deposit feeder	Bode (1989)
Aequipecten opercularis	Suspension feeder	Dominici (2001), Kiorboe and Mohlenberg (1981), Tillin *et al.* (2006), Lehane and Davenport (2002), Thouzeau *et al.* (1996), Hunt (1925)
Alvania beani	Detritivore	Koutsoubas *et al.* (2000), Graham (1988)
Alvania punctura	Detritivore	Graham (1988), Fretter and Montgomery (1968)
Alvania semistriata	Detritivore	Graham (1988), Antoniadou *et al.* (2005)
Ammonicera rota	Grazer	Rueda *et al.* (2009), Graham (1988)
Amphilochus manudens	Commensal; deposit feeder	Duffy (1990), O'Gorman (2010a), Chapman (2007)
Anomia ephippium	Suspension feeder	Dominici (2001), Rueda *et al.* (2009), Koulouri *et al.* (2006), Bramanti *et al.* (2003)
Aora gracilis	Tube feeder	Dixon and Moore (1997), Taylor and Brown (2006)
Apherusa bispinosa	Grazer; deposit feeder	O'Gorman (2010a), Bradstreet and Cross (1982), Werner (1997), Scott *et al.* (1999), Poltermann (2001), Johnson and Attramadal (1982)
Apseudes latreillei	Suspension feeder; scavenger	O'Gorman (2010a), Johnson and Attramadal (1982), Kudinova-Pasternak (1991), Dennell (1937), Blazewicz-Paszkowycz and Ligowski (2002), Drumm (2005), Holdich and Jones (1983), Grall *et al.* (2006)
Apseudes talpa	Suspension feeder; scavenger	O'Gorman (2010a), Johnson and Attramadal (1982), Kudinova-Pasternak (1991), Dennell (1937), Blazewicz-Paszkowycz and Ligowski (2002), Drumm (2005), Holdich and Jones (1983)

(*continued*)

Table S15 (*continued*)

Taxa	Feeding mode	References
Ascidiella aspersa	Suspension feeder	Tillin *et al.* (2006), Hunt (1925), Hily (1991), Pascoe *et al.* (2007), Randlov and Riisgard (1979), Millar (1970)
Asterina phylactica	Omnivore; scavenger	Crump and Emson (1983), Emson and Crump (1984)
Bittium reticulatum	Grazer; deposit feeder	Dominici (2001), Rueda *et al.* (2009), Koulouri *et al.* (2006), Koutsoubas *et al.* (2000), Graham (1988), Borja (1986), Fernandez *et al.* (1988), Salgeback and Savazzi (2006), Morton (1967), Fretter (1951)
Boreotrophon truncatus	Deposit feeder; carnivore	Graham (1988)
Buccinum undatum	Scavenger; carnivore	Tillin *et al.* (2006), Graham (1988), O'Gorman (2010a), Grall *et al.* (2006), Scolding *et al.* (2007), Taylor (1978), Hancock (1960), Nielsen (1975); Himmelman and Hamel (1993), Ramsay *et al.* (1998)
Calanoida	Grazer; suspension feeder; carnivore	Bradstreet and Cross (1982), Marshall and Orr (1955), Turner (2000, 2004), Marshall (1973), Mauchline (1998), Kerfoot (1977, 1978), Montagna (1984)
Callopora lineata	Suspension feeder	Hayward and Ryland (1977)
Caprella acanthifera	Scavenger (scraper)	Guerra-García and Tierno de Figueroa (2009), Caine (1974), Guerra-Garcia *et al.* (2002)
Caprella equilibra	Grazer; scavenger; carnivore	Guerra-García and Tierno de Figueroa (2009), Guerra-Garcia *et al.* (2002), Keith (1969), Alan (1970)
Caprella linearis	Detritivore	Guerra-García and Tierno de Figueroa (2009), Agrawal (1964)
Ceradocus semiserratus	Detritivore	O'Gorman (2010b)
Cerastoderma edule	Suspension feeder	Kiorboe and Mohlenberg (1981), Lehane and Davenport (2002), Hawkins *et al.* (1990), Kamermans (1994), Brown (1976), Leguerrier *et al.* (2003)
Cerithiopsis tubercularis	Deposit feeder	Rueda *et al.* (2009), Morton (1967), Fretter (1951), Fretter and Manly (1979)
Chironomidae spp.	Grazer	Goldfinch and Carman (2000), Maasri *et al.* (2008), Robles and Cubit (1981)
Chlamys varia	Suspension feeder	Rueda *et al.* (2009), Koulouri *et al.* (2006), Grall *et al.* (2006), Beninger *et al.* (1991)
Circulus striatus	Grazer; scavenger	Graham (1988), Fretter (1956)
Clathrina coriacea	Suspension feeder	Bidder (1920)
Cliona celata	Suspension feeder	Hunt (1925)
Coriandria fulgida	Detritivore	Graham (1988)

Crassicorophium bonnellii	Tube feeder	Dixon and Moore (1997), Shillaker and Moore (1987a,b), Foster-Smith and Shillaker (1977), Crawford (1937), Enequist (1949)
Crassicorophium crassicorne	Tube feeder	O'Gorman (2010a), Dixon and Moore (1997), Shillaker and Moore (1987a,b), Foster-Smith and Shillaker (1977), Crawford (1937)
Crisia denticulata	Suspension feeder	O'Gorman (2010a), Riisgard and Manriquez (1997), Nielsen and Riisgard (1998), Hayward and Ryland (1985)
Crisia eburnea	Suspension feeder	Riisgard and Manriquez (1997), Nielsen and Riisgard (1998), Hayward and Ryland (1985)
Cumacea A	Grazer; deposit feeder	Blazewicz-Paszkowycz and Ligowski (2002), Dennell (1934), Foxon (1936), Dixon (1944), O'Gorman (2010c)
Cuthona sp.	Carnivore	Miller (1961)
Cyclopoida	Grazer; suspension feeder; carnivore	Turner (2000, 2004), Marshall (1973), Kerfoot (1977, 1978), Montagna (1984)
Cyprid larvae	Non-feeding larval stage	Jarrett and Pechenik (1997)
Cythere lutea	Grazer; deposit feeder	Montagna (1984), Athersuch *et al.* (1989)
Dexamine spinosa	Grazer	Greze *et al.* (1968), Strong *et al.* (2009)
Dexamine thea	Grazer	O'Gorman (2010a), Greze *et al.* (1968), Strong *et al.* (2009)
Disporella hispida	Suspension feeder	Riisgard and Manriquez (1997), Nielsen and Riisgard (1998), Hayward and Ryland (1985)
Dysidea fragilis	Suspension feeder	Grall *et al.* (2006)
Elasmopus rapax	Detritivore	O'Gorman (2010a), Keith (1969), Zimmerman *et al.* (1979), Duffy and Hay (2000), Nelson (1979)
Electra pilosa	Suspension feeder	Hayward and Ryland (1977), Riisgard and Manriquez (1997), Best and Thorpe (1986a,b), Jebram (1981), Okamura (1988)
Elysia viridis	Grazer	Miller (1961), Gallop *et al.* (1980), Jensen (1989), Williams and Walker (1999)
Emarginula fissura	Grazer; scavenger; carnivore	Dominici (2001), Graham (1939, 1988), Grall *et al.* (2006)
Epilepton clarkiae	Suspension feeder	Jespersen *et al.* (2007), Graham (1955)
Epitonium clathrus	Parasite	Graham (1988), O'Gorman (2010a), Perron (1978), Robertson (1963), Ankel (1937)
Ericthonius brasiliensis	Tube feeder	Duffy (1990), O'Gorman (2010a), Dixon and Moore (1997), Hughes (1975), Zavattari (1920)
Ericthonius punctatus	Tube feeder	Dixon and Moore (1997), Hughes (1975), Bradshaw *et al.* (2003)
Eubranchus farrani	Carnivore	Koulouri *et al.* (2006), Miller (1961)
Exogone gemmifera	Grazer; Deposit feeder	Neumann *et al.* (1970), Fauchald and Jumars (1979), Rasmussen (1973)

(*continued*)

Table S15 (*continued*)

Taxa	Feeding mode	References
Foraminifera A	Deposit feeder	O'Gorman (2010c), Austin *et al.* (2005), Murray (2006), Suhr *et al.* (2003), Ward *et al.* (2003), Murray (1963), Sandon (1932)
Foraminifera B	Deposit feeder	O'Gorman (2010c), Austin *et al.* (2005), Murray (2006), Suhr *et al.* (2003), Ward *et al.* (2003), Murray (1963), Sandon (1932)
Foraminifera C	Deposit feeder	O'Gorman (2010c), Austin *et al.* (2005), Murray (2006), Suhr *et al.* (2003), Ward *et al.* (2003), Murray (1963), Sandon (1932)
Foraminifera D	Deposit feeder	O'Gorman (2010c), Austin *et al.* (2005), Murray (2006), Suhr *et al.* (2003), Ward *et al.* (2003), Murray (1963), Sandon (1932)
Foraminifera E	Deposit feeder	O'Gorman (2010c), Austin *et al.* (2005), Murray (2006), Suhr *et al.* (2003), Ward *et al.* (2003), Murray (1963), Sandon (1932)
Foraminifera F	Deposit feeder	O'Gorman (2010c), Austin *et al.* (2005), Murray (2006), Suhr *et al.* (2003), Ward *et al.* (2003), Murray (1963), Sandon (1932)
Foraminifera G	Deposit feeder	O'Gorman (2010c), Austin *et al.* (2005), Murray (2006); Suhr *et al.* (2003), Ward *et al.* (2003), Murray (1963), Sandon (1932)
Galathea squamifera	Deposit feeder; carnivore	Nicol (1933), Samuelson (1970), DeGrave and Turner (1997), O'Gorman (Unpublished)
Gammaropsis maculata	Deposit feeder	O'Gorman (2010a), Dixon and Moore (1997), Enequist (1949)
Gammarus locusta	Grazer; deposit feeder	Duffy (1990), O'Gorman (2010a), Poltermann (2001), Greze *et al.* (1968), Zimmerman *et al.* (1979), Smith *et al.* (1982)
Gammarus zaddachi	Grazer; deposit feeder	Duffy (1990), O'Gorman (2010a), Poltermann (2001), Greze *et al.* (1968), Zimmerman *et al.* (1979), Smith *et al.* (1982)
Gastropod A	Deposit feeder	O'Gorman (2010d)
Gibbula umbilicalis	Grazer; deposit feeder	Koulouri *et al.* (2006), Graham (1988), O'Gorman (2010a), Jacobs *et al.* (1983), Prathep *et al.* (2003)
Halacarellus basteri	Scavenger	Bartsch (2006), MacQuitty (1984)
Harpacticoida	Grazer; suspension feeder	Turner (2000, 2004), Marshall (1973), Montagna (1984)
Hiatella arctica	Suspension feeder	Dominici (2001), Rueda *et al.* (2009), Koulouri *et al.* (2006), Grall *et al.* (2006), Aitken and Fournier (1993)
Hydrozoa	Carnivore	Christensen (1967), Mills (1976), Cornelius (1995)
Hymenopteran larvae	Possible terrestrial input	

Iothia fulva	Detritivore	Graham (1988)
Idotea A	Grazer; scavenger; carnivore	O'Gorman (2010b), Prathep *et al.* (2003), Sekiguchi (1982), Naylor (1955), Mistri *et al.* (2001), Goff and Cole (1976), Jormalainen *et al.* (2001), Naylor (1972)
Idotea B	Grazer; scavenger; carnivore	O'Gorman (2010b), Sekiguchi (1982), Naylor (1955), Mistri *et al.* (2001), Goff and Cole (1976), Jormalainen *et al.* (2001), Naylor (1972)
Janua pagenstecheri	Suspension feeder	O'Gorman (2010b), Fauchald and Jumars (1979), Orton (1914), Yonge (1928), Day (1967)
Lasaea rubra	Suspension feeder	McQuiston (1969), Ballantine and Morton (1956), Morton (1956)
Lembos websteri	Tube feeder	Dixon and Moore (1997), Shillaker and Moore (1987a,b), Foster-Smith and Shillaker (1977), Zimmerman *et al.* (1979)
Leptocheirus tricristatus	Tube feeder	O'Gorman (2010a), Dixon and Moore (1997), Goodhart (1939)
Leptochelia savignyi	Grazer; deposit feeder	O'Gorman (2010a), Johnson and Attramadal (1982), Kudinova-Pasternak (1991), Dennell (1937), Blazewicz-Paszkowycz and Ligowski (2002), Drumm (2005), Holdich and Jones (1983), Lewis (1998)
Leptocythere pellucida	Grazer; deposit feeder	Athersuch *et al.* (1989)
Leptomysis lingvura	Deposit feeder	Dauby (1995), Tattersall and Tattersall (1951), Mauchline (1980), Kobusch (1998)
Loxoconcha rhomboidea	Grazer; deposit feeder	Montagna (1984), Athersuch *et al.* (1989), Athersuch and Whittaker (1976)
Lysianassa ceratina	Scavenger; deposit feeder	Grall *et al.* (2006), Conlan (1994)
Macoma balthica	Suspension feeder; deposit feeder	Kamermans (1994), Leguerrier *et al.* (2003), Hummel (1985), Bradfield and Newell (1961), Bubnova (1972)
Melita palmata	Grazer; deposit feeder	Zimmerman *et al.* (1979), Sekiguchi (1982), Mistri *et al.* (2001), Conlan (1994)
Microdeutopus anomalus	Tube feeder	O'Gorman (2010a), Dixon and Moore (1997), Heckscher *et al.* (1996), Jimenez *et al.* (1996)
Microprotopus maculatus	Grazer; scavenger; carnivore	Schellenberg (1929)
Modiolula phaseolina	Suspension feeder	O'Gorman (2010b), Brown (1976), Prathep *et al.* (2003), Lesser *et al.* (1994), Navarro and Thompson (1996), Wildish and Kristmanson (1984)
Monia patelliformis	Suspension feeder	Dominici (2001), Rueda *et al.* (2009), Bramanti *et al.* (2003), O'Gorman (2010b)
Munna kroyeri	Grazer; deposit feeder	O'Gorman (2010c), Sekiguchi (1982), Naylor (1955), Mistri *et al.* (2001), Goff and Cole (1976), Jormalainen *et al.* (2001), Naylor (1972), Kussakin (1926)
Musculus discors	Suspension feeder	Koulouri *et al.* (2006), O'Gorman (2010b), Aitken and Fournier (1993), Larsen *et al.* (2001)
Mytilus edulis	Suspension feeder	Kiorboe and Mohlenberg (1981), Hawkins *et al.* (1990); Kamermans (1994), Beninger *et al.* (1991), Davenport *et al.* (2000), Ward *et al.* (1993)
Nannastacus unguiculatus	Grazer; deposit feeder	O'Gorman (2010a), Blazewicz-Paszkowycz and Ligowski (2002), Dennell (1934), Foxon (1936), Dixon (1944)

(*continued*)

Table S15 (*continued*)

Taxa	Feeding mode	References
Nematoda spp.	Deposit feeder	Montagna (1984), Britton (1993), Hyman (1951), Platt *et al.* (1983)
Nereis sp.	Grazer; scavenger; carnivore	Barrio Frojan *et al.* (2005), Goerke (1971), Gardiner (1903)
Odostomia plicata	Parasite	Rueda *et al.* (2009), Graham (1988), Fretter and Graham (1949)
Omalogyra atomus	Pendulum feeder	Graham (1988), Fretter (1948, 1951)
Onoba semicosta	Grazer; deposit feeder	Graham (1988)
Ophiothrix fragilis	Grazer; suspension feeder	Tillin *et al.* (2006), Allen (1998), Warner and Woodley (1975), Roushdy and Hansen (1960), Davoult and Gounin (1995), Vevers (1956)
Ophiura ophiura	Grazer; scavenger; carnivore	Tillin *et al.* (2006), Ramsay *et al.* (1998), O'Gorman (unpublished), Tyler (1977)
Ostracod A	Grazer; deposit feeder	Montagna (1984), Athersuch *et al.* (1989), O'Gorman (2010d), Cannon (1933)
Ostracod B	Grazer; deposit feeder	Montagna (1984), Athersuch *et al.* (1989), O'Gorman (2010d), Cannon (1933)
Ostracod C	Grazer; deposit feeder	Montagna (1984), Athersuch *et al.* (1989), O'Gorman (2010d), Cannon (1933)
Paradoxostoma variabile	Grazer; deposit feeder	Montagna (1984), Athersuch *et al.* (1989), Horne and Whittaker (1985)
Parvicardium exiguum	Suspension feeder	Rueda *et al.* (2009), Koulouri *et al.* (2006)
Parvicardium ovale	Suspension feeder	Rueda *et al.* (2009), Koulouri *et al.* (2006), O'Gorman (2010a), Leguerrier *et al.* (2003)
Parvicardium scabrum	Suspension feeder	Rueda *et al.* (2009), Koulouri *et al.* (2006)
Perinereis cultrifera	Grazer; scavenger	Fauchald and Jumars (1979), Barrio Frojan *et al.* (2005), Goerke (1971), Cazaux (1968), Yonge (1954)
Phtisica marina	Scavenger; carnivore	Guerra-García and Tierno de Figueroa (2009), Caine (1974), Guerra-Garcia *et al.* (2002), Bradshaw *et al.* (2003), Costa (1960)
Phyllodocidae sp.	Scavenger; carnivore	Fauchald and Jumars (1979), Rasmussen (1973), Barrio Frojan *et al.* (2005), Gardiner (1903), Pearson (1971)
Pilumnus hirtellus	Grazer; scavenger; carnivore	Grall *et al.* (2006), O'Gorman (unpublished), Prathep *et al.* (2003), Kyomo (1999), Lobb (1972)
Platynereis dumerilii	Grazer; scavenger; carnivore	Hunt (1925), Fauchald and Jumars (1979), Rasmussen (1973), Day (1967), Barrio Frojan *et al.* (2005), Goerke (1971)
Pomatoceros lamarcki	Suspension feeder	Hunt (1925), O'Gorman (2010a), Fauchald and Jumars (1979), Orton (1914), Yonge (1928), Pearson (1971)
Pomatoceros triqueter	Suspension feeder	Hunt (1925), O'Gorman (2010a), Fauchald and Jumars (1979), Orton (1914), Yonge (1928), Pearson (1971)

Pontocypris mytiloides	Grazer; deposit feeder	Montagna (1984), Athersuch *et al.* (1989)
Pseudoparatanais batei	Grazer; deposit feeder	O'Gorman (2010a), Johnson and Attramadal (1982), Kudinova-Pasternak (1991), Dennell (1937), Blazewicz-Paszkowycz and Ligowski (2002), Drumm (2005), Holdich and Jones (1983)
Retusa truncatula	Carnivore	Koulouri *et al.* (2006), O'Gorman (2010a), Berry (1994), Malaquias *et al.* (2009), Chaban (2000), Smith (1967)
Rissoa parva	Epiphytic grazer; detritivore	Graham (1988), Fretter and Montgomery (1968), Prathep *et al.* (2003), Fretter and Graham (1981), Wigham (1976), Fretter and Graham (1978)
Rissoa sarsi	Epiphytic grazer; detritivore	Graham (1988), Fretter and Montgomery (1968), Fretter and Graham (1978, 1981)
Rissoella diaphana	Epiphytic grazer; deposit feeder	Rueda *et al.* (2009), Graham (1988), Fretter (1948, 1951), Fretter and Graham (1981)
Rissoella opalina	Epiphytic grazer; deposit feeder	Graham (1988), Fretter (1948, 1951), Fretter and Graham (1981)
Sabella pavonina	Suspension feeder	Hunt (1925), Fauchald and Jumars (1979), Day (1967), Barrio Frojan *et al.* (2005), Gardiner (1903), Nicol (1930)
Sagitta elegans	Suspension feeder; carnivore	Solov'ev and Kosobokova (2003), Batistic *et al.* (2003), Tonnesson and Tiselius (2005), Pearre (1976), Sullivan (1980), Brodeur and Terazaki (1999), Terazaki (1993, 2004), Alvarez-Cadena (1993)
Scrupocellaria spp.	Suspension feeder	Hayward and Ryland (1977), Riisgard and Manriquez (1997)
Semibalanus balanoides	Suspension feeder	Lake *et al.* (2006), Trager *et al.* (1990), Bertness *et al.* (1991), Sanford *et al.* (1994), Barnes (1963)
Semicytherura nigrescens	Grazer; deposit feeder	Athersuch *et al.* (1989), Whittaker (1974)
Serpulid A	Suspension feeder	Hunt (1925), O'Gorman (2010a), Fauchald and Jumars (1979), Orton (1914), Yonge (1928), Day (1967), Pearson (1971)
Siriella armata	Carnivore	Tattersall and Tattersall (1951), Mauchline (1980), Kobusch (1998), De Jong-Moreau *et al.* (2001), Hobson and Chess (1976)
Skenea serpuloides	Detritivore	Graham (1988)
Spirorbis A	Suspension feeder	O'Gorman (2010a), Fauchald and Jumars (1979), Orton (1914), Yonge (1928), Day (1967)
Spirorbis B	Suspension feeder	O'Gorman (2010a), Fauchald and Jumars (1979), Orton (1914), Yonge (1928), Day (1967)

(*continued*)

Table S15 (*continued*)

Taxa	Feeding mode	References
Stenothoe marina	Commensal; suspension feeder	Chapman (2007), O'Gorman (2010b), Bradshaw *et al.* (2003)
Syllidae A	Grazer; scavenger; carnivore	Neumann *et al.* (1970), Fauchald and Jumars (1979), Day (1967), Barrio Frojan *et al.* (2005)
Syllidae B	Grazer; scavenger; carnivore	Neumann *et al.* (1970), Fauchald and Jumars (1979), Day (1967), Barrio Frojan *et al.* (2005)
Tapes aureus	Suspension feeder	O'Gorman (2010a), Leguerrier *et al.* (2003), Graham (1955)
Tectura virginea	Grazer; scavenger; carnivore	Graham (1988)
Tomopteris helgolandica	Scavenger; carnivore	Fauchald and Jumars (1979), Lebour (1923), Hartmann-Schroder (1971)
Tritaeta gibbosa	Grazer	O'Gorman (2010b), Greze *et al.* (1968), Strong *et al.* (2009), Peacock (1971)
Tryphosella sarsi	Scavenger; deposit feeder	Grall *et al.* (2006), O'Gorman (2010b), Enequist (1949), Conlan (1994)
Tubulipora liliacea	Suspension feeder	Riisgard and Manriquez (1997), Nielsen and Riisgard (1998), Hayward and Ryland (1985)
Turbellaria A	Scavenger	Britton (1993), Hyman (1951)
Typosyllis prolifera	Grazer; scavenger; carnivore	Neumann *et al.* (1970), Fauchald and Jumars (1979), Day (1967), Barrio Frojan *et al.* (2005)
Vitreolina philippi	Parasite	Rueda *et al.* (2009), Graham (1988), Campani (2007), Warén (1983)
Xestoleberis aurantia	Grazer; deposit feeder	Montagna (1984), Athersuch *et al.* (1989), Whittaker (1978)
Algae	Basal resource	
Bacteria	Basal resource	
Cladocerans	Basal resource	
CPOM	Basal resource	
Diatoms	Basal resource	
FPOM	Basal resource	
Microphytobenthos	Basal resource	
Phytoplankton	Basal resource	

APPENDIX IV. DETAILS OF THE 144 MESOCOSM FOOD WEBS DESCRIBED IN THE STUDY

Figure S1 Replicate food webs for the W^+S^+ treatment.

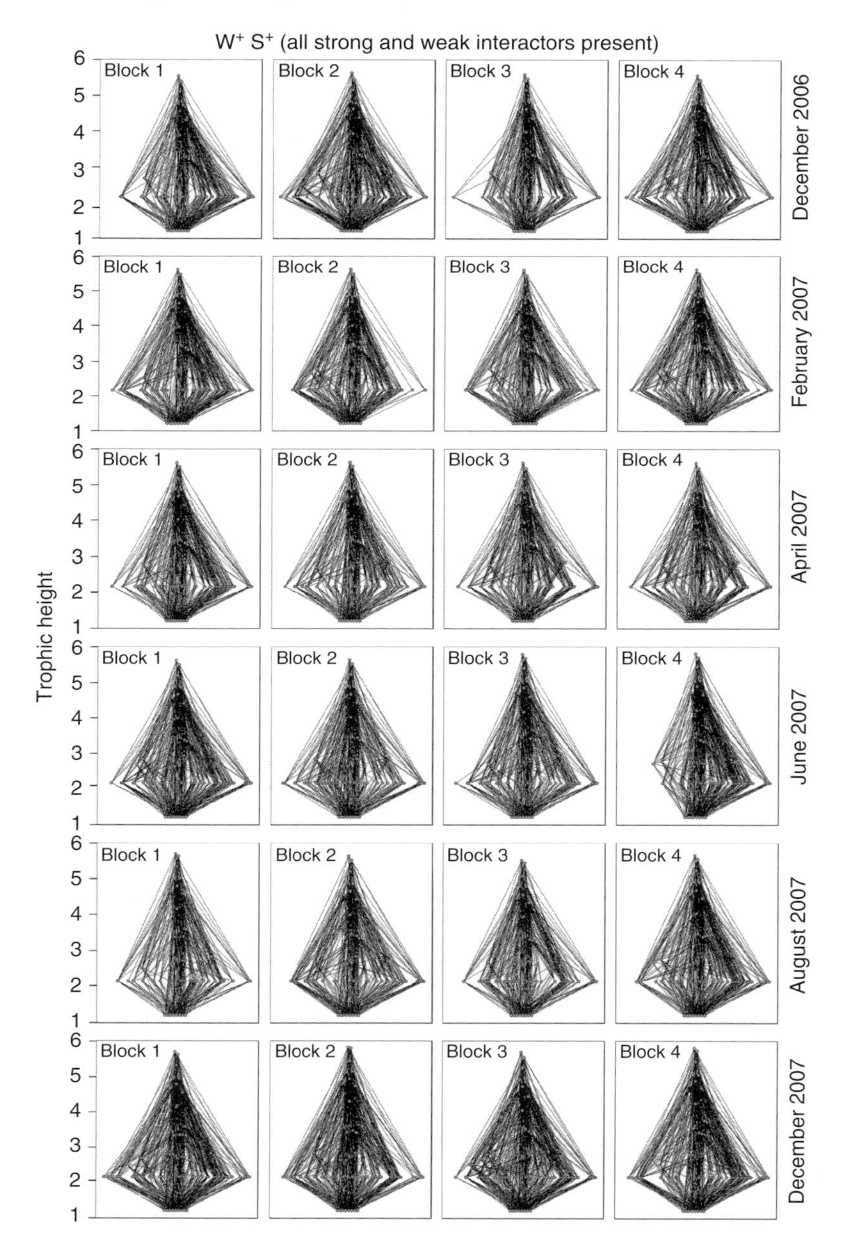

Table S16 Species present in each of the food webs shown in Figure S1 above (grey squares indicate species presence in a web; white squares indicate species absence from a web)

	Dec 06				Feb 07				Apr 07				Jun 07				Aug 07				Dec 07			
Taxa	1	2	3	4	1	2	3	4	1	2	3	4	1	2	3	4	1	2	3	4	1	2	3	4
Carcinus maenas	■	■	■	■	■	■	■	■	■	■	■	■	■	■	■	■	■	■	■	■	■	■	■	■
Ctenolabrus rupestris	■	■	■	■	■	■	■	■	■	■	■	■	■	■	■	■	■	■	■	■	■	■	■	■
Gaidropsarus mediterraneus	■	■	■	■	■	■	■	■	■	■	■	■	■	■	■	■	■	■	■	■	■	■	■	■
Gobius niger	■	■	■	■	■	■	■	■	■	■	■	■	■	■	■	■	■	■	■	■	■	■	■	■
Gobius paganellus	■	■	■	■	■	■	■	■	■	■	■	■	■	■	■	■	■	■	■	■	■	■	■	■
Marthasterias glacialis	■	■	■	■	■	■	■	■	■	■	■	■	■	■	■	■	■	■	■	■	■	■	■	■
Necora puber	■	■	■	■	■	■	■	■	■	■	■	■	■	■	■	■	■	■	■	■	■	■	■	■
Palaemon serratus	■	■	■	■	■	■	■	■	■	■	■	■	■	■	■	■	■	■	■	■	■	■	■	■
Paracentrotus lividus	■	■	■	■	■	■	■	■	■	■	■	■	■	■	■	■	■	■	■	■	■	■	■	■
Taurulus bubalis	■	■	■	■	■	■	■	■	■	■	■	■	■	■	■	■	■	■	■	■	■	■	■	■
Gobiusculus flavescens	■	■	■	■	■	■	■	■	■	■	■	■	■	■	■	■	■	■	■	■	■	■	■	■
Pomatoschistus pictus	■	■	■	■	■	■	■	■	■	■	■	■	■	■	■	■	■	■	■	■	■	■	■	■
Abra alba		■	■	■							■										■			
Acanthochitona crinitus														■	■						■			
Aequipecten opercularis						■		■					■										■	
Alvania beani		■		■	■			■	■	■		■	■	■							■	■	■	■
Alvania semistriata																					■	■		■
Anomia ephippium	■	■		■	■	■	■	■			■						■	■	■	■	■	■		■
Aora gracilis	■		■		■	■	■	■	■						■			■	■					
Apherusa bispinosa	■	■																						
Apseudes latreillei		■								■	■		■		■			■			■	■	■	■
Apseudes talpa				■	■	■						■	■											
Ascidiella aspersa						■	■	■			■							■						■
Asterina phylactica																						■		
Bittium reticulatum	■	■	■	■	■	■	■	■	■	■	■	■	■	■	■	■	■	■	■	■	■	■	■	■
Buccinum undatum																		■				■	■	
Callopora lineata										■	■						■					■		
Caprella equilibra																							■	
Ceradocus semiserratus		■												■										
Cerithiopsis tubercularis	■	■	■	■	■	■	■	■	■	■	■	■	■	■	■		■		■	■	■	■	■	■
Chironomidae spp.										■			■					■						
Chlamys varia			■	■				■	■			■									■	■	■	
Clathrina coriacea						■			■					■	■			■	■	■	■	■	■	
Coriandria fulgida				■																				
Crassicorophium bonnellii																			■					
Crassicorophium crassicorne	■	■	■	■	■	■	■	■	■	■	■	■	■	■	■	■	■	■	■	■	■	■	■	■
Crisia denticulata											■	■		■							■	■		
Cumacea A													■	■										
Cyclopoida		■		■	■	■	■	■	■	■	■	■	■	■	■			■		■		■	■	■
Cyprid larvae							■				■													
Dexamine spinosa			■		■		■						■	■	■	■		■	■	■	■	■	■	■
Disporella hispida											■				■					■		■		
Dysidea fragilis																							■	
Elasmopus rapax		■			■									■										
Electra pilosa																							■	■
Elysia viridis	■							■	■					■					■	■			■	■
Epilepton clarkiae	■	■	■	■	■	■	■	■	■	■	■	■	■	■	■	■	■		■	■	■	■	■	■
Epitonium clathrus											■													
Ericthonius brasiliensis		■																					■	
Ericthonius punctatus																					■		■	
Eubranchus farrani	■		■	■	■																			
Exogene gemmifera	■	■	■	■		■		■	■	■	■		■	■	■	■		■	■	■	■	■	■	
Foraminifera A	■	■	■	■	■	■		■			■	■	■	■		■				■		■		
Foraminifera B		■	■	■		■	■	■	■		■		■	■		■		■	■		■	■	■	
Foraminifera F		■	■	■			■		■	■	■	■	■	■	■	■	■	■	■	■	■	■	■	■
Foraminifera G	■	■	■			■	■	■	■	■	■	■	■	■	■	■				■		■		
Galathea squamifera																■					■	■	■	■
Gammaropsis maculata	■	■	■	■																				
Gammarus locusta						■	■		■															
Gibbula umbilicalis									■	■	■			■					■		■	■		

Table S17 Table S16 continued

	Dec 06				Feb 07				Apr 07				Jun 07				Aug 07				Dec 07			
Taxa	1	2	3	4	1	2	3	4	1	2	3	4	1	2	3	4	1	2	3	4	1	2	3	4
Halacarellus basteri																								
Harpacticoida																								
Hiatella arctica																								
Hymenopteran A																								
Isopod A																								
Isopod B																								
Janua pagenstecheri																								
Lembos websteri																								
Leptocythere pellucida																								
Leptomysis lingvura																								
Loxoconcha rhomboidea																								
Lysianassa cerafina																								
Melita palmata																								
Microdeutopus anomalus																								
Microprotopus maculatus																								
Modiolula phaseolina																								
Monia patelliformis																								
Munna kroyeri																								
Musculus discors																								
Nannastacus unguiculatus																								
Nematoda spp.																								
Nereis A																								
Odostomia plicata																								
Omalogyra atomus																								
Onoba semicosta																								
Ophiothrix fragilis																								
Ophiura ophiura																								
Ostracod A																								
Paradoxostoma variabile																								
Parvicardium exiguum																								
Parvicardium ovale																								
Parvicardium scabrum																								
Perinereis cultrifera																								
Phyllodocid A																								
Pilumnus hirtellus																								
Platynereis dumerili																								
Pomatoceros lamarcki																								
Pomatoceros triqueter																								
Pontocypris mytiloides																								
Pseudoparatanais batei																								
Retusa truncatula																								
Rissoa parva																								
Rissoa sarsi																								
Rissoella diaphana																								
Rissoella opalina																								
Sabella pavonina																								
Sagitta elegans																								
Scrupocellaria spp.																								
Semibalanus balanoides																								
Semicytherura nigrescens																								
Siriella armata																								
Skenea serpuloides																								
Spirorbis A																								
Stenothoe marina																								
Syllidae A																								
Syllidae B																								
Tapes aureus																								
Tomopteris helgolandica																								
Tryphosella sarsi																								
Tubulipora liliacea																								
Turbellaria A																								
Typosyllis prolifera																								
Vitreolina philippii																								
Xestoleberis aurantia																								

Figure S2 Replicate food webs for the $W^{-2}S^{+}$ treatment.

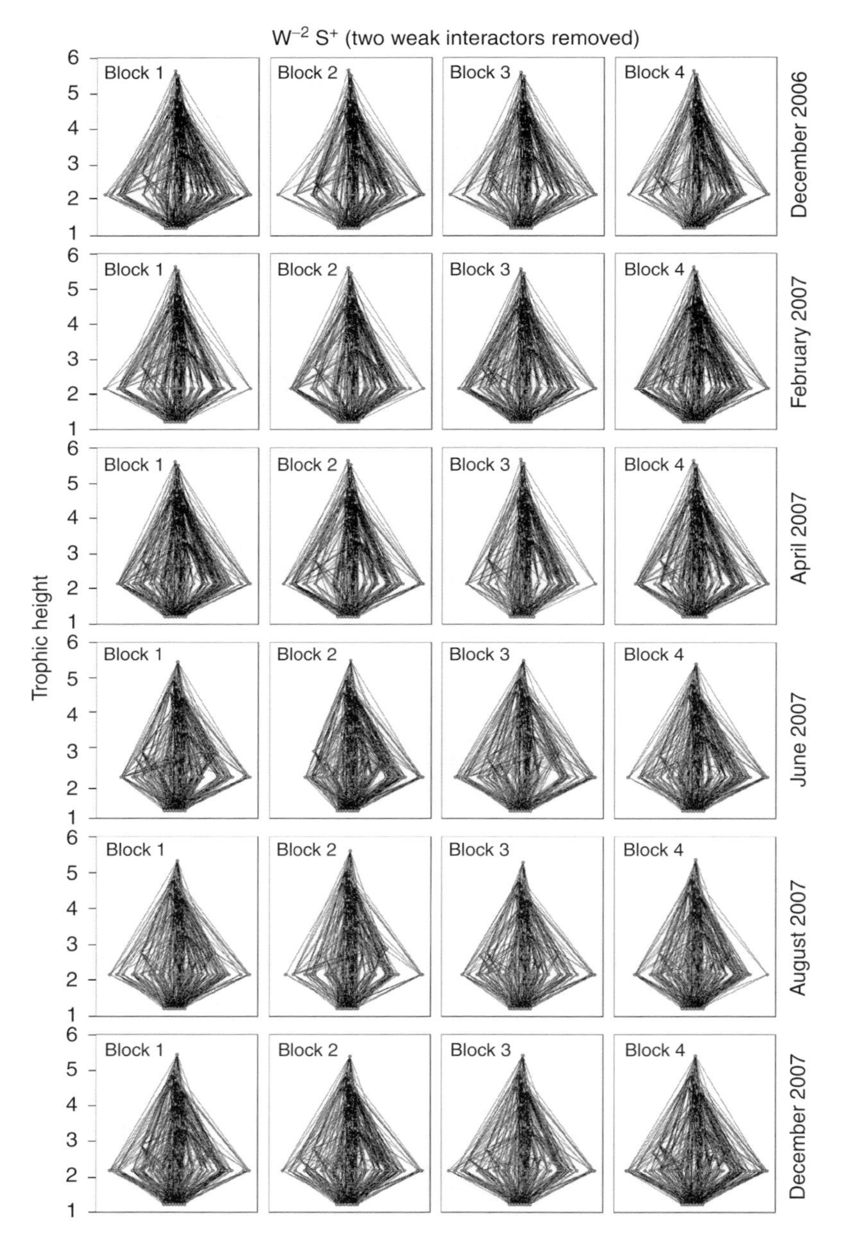

Table S18 Species present in each of the food webs shown in Figure S1 above (grey squares indicate species presence in a web; white squares indicate species absence from a web)

	Dec 06				Feb 07				Apr 07				Jun 07				Aug 07				Dec 07			
Taxa	1	2	3	4	1	2	3	4	1	2	3	4	1	2	3	4	1	2	3	4	1	2	3	4
Carcinus maenas	■	■	■	■	■	■	■	■	■	■	■	■	■	■	■	■	■	■	■	■	■	■	■	■
Ctenolabrus rupestris	■	■	■	■	■	■	■	■	■	■	■	■	■	■	■	■	■	■	■	■	■	■	■	■
Gaidropsarus mediterraneus	■	■	■	■	■	■	■	■	■	■	■	■	■	■	■	■	■	■	■	■	■	■	■	■
Gobius niger	■	■	■	■	■	■	■	■	■	■	■	■	■	■	■	■	■	■	■	■	■	■	■	■
Gobius paganellus	■	■	■	■	■	■	■	■	■	■	■	■												
Marthasterias glacialis	■	■	■	■	■	■	■	■	■	■	■	■	■	■	■	■	■	■	■	■	■	■	■	■
Necora puber	■	■	■	■	■	■	■	■	■	■	■	■												
Palaemon serratus	■	■	■	■	■	■	■	■	■	■	■	■	■	■	■	■	■	■	■	■	■	■	■	■
Paracentrotus lividus	■	■	■	■	■	■	■	■	■	■	■	■	■	■	■	■	■	■	■	■	■	■	■	■
Taurulus bubalis	■	■	■	■	■	■	■	■	■	■	■	■	■	■	■	■	■	■	■	■	■	■	■	■
Gobiusculus flavescens	■	■	■	■	■	■	■	■	■	■	■	■	■	■	■	■	■	■	■	■	■	■	■	■
Pomatoschistus pictus	■	■	■	■	■	■	■	■	■	■	■	■	■	■	■	■	■	■	■	■	■	■	■	■
Abra alba	■	■	■		■																		■	
Acanthochitona crinitus																								■
Aequipecten opercularis				■			■						■						■					■
Alvania beani	■			■						■					■	■	■				■	■	■	■
Alvania semistriata							■								■			■			■			■
Anomia ephippium	■		■		■	■	■	■	■	■		■			■		■		■	■	■	■	■	■
Aora gracilis		■	■		■	■	■	■	■					■						■				
Apseudes latreillei										■										■	■		■	■
Apseudes talpa						■	■	■	■	■														
Ascidiella aspersa	■				■	■	■	■				■	■			■	■		■			■		
Bittium reticulatum	■	■	■	■	■	■	■	■	■	■	■	■	■	■	■	■	■		■	■	■	■	■	■
Buccinum undatum														■										
Callopora lineata									■															■
Caprella acanthifera																■								
Caprella equilibra			■																					
Caprella linearis															■									
Ceradocus semiserratus				■				■		■														
Cerithiopsis tubercularis	■		■		■		■	■	■			■	■	■	■	■	■	■	■	■	■	■	■	■
Chironomidae spp.																			■					
Chlamys varia								■									■				■	■		
Clathrina coriacea									■		■	■	■	■	■	■	■	■	■	■	■	■	■	■
Coriandria fulgida		■												■				■						
Crassicorophium bonnellii												■									■			
Crassicorophium crassicorne	■	■	■	■	■	■	■	■	■	■	■	■	■	■	■	■	■	■	■	■	■	■	■	■
Crisia denticulata																				■	■			■
Cumacea A					■									■										
Cyclopoida	■	■		■	■	■	■	■	■	■	■	■	■		■	■		■						
Cyprid larvae											■	■												
Dexamine spinosa								■	■				■	■	■	■	■			■	■	■	■	■
Dexamine thea											■													
Dysidea fragilis													■	■	■			■						■
Elasmopus rapax																	■	■						
Electra pilosa																				■				
Elysia viridis	■	■				■									■			■			■	■	■	■
Emarginula fissura																■								
Epilepton clarkiae	■	■	■	■	■	■	■	■	■	■	■	■	■	■	■	■	■	■	■	■	■	■	■	■
Ericthonius brasiliensis																							■	
Ericthonius punctatus																						■		
Eubranchus farrani	■				■									■										
Exogene gemmifera		■			■	■	■	■	■	■			■		■	■				■		■		
Foraminifera A	■	■			■	■	■	■	■	■	■	■	■	■		■	■	■	■		■	■	■	■
Foraminifera B	■		■	■	■	■		■	■	■	■	■	■	■		■	■	■	■	■	■	■	■	■
Foraminifera C	■																							
Foraminifera E											■										■			■
Foraminifera F				■			■		■	■	■	■	■	■	■	■	■	■	■	■	■	■	■	■
Foraminifera G			■		■	■	■	■	■	■	■	■	■	■				■		■	■			■
Galathea squamifera																■					■	■		
Gammaropsis maculata	■	■	■	■	■		■																	

Table S19 Table S18 continued

	Dec 06				Feb 07				Apr 07				Jun 07				Aug 07				Dec 07			
Taxa	1	2	3	4	1	2	3	4	1	2	3	4	1	2	3	4	1	2	3	4	1	2	3	4
Gammarus locusta																								
Gibbula umbilicalis																								
Halacarellus basteri																								
Harpacticoida																								
Hiatella arctica																								
Isopod A																								
Isopod B																								
Janua pagenstecheri																								
Lembos websteri																								
Leptocheirus tricristatus																								
Leptocythere pellucida																								
Leptomysis lingvura																								
Loxoconcha rhomboidea																								
Lysianassa ceratina																								
Melita palmata																								
Microdeutopus anomalus																								
Microprotopus maculatus																								
Modiolula phaseolina																								
Monia patelliformis																								
Munna kroyeri																								
Musculus discors																								
Mytilus edulis																								
Nannastacus unguiculatus																								
Nematoda spp.																								
Nereis A																								
Odostomia plicata																								
Omalogyra atomus																								
Onoba semicosta																								
Ophiothrix fragilis																								
Ophiura ophiura																								
Ostracod A																								
Paradoxostoma variabile																								
Parvicardium exiguum																								
Parvicardium ovale																								
Parvicardium scabrum																								
Perinereis cultrifera																								
Phyllodocid A																								
Pilumnus hirtellus																								
Platynereis dumerili																								
Pomatoceros lamarcki																								
Pomatoceros triqueter																								
Pontocypris mytiloides																								
Pseudoparatanais batei																								
Retusa truncatula																								
Rissoa parva																								
Rissoa sarsi																								
Rissoella diaphana																								
Rissoella opalina																								
Sagitta elegans																								
Scrupocellaria spp.																								
Semibalanus balanoides																								
Semicytherura nigrescens																								
Siriella armata																								
Skenea serpuloides																								
Stenothoe marina																								
Syllidae A																								
Syllidae B																								
Tapes aureus																								
Tomopteris helgolandica																								
Tryphosella sarsi																								
Tubulipora liliacea																								
Turbellaria A																								
Typosyllis prolifera																								
Xestoleberis aurantia																								

Figure S3 Replicate food webs for the $W^{-3}S^{+}$ treatment.

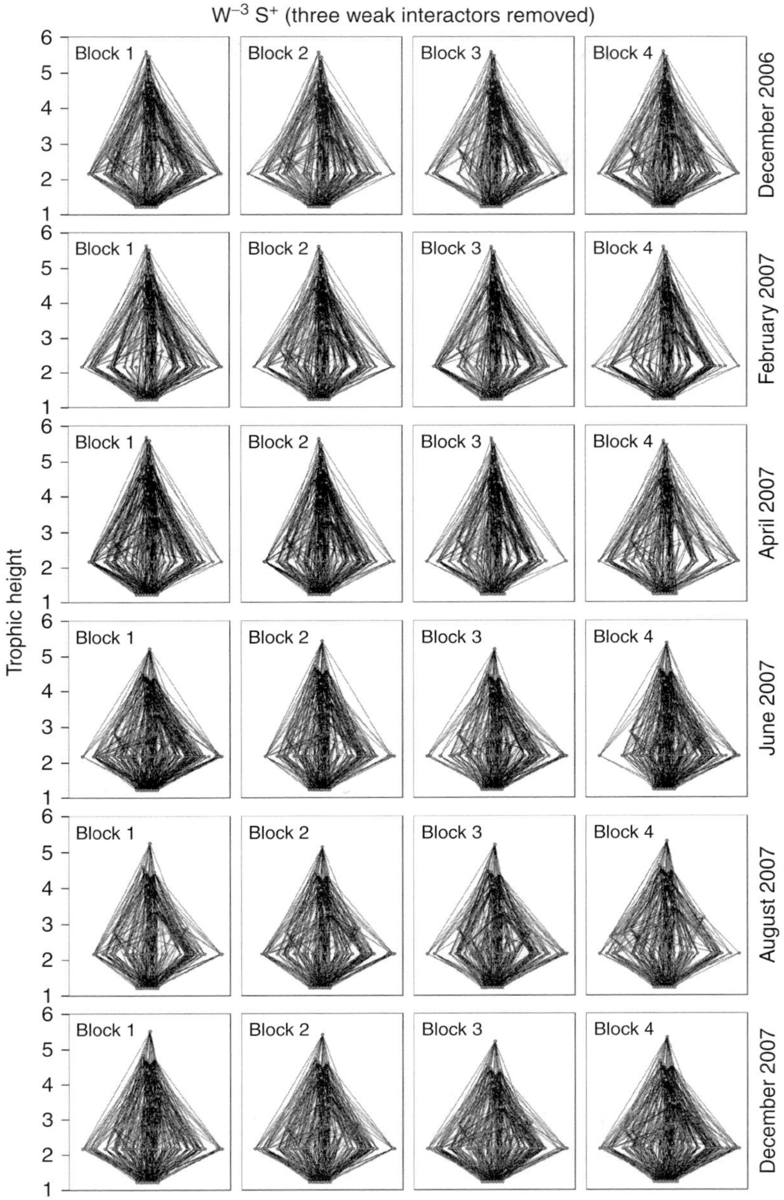

Table S20 Species present in each of the food webs shown in Figure S1 above (grey squares indicate species presence in a web; white squares indicate species absence from a web)

	Dec 06				Feb 07				Apr 07				Jun 07				Aug 07				Dec 07			
Taxa	1	2	3	4	1	2	3	4	1	2	3	4	1	2	3	4	1	2	3	4	1	2	3	4
Carcinus maenas	■	■	■	■	■	■	■	■	■	■	■	■												
Ctenolabrus rupestris	■	■	■	■	■	■	■	■	■	■	■	■	■	■	■	■	■	■	■	■	■	■	■	■
Gaidropsarus mediterraneus	■	■	■	■	■	■	■	■	■	■	■	■	■	■	■	■	■	■	■	■	■	■	■	■
Gobius niger	■	■	■	■	■	■	■	■	■	■	■	■	■	■	■	■	■	■	■	■	■	■	■	■
Gobius paganellus	■	■	■	■	■	■	■	■	■	■	■	■												
Marthasterias glacialis	■	■	■	■	■	■	■	■	■	■	■	■	■	■	■	■	■	■	■	■	■	■	■	■
Necora puber	■	■	■	■	■	■	■	■	■	■	■	■												
Palaemon serratus	■	■	■	■	■	■	■	■	■	■	■	■	■	■	■	■	■	■	■	■	■	■	■	■
Paracentrotus lividus	■	■	■	■	■	■	■	■	■	■	■	■	■	■	■	■	■	■	■	■	■	■	■	■
Taurulus bubalis	■	■	■	■	■	■	■	■	■	■	■	■	■	■	■	■	■	■	■	■	■	■	■	■
Gobiusculus flavescens	■	■	■	■	■	■	■	■	■	■	■	■	■	■	■	■	■	■	■	■	■	■	■	■
Pomatoschistus pictus	■	■	■	■	■	■	■	■	■	■	■	■	■	■	■	■	■	■	■	■	■	■	■	■
Abra alba		■						■																
Aequipecten opercularis	■	■				■																		■
Alvania beani			■	■	■	■					■	■	■		■	■			■	■	■	■	■	■
Alvania semistriata															■						■	■	■	■
Amphilochus manudens								■																
Anomia ephippium	■	■	■	■	■		■		■			■						■	■	■	■	■	■	■
Aora gracilis	■			■	■	■								■			■	■	■					
Apseudes latreillei			■				■	■	■	■											■	■	■	■
Apseudes talpa				■		■		■	■	■	■	■	■											
Ascidiella aspersa					■	■	■	■			■		■		■		■			■	■		■	
Asterina phylactica																					■			
Bittium reticulatum	■	■	■	■	■	■	■	■	■	■	■	■	■	■	■	■	■	■	■	■	■	■	■	■
Buccinum undatum														■		■								
Calanoida																				■				
Callopora lineata													■											
Caprella equilibra	■																							
Caprella linearis							■						■						■					
Ceradocus semiserratus													■				■							
Cerithiopsis tubercularis			■	■	■	■		■	■	■		■	■	■	■	■	■	■	■	■	■	■	■	■
Chironomidae spp.																							■	
Chlamys varia	■	■							■	■							■				■	■	■	■
Circulus striatus									■			■												
Clathrina coriacea	■				■				■	■		■	■	■	■	■						■	■	■
Coriandria fulgida	■	■																■	■			■	■	
Crassicorophium bonnellii									■	■									■		■			
Crassicorophium crassicorne	■	■	■	■	■	■	■	■	■	■	■		■	■	■	■	■	■	■	■	■	■	■	■
Crisia denticulata									■		■	■								■	■			■
Cumacea A													■			■						■		
Cuthona A													■											
Cyclopoida	■	■		■	■	■	■	■	■	■	■	■	■	■			■	■		■		■	■	■
Cyprid larvae											■	■												
Dexamine spinosa	■										■		■	■	■	■		■		■	■	■	■	■
Disporella hispida								■			■				■					■	■			
Dysidea fragilis																			■					
Elasmopus rapax																	■							
Electra pilosa														■					■		■	■	■	
Elysia viridis				■									■	■		■		■			■	■	■	■
Epilepton clarkiae	■	■	■	■	■	■	■	■	■	■	■	■	■	■	■	■	■	■	■	■	■	■	■	■
Epitonium clathrus												■												
Eubranchus farrani					■							■												
Exogene gemmifera						■		■	■	■		■	■	■		■	■			■	■	■	■	■
Foraminifera A		■	■	■						■			■		■	■		■	■	■		■	■	■
Foraminifera B			■	■		■	■		■	■	■	■	■	■		■	■		■	■	■	■	■	
Foraminifera C		■	■	■																				
Foraminifera F	■		■	■					■	■	■	■	■	■	■	■	■	■	■	■	■	■	■	■
Foraminifera G	■		■	■		■	■	■	■	■	■	■	■		■	■	■	■		■	■		■	
Galathea squamifera																■					■			■
Gammaropsis maculata	■	■	■	■		■																		

Table S21 Table S20 continued

	Dec 06				Feb 07				Apr 07				Jun 07				Aug 07				Dec 07			
Taxa	1	2	3	4	1	2	3	4	1	2	3	4	1	2	3	4	1	2	3	4	1	2	3	4
Gammarus locusta																								
Gibbula umbilicalis																								
Halacarellus basteri																								
Harpacticoida																								
Hiatella arctica																								
Hymenopteran A																								
Isopod A																								
Isopod B																								
Janua pagenstecheri																								
Lembos websteri																								
Leptochelia savignyi																								
Leptocythere pellucida																								
Leptomysis lingvura																								
Loxoconcha rhomboidea																								
Lysianassa ceratina																								
Melita palmata																								
Microdeutopus anomalus																								
Microprotopus maculatus																								
Modiolula phaseolina																								
Monia patelliformis																								
Munna kroyeri																								
Musculus discors																								
Mytilus edulis																								
Nannastacus unguiculatus																								
Nematoda spp.																								
Nereis A																								
Odostomia plicata																								
Omalogyra atomus																								
Onoba semicosta																								
Ophiothrix fragilis																								
Ophiura ophiura																								
Ostracod A																								
Paradoxostoma variabile																								
Parvicardium exiguum																								
Parvicardium ovale																								
Parvicardium scabrum																								
Perinereis cultrifera																								
Phyllodocid A																								
Pilumnus hirtellus																								
Platynereis dumerili																								
Pomatoceros lamarcki																								
Pomatoceros triqueter																								
Pontocypris mytiloides																								
Pseudoparatanais batei																								
Retusa truncatula																								
Rissoa parva																								
Rissoa sarsi																								
Rissoella diaphana																								
Rissoella opalina																								
Sabella pavonina																								
Sagitta elegans																								
Scrupocellaria spp.																								
Semibalanus balanoides																								
Semicytherura nigrescens																								
Siriella armata																								
Skenea serpuloides																								
Stenothoe marina																								
Syllidae A																								
Syllidae B																								
Tapes aureus																								
Tomopteris helgolandica																								
Tryphosella sarsi																								
Tubulipora liliacea																								
Turbellaria A																								
Typosyllis prolifera																								
Xestoleberis aurantia																								

Figure S4 Replicate food webs for the W^+S^{-2} treatment.

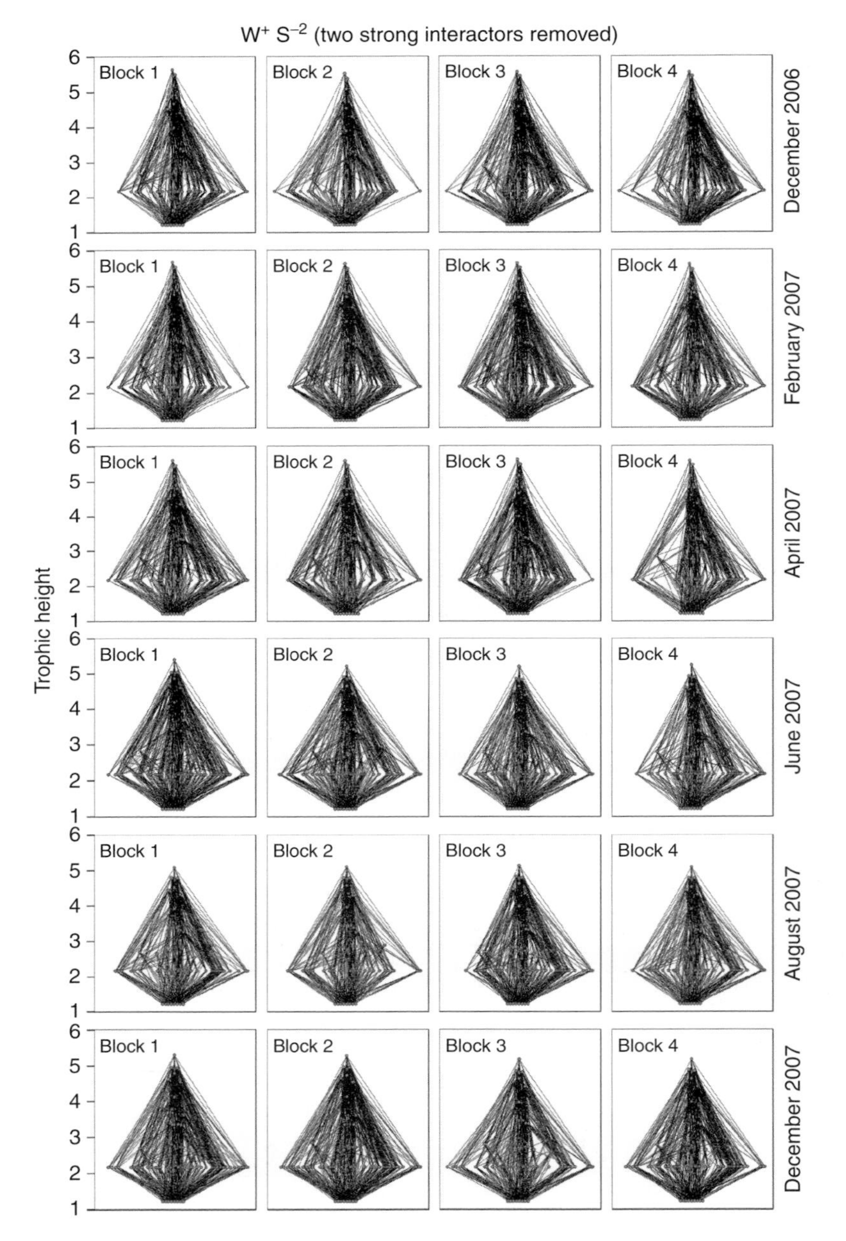

Table S22 Species present in each of the food webs shown in Figure S1 above (grey squares indicate species presence in a web; white squares indicate species absence from a web)

	Dec 06				Feb 07				Apr 07				Jun 07				Aug 07				Dec 07			
Taxa	1	2	3	4	1	2	3	4	1	2	3	4	1	2	3	4	1	2	3	4	1	2	3	4
Carcinus maenas	■	■	■	■	■	■	■	■	■	■	■	■	■	■	■	■	■	■	■	■	■	■	■	■
Ctenolabrus rupestris	■	■	■	■	■	■	■	■	■	■	■	■	■	■	■	■	■	■	■	■	■	■	■	■
Gaidropsarus mediterraneus	■	■	■	■	■	■	■	■	■	■	■	■												
Gobius niger	■	■	■	■	■	■	■	■	■	■	■	■	■	■	■	■	■	■	■	■	■	■	■	■
Gobius paganellus	■	■	■	■	■	■	■	■	■	■	■	■	■	■	■	■	■	■	■	■	■	■	■	■
Marthasterias glacialis	■	■	■	■	■	■	■	■	■	■	■	■												
Necora puber	■	■	■	■	■	■	■	■	■	■	■	■	■	■	■	■	■	■	■	■	■	■	■	■
Palaemon serratus	■	■	■	■	■	■	■	■	■	■	■	■	■	■	■	■	■	■	■	■	■	■	■	■
Paracentrotus lividus	■	■	■	■	■	■	■	■	■	■	■	■	■	■	■	■	■	■	■	■	■	■	■	■
Taurulus bubalis	■	■	■	■	■	■	■	■	■	■	■	■	■	■	■	■	■	■	■	■	■	■	■	■
Gobiusculus flavescens	■	■	■	■	■	■	■	■	■	■	■	■	■	■	■	■	■	■	■	■	■	■	■	■
Pomatoschistus pictus	■	■	■	■	■	■	■	■	■	■	■	■	■	■	■	■	■	■	■	■	■	■	■	■
Abra alba		■	■	■																			■	
Acanthochitona crinitus														■										
Aequipecten opercularis						■					■								■					
Alvania beani					■				■				■	■							■	■	■	■
Alvania semistriata																					■	■	■	■
Anomia ephippium		■	■	■			■	■	■	■	■	■	■	■	■		■	■		■	■	■	■	■
Aora gracilis	■	■		■	■	■	■		■		■				■			■	■	■	■			
Apherusa bispinosa			■			■																		
Apseudes latreillei		■												■							■	■	■	■
Apseudes talpa					■	■		■		■	■													
Ascidiella aspersa		■			■		■	■		■	■		■		■		■		■			■		
Bittium reticulatum	■	■	■	■	■	■	■	■	■	■		■	■	■	■	■	■		■		■	■	■	■
Buccinum undatum																					■			■
Calanoida																				■				
Callopora lineata										■		■										■		
Caprella linearis																				■				
Ceradocus semiserratus		■	■					■																
Cerithiopsis tubercularis	■		■	■	■	■	■		■				■	■	■			■	■	■	■	■	■	■
Chironomidae spp.																			■					
Chlamys varia	■		■		■			■		■								■			■	■	■	
Circulus striatus						■					■													
Clathrina coriacea			■			■			■			■	■	■	■	■	■	■	■	■	■	■		■
Coriandria fulgida	■						■						■									■		
Crassicorophium bonnellii								■			■		■						■					■
Crassicorophium crassicorne	■	■	■	■	■	■	■	■	■	■	■		■	■	■	■	■	■	■	■	■	■	■	■
Crisia denticulata										■											■		■	■
Cumacea A											■			■										
Cyclopoida	■				■	■	■	■	■	■	■	■	■	■		■	■	■						■
Cyprid larvae										■		■												
Dexamine spinosa						■							■	■	■	■					■	■	■	■
Dexamine thea						■				■														■
Disporella hispida													■						■			■		
Dysidea fragilis																		■					■	
Elasmopus rapax		■						■		■									■					
Electra pilosa															■				■					
Elysia viridis			■		■		■						■	■	■	■			■	■	■	■	■	■
Emarginula fissura														■			■							
Epilepton clarkiae	■	■	■	■	■	■	■	■	■	■	■	■	■	■	■	■	■	■	■	■	■	■	■	■
Epitonium clathrus						■					■		■											
Ericthonius brasiliensis													■											
Eubranchus farrani	■	■	■	■			■																	
Exogene gemmifera	■	■	■		■	■	■	■	■	■	■		■	■	■	■					■	■	■	■
Foraminifera A	■	■	■	■	■	■	■	■	■	■	■	■	■	■	■	■	■	■		■	■	■		■
Foraminifera B			■	■	■		■	■	■	■		■	■	■				■	■			■	■	■
Foraminifera C			■																					
Foraminifera F			■		■		■		■	■	■	■	■	■	■	■	■	■	■	■	■	■	■	■
Foraminifera G	■		■		■	■	■	■	■	■	■	■	■	■	■	■		■		■		■		
Galathea squamifera																					■		■	

Table S23 Table S22 continued

	Dec 06				Feb 07				Apr 07				Jun 07				Aug 07				Dec 07			
Taxa	1	2	3	4	1	2	3	4	1	2	3	4	1	2	3	4	1	2	3	4	1	2	3	4
Gammaropsis maculata																								
Gammarus locusta																								
Gibbula umbilicalis																								
Halacarellus basteri																								
Harpacticoida																								
Hiatella arctica																								
Hymenopteran A																								
Isopod A																								
Isopod B																								
Janua pagenstecheri																								
Lembos webstri																								
Leptocheirus tricristatus																								
Leptochelia savignyi																								
Leptocythere pellucida																								
Leptomysis lingvura																								
Loxoconcha rhomboidea																								
Lysianassa ceratina																								
Melita palmata																								
Microdeutopus anomalus																								
Microprotopus maculatus																								
Modiolula phaseolina																								
Monia patelliformis																								
Munna kroyeri																								
Musculus discors																								
Mytilus edulis																								
Nannastacus unguiculatus																								
Nematoda spp.																								
Nereis A																								
Odostomia plicata																								
Omalogyra atomus																								
Onoba semicosta																								
Ophiothrix fragilis																								
Ophiura ophiura																								
Ostracod A																								
Paradoxostoma variabile																								
Parvicardium exiguum																								
Parvicardium ovale																								
Parvicardium scabrum																								
Perinereis cultrifera																								
Phyllodocid A																								
Platynereis dumerili																								
Pomatoceros lamarcki																								
Pomatoceros triqueter																								
Pontocypris mytiloides																								
Pseudoparatanais batei																								
Retusa truncatula																								
Rissoa parva																								
Rissoa sarsi																								
Rissoella diaphana																								
Rissoella opalina																								
Sabella pavonina																								
Sagitta elegans																								
Scrupocellaria spp.																								
Semibalanus balanoides																								
Semicytherura nigrescens																								
Siriella armata																								
Skenea serpuloides																								
Stenothoe marina																								
Syllidae A																								
Syllidae B																								
Tapes aureus																								
Tomopteris helgolandica																								
Tryphosella sarsi																								
Tubulipora liliacea																								
Turbellaria A																								
Typosyllis prolifera																								
Xestoleberis aurantia																								

Figure S5 Replicate food webs for the $W^{+}S^{-3}$ treatment.

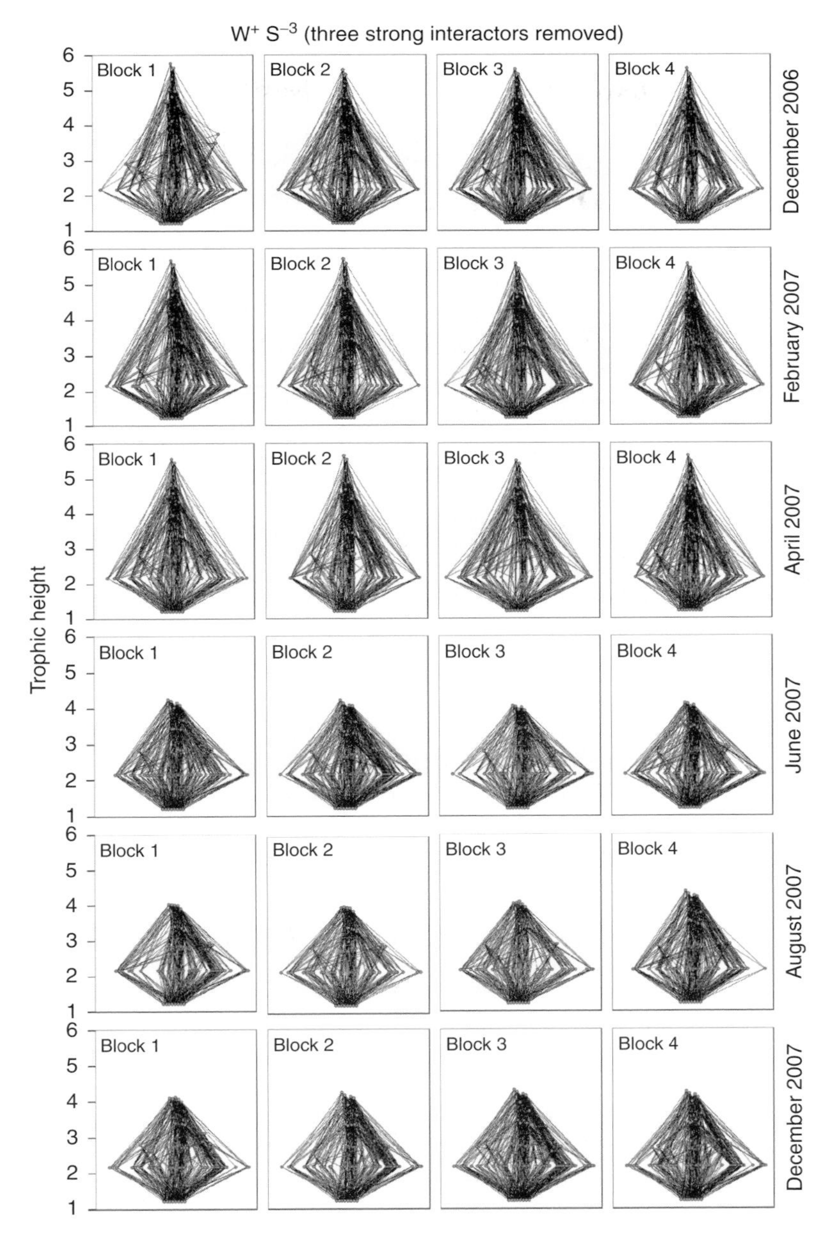

Table S24 Species present in each of the food webs shown in Figure S1 above (grey squares indicate species presence in a web; white squares indicate species absence from a web)

	Dec 06				Feb 07				Apr 07				Jun 07				Aug 07				Dec 07			
Taxa	1	2	3	4	1	2	3	4	1	2	3	4	1	2	3	4	1	2	3	4	1	2	3	4
Carcinus maenas	■	■	■	■	■	■	■	■	■	■	■	■	■	■	■	■	■	■	■	■	■	■	■	■
Ctenolabrus rupestris	■	■	■	■	■	■	■	■	■	■	■	■	■	■	■	■	■	■	■	■	■	■	■	■
Gaidropsarus mediterraneus	■	■	■	■	■	■	■	■	■	■	■	■												
Gobius niger	■	■	■	■	■	■	■	■	■	■	■	■	■	■	■	■	■	■	■	■	■	■	■	■
Gobius paganellus	■	■	■	■	■	■	■	■	■	■	■	■	■	■	■	■	■	■	■	■	■	■	■	■
Marthasterias glacialis	■	■	■	■	■	■	■	■	■	■	■	■												
Necora puber	■	■	■	■	■	■	■	■	■	■	■	■	■	■	■	■	■	■	■	■	■	■	■	■
Palaemon serratus	■	■	■	■	■	■	■	■	■	■	■	■												
Paracentrotus lividus	■	■	■	■	■	■	■	■	■	■	■	■	■	■	■	■	■	■	■	■	■	■	■	■
Taurulus bubalis	■	■	■	■	■	■	■	■	■	■	■	■	■	■	■	■	■	■	■	■	■	■	■	■
Gobiusculus flavescens	■	■	■	■	■	■	■	■	■	■	■	■	■	■	■	■	■	■	■	■	■	■	■	■
Pomatoschistus pictus	■	■	■	■	■	■	■	■	■	■	■	■	■	■	■	■	■	■	■	■	■	■	■	■
Abra alba	■						■				■													
Aequipecten opercularis							■																	■
Alvania beani		■	■		■	■			■					■	■	■		■			■	■	■	■
Alvania semistriata																					■		■	■
Anomia ephippium		■	■	■	■			■	■		■		■	■	■	■	■	■	■	■	■	■	■	■
Aora gracilis			■	■	■	■	■					■		■				■	■					
Apseudes latreillei		■	■											■		■				■		■	■	■
Apseudes talpa	■				■	■	■	■		■													■	
Ascidiella aspersa					■	■	■	■		■	■						■							
Bittium reticulatum	■	■	■	■	■	■	■	■	■	■	■	■	■	■	■	■		■	■	■	■	■	■	■
Buccinum undatum														■		■								■
Callopora lineata													■			■			■				■	
Ceradocus semiserratus			■	■			■	■									■							
Cerithiopsis tubercularis		■	■			■	■		■		■	■	■	■		■		■	■		■	■	■	■
Chlamys varia	■	■							■		■		■		■						■	■	■	
Circulus striatus									■	■		■	■											
Clathrina coriacea			■		■				■	■			■	■				■	■		■	■	■	■
Crassicorophium bonnellii				■								■								■	■			■
Crassicorophium crassicorne	■	■	■	■	■	■	■	■	■	■	■	■	■	■	■	■	■	■	■	■	■	■	■	■
Crisia denticulata											■									■				■
Cumacea A						■				■		■	■		■									
Cyclopoida	■	■	■	■	■	■	■	■	■	■	■	■	■			■	■	■	■	■				
Cyprid larvae												■												
Dexamine spinosa		■								■			■	■	■	■			■	■	■	■	■	■
Dexamine thea																		■						
Disporella hispida																					■			
Dysidea fragilis															■				■	■	■		■	■
Elasmopus rapax				■				■																
Electra pilosa													■											
Elysia viridis														■	■					■	■	■	■	■
Epilepton clarkiae	■	■	■	■	■	■	■	■	■	■	■	■	■	■	■	■	■	■	■	■	■	■	■	■
Ericthonius brasiliensis												■												
Eubranchus farrani	■	■				■		■					■											
Exogene gemmifera	■	■	■		■	■	■		■	■			■	■	■	■		■		■	■		■	
Foraminifera A	■	■	■	■		■	■	■	■	■		■	■	■		■	■	■	■				■	
Foraminifera B	■	■	■	■	■			■	■	■		■	■	■	■	■	■	■	■	■				
Foraminifera F			■			■	■	■	■	■	■	■	■	■	■	■	■	■	■	■	■	■	■	■
Foraminifera G	■	■	■		■	■	■	■	■	■		■	■				■	■	■	■	■			
Galathea squamifera																■						■	■	■
Gammaropsis maculata	■	■	■	■																				
Gammarus locusta						■						■												
Gammarus zaddachi					■																			
Gibbula umbilicalis												■									■		■	■
Halacarellus basteri		■																						
Harpacticoida	■	■	■	■	■	■	■	■	■	■	■	■	■	■	■	■	■	■	■	■	■	■	■	■
Hiatella arctica	■	■	■	■	■	■	■	■	■	■	■	■	■	■	■	■	■	■		■	■	■	■	■
Hydrozoa	■																							
Isopod A	■	■	■									■		■		■								

Table S25 Table S24 continued

	Dec 06				Feb 07				Apr 07				Jun 07				Aug 07				Dec 07			
Taxa	1	2	3	4	1	2	3	4	1	2	3	4	1	2	3	4	1	2	3	4	1	2	3	4
Isopod B																								
Janua pagenstecheri																								
Lembos websteri																								
Leptocythere pellucida																								
Loxoconcha rhomboidea																								
Lysianassa ceratina																								
Melita palmata																								
Microdeutopus anomalus																								
Microprotopus maculatus																								
Modiolula phaseolina																								
Monia patelliformis																								
Munna kroyeri																								
Musculus discors																								
Mytilus edulis																								
Nannastacus unguiculatus																								
Nematoda spp.																								
Nereis A																								
Odostomia plicata																								
Omalogyra atomus																								
Onoba semicosta																								
Ophiothrix fragilis																								
Ophiura ophiura																								
Ostracod A																								
Paradoxostoma variabile																								
Parvicardium exiguum																								
Parvicardium ovale																								
Parvicardium scabrum																								
Perinereis cultrifera																								
Phyllodocid A																								
Pilumnus hirtellus																								
Platynereis dumerili																								
Pomatoceros lamarcki																								
Pomatoceros triqueter																								
Pontocypris mytiloides																								
Pseudoparatanais batei																								
Retusa truncatula																								
Rissoa parva																								
Rissoa sarsi																								
Rissoella diaphana																								
Rissoella opalina																								
Sabella pavonina																								
Sagitta elegans																								
Scrupocellaria spp.																								
Semibalanus balanoides																								
Semicytherura nigrescens																								
Siriella armata																								
Skenea serpuloides																								
Spirorbis B																								
Stenothoe marina																								
Syllidae A																								
Syllidae B																								
Tapes aureus																								
Tomopteris helgolandica																								
Tryphosella sarsi																								
Tubulipora liliacea																								
Turbellaria A																								
Typosyllis prolifera																								
Xestoleberis aurantia																								

Figure S6 Replicate food webs for the W^-S^- treatment.

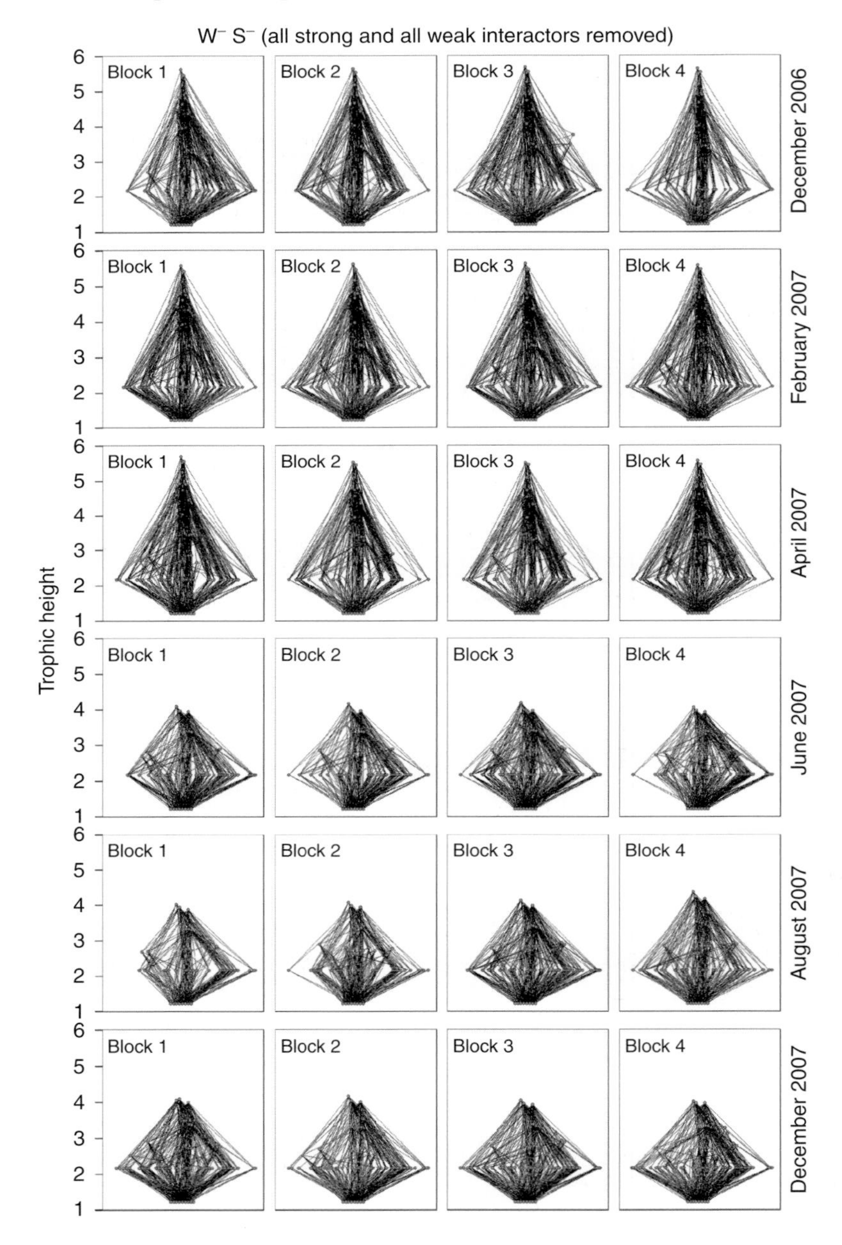

Table S26 Species present in each of the food webs shown in Figure S1 above (grey squares indicate species presence in a web; white squares indicate species absence from a web)

	Dec 06				Feb 07				Apr 07				Jun 07				Aug 07				Dec 07			
Taxa	1	2	3	4	1	2	3	4	1	2	3	4	1	2	3	4	1	2	3	4	1	2	3	4
Carcinus maenas	■	■	■	■	■	■	■	■	■	■	■	■												
Ctenolabrus rupestris	■	■	■	■	■	■	■	■	■	■	■	■	■	■	■	■	■	■	■	■	■	■	■	■
Gaidropsarus mediterraneus	■	■	■	■	■	■	■	■	■	■	■	■												
Gobius niger	■	■	■	■	■	■	■	■	■	■	■	■	■	■	■	■	■	■	■	■	■	■	■	■
Gobius paganellus	■	■	■	■	■	■	■	■	■	■	■	■												
Marthasterias glacialis	■	■	■	■	■	■	■	■	■	■	■	■												
Necora puber	■	■	■	■	■	■	■	■	■	■	■	■												
Palaemon serratus	■	■	■	■	■	■	■	■	■	■	■	■												
Paracentrotus lividus	■	■	■	■	■	■	■	■	■	■	■	■	■	■	■	■	■	■	■	■	■	■	■	■
Taurulus bubalis	■	■	■	■	■	■	■	■	■	■	■	■	■	■	■	■	■	■	■	■	■	■	■	■
Gobiusculus flavescens	■	■	■	■	■	■	■	■	■	■	■	■	■	■	■	■	■	■	■	■	■	■	■	■
Pomatoschistus pictus	■	■	■	■	■	■	■	■	■	■	■	■	■	■	■	■	■	■	■	■	■	■	■	■
Abra alba			■	■		■		■																
Acanthochitona crinitus																								■
Aequipecten opercularis								■					■				■							
Alvania beani						■		■	■	■		■		■	■	■		■		■	■	■	■	■
Alvania semistriata									■		■										■	■	■	■
Anomia ephippium		■	■		■	■	■	■		■	■	■							■	■	■		■	■
Aora gracilis	■	■	■		■	■	■								■		■	■	■	■				
Apseudes latreillei									■						■				■		■	■	■	■
Apseudes talpa					■	■	■	■	■	■		■	■											
Ascidiella aspersa					■		■	■					■								■			
Bittium reticulatum	■	■	■	■	■	■	■	■	■	■	■	■	■	■	■	■	■	■	■	■	■	■	■	■
Buccinum undatum														■	■									
Callopora lineata			■		■																		■	■
Ceradocus semiserratus							■		■	■														
Cerithiopsis tubercularis			■	■	■		■	■	■			■	■		■			■	■	■	■	■	■	■
Chironomidae spp.															■	■			■					
Chlamys varia		■	■	■	■	■			■		■	■			■					■	■	■	■	■
Circulus striatus					■						■	■						■						
Clathrina coriacea					■		■	■	■							■		■			■	■	■	■
Coriandria fulgida															■	■			■					
Crassicorophium bonnellii												■									■			
Crassicorophium crassicorne	■	■	■	■	■	■	■	■	■	■		■	■	■	■	■	■	■	■	■	■	■	■	■
Crisia denticulata										■	■													
Cumacea A											■											■		
Cyclopoida	■	■	■	■	■	■	■	■	■	■	■	■	■	■	■				■	■			■	
Cyprid larvae									■		■													
Dexamine spinosa													■	■	■	■			■		■	■	■	■
Disporella hispida															■					■	■		■	
Dysidea fragilis																		■			■			
Elasmopus rapax						■			■	■														
Elysia viridis															■	■	■				■	■	■	■
Emarginula fissura																				■				
Epilepton clarkiae	■	■	■	■	■	■	■	■	■	■	■	■	■	■	■	■	■	■	■	■	■	■	■	■
Epitonium clathrus																		■						
Ericthonius brasiliensis																					■	■		
Eubranchus farrani		■			■																			
Exogene gemmifera	■	■	■	■	■		■	■	■	■		■	■	■	■	■			■				■	■
Foraminifera A	■	■	■	■		■	■	■	■		■	■	■	■	■	■	■	■	■	■	■			
Foraminifera B		■				■	■	■	■		■	■	■	■	■			■	■	■	■	■	■	
Foraminifera C			■																					
Foraminifera F	■	■				■		■	■	■	■	■	■	■	■	■	■	■	■	■	■	■	■	■
Foraminifera G		■			■	■	■	■	■		■	■		■	■	■		■		■	■		■	
Galathea squamifera																	■				■			
Gammaropsis maculata	■	■	■		■																			
Gammarus locusta					■	■						■										■		
Gibbula umbilicalis										■		■		■	■	■		■						■
Halacarellus basteri					■				■															
Harpacticoida	■	■	■	■	■	■	■	■	■	■	■	■	■	■	■	■	■	■	■	■	■	■	■	■

Table S27 Table S26 continued

	Dec 06				Feb 07				Apr 07				Jun 07				Aug 07				Dec 07			
Taxa	1	2	3	4	1	2	3	4	1	2	3	4	1	2	3	4	1	2	3	4	1	2	3	4
Hiatella arctica																								
Hydrozoa																								
Hymenopteran A																								
Isopod A																								
Isopod B																								
Janua pagenstecheri																								
Lembos websteri																								
Leptocheirus tricristatus																								
Leptocythere pellucida																								
Leptomysis lingvura																								
Loxoconcha rhomboidea																								
Lysianassa ceratina																								
Melita palmata																								
Microdeutopus anomalus																								
Microprotopus maculatus																								
Modiolula phaseolina																								
Monia patelliformis																								
Munna kroyeri																								
Musculus discors																								
Mytilus edulis																								
Nannastacus unguiculatus																								
Nematoda spp.																								
Nereis A																								
Omalogyra atomus																								
Onoba semicosta																								
Ophiothrix fragilis																								
Ophiura ophiura																								
Ostracod A																								
Paradoxostoma variabile																								
Parvicardium exiguum																								
Parvicardium ovale																								
Parvicardium scabrum																								
Perinereis cultrifera																								
Phyllodocid A																								
Pilumnus hirtellus																								
Platynereis dumerili																								
Pomatoceros lamarcki																								
Pomatoceros triqueter																								
Pontocypris mytiloides																								
Pseudoparatanais batei																								
Retusa truncatula																								
Rissoa parva																								
Rissoa sarsi																								
Rissoella diaphana																								
Rissoella opalina																								
Sabella pavonina																								
Sagitta elegans																								
Scrupocellaria spp.																								
Semibalanus balanoides																								
Semicytherura nigrescens																								
Siriella armata																								
Skenea serpuloides																								
Stenothoe marina																								
Syllidae A																								
Syllidae B																								
Tapes aureus																								
Tectura virginia																								
Tomopteris helgolandica																								
Tryphosella sarsi																								
Tubulipora liliacea																								
Turbellaria A																								
Xestoleberis aurantia																								

APPENDIX V. LISTS OF SPECIES COMPRISING EACH OF THE COMPOSITE WEBS SHOWN IN FIGURE 4

Table S28 List of species comprising each of the composite webs shown in Figure 4

	W^+S^+				$W^{-2}S^+$				$W^{-3}S^+$				W^+S^{-2}				W^+S^{-3}				W^-S^-			
	Pre		Post		Pre		Post		Pre		Post		Pre		Post		Pre		Post		Pre		Post	
Species	TH	TL	TH	TL	TH	TL	TH	TL	TH	TL	TH	TL	TH	TL	TH	TL	TH	TL	TH	TL	TH	TL	TH	TL
Carcinus maenas	4.7	30	5.1	30	4.7	31	5.0	28	4.7	29			4.7	32	5.0	30	4.7	31	3.4	28	4.7	29		
Ctenolabrus rupestris	4.9	50	5.2	54	4.9	48	5.1	53	4.9	47	5.0	55	4.9	53	5.1	52	4.9	52	4.3	49	4.9	51	4.3	51
Gaidropsarus mediterraneus	5.5	37	5.9	33	5.5	34	5.7	35	5.5	35	5.8	35	5.5	37			5.6	33			5.6	33		
Gobius niger	4.4	55	4.7	60	4.4	55	4.7	63	4.4	55	5.0	62	4.4	59	4.7	62	4.4	57	4.3	54	4.4	53	4.3	57
Gobius paganellus	4.8	64	5.2	64	4.8	63			4.8	64			4.8	66	5.1	65	4.9	65	4.3	60	4.9	61		
Marthasterias glacialis	5.7	17	6.0	23	5.7	18	5.3	20	5.7	19	4.7	21	5.6	20			5.7	20			5.7	19		
Necora puber	5.2	19	5.5	21	5.2	20			5.2	19			5.2	21	5.5	19	5.3	21	4.4	19	5.3	19		
Palaemon serratus	3.9	46	4.3	46	4.0	44	4.3	44	4.0	44	4.5	47	4.0	47	4.3	45	4.0	44			4.0	42		
Paracentrotus lividus	2	5	2	5	2	5	2	4	2	5	2	3	2	5	2	4	2	5	2	4	2	5	2	2
Taurulus bubalis	4.5	39	4.8	42	4.5	33	4.8	40	4.5	38	5.0	42	4.5	39	4.8	39	4.6	36	4.3	37	4.6	33	4.3	37
Gobiusculus flavescens	3.3	33	3.4	34	3.3	34	3.4	34	3.3	34	3.5	34	3.3	33	3.4	33	3.4	33	3.3	27	3.4	30	3.4	30
Pomatoschistus pictus	3.3	52	3.6	54	3.3	52	3.5	52	3.3	53	3.8	53	3.3	53	3.7	52	3.3	51	3.7	48	3.4	49	3.6	49
Abra alba	2	7	2	9	2	7	2	8	2	7			2	7	2	8	2	7			2	7		
Acanthochitona crinitus			2	5			2	5							2	5							2	5
Aequipecten opercularis	2.6	14	2.6	13	2.6	14	2.5	10	2.6	14	2.7	12	2.6	14	2.7	12	2.6	14	2.5	11	2.6	14	2.5	9
Alvania beani	2	7	2	8	2	7	2	7	2	7	2	7	2	7	2	6	2	7	2	6	2	7	2	5
Alvania semistriata			2	5	2	4	2	4			2	4			2	4			2	5	2	4	2	4
Amphilochus manudens									2	4														10
Anomia ephippium	2	15	2	15	2	15	2	13	2	15	2	12	2	15	2	14	2	15	2	13	2	15	2	

(continued)

Table S28 (*continued*)

Species	$W^{+}S^{+}$				$W^{-2}S^{+}$				$W^{-3}S^{+}$				$W^{+}S^{-2}$				$W^{+}S^{-3}$				$W^{-}S^{-}$			
	Pre		Post		Pre		Post		Pre		Post		Pre		Post		Pre		Post		Pre		Post	
	TH	TL	TH	TL	TH	TL	TH	TL	TH	TL	TH	TL	TH	TL	TH	TL	TH	TL	TH	TL	TH	TL	TH	TL
Aora gracilis	2.9	12	3.1	13	2	11	3.1	12	2	11	2	10	2	11	2.9	12	2.9	12	2	11	2	11	2.9	11
Apherusa bispinosa	2	6											2	6										
Apseudes latreillei	2	8	2	8	2	8	2	7	2	8	2	7	2	8	2	8	2	8	2	7	2	8	2	6
Apseudes talpa	2	6	2	6	2	6			2	6	2	6	2	6			2	6	2	6	2	6	2	6
Ascidiella aspersa	2	7	2	8	2	7	2	7	2	7	2	7	2	7	2	8	2	7	2	8	2	7	2	7
Asterina phylactica			3.5	12							3.8	12												
Bittium reticulatum	2.7	16	2.8	19	2.8	18	2.8	19	2.8	17	2.8	18	2.8	17	2.8	16	2.7	16	2.8	16	2.8	17	2.8	15
Buccinum undatum			3.5	8			3.5	8			3.8	9			3.8	8			3.5	9			3.5	9
Calanoida											2	19			2	18								
Callopora lineata	2	5	2	5	2	5	2	5			2	5	2	5	2	5			2	5	2	5	2	5
Caprella acanthifera							2	7			2	7												
Caprella equilibra			2.8	9	2.8	9			2.8	9														
Caprella linearis							2	6	2	7	2	6			2	6								
Ceradocus semiserratus	2	3	2	3	2	3					2	3	2	3			2	3	2	3	2	3		
Cerithiopsis tubercularis	2	10	2.5	13	2	10	2.5	12	2	10	2.5	12	2	10	2.5	10	2	10	2.5	10	2	10	2.5	9
Chironomidae spp.			2	8			2	7			2	7			2	8							2	6
Chlamys varia	2	6	2	10	2	10	2	8	2	10	2	8	2	10	2	9	2	10	2	9	2	10	2	7
Circulus striatus									2	6			2	6			2	6	2	5	2	6	2	4
Clathrina coriacea	2	10	2	7	2	7	2	9	2	7	2	8	2	7	2	9	2	8	2	8	2	7	2	10
Coriandria fulgida	2	8			2	10	2	10	2	10	2	9	2	10	2	10							2	8
Crassicorophium bonnellii			2	12	2	13	2	11	2	13	2	11	2	13	2	11	2	13	2	11	2	13	2	10

Crassicorophium crassicorne	2	10	2	14	2	15	2	13	2	15	2	13	2	15	2	13	2	15	2	12	2	15	2	11
Crisia denticulata	2	15	2	11			2	10	2	11	2	10	2	11	2	11	2	10	2	10	2	11		
Cumacea A			2	5	2	5	2	5			2	5	2	5	2	5	2	5	2	5	2	5	2	5
Cuthona A											2	1												
Cyclopoida	2	17	2	19	2	17	2	17	2	18	2.5	19	2	17	2.5	17	2	17	2	17	2	16	2	16
Dexamine spinosa	2	11	2	7	2	7	2	6	2	7	2	6	2	7	2	6	2	7	2	6			2	5
Dexamine thea					2	4							2	4	2	3			2	3				
Disporella hispida	2	7	2	4					2	4	2	4			2	4			2	4			2	4
Dysidea fragilis			2	11			2	13			2	12			2	13			2	12			2	14
Elasmopus rapax	2	4	2	6			2	5			2	5	2	6	2	5	2	6			2	6		
Electra pilosa	2	6	2	6			2	6			2	6			2	6			2	6				
Elysia viridis	2	3	2	3	2	3	2	3	2	3	2	3	2	3	2	2			2	2			2	2
Emarginula fissura							2.7	7							2.7	7							2.7	6
Epilepton clarkiae	2	9	2	11	2	9	2	10	2	9	2	10	2	9	2	9	2	9	2	10	2	9	2	9
Epitonium clathrus	2	1							2	1			2	1	2	1							2	1
Ericthonius brasiliensis	2.4	15	2.4	15			2.4	14							2.4	14	2.4	15					2.4	13
Ericthonius punctatus			2.4	15			2.4	14																
Eubranchus farrani	2	1			2	1	2	1	2	1			2	1			3.7	2	2	3	3.7	2		
Exogone gemmifera	2	5	2	6	2	5	2	6	2	5	2	6	2	5	2	6	2	5	2	5	2	5	2	5
Foraminifera A	2	9	2	11	2	9	2	10	2	9	2	9	2	9	2	11	2	9	2	10	2	9	2	8
Foraminifera B	2	9	2	11	2	9	2	9	2	9	2	9	2	9	2	11	2	9	2	10	2	9	2	8
Foraminifera C					2	5			2	5			2	5							2	5		
Foraminifera E					2	6	2	8																
Foraminifera F	2	8	2	10	2	8	2	10	2	8	2	9	2	8	2	10	2	8	2	10	2	8	2	9
Foraminifera G	2	6	2	8	2	6	2	7	2	6	2	7	2	6	2	8	2	6	2	8	2	6	2	8
Galathea squamifera			3.4	20			3.4	18			3.7	20			3.6	20			3.5	20			3.4	20
Gammaropsis maculata	2	5			2	5			2	5			2	5	2	4	2	5			2	5		
Gammarus locusta	2	8			2	8	2	7	2	8			2	8	2	7	2	8			2	8	2	6
Gammarus zaddachi																	2	6						

(*continued*)

Table S28 (*continued*)

Species	W^+S^+				$W^{-2}S^+$				$W^{-3}S^+$				W^+S^{-2}				W^+S^{-3}				W^-S^-			
	Pre		Post		Pre		Post		Pre		Post		Pre		Post		Pre		Post		Pre		Post	
	TH	TL	TH	TL	TH	TL	TH	TL	TH	TL	TH	TL	TH	TL	TH	TL	TH	TL	TH	TL	TH	TL	TH	TL
Gibbula umbilicalis	2.5	8	2.7	10	2.5	8	2.7	8			2.7	8	2.5	8	2.7	8	2.5	8	2.7	8	2.5	8	2.7	6
Halacarellus basteri	2	5	2	5	2	5			2	5	2	5	2	5			2	5			2	5		
Harpacticoida	2	22	2	25	2	22	2	22	2	23	2	23	2	22	2	21	2	23	2	22	2	22	2	20
Hiatella arctica	2	11	2	13	2	11	2	11	2	11	2	10	2	11	2	12	2	11	2	12	2	11	2	9
Hydrozoa																	2.8	11			2.8	10		
Idotea A	2.7	8	2.8	9	2.5	7	2.8	8	2.7	8	2.8	8	2.7	8	2.8	8	2.7	8	2.8	8	2.7	8	2.5	5
Idotea B	2.7	9	2.8	10	2.5	8	2.8	9	2.7	9	2.8	9	2.7	9	2.8	9			2.8	8	2.7	9	2.5	5
Janua pagenstecheri	2	8	2	9	2	8	2	8	2	8	2	8	2	8	2	8	2	8	2	8	2	8	2	7
Lembos websteri	2.6	17	2.6	17	2.6	17	2.6	16	2.6	17	2.6	16	2.6	17	2.6	16	2.6	17	2.6	16	2.6	17	2.6	15
Leptocheirus tricristatus							2	7					2	7	2	7					2	7	2	7
Leptochelia savignyi											2	6			2	7								
Leptocythere pellucida	2	5	2	5			2	5			2	5			2	5			2	5			2	5
Leptomysis lingvura	2	7			2	7	2	7	2	7	2	7	2	7	2	6					2	7		
Loxoconcha rhomboidea	2	12	2	12	2	12	2	11	2	12	2	10	2	12	2	12	2	12	2	11	2	12	2	9
Lysianassa ceratina	2	5	2	5	2	5	2	5	2	5	2	5			2	4	2	5	2	4	2	5	2	4
Melita palmata	2	9			2	9					2	7	2	9			2	9	2	7	2	9	2	5
Microdeutopus anomalus	2	11	2	11	2	11	2	10	2	11	2	10	2	11	2	10	2	11	2	10	2	11	2	9
Microprotopus maculatus	2	5			2	5			2	5			2	5			2	5	2	4	2	5		
Modiolula phaseolina	2.7	16	2.7	21	2.7	16	2.7	18	2.7	16	3	19	2.7	16	3	19	2.7	16	2.7	18	2.7	16	2.7	15
Monia patelliformis			2	13			2	11			2	10	2	13	2	12	2	13	2	12			2	9

Munna kroyeri	2	8	2	8	2	8			2	8	2	7	2	8	2	7	2	8	2	6	2	8		
Musculus discors	2.7	16	2.7	21	2.7	16	2.7	18	2.7	16	3	19	2.7	16	3	19	2.7	16	2.7	18	2.7	16	2.7	15
Mytilus edulis					2.7	16			2.7	16	3	19	2.7	16			2.7	16	2.7	18			2.7	15
Nannastacus unguiculatus	2.5	6	2.7	7	2.5	6	2.7	7	2.5	6	2.7	7	2.5	6	2.7	7	2.5	6	2.7	7	2.5	6	2.7	7
Nematoda spp.	2	13	2	15	2	12	2	13	2	12	2	11	2	12	2	12	2	14	2	10	2	13	2	10
Nereis A	2	9			2	9			2	9	2	10	2	9			2	9			2	9		
Odostomia plicata			3	2	3	2			3	2	3	1			3	2			3	1				
Omalogyra atomus	2	3	2	3			2	2			2	2			2	3			2	3			2	2
Onoba semicosta	2.4	8	2.6	10			2.6	9	2.4	8	2.6	9			2.6	9	2.4	8	2.6	9			2.6	9
Ophiothrix fragilis			2	14			2	12	2	12	2	12	2	12					2	12	2	12	2	11
Ophiura ophiura			3.4	25			3.3	23			3.7	25			3.6	22			3.5	24			3.5	23
Ostracod A	2	8	2	8	2	8	2	7	2	8	2	6	2	8	2	8	2	8	2	8	2	8	2	6
Paradoxostoma variabile	2	10	2	10	2	10	2	9	2	10	2	8	2	10	2	10	2	10	2	10	2	10	2	8
Parvicardium exiguum	2	11	2	14	2	11	2	12	2	11	2	11	2	11	2	13	2	11	2	14	2	11	2	11
Parvicardium ovale	2	12	2	15	2	12	2	13	2	12	2	12	2	12	2	14	2	12	2	14	2	12	2	11
Parvicardium scabrum	2	12	2	15	2	12	2	13	2	12	2	12	2	12	2	14	2	12	2	14	2	12	2	11
Perinereis cultrifera			2	10			2	9			2	9			2	9			2	8			2	7
Phyllodocid A	3.2	8	3.2	9	2.8	8	2.8	9	3.2	8	3.1	10	3.2	8	3.2	9	2.8	8	3.2	9	3.2	8	3.2	9
Pilumnus hirtellus			3.8	27			3.7	26			4.1	26							3.9	24			3.9	23
Platynereis dumerili	2.6	11	2.6	12	2	10	2	10	2.6	11	2.6	11	2.6	11	2.6	11	2	10	2.6	10	2.6	11	2.6	9
Pomatoceros lamarcki	2	12	2	14	2	13			2	13			2	12	2	13	2	12			2	12	2	9
Pomatoceros triqueter	2	12	2	14	2	13	2	11	2	13	2	11	2	12	2	13	2	12	2	13	2	12	2	9
Pontocypris mytiloides	2	7	2	7	2	7	2	7	2	7	2	7	2	7	2	7	2	7	2	6	2	7	2	6
Pseudoparatanais batei	2	7	2	7	2	7	2	6	2	7	2	6	2	7	2	7	2	7	2	6			2	5
Retusa truncatula			3	4			3	5			3	4			3	4			3	4			3	4
Rissoa parva	2	13	2.4	16	2	13	2.4	15	2	13	2.4	14	2	13	2.4	15	2	13	2.4	15	2	13	2.4	13
Rissoa sarsi	2	13	2	15	2	13	2	14	2	13	2	13	2	13	2	14	2	13	2	14	2	13	2	12
Rissoella diaphana	2	12	2	13	2	12	2	12			2	11	2	12	2	12	2	12	2	12			2	10

(*continued*)

Table S28 (*continued*)

Species	W^+S^+				$W^{-2}S^+$				$W^{-3}S^+$				W^+S^{-2}				W^+S^{-3}				W^-S^-			
	Pre		Post		Pre		Post		Pre		Post		Pre		Post		Pre		Post		Pre		Post	
	TH	TL	TH	TL	TH	TL	TH	TL	TH	TL	TH	TL	TH	TL	TH	TL	TH	TL	TH	TL	TH	TL	TH	TL
Rissoella opalina	2	11	2	12	2	11	2	11	2	11	2	11	2	11	2	11	2	11	2	11	2	11	2	10
Sabella pavonina	2	7	2	9					2	7	2	9	2	7	2	9			2	9	2	7	2	9
Sagitta elegans	2.8	11	2.8	12	2.8	11	2.8	11	2.8	11	3.1	12	2.8	11	3.1	13	2.8	11	2.8	11	2.8	11	2.8	10
Scrupocellaria spp.	2	11	2	11	2	11	2	10	2	11	2	10	2	11	2	11	2	10	2	10	2	11	2	9
Semibalanus balanoides	2.8	8	2.8	8			2.8	8	2.8	8	3.0	8			3.0	8	2.8	8	2.8	7	2.8	8	2.8	6
Semicytherura nigrescens	2	6			2	6			2	6			2	6			2	6			2	6		
Siriella armata	2.7	10	2.7	10			2.7	10			3	10	2.7	10					2.7	8			2.7	8
Skenea serpuloides	2	3			2	3	2	2	2	3	2	2	2	3	2	3	2	3	2	3	2	3	2	2
Spirorbis A	2	5																						
Spirorbis B																	2	4						
Stenothoe marina	2	6	2	6	2	6	2	6	2	6	2	6	2	6	2	6	3.4	7	2	6	3.4	7	2	6
Syllidae A	2	2	2	3	2	2	2	3	2	2	2	3	2	2	2	3	3.7	3	2	3	3.7	3	2	3
Syllidae B	2	2	2	3	2	2	2	3	2	2	2	3	2	2	2	3	3.7	3	2	3	3.7	3	2	3
Tapes aureus			2	6	2	6	2	6	2	6	2	6	2	6	2	5	2	6	2	5	2	6	2	5
Tectura virginia																							2.6	8
Tomopteris helgolandica			3.6	3			3.6	3			3.9	3			3.9	3			3.6	3			3.6	3
Tryphosella sarsi			2	2			2	2			2	2			2	2	2	2	2	2			2	2
Tubulipora liliacea	2	4	2	4	2	4	2	4	2	4	2	4	2	4	2	4	2	4	2	4			2	4
Turbellaria A	2	5	2	5			2	5	2	5	2	5	2	5	2	4	2	5	2	4	2	5	2	4
Typosyllis prolifera	2	2			2	2			2	2			2	2			3.7	3	2	3				
Vitreolina philippi			4.3	3																				
Xestoleberis aurantia	2	11	2	11	2	11	2	10	2	11	2	9	2	11	2	11	2	11	2	10	2	11	2	8

Algae	1	42	1	45	1	40	1	46	1	40	1	46	1	44	1	46	1	39	1	44	1	35	1	45
Bacteria	1	45	1	51	1	43	1	49	1	48	1	50	1	47	1	50	1	45	1	46	1	45	1	50
Cladocerans	1	5	1	7	1	5	1	7	1	5	1	8	1	6	1	6	1	6	1	7	1	5	1	6
CPOM	1	46	1	49	1	41	1	50	1	39	1	52	1	44	1	48	1	43	1	44	1	40	1	44
Diatoms	1	51	1	60	1	50	1	60	1	50	1	61	1	55	1	59	1	51	1	56	1	51	1	59
FPOM	1	77	1	82	1	72	1	83	1	74	1	86	1	78	1	85	1	75	1	77	1	73	1	80
Microphytobenthos	1	39	1	41	1	37	1	43	1	36	1	41	1	39	1	45	1	38	1	38	1	33	1	44
Phytoplankton	1	33	1	39	1	29	1	35	1	33	1	36	1	34	1	37	1	33	1	35	1	33	1	36
Cyprid larvae	1	7			1	8			1	8			1	8			1	8			1	8		
Hymenopteran larvae			1	1							1	1			1	1							1	1

The trophic height (TH) and number of trophic links (TL) for each species are also provided.

REFERENCES

Agrawal, V.P. (1964). Functional morphology of the stomach of *Caprella linearis* (L.). *Curr. Sci.* **33**, 618–619.

Aitken, A.E., and Fournier, J. (1993). Macrobenthos communities of Cambridge, McBeth and Itirbilung Fjords, Baffin Island, Northwest-territories, Canada. *Arctic* **46**, 60–71.

Alan, R. (1970). The Feeding of *Caprella equilibra* Say, 1818 (Amphipoda: Crustacea). PhD Thesis, San Diego State College, pp. 1–91.

Allen, J.R. (1998). Suspension feeding in the brittle-star *Ophiothrix fragilis*: Efficiency of particle retention and implications for the use of encounter-rate models. *Mar. Biol.* **132**, 383–390.

Alvarez-Cadena, J.N. (1993). Feeding of the chaetognath *Sagitta elegans* Verrill. *Estuar. Coast. Shelf Sci.* **36**, 195–206.

Amouroux, J.M., Gremare, A., and Amouroux, J. (1988). Modelling of consumption and assimilation in Abra alba Mollusca Bivalvia. *Mar. Ecol. Prog. Ser.* **51**, 87–98.

Amouroux, J.M., Gremare, A., and Cahet, G. (1991). Experimental study of the interactions between a natural c-14 radiolabeled sediment and a deposit-feeding bivalve—Abra alba. *Oceanol. Acta* **14**, 589–597.

Ankel, W.E. (1937). Beobachtungen an Prosobranchiern der schwedischen westkuste. *Ark. Zool.* **30**, 1–28.

Antoniadou, C., Koutsoubas, D., and Chintiroglou, C.C. (2005). Mollusca fauna from infralittoral hard substrate assemblages in the North Aegean Sea. *Belg. J. Zool.* **135**, 119–126.

Athersuch, J., and Whittaker, J.E. (1976). On Loxoconcha rhomboidea. *Stereo-Atlas Ostracod Shell.* **3**, 81–90.

Athersuch, J., Horne, D.J., and Whittaker, J.E. (1989). Marine and Brackish Water Ostracods (Superfamilies Cypridacea and Cytheracea). London.

Austin, H.A., Austin, W.E., and Paterson, D.M. (2005). Extracellular cracking and content removal of the benthic diatom *Pleurosigma angulatum* (Quekett) by the benthic foraminifera *Haynesina germanica* (Ehrenberg). *Mar. Micropaleontol.* **57**, 68–73.

Bachelet, G., and Cornet, M. (1981). Some data on the life-history of Abra alba (Mollusca, Bivalvia) in the Southern part of the Bay of Biscay. *Ann. Inst. Oceanogr.* **57**, 111–123.

Baird, D., and Ulanowicz, R.E. (1989). The seasonal dynamics of the Chesapeake Bay ecosystem. *Ecol. Monogr.* **59**, 329–364.

Ballantine, D., and Morton, J.E. (1956). Filtering, feeding, and digestion in the Lamellibranch *Lasaea rubra. J. Mar. Biol. Assoc. UK* **35**, 241–274.

Barnes, R.D. (1963). Invertebrate Zoology. Saunders.

Barrio Frojan, C.R.S., Hawkins, L.E., Aryuthaka, C., Nimsantijaroen, S., Kendall, M.A., and Paterson, G.L.J. (2005). Patterns of polychaete communities in tropical sedimentary habitats: A case study in south-western Thailand. *Raffles Bull. Zool.* **53**, 1–11.

Bartsch, I. (2006). Halacaroidea (Acari): A guide to marine genera. *Org. Divers. Evol.* **6**.

Batistic, M., Mikus, J., and Njire, J. (2003). Chaetognaths in the South Adriatic: Vertical distribution and feeding. *J. Mar. Biol. Assoc. UK* **83**, 1301–1306.

Beninger, P.G., Lepennec, M., and Donval, A. (1991). Mode of particle ingestion in 5 species of suspension-feeding bivalve Mollusks. *Mar. Biol.* **108**, 255–261.

Berry, A.J. (1994). Foraminiferan prey in the annual life-cycle of the predatory opisthobranchs gastropod *Retusa obtusa* (Montagu). *Estuar. Coast. Shelf Sci.* **38**, 603–612.

Bersier, L.F., and Sugihara, G. (1997). Scaling regions for food web properties. *Proc. Natl. Acad. Sci. USA* **94**, 1247–1251.

Bertness, M.D., Gaines, S.D., Bermudez, D., and Sanford, E. (1991). Extreme spatial variation in the growth and reproductive output of the acorn barnacle *Semibalanus balanoides*. *Mar. Ecol. Prog. Ser.* **75**, 91–100.

Best, M.A., and Thorpe, J.P. (1986a). Effects of food particle concentration on feeding current velocity in 6 species of marine Bryozoa. *Mar. Biol.* **93**, 255–262.

Best, M.A., and Thorpe, J.P. (1986b). Feeding-current interactions and competition for food among the Bryozoan epiphytes of *Fucus serratus*. *Mar. Biol.* **93**, 371–375.

Bidder, G.P. (1920). Notes on the physiology of sponges. *J. Linn. Soc. Lond. Zool.* **13**, 315–326.

Blackburn, T.M., and Gaston, K.J. (1997). A critical assessment of the form of the interspecific relationship between abundance and body size in animals. *J. Anim. Ecol.* **66**, 233–249.

Blazewicz-Paszkowycz, M., and Ligowski, R. (2002). Diatoms as food source indicator for some Antarctic Cumacea and Tanaidacea (Crustacea). *Antarct. Sci.* **14**, 11–15.

Bode, A. (1989). Production of the intertidal chiton *Acanthochitona crinita* within a community of Corallina elongata (Rhodophyta). *J. Molluscan Stud.* **55**, 37–44.

Borgmann, U. (1987). Models on the slope of, and biomass flow up, the biomass size spectrum. *Can. J. Fish. Aquat. Sci.* **44**, 136–140.

Borja, A. (1986). Feeding and spatial distribution in 3 gastropod mollusks—Rissoa parva (Dacosta), Barleeia unifasciata (Montagu) and *Bittium reticulatum* (Dacosta). *Cah. Biol. Mar.* **27**, 69–75.

Borrvall, C., and Ebenman, B. (2006). Early onset of secondary extinctions in ecological communities following the loss of top predators. *Ecol. Lett.* **9**, 435–442.

Bradfield, A.E., and Newell, G.E. (1961). The behaviour of *Macoma balthca* (L.). *J. Mar. Biol. Assoc. UK* **41**, 81–87.

Bradshaw, C., Collins, P., and Brand, A.R. (2003). To what extent does upright sessile epifauna affect benthic biodiversity and community composition? *Mar. Biol.* **143**, 783–791.

Bradstreet, M.S.W., and Cross, W.E. (1982). Trophic relationships at high arctic ice edges. *Arctic* **35**, 1–12.

Bramanti, L., Magagnini, G., and Santangelo, G. (2003). Settlement and recruitment: The first stages in the life cycle of two epibenthic suspension feeders (*Corallium rubrum* and *Anomia ephippium*). *Ital. J. Zool.* **70**, 175–178.

Briand, F., and Cohen, J.E. (1987). Environmental correlates of food chain length. *Science* **238**, 956–960.

Britton, J.C. (1993). In: *The Marine Biology of the South China Sea: Proceedings of the First International Conference on the Marine Biology of Hong Kong and the South China Sea, Hong Kong, 28 October–3 November 1990* (Ed. by B. Morton), pp. 357–391. Hong Kong University Press.

Brodeur, R.D., and Terazaki, M. (1999). Springtime abundance of chaetognaths in the shelf region of the northern Gulf of Alaska, with observations on the vertical distribution and feeding of *Sagitta elegans*. *Fish. Oceanogr.* **8**, 93–103.

Brose, U., Jonsson, T., Berlow, E.L., Warren, P., Banasek-Richter, C., Bersier, L.F., Blanchard, J.L., Brey, T., Carpenter, S.R., Blandenier, M.F.C., Cushing, L.,

Dawah, H.A., *et al.* (2006a). Consumer–resource body-size relationships in natural food webs. *Ecology* **87**, 2411–2417.
Brose, U., Williams, R.J., and Martinez, N.D. (2006b). Allometric scaling enhances stability in complex food webs. *Ecol. Lett.* **9**, 1228–1236.
Brown, R.A. (1976). Queens University of Belfast.
Brown, J.H. (1995). Macroecology. University Chicago Press.
Brown, J.H., and Gillooly, J.F. (2003). Ecological food webs: High-quality data facilitate theoretical unification. *Proc. Natl. Acad. Sci. USA* **100**, 1467–1468.
Brown, J.H., and Nicoletto, P.F. (1991). Spatial scaling of species composition—Body masses of North-American land mammals. *Am. Nat.* **138**, 1478–1512.
Bruno, J.F., and O'Connor, M.I. (2005). Cascading effects of predator diversity and omnivory in a marine food web. *Ecol. Lett.* **8**, 1048–1056.
Bubnova, N.P. (1972). The nutrition of the detritus-feeding mollusks *Macoma balthica* (L.) and *Portlandia arctica* (Gray) and their influence on bottom sedirnents. *Okeanologiya, Moscow* **12**, 899–905.
Byrnes, J., Stachowicz, J.J., Hultgren, K.M., Hughes, A.R., Olyarnik, S.V., and Thornber, C.S. (2006). Predator diversity strengthens trophic cascades in kelp forests by modifying herbivore behaviour. *Ecol. Lett.* **9**, 61–71.
Caine, E.A. (1974). Comparative functional morphology of feeding in three species of caprellids. *J. Exp. Mar. Biol. Ecol.* **15**, 81–96.
Calder, W.A. (1984). Size, Function, and Life History. Harvard University Press, Cambridge, MA.
Campani, E. (2007). Eulimidae Mediterranei: generalità e caratteri diagnostici. *Atti Primo Convegno Malacologico Pontino* 1–29.
Cannon, H. (1933). On the feeding mechanism of certain marine Ostracods. *Trans. R. Soc. Edinb.* **57**, 739–764.
Carpenter, S.R., Kitchell, J.F., and Hodgson, J.R. (1985). Cascading trophic interactions and lake productivity. *Bioscience* **35**, 634–639.
Cazaux, C. (1968). Etude morphologique du developpement larvaire d'annelides polychetes (Basin d'Arcachon). I. Aphroditidae, Chrysopetalidae. *Arch. Zool. Exp. Gen.* **109**, 477–543.
Chaban, E.M. (2000). Some materials for revision of opisthobranchs of the family Retusidae (Mollusca: Cephalaspidea). *Proc. Zool. Inst. RAS* **286**, 175.
Chapman, J.W. (2007). In: *The Light and Smith Manual: Intertidal Invertebrates from Central California to Oregon* (Ed. by J.T. Carlton), pp. 545–618. University of California Press, California.
Christensen, H.E. (1967). Ecology of *Hydractinia echinata* (Fleming) (Hydroidea, Athecata). I. Feeding biology. *Ophelia* **4**, 245–275.
Christian, R.R., and Luczkovich, J.J. (1999). Organizing and understanding a winter's seagrass foodweb network through effective trophic levels. *Ecol. Model.* **117**, 99–124.
Cohen, J.E., and Newman, C.M. (1991). Community area and food chain length—Theoretical predictions. *Am. Nat.* **138**, 1542–1554.
Cohen, J.E., Briand, F., and Newman, C.M. (1986). A stochastic-theory of community food webs. 3. Predicted and observed lengths of food-chains. *Proc. R. Soc. Lond. Ser. B Biol. Sci.* **228**, 317–353.
Cohen, J.E., Beaver, R.A., Cousins, S.H., Deangelis, D.L., Goldwasser, L., Heong, K.L., Holt, R.D., Kohn, A.J., Lawton, J.H., Martinez, N., O'Malley, R., Page, L.M., *et al.* (1993a). Improving food webs. *Ecology* **74**, 252–258.
Cohen, J.E., Pimm, S.L., Yodzis, P., and Saldana, J. (1993b). Body sizes of animal predators and animal prey in food webs. *J. Anim. Ecol.* **62**, 67–78.

Cohen, J.E., Jonsson, T., and Carpenter, S.R. (2003). Ecological community description using the food web, species abundance, and body size. *Proc. Natl. Acad. Sci. USA* **100**, 1781–1786.

Coll, M., Lotze, H.K., and Romanuk, T.N. (2008). Structural degradation in Mediterranean Sea food webs: Testing ecological hypotheses using stochastic and mass-balance modelling. *Ecosystems* **11**, 939–960.

Conlan, K.E. (1994). Amphipod crustaceans and environmental disturbance: A review. *J. Nat. Hist.* **28**, 519–554.

Cornelius, P.F.S. (1995). North-West European Thecate Hydroids and their Medusae. London.

Costa, A. (1960). Note préliminaire sur l'éthologie alimentaire de deux Caprellides de la Rade de Villefranche sur Mer. *Trav. Station Zool. Russe Villefranche* **19**, 103–105.

Costello, M.J. (1992). Abundance and spatial overlap of Gobies (Gobiidae) in Lough-Hyne, Ireland. *Environ. Biol. Fish.* **33**, 239–248.

Crawford, G.I. (1937). A review of the amphipod genus *Corophium*, with notes on the British species. *J. Mar. Biol. Assoc. UK* **21**, 589–630.

Crook, A.C., Long, M., and Barnes, D.K.A. (2000). Quantifying daily migration in the sea urchin *Paracentrotus lividus*. *J. Mar. Biol. Assoc. UK* **80**, 177–178.

Crump, R.G., and Emson, R.H. (1983). The natural history life history and ecology of the 2 British UK species of Asterina. *Field Stud.* **5**, 867–882.

Cyr, H., Downing, J.A., and Peters, R.H. (1997). Density-body size relationships in local aquatic communities. *Oikos* **79**, 333–346.

Damuth, J. (1981). Population-density and body size in mammals. *Nature* **290**, 699–700.

Damuth, J. (1987). Interspecific allometry of population-density in mammals and other animals—The independence of body-mass and population energy-use. *Biol. J. Linn. Soc.* **31**, 193–246.

Dauby, P.A. (1995). A delta-13C study of the feeding habits in four Mediterranean *Leptomysis* species (Crustacea: Mysidacea). *Mar. Ecol.* **16**, 93–102.

Davenport, J., Smith, R., and Packer, M. (2000). Mussels *Mytilus edulis*: Significant consumers and destroyers of mesozooplankton. *Mar. Ecol. Prog. Ser.* **198**, 131–137.

Davoult, D., and Gounin, F. (1995). Suspension-feeding activity of a dense *Ophiothrix fragilis* (Abildgaard) population at the water–sediment interface—Time coupling of food availability and feeding-behavior of the species. *Estuar. Coast. Shelf Sci.* **41**, 567–577.

Dawah, H.A., Hawkins, B.A., and Claridge, M.F. (1995). Structure of the parasitoid communities of grass-feeding chalcid wasps. *J. Anim. Ecol.* **64**, 708–720.

Day, J.H. (1967). A Monograph on the Polychaeta of Southern Africa. British Museum (Natural History), London.

De Jong-Moreau, L., Casnova, B., and Casanova, J.P. (2001). Detailed comparative morphology of the peri-oral structures of the Mysidacea and Euphausiacea (Crustacea): An indication for the food preference. *J. Mar. Biol. Assoc. UK* **81**, 235–241.

DeGrave, S., and Turner, J.R. (1997). Activity rhythms of the squat lobsters, *Galathea squamifera* and G-strigosa (Crustacea: Decapoda: Anomura) in south-west Ireland. *J. Mar. Biol. Assoc. UK* **77**, 273–276.

Dekker, R. (1989). The macrozoobenthos of the subtidal western Dutch Wadden sea. 1. Biomass and species richness. *Neth. J. Sea Res.* **23**, 57–68.

Dennell, R. (1934). The feeding mechanism of the Cumacean Crustacean Diastylis bradyi. *Trans. R. Soc. Edinb.* **58**, 125–142.

Dennell, R. (1937). On the feeding mechanism of the Apseudes Talpa, and the evolution of the Peracaridan feeding mechanisms. *Trans. R. Soc. Edinb.* **59**, 57–78.

Dirzo, R., and Raven, P.H. (2003). Global state of biodiversity and loss. *Annu. Rev. Environ. Resour.* **28**, 137–167.

Dixon, A.Y. (1944). Notes on certain aspects of the biology of *Cumopsis goodsiri* (Van Beneden) and some other cumaceans in relation to their environment. *J. Mar. Biol. Assoc. UK* **26**, 61–71.

Dixon, I.M.T., and Moore, P.G. (1997). A comparative study on the tubes and feeding behaviour of eight species of corophioid amphipoda and their bearing on phylogenetic relationships within the Corophioidea. *Philos. Trans. R. Soc. Lond. B Biol. Sci.* **352**, 93–112.

Dominici, S. (2001). Taphonomy and paleoecology of shallow marine macrofossil assemblages in a collisional setting (late Pliocene-early Pleistocene, western Emilia, Italy). *Palaios* **16**, 336–353.

Drumm, D.T. (2005). Comparison of feeding mechanisms, respiration, and cleaning behavior in two kalliapseudids, *Kalliapseudes macsweenyi and Psammokalliapseudes granulosus* (Peracarida: Tanaidacea). *J. Crust. Biol.* **25**, 203–211.

Duffy, J.E. (1990). Amphipods on seaweeds: Partners or pests? *Oecologia* **83**, 267–276.

Duffy, J.E. (2003). Biodiversity loss, trophic skew and ecosystem functioning. *Ecol. Lett.* **6**, 680–687.

Duffy, J.E., and Hay, M.E. (2000). Strong impacts of grazing amphipods on the organization of a benthic community. *Ecol. Monogr.* **70**, 237–263.

Dunne, J.A., Williams, R.J., and Martinez, N.D. (2002a). Food-web structure and network theory: The role of connectance and size. *Proc. Natl. Acad. Sci. USA* **99**, 12917–12922.

Dunne, J.A., Williams, R.J., and Martinez, N.D. (2002b). Network structure and biodiversity loss in food webs: Robustness increases with connectance. *Ecol. Lett.* **5**, 558–567.

Ebling, F.J., Kitching, J.A., Purchon, R.D., and Bassindale, R. (1948). The ecology of the Lough Ine rapids with special reference to water currents. 2. The fauna of the Saccorhiza canopy. *J. Anim. Ecol.* **17**, 223–244.

Ebling, F.J., Kitching, J.A., Muntz, L., and Taylor, C.M. (1964). The ecology of Lough Ine. 13. Experimental-observations of the destruction of *Mytilus-edulis* and *Nucella-lapillus* by crabs. *J. Anim. Ecol.* **33**, 73–82.

Ebling, F.J., Hawkins, A.D., Kitching, J.A., Muntz, L., and Pratt, V.M. (1966). Ecology of Lough Ine. 16. Predation and diurnal migration in *Paracentrotus* community. *J. Anim. Ecol.* **35**, 559–566.

Elton, C.S. (1927). Animal Ecology. Sidgwick & Jackson, London.

Emmerson, M.C., and Raffaelli, D. (2004). Predator–prey body size, interaction strength and the stability of a real food web. *J. Anim. Ecol.* **73**, 399–409.

Emmerson, M.E., Montoya, J.M., and Woodward, G. (2005). Body size, interaction strength and food web dynamics. In: *Dynamic Food Webs: Multispecies Assemblages, Ecosystem Development, and Environmental Change* (Ed. by P.C. De Ruiter, V. Wolters and J.C. Moore). Academic Press, Amsterdam.

Emson, R.H., and Crump, R.G. (1984). Comparative studies on the ecology of *Asterina gibbosa* and *Asterina phylactica* at Lough Ine. *J. Mar. Biol. Assoc. UK* **64**, 35–53.

Enequist, P. (1949). Studies on the soft-bottom amphipods of the Skagerrak. *Zool Bidr Uppsala* **28**, 299–492.
Enquist, B.J., Brown, J.H., and West, G.B. (1998). Allometric scaling of plant energetics and population density. *Nature* **395**, 163–165.
Estes, J.A., and Palmisano, J.F. (1974). Sea otters—Their role in structuring near-shore communities. *Science* **185**, 1058–1060.
Fahrig, L. (2003). Effects of habitat fragmentation on biodiversity. *Annu. Rev. Ecol. Evol. Syst.* **34**, 487–515.
Fauchald, K., and Jumars, P.A. (1979). The diet of worms—A study of polychaete feeding guilds. *Oceanogr. Mar. Biol. Ann. Rev.* **17**, 193–284.
Fernandez, E., Anadon, R., and Fernandez, C. (1988). Life histories and growth of the gastropods *Bittium reticulatum* and *Barleeia unifasciata* inhabiting the seaweed *Gelidium latifolium*. *J. Molluscan Stud.* **54**, 119–129.
Finke, D.L., and Denno, R.F. (2005). Predator diversity and the functioning of ecosystems: The role of intraguild predation in dampening trophic cascades. *Ecol. Lett.* **8**, 1299–1306.
Foster-Smith, R.L., and Shillaker, R.O. (1977). Tube irrigation by *Lembos websteri* Bate and *Corophium bonnelli* Milne Edwards (Crustacea: Amphipoda). *J. Exp. Mar. Biol. Ecol.* **26**, 289–296.
Foxon, G.E.H. (1936). Notes on the natural history of certain sand-dwelling Cumacea. *Ann. Mag. Nat. Hist.* **17**, 377–393.
Fretter, V. (1948). The structure and life history of some minute Prosobranchs of rock pools—*Skeneopsis planorbis* (Fabricius), *Omalogyra atomus* (Philippi), *Rissoella diaphana* (Alder) and *Rissoella opalina* (Jeffreys). *J. Mar. Biol. Assoc. UK* **27**, 597–632.
Fretter, V. (1951). Observations on the life history and functional morphology of *Cerithiopsis tubercularis* (Montagu) and *Triphora perversa* (l). *J. Mar. Biol. Assoc. UK* **29**, 567–586.
Fretter, V. (1956). The anatomy of the prosobranch *Circulus striatus* (Philippi) and a review of its systematic position. *Proc. Zool. Soc. Lond.* **126**, 369–381.
Fretter, V., and Graham, A. (1949). The structure and mode of life of the Pyramidellidae, parasitic opisthobranchs. *J. Mar. Biol. Assoc. UK* **28**, 493–532.
Fretter, V., and Graham, A. (1978). The Prosobranch Molluscs of Britain and Denmark Part 4 Marine Rissoacea. *J. Molluscan Stud. Suppl.* **6**, 153–241.
Fretter, V., and Graham, A. (1981). The Prosobranch Molluscs of Britain and Denmark. Part 4. Marine Rissoacea. *J. Molluscan Stud. Suppl.* **6**, 153–241.
Fretter, V., and Manly, R. (1979). Observations on the biology of some sublittoral prosobranchs. *J. Molluscan Stud.* **45**, 209–218.
Fretter, V., and Montgomery, M.C. (1968). The treatment of food by prosobranch veligers. *J. Mar. Biol. Assoc. UK* **48**, 499–520.
Gallop, A., Bartrop, J., and Smith, D.C. (1980). The biology of chloroplast acquisition by *Elysia viridis*. *Proc. R. Soc. Lond. B Biol. Sci.* **207**, 335–349.
Gardiner, J.S. (1903). In: *The Fauna and Geography of the Maldive and Laccadive Archipelagoes*. pp. 313–346. Cambridge University Press, Cambridge.
Goerke, H. (1971). Die Ernährungsweise der Nereis-Arten (Polychaeta, Nereidae) der Deutschen Küsten. *Veröff. Inst. Meeresforsch. Bremerhav.* **13**, 1–50.
Goff, L., and Cole, K. (1976). The biology of *Harveyella mirabilis* (Crytonemiales, Rhodophyceae). III. Spore germination and subsequent development within the host *Odonthalia floccosa* (Ceramiales, Rhodophyceae). *Can. J. Bot.* **54**, 268–280.
Goldfinch, A.C., and Carman, K.R. (2000). Chironomid grazing on benthic microalgae in a Louisiana salt marsh. *Estuaries* **23**, 536–547.

Goodhart, C.B. (1939). Notes on the bionomics of the tubebuilding amphipod Leptocheirus pilosus Zaddach. *J. Mar. Biol. Assoc. UK* **23**, 311–325.
Graham, A. (1939). On the structure of the alimentary canal of style-bearing prosobranchs. *Proc. Zool. Soc. Lond.* **109**, 75–112.
Graham, A. (1955). Molluscan diets. *J. Molluscan Stud.* **31**, 144–158.
Graham, A. (1988). Molluscs: Prosobranch and Pyramidellid Gastropods. London.
Grall, J., Le Loch, F., Guyonnet, B., and Riera, P. (2006). Community structure and food web based on stable isotopes (delta N-15 and delta C-13) analysis of a North Eastern Atlantic Maerl bed. *J. Exp. Mar. Biol. Ecol.* **338**, 1–15.
Greze, V.N., Baldina, E.M., and Bileva, O.K. (1968). Production of planktonic copepods in the neritic zone of the Black Sea. *Okeanologiya* **8**, 1066–1070.
Griffiths, D. (1992). Size, abundance, and energy use in communities. *J. Anim. Ecol.* **61**, 307–315.
Griffiths, D. (1998). Sampling effort, regression method, and the shape and slope of size-abundance relations. *J. Anim. Ecol.* **67**, 795–804.
Guerra-García, J.M., and Tierno de Figueroa, J.M. (2009). What do caprellids (Crustacea: Amphipoda) feed on? *Mar. Biol.* **156**, 1881–1890.
Guerra-Garcia, J.M., Corzo, J., and Garcia-Gomez, J.C. (2002). Clinging behaviour of the caprellidea (amphipoda) from the Strait of Gibraltar. *Crustaceana* **75**, 41–50.
Hall, S.J., and Raffaelli, D.G. (1991). Food web patterns: Lessons from a species-rich food web. *J. Anim. Ecol.* **60**, 823–842.
Hall, R.O., Wallace, J.B., and Eggert, S.L. (2000). Organic matter flow in stream food webs with reduced detrital resource base. *Ecology* **81**, 3445–3463.
Hancock, D.A. (1960). The ecology of molluscan enemies of the edible mollusc. *Proc. Malacol. Soc. Lond.* **34**, 134–143.
Harmon, J.P., Moran, N.A., and Ives, A.R. (2009). Species response to environmental change: Impacts of food web interactions and evolution. *Science* **323**, 1347–1350.
Harrison, R.D. (2000). Repercussions of El Nino: Drought causes extinction and the breakdown of mutualism in Borneo. *Proc. R. Soc. Lond. Ser. B Biol. Sci.* **267**, 911–915.
Hartmann-Schroder, G. (1971). Annelida, Borstenwürmer, Polychaeta. Die tierwelt Deutschlands. Gustav Fischer Verlag.
Havens, K. (1992). Scale and structure in natural food webs. *Science* **257**, 1107–1109.
Hawkins, A.J.S., Navarro, E., and Iglesias, J.I.P. (1990). Comparative allometries of gut-passage time, gut content and metabolic fecal loss in *Mytilus edulis* and *Cerastoderma edule*. *Mar. Biol.* **105**, 197–204.
Hayward, P.J., and Ryland, J.S. (1977). British Anascan Bryozoans: Cheilostomata, Anasca: Keys and Notes for the Identification of the Species. London.
Hayward, P.J., and Ryland, J.S. (1985). Cyclostome Bryozoans: Keys and Notes for the Identification of the Species. London.
Hayward, P.J., and Ryland, J.S. (1995). Handbook of the Marine Fauna of North-West Europe. Oxford University Press, New York.
Heckscher, E., Hauxwell, J., Jimenez, E.G., Rietsma, C., and Valiela, I. (1996). Selectivity by the herbivorous amphipod *Microdeutopus gryllotalpa* among five species of macroalgae. *Biol. Bull. (Woods Hole)* **191**, 324–326.
Hily, C. (1991). Is the activity of benthic suspension feeders a factor controlling water quality in the Bay of Brest France. *Mar. Ecol. Prog. Ser.* **69**, 179–188.
Himmelman, J.H., and Hamel, J.R. (1993). Diet, behavior and reproduction of the whelk *Buccinum undatum* in the Northern Gulf of St Lawrence, Eastern Canada. *Mar. Biol.* **116**, 423–430.

Hobson, E.S., and Chess, J.R. (1976). Trophic interactions among fish and zooplankters near shore at Santa Catalina Island, California. *Fish. Bull. Natl. Ocean. Atmos. Admin. Washington, DC* **74**, 567–598.

Holdich, D.M., and Jones, J.A. (1983). Tanaids: Keys and Notes for the Identification of the Species. London.

Holling, C.S. (1992). Cross-scale morphology, geometry, and dynamics of ecosystems. *Ecol. Monogr.* **62**, 447–502.

Holmes, J.M.C., and O'Connor, J.P. (1990). Collecting marine crustaceans with a light trap. In: *The Ecology of Lough Hyne: Proceedings of a Conference, 4–5 September, 1990* (Ed. by A.A. Myers, C. Little, M.J. Costello and J.C. Patridge). Royal Irish Academy, Dublin.

Holt, R.D. (1985). Population-dynamics in 2-patch environments—Some anomalous consequences of an optimal habitat distribution. *Theor. Popul. Biol.* **28**, 181–208.

Horne, D.J., and Whittaker, J.E. (1985). A revision of the genus Paradoxostoma Fischer (Crustacea, Ostracoda) in British waters. *Zool. J. Linn. Soc.* **85**, 131–203.

Hughes, R.G. (1975). The distribution of epizoites on the hydroid *Nemertesia antennina*. *J. Mar. Biol. Assoc. UK* **55**, 275–294.

Hummel, H. (1985). Food-intake of *Macoma balthica* (Mollusca) in relation to seasonal-changes in its potential food on a tidal flat in the Dutch Wadden Sea. *Neth. J. Sea Res.* **19**, 52–76.

Hunt, O.D. (1925). The food of the bottom fauna of the Plymouth fishing grounds. *J. Mar. Biol. Assoc. UK* **13**, 560–599.

Hutchinson, G.E. (1957). Population studies—Animal ecology and demography—Concluding remarks. *Cold Spring Harbor Symp. Quant. Biol.* **22**, 415–427.

Hutchinson, G.E. (1959). Homage to Santa Rosalia or why are there so many kinds of animals. *Am. Nat.* **93**, 145–159.

Hutchinson, G.E., and Macarthur, R.H. (1959). A theoretical ecological model of size distributions among species of animals. *Am. Nat.* **93**, 117–125.

Hyman, L.H. (1951). The invertebrates. II. Platyhelminthes and Rhychocoela, The Acoelomate Bilateria. McGraw Hill, New York.

Ings, T.C., Montoya, J.M., Bascompte, J., Bluthgen, N., Brown, L., Dormann, C.F., Edwards, F., Figueroa, D., Jacob, U., Jones, J.I., Lauridsen, R.B., Ledger, M.E., *et al.* (2009). Ecological networks—Beyond food webs. *J. Anim. Ecol.* **78**, 253–269.

Jackson, J.B.C., Kirby, M.X., Berger, W.H., Bjorndal, K.A., Botsford, L.W., Bourque, B.J., Bradbury, R.H., Cooke, R., Erlandson, J., Estes, J.A., Hughes, T.P., Kidwell, S., *et al.* (2001). Historical overfishing and the recent collapse of coastal ecosystems. *Science* **293**, 629–638.

Jacob, U. (2005). Trophic Dynamics of Antarctic Shelf Ecosystems—Food Webs and Energy Flow Budgets. University of Bremen, PhD Thesis.

Jacobs, R., Hermelink, P.M., and Vangeel, G. (1983). Epiphytic algae on eelgrass at Roscoff, France. *Aquat. Bot.* **15**, 157–173.

Jarrett, J.N., and Pechenik, J.A. (1997). Temporal variation in cyprid quality and juvenile growth capacity for an intertidal barnacle. *Ecology* **78**, 1262–1265.

Jebram, D. (1981). Influences of the food on the colony forms of Electra pilosa Bryozoa Cheilostomata. *Zool. Jahrb. Abt. Syst. Oekolog. Geogr. Tiere* **108**, 1–14.

Jennings, S., Pinnegar, J.K., Polunin, N.V.C., and Boon, T.W. (2001). Weak cross-species relationships between body size and trophic level belie powerful size-based trophic structuring in fish communities. *J. Anim. Ecol.* **70**, 934–944.

Jennings, S., Pinnegar, J.K., Polunin, N.V.C., and Warr, K.J. (2002). Linking size-based and trophic analyses of benthic community structure. *Mar. Ecol. Prog. Ser.* **226**, 77–85.

Jensen, K.R. (1989). Learning as a factor in diet selection by *Elysia viridis* (Montagu) (Opisthobranchia). *J. Molluscan Stud.* **55**, 79–88.
Jespersen, A., Lutzen, J., and Oliver, P.G. (2007). Morphology, biology and systematic position of Epilepton clarkiae (Clark, 1852) (Galeommatoidea: Montacutidae) a bivalve commensal with sipunculans. *J. Conchol.* **39**, 391–401.
Jimenez, E.G., Hauxwell, J., Heckscher, E., Rietsma, C., and Valiela, I. (1996). Selection of nitrogen-enriched macroalgae (Cladophora vagabunda and Gracilaria tikvahiae) by the herbivorous amphipod *Microdeutopus gryllotalpa. Biol. Bull. (Woods Hole)* **191**, 323–324.
Johnson, S.B., and Attramadal, Y.G. (1982). A functional–morphological model of Tanais cavolinii Milne-Edwards (Crustacea, Tanaidacea) adapted to a tubicolous life-strategy. *Sarsia* **67**, 29–42.
Johnson, M.P., Costello, M.J., and O'Donnell, D. (1995). The nutrient economy of a marine inlet Lough-Hyne, south-west Ireland. *Ophelia* **41**, 137–151.
Jonsson, T., and Ebenman, B. (1998). Effects of predator–prey body size ratios on the stability of food chains. *J. Theor. Biol.* **193**, 407–417.
Jonsson, T., Cohen, J.E., and Carpenter, S.R. (2005). Food webs, body size, and species abundance in ecological community description. *Adv. Ecol. Res.* **36**, 1–84.
Jormalainen, V., Honkanen, T., and Heikkila, N. (2001). Feeding preferences and performance of a marine isopod on seaweed hosts: Cost of habitat specialization. *Mar. Ecol. Prog. Ser.* **220**, 219–230.
Kamermans, P. (1994). Similarity in food source and timing of feeding in deposit-feeding and suspension-feeding bivalves. *Mar. Ecol. Prog. Ser.* **104**, 63–75.
Kaunzinger, C.M.K., and Morin, P.J. (1998). Productivity controls food-chain properties in microbial communities. *Nature* **395**, 495–497.
Keith, D.E. (1969). Aspects of feeding in *Caprella californica* Stimpson and *Caprella equilibra* Say (Amphipoda). *Crustaceana* **16**, 119–124.
Kerfoot, W.C. (1977). Implications of copepod predation. *Limnol. Oceanogr.* **22**, 316–325.
Kerfoot, W.C. (1978). Combat between predatory copepods and their prey: Cyclops, Epischura, and Bosmina. *Limnol. Oceanogr.* **23**, 1089–1102.
Kiorboe, T., and Mohlenberg, F. (1981). Particle selection in suspension-feeding bivalves. *Mar. Ecol. Prog. Ser.* **5**, 291–296.
Kitching, J.A., and Ebling, F.J. (1961). The ecology of Lough Ine. 11. The control of algae by *Paracentrotus lividus* (Echinoidea). *J. Anim. Ecol.* **30**, 373–383.
Kitching, J.A., and Thain, V.M. (1983). The ecology of Lough Ine. 22. The ecological impact of the sea-urchin *Paracentrotus lividus* (Lamarck) in Lough Ine, Ireland. *Philos. Trans. R. Soc. Lond., Ser. B, Biol. Sci.* **300**, 513–552.
Kitching, J.A., Sloane, J.F., and Ebling, F.J. (1959). The ecology of Lough Ine. 8. Mussels and their predators. *J. Anim. Ecol.* **28**, 331–341.
Kleiber, M. (1947). Body size and metabolic rate. *Physiol. Rev.* **27**, 511–541.
Kobusch, W. (1998). The foregut of the Mysida (Crustacea, Peracarida) and its phylogenetic relevance. *Philos. Trans. R. Soc. Lond. B Biol. Sci.* **353**, 559–581.
Koulouri, P., Dounas, C., Arvanitidis, C., Koutsoubas, D., and Eleftheriou, A. (2006). Molluscan diversity along a Mediterranean soft bottom sublittoral ecotone. *Sci. Mar.* **70**, 573–583.
Koutsoubas, D., Arvanitidis, C., Dounas, C., and Drummond, L. (2000). Community structure and dynamics of the molluscan fauna in a Mediterranean lagoon (Gialova lagoon, SW Greece). *Belg. J. Zool.* **130**, 135–142.
Kudinova-Pasternak, R.K. (1991). Troficheskie gruppy Tanaidacea (Crustacea, Peracarida). [Trophic groups of Tanaidacea (Crustacea, Peracarida)]. *Zool. Zh.* **70**, 30–37.

Kussakin, O.G. (1926). On the fauna of Munnidae (Isopoda, Asellota) from the Far-Eastern seas of the USSR. *Tr. Zool. Inst. Akad. Nauk. SSSR* **30**, 66–109.
Kyomo, J. (1999). Feeding patterns, habits and food storage in *Pilumnus vespertilio* (Brachyura: Xanthidae). *Bull. Mar. Sci.* **65**, 381–389.
Lake, B., Johnson, A.S., and Mauck, R.A. (2006). Influence of orientation and flow speed on feeding behavior and metabolism of the barnacle *Semibalanus balanoides*. *Integr. Comp. Biol.* **46**, 10.1093/icb/icl0560.
Larsen, T.S., Kristensen, J.A., Asmund, G., and Bjerregaard, P. (2001). Lead and zinc in sediments and biota from Maarmorilik, West Greenland: An assessment of the environmental impact of mining wastes on an Arctic fjord system. *Environ. Pollut.* **114**, 275–283.
Lawler, S.P., and Morin, P.J. (1993). Food-web architecture and population-dynamics in laboratory microcosms of protists. *Am. Nat.* **141**, 675–686.
Layer, K., Riede, J.O., Hildrew, A.G., and Woodward, G. (2010). Food web structure and stability in 20 streams across a wide pH gradient. *Adv. Ecol. Res.* **42**, 265–299.
Lebour, M. (1923). The food of plankton organisms. II. *J. Mar. Biol. Assoc. UK* **13**, 70–92.
Leguerrier, D., Niquil, N., Boileau, N., Rzeznik, J., Sauriau, P.G., Le Moine, O., and Bacher, C. (2003). Numerical analysis of the food web of an intertidal mudflat ecosystem on the Atlantic coast of France. *Mar. Ecol. Prog. Ser.* **246**, 17–37.
Lehane, C., and Davenport, J. (2002). Ingestion of mesozooplankton by three species of bivalve; *Mytilus edulis*, *Cerastoderma edule* and *Aequipecten opercularis*. *J. Mar. Biol. Assoc. UK* **82**, 615–619.
Lesser, M.P., Witman, J.D., and Sebens, K.P. (1994). Effects of flow and seston availability on scope for growth of benthic suspension-feeding invertebrates from the Gulf of Maine. *Biol. Bull.* **187**, 319–335.
Lewis, J.B. (1998). Occurrence and distribution of the tanaid crustacean *Leptochelia savignyi* on the calcareous hydrozoan *Millepora complanata*. *Bull. Mar. Sci.* **63**, 629–632.
Lincoln, R.J. (1979). British Marine Amphipoda: Gammaridea. British Museum (Natural History).
Link, J. (2002). Does food web theory work for marine ecosystems? *Mar. Ecol. Prog. Ser.* **230**, 1–9.
Little, C. (1990). Ecology of the rocky intertidal. In: *The ecology of Lough Hyne: Proceedings of a Conference, 4–5 September, 1990* (Ed. by A.A. Myers, C. Little, M.J. Costello and J.C. Patridge). Royal Irish Academy, Dublin.
Lobb, S.M. (1972) University of Reading, p. 209.
Loreau, M., Naeem, S., Inchausti, P., Bengtsson, J., Grime, J.P., Hector, A., Hooper, D.U., Huston, M.A., Raffaelli, D., Schmid, B., Tilman, D., and Wardle, D.A. (2001). Biodiversity and ecosystem functioning: Current knowledge and future challenges. *Science* **294**, 804–808.
Maasri, A., Fayolle, S., Gandouin, E., Garnier, R., and Franquet, E. (2008). Epilithic chironomid larvae and water enrichment: Is larval distribution explained by epilithon quantity or quality? *J. North Am. Benthol. Soc.* **27**, 38–51.
MacQuitty, M. (1984). In: *Acarology VI* (Ed. by D.A. Griffiths and C.E. Bowman), pp. 571–580. Halsted Press, New York.
Malaquias, M.A.E., Berecibar, E., and Reid, D.G. (2009). Reassessment of the trophic position of Bullidae (Gastropoda: Cephalaspidea) and the importance of diet in the evolution of cephalaspidean gastropods. *J. Zool.* **277**, 88–97.
Marshall, S.M. (1973). Respiration and feeding in copepods. *Adv. Mar. Biol.* **11**, 57–120.

Marshall, S.M., and Orr, A.P. (1955). The biology of a marine copepod *Calanus finmarchicus* (Gunnerus). Oliver & Boyd, Edinburgh.

Martinez, N.D. (1991). Artifacts or attributes—Effects of resolution on the little-rock lake food web. *Ecol. Monogr.* **61**, 367–392.

Martinez, N.D. (1992). Constant connectance in community food webs. *Am. Nat.* **139**, 1208–1218.

Martinez, N.D. (1993). Effects of resolution on food web structure. *Oikos* **66**, 403–412.

Mauchline, J. (1980). The biology of mysids and euphausiids. *Adv. Mar. Biol.* **18**, 13–63.

Mauchline, J. (1998). The Biology of Calanoid Copepods. Elsevier.

May, R.M. (1975). Patterns of species abundance and diversity. In: *Ecology of Species and Communities* (Ed. by M. Cody and J.M. Diamond). Harvard University Press, Cambridge, MA.

May, R.M. (1988). How many species are there on Earth? *Science* **241**, 1441–1449.

McAllen, R., Davenport, J., Bredendieck, K., and Dunne, D. (2009). Seasonal structuring of a benthic community exposed to regular hypoxic events. *J. Exp. Mar. Biol. Ecol.* **368**, 67–74.

McCann, K., Hastings, A., and Huxel, G.R. (1998). Weak trophic interactions and the balance of nature. *Nature* **395**, 794–798.

McCann, K.S., Rasmussen, J.B., and Umbanhowar, J. (2005). The dynamics of spatially coupled food webs. *Ecol. Lett.* **8**, 513–523.

McLaughlin, O.B., Jonsson, T., and Emmerson, M.C. (2010). Temporal variability in predator–prey relationships of a forest floor food web. *Adv. Ecol. Res.* **42**, 171–264.

McMahon, T.A., and Bonner, J.T. (1983). On Size and Life. Scientific American Library, New York.

McQuiston, R.W. (1969). Cyclic activity in the digestive diverticula of *Lasaea rubra* (Montagu) (Blvalvia; Eulamellibranchia). *Proc. Malacol. Soc. Lond.* **38**, 483–492.

Memmott, J., Martinez, N.D., and Cohen, J.E. (2000). Predators, parasitoids and pathogens: Species richness, trophic generality and body sizes in a natural food web. *J. Anim. Ecol.* **69**, 1–15.

Millar, R.H. (1970). British Ascidians, Tunicata: Ascidiacea. London.

Miller, M.C. (1961). Distribution and food of the Nudibranchiate Mollusca of the South of the Isle of Man. *J. Anim. Ecol.* **30**, 95–116.

Mills, C.E. (1976). Podocoryne selena, a New Species of Hydroid from the Gulf of Mexico, and a Comparison with *Hydractinia echinata. Biol. Bull.* **151**, 214–224.

Minchin, D. (1987). Fishes of the Lough Hyne marine reserve. *J. Fish Biol.* **31**, 343–352.

Mistri, M., Fano, E.A., and Rossi, R. (2001). Redundancy of macrobenthos from lagoonal habitats in the Adriatic Sea. *Mar. Ecol. Prog. Ser.* **215**, 289–296.

Montagna, P.A. (1984). Insitu measurement of meiobenthic grazing rates on sediment bacteria and edaphic diatoms. *Mar. Ecol. Prog. Ser.* **18**, 119–130.

Montoya, J.M., and Sole, R.V. (2003). Topological properties of food webs: From real data to community assembly models. *Oikos* **102**, 614–622.

Montoya, J.M., Emmerson, M.C., and Woodward, G. (2005). Perturbations and indirect effects in complex food webs. In: *Dynamic Food Webs: Multispecies Assemblages, Ecosystem Development, and Environmental Change* (Ed. by P.C. De Ruiter, V. Wolters and J.C. Moore). Academic Press, Amsterdam.

Montoya, J.M., Woodward, G., Emmerson, M.C., and Solé, R.V. (2009). Press perturbations and indirect effects in real food webs. *Ecology* **90**, 2426–2433.

Morton, J.E. (1956). The tidal rhythm and action of the digestive system of the Lamellibranch *Lasaea rubra*. *J. Mar. Biol. Assoc. UK* **35**, 563–586.
Morton, J.E. (1967). Molluscs. Hutchinson University Library.
Muntz, L., Ebling, F.J., and Kitching, J.A. (1965). The ecology of Lough Ine. 14. Predatory activity of large crabs. *J. Anim. Ecol.* **34**, 315–329.
Murray, J.W. (1963). Ecological experiments on Foraminiferida. *J. Mar. Biol. Assoc. UK* **43**, 621–643.
Murray, J.W. (2006). Ecology and Applications of Benthic Foraminifera. Cambridge University Press.
Navarro, J.M., and Thompson, R.J. (1996). Physiological energetics of the horse mussel *Modiolus modiolus* in a cold ocean environment. *Mar. Ecol. Prog. Ser.* **138**, 135–148.
Naylor, E. (1955). The diet and feeding mechanism of Idotea. *J. Mar. Biol. Assoc. UK* **34**, 347–355.
Naylor, E.E. (1972). British Marine Isopods: Keys and Notes for the Identification of the Species. London.
Nelson, W.G. (1979). An analysis of structural pattern in an eelgrass (*Zostera marina* L.) amphipod community. *J. Exp. Mar. Biol. Ecol.* **39**, 231–264.
Neumann, A.C., Gebelein, C.D., and Scoffin, T.P. (1970). Composition, structure and erodability of subtidal mats, Abaco, Bahamas. *J. Sed. Petrol.* **40**, 274–297.
Neutel, A.M., Heesterbeek, J.A.P., and de Ruiter, P.C. (2002). Stability in real food webs: Weak links in long loops. *Science* **296**, 1120–1123.
Nicol, E.A.T. (1930). The feeding mechanism, formation of the tube, and physiology of digestion in *Sabella pavonina*. *Trans. R. Soc. Edinb.* **56**, 537–600.
Nicol, E.A.T. (1933). The feeding habits of Galatheidea. *J. Mar. Biol. Assoc. UK* **18**, 87–106.
Nielsen, C. (1975). Observations on *Buccinum undatum* L. attacking bivalves and on prey responses, with a short review on attack methods of other prosobranchs. *Ophelia* **13**, 87–108.
Nielsen, C., and Riisgard, H.U. (1998). Tentacle structure and filter-feeding in *Crisia eburnea* and other cyclostomatous bryozoans, with a review of upstream-collecting mechanisms. *Mar. Ecol. Prog. Ser.* **168**, 163–186.
Norton, T.A., Hiscock, K., and Kitching, J.A. (1977). Ecology of Lough Ine. 20. Laminaria Forest at Carrigathorna. *J. Ecol.* **65**, 919–941.
O'Gorman, E.J. (2010a). Inferred from Available Genus Information.
O'Gorman, E.J. (2010b). Inferred from Available Family Information.
O'Gorman, E.J. (2010c). Inferred from Available Order Information.
O'Gorman, E.J. (2010d). Inferred from Available Class Information.
O'Gorman, E.J. Personal observations during sampling of the mesocosm cages used in the experiment, unpublished.
O'Gorman, E.J., and Emmerson, M.C. (2010). Manipulating interaction strengths and the consequences for trivariate patterns in a marine food web. *Adv. Ecol. Res.* **42**, 301–419.
O'Gorman, E.J., and Emmerson, M.C. (2009). Perturbations to trophic interactions and the stability of complex food webs. *Proc. Natl. Acad. Sci. USA* **106**, 13393–13398.
O'Gorman, E., Enright, R., and Emmerson, M. (2008). Predator diversity enhances secondary production and decreases the likelihood of trophic cascades. *Oecologia* **158**, 557–567.

O'Gorman, E.J., Jacob, U., Jonsson, T., and Emmerson, M.C. (2010). Interaction strength, food web topology and the relative importance of species in food webs. *J. Anim. Ecol.* **79**, 682–692.

Okamura, B. (1988). The influence of neighbors on the feeding of an epifaunal Bryozoan. *J. Exp. Mar. Biol. Ecol.* **120**, 105–123.

Olesen, J.M., Dupont, Y.L., O'Gorman, E.J., Ings, T.C., Layer, K., Melián, C.J., Troejelsgaard, K., Pichler, D.E., Rasmussen, C., and Woodward, G. (2010). From Broadstone to Zackenberg: Space, time and hierarchies in ecological networks. *Adv. Ecol. Res.* **42**, 1–69.

Opitz, S. (1993). A quantitative model of the trophic interactions in a Caribbean coral reef ecosystem. In: *Trophic Models of Aquatic Ecosystems* (Ed. by V. Christensen and D. Pauly). ICLARM Conference Proceedings 26.

Orton, J.H. (1914). On ciliary mechanisms in brachiopods and some polychaetes, with a comparison of the ciliary mechanisms on the gills of molluscs, Protochordata, brachiopods, and cryptocephalous polychaetes, and an account of the endostyle of Crepidula and its allies. *J. Mar. Biol. Assoc. UK* **10**, 283–311.

Pascoe, P.L., Parry, H.E., and Hawkins, A.J.S. (2007). Dynamic filter-feeding responses in fouling organisms. *Aquat. Biol.* **1**, 177–185.

Peacock, N. (1971) University of Wales, MSc thesis.

Pearre, S.J. (1976). Vertical migration and feeding of *Sagitta elegans* Verrill. *Ecology* **54**, 300–314.

Pearson, T.H. (1971). Studies on the ecology of the macrobenthic fauna of Lochs Linnhe and Eil, west coast of Scotland. II. Analysis of the macrobenthic fauna by comparison of feeding groups. *Vie Milieu* **1**, 53–91.

Perron, F. (1978). Habitat and feeding-behavior of Wentletrap Epitonium greenlandicum. *Malacologia* **17**, 63–72.

Petchey, O.L., McPhearson, P.T., Casey, T.M., and Morin, P.J. (1999). Environmental warming alters food-web structure and ecosystem function. *Nature* **402**, 69–72.

Peters, R.H. (1983). The Ecological Implications of Body Size. Cambridge University Press, Cambridge.

Pimm, S.L. (1982). Food Webs. University of Chicago Press, Chicago.

Pimm, S.L., and Lawton, J.H. (1977). Number of trophic levels in ecological communities. *Nature* **268**, 329–331.

Pimm, S.L., Russell, G.J., Gittleman, J.L., and Brooks, T.M. (1995). The future of biodiversity. *Science* **269**, 347–350.

Pimm, S., Raven, P., Peterson, A., Sekercioglu, C.H., and Ehrlich, P.R. (2006). Human impacts on the rates of recent, present, and future bird extinctions. *Proc. Natl. Acad. Sci. USA* **103**, 10941–10946.

Platt, H.M., Warwick, R.M., and Somerfield, P.J. (1983). Free-Living Marine Nematodes: Pictorial Key to World Genera and Notes for the Identification of British Species. London.

Polis, G.A. (1991). Complex trophic interactions in deserts—An empirical critique of food-web theory. *Am. Nat.* **138**, 123–155.

Polis, G.A., Sears, A.L.W., Huxel, G.R., Strong, D.R., and Maron, J. (2000). When is a trophic cascade a trophic cascade? *Trends Ecol. Evol.* **15**, 473–475.

Poltermann, M. (2001). Arctic sea ice as feeding ground for amphipods: Food sources and strategies. *Polar Biol.* **24**, 89–96.

Post, D.M. (2002). The long and short of food-chain length. *Trends Ecol. Evol.* **17**, 269–277.

Post, D.M., Pace, M.L., and Hairston, N.G. (2000). Ecosystem size determines food-chain length in lakes. *Nature* **405**, 1047–1049.

Prathep, A., Marrs, R.H., and Norton, T.A. (2003). Spatial and temporal variations in sediment accumulation in an algal turf and their impact on associated fauna. *Mar. Biol.* **142**, 381–390.
Pulliam, H.R. (1988). Sources, sinks, and population regulation. *Am. Nat.* **132**, 652–661.
Purvis, A., Jones, K.E., and Mace, G.M. (2000). Extinction. *Bioessays* **22**, 1123–1133.
Raffaelli, D. (2002). Ecology—From Elton to mathematics and back again. *Science* **296**, 1035–1036.
Ramsay, K., Kaiser, M.J., and Hughes, R.N. (1998). Responses of benthic scavengers to fishing disturbance by towed gears in different habitats. *J. Exp. Mar. Biol. Ecol.* **224**, 73–89.
Randlov, A., and Riisgard, H.U. (1979). Efficiency of particle retention and filtration-rate in 4 species of ascidians. *Mar. Ecol. Prog. Ser.* **1**, 55–59.
Rasmussen, E. (1973). Systematics and ecology of the Isefjord marine fauna (Denmark). With a survey of the eelgrass (Zostera) vegetation and its communities. *Ophelia* **11**, 1–505.
Rawlinson, K.A., Davenport, J., and Barnes, D.K.A. (2004). Vertical migration strategies with respect to advection and stratification in a semi-enclosed lough: A comparison of mero- and holozooplankton. *Mar. Biol.* **144**, 935–946.
Reagan, D.P., and Waide, R.B. (1996). The food web of a tropical rain forest. University of Chicago Press, London.
Rees, T.K. (1935). The marine algae of Lough Ine. *J. Ecol.* **23**, 69–133.
Rejmanek, M., and Stary, P. (1979). Connectance in real biotic communities and critical-values for stability of model ecosystems. *Nature* **280**, 311–313.
Renouf, L.P.W. (1931). Preliminary work of a new biological field station (Lough Ine, Co. Cork, I.F.S.). *J. Ecol.* **19**, 410–438.
Reuman, D.C., and Cohen, J.E. (2005). Estimating relative energy fluxes using the food web, species abundance, and body size. *Adv. Ecol. Res.* **36**, 137–182.
Riede, J.O., Rall, B.C., Banasek-Richter, C., Navarrete, S.A., Wieters, E.A., and Brose, U. (2010). Scaling of food-web properties with diversity and complexity across ecosystems. *Adv. Ecol. Res.* **42**, 139–170.
Riisgard, H.U., and Manriquez, P. (1997). Filter-feeding in fifteen marine ectoprocts (Bryozoa): Particle capture and water pumping. *Mar. Ecol. Prog. Ser.* **154**, 223–239.
Rinaldo, A., Maritan, A., Cavender-Bares, K.K., and Chisholm, S.W. (2002). Cross-scale ecological dynamics and microbial size spectra in marine ecosystems. *Proc. R. Soc. Lond., Ser. B Biol. Sci.* **269**, 2051–2059.
Robertson, R. (1963). Wentletraps (Epitoniidae) feeding on sea anemones and corals. *Proc. Malacol. Soc. Lond.* **35**, 51–63.
Robles, C.D., and Cubit, J. (1981). Influence of biotic factors in an upper inter-tidal community—Dipteran larvae grazing on algae. *Ecology* **62**, 1536–1547.
Rogers, S.I. (1990). Collecting marine crustaceans with a light trap. In: *The ecology of Lough Hyne: Proceedings of a Conference, 4–5 September, 1990* (Ed. by A.A. Myers, C. Little, M.J. Costello and J.C. Patridge). Royal Irish Academy, Dublin.
Rosenberg, R. (1993). Suspension-feeding in Abra alba (Mollusca). *Sarsia* **78**, 119–121.
Roushdy, H.M., and Hansen, V.K. (1960). Ophiuroids feeding on phytoplankton. *Nature* **188**, 517–518.
Rueda, J.L., Gofas, S., Urra, J., and Salas, C. (2009). A highly diverse molluscan assemblage associated with eelgrass beds (*Zostera marina* L.) in the Alboran Sea: Micro-habitat preference, feeding guilds and biogeographical distribution. *Sci. Mar.* **73**, 679–700.

Sala, O.E., Chapin, F.S., Armesto, J.J., Berlow, E., Bloomfield, J., Dirzo, R., Huber-Sanwald, E., Huenneke, L.F., Jackson, R.B., Kinzig, A., Leemans, R., Lodge, D. M., *et al.* (2000). Biodiversity—Global biodiversity scenarios for the year 2100. *Science* **287**, 1770–1774.

Salgeback, J., and Savazzi, E. (2006). Constructional morphology of cerithiform gastropods. *Paleontol. Res.* **10**, 233–259.

Samuelson, T.J. (1970). The biology of six species of Anomura. (Crustacea, Decapoda) from Raunefjorden, Western Norway. *Sarsia* **45**, 25–52.

Sandon, H. (1932). The Food of Protozoa. Misr-Sokkar Press, Cairo.

Sanford, E., Bermudez, D., Bertness, M.D., and Gaines, S.D. (1994). Flow, food-supply and acorn barnacle population-dynamics. *Mar. Ecol. Prog. Ser.* **104**, 49–62.

Schellenberg, R.O. (1929). Korperbau und Grabweise einiger Amphipoden. *Zool. Anz.* **85**, 186–190.

Schmid-Araya, J.M., Hildrew, A.G., Robertson, A., Schmid, P.E., and Winterbottom, J. (2002). The importance of meiofauna in food webs: Evidence from an acid stream. *Ecology* **83**, 1271–1285.

Schmidt-Nielsen, K. (1984). Scaling: Why is Animal Size so Important? Cambridge University Press, Cambridge.

Schoener, T.W. (1974). Resource partitioning in ecological communities. *Science* **185**, 27–39.

Scolding, J.W.S., Richardson, C.A., and Luckenbach, M.J. (2007). Predation of cockles (*Cerastoderma edule*) by the whelk (*Buccinum undatum*) under laboratory conditions. *J. Molluscan Stud.* **73**, 333–337.

Scott, C.L., Falk-Petersen, S., Sargent, J.R., Hop, H., Lonne, O.J., and Poltermann, M. (1999). Lipids and trophic interactions of ice fauna and pelagic zooplankton in the marginal ice zone of the Barents Sea. *Polar Biol.* **21**, 65–70.

Seabloom, E.W., Williams, J.W., Slayback, D., Stoms, D.M., Viers, J.H., and Dobson, A.P. (2006). Human impacts, plant invasion, and imperiled, plant species in California. *Ecol. Appl.* **16**, 1338–1350.

Sekiguchi, H. (1982). Scavenging amphipods and isopods attacking the spiny lobster caught in a gill-net. *Rep. Fish. Res. Lab. Mie Univ.* **3**, 21–30.

Shillaker, R.O., and Moore, P.G. (1987a). The feeding habits of the amphipods *Lembos websteri* (Bate) and *Corophium bonnellii* (Milne Edwards). *J. Exp. Mar. Biol. Ecol.* **110**, 93–112.

Shillaker, R.O., and Moore, P.G. (1987b). Tube-emergence behavior in the amphipods *Lembos websteri* (B)ate and *Corophium bonnellii* (Milne Edwards). *J. Exp. Mar. Biol. Ecol.* **111**, 231–242.

Smith, S.T. (1967). The development of *Retusa obtusa* (Montagu) (Gastropoda, Opisthobranchia). *Can. J. Zool.* **45**, 737–764.

Smith, G.A., Nickels, J.S., Davis, W.M., Martz, R.F., Findlay, R.H., and White, D. C. (1982). Perturbations in the biomass, metabolic activity, and community structure of the estuarine detrital microbiota: Resource partitioning in amphipod grazing. *J. Exp. Mar. Biol. Ecol.* **64**, 125–144.

Solov'ev, K.A., and Kosobokova, K.N. (2003). Feeding of the chaetognaths *Parasagitta elegans* Verrill (Chaetognatha) in the White Sea. *Oceanology* **43**, 524–531.

Spencer, M., and Warren, P.H. (1996). The effects of habitat size and productivity on food web structure in small aquatic microcosms. *Oikos* **75**, 419–430.

Sterner, R.W., Bajpai, A., and Adams, T. (1997). The enigma of food chain length: Absence of theoretical evidence for dynamic constraints. *Ecology* **78**, 2258–2262.

Stork, N.E., and Blackburn, T.M. (1993). Abundance, body-size and biomass of arthropods in tropical forest. *Oikos* **67**, 483–489.

Stromgren, T., Nielsen, M.V., and Reiersen, L.O. (1993). The effect of hydrocarbons and drilling-fluids on the fecal pellet production of the deposit feeder Abra alba. *Aquat. Toxicol.* **24**, 275–286.

Strong, J.A., Maggs, C.A., and Johnson, M.P. (2009). The extent of grazing release from epiphytism for Sargassum muticum (Phaeophyceae) within the invaded range. *J. Mar. Biol. Assoc. UK* **89**, 303–314.

Sugihara, G., Schoenly, K., and Trombla, A. (1989). Scale invariance in food web properties. *Science* **245**, 48–52.

Suhr, S.B., Pond, D.W., Gooday, A.J., and Smith, C.R. (2003). Selective feeding by benthic foraminifera on phytodetritus on the western Antarctic Peninsula shelf: Evidence from fatty acid biomarker analysis. *Mar. Ecol. Prog. Ser.* **262**, 153–162.

Sullivan, B.K. (1980). In situ feeding behaviour of *Sagitta elegans* and *Eukrohinia hamata* (Chaetognatha) in relation to the vertical distribution and abundance of prey at ocean station "P". *Limnol. Oceanogr.* **25**, 317–326.

Tagmouti-Talha, F., Moutaouakkil, A., Taib, N., Mikou, A., Talbi, M., Fellat-Zerrouk, K., and Blaghen, M. (2000). Detection of paralytic and diarrhetic shellfish toxins in Moroccan cockles (*Acanthocardia tuberculata*). *Bull. Environ. Contam. Toxicol.* **65**, 707–716.

Tattersall, O.S., and Tattersall, W.M. (1951). The British Mysidacea. Ray Society, London.

Tavares-Cromar, A.F., and Williams, D.D. (1996). The importance of temporal resolution in food web analysis: Evidence from a detritus-based stream. *Ecol. Monogr.* **66**, 91–113.

Taylor, J.D. (1978). Diet of *Buccinum undatum* and Neptunea antiqua (Gastropoda Buccinidae). *J. Conchol.* **29**, 309–318.

Taylor, R.B., and Brown, P.J. (2006). Herbivory in the gammarid amphipod Aora typica: Relationships between consumption rates, performance and abundance across ten seaweed species. *Mar. Biol. (Berlin)* **149**, 455–463.

Terazaki, M. (1993). Seasonal-variation and life-history of the pelagic Chaetognatha, *Sagitta elegans* Verrill, in Toyama Bay, Southern Japan Sea. *J. Plankton Res.* **15**, 703–714.

Terazaki, M. (2004). Life history of the chaetognath *Sagitta elegans* in the World's oceans. *Coast. Mar. Sci.* **29**, 1–12.

Thompson, R.M., and Townsend, C.R. (1999). The effect of seasonal variation on the community structure and food-web attributes of two streams: Implications for food-web science. *Oikos* **87**, 75–88.

Thouzeau, G., Jean, F., and Del Amo, Y. (1996). Sedimenting phytoplankton as a major food source for suspension-feeding queen scallops (*Aequipecten opercularis* L.) off Roscoff (western English Channel)? *J. Shellfish Res.* **15**, 504–505.

Tillin, H.M., Hiddink, J.G., Jennings, S., and Kaiser, M.J. (2006). Chronic bottom trawling alters the functional composition of benthic invertebrate communities on a sea-basin scale. *Mar. Ecol. Prog. Ser.* **318**, 31–45.

Tokeshi, M. (1993). Species abundance patterns and community structure. *Adv. Ecol. Res.* **24**, 111–186.

Tonnesson, K., and Tiselius, P. (2005). Diet of the chaetognaths *Sagitta setosa* and *S. elegans* in relation to prey abundance and vertical distribution. *Mar. Ecol. Prog. Ser.* **289**, 177–190.

Townsend, C.R., Thompson, R.M., McIntosh, A.R., Kilroy, C., Edwards, E., and Scarsbrook, M.R. (1998). Disturbance, resource supply, and food-web architecture in streams. *Ecol. Lett.* **1**, 200–209.

Trager, G.C., Hwang, J.S., and Strickler, J.R. (1990). Barnacle suspension-feeding in variable flow. *Mar. Biol.* **105**, 117–127.

Turner, J.T. (2000). In: *Proceedings of the International Symposium on Marine Biology in Taiwan—Crustacean and Zooplankton Taxonomy, Ecology and Living Resources, 26–27 May 1998*. National Taiwan Museum Special Publication Series, pp. 37–57.

Turner, J.T. (2004). The importance of small planktonic copepods and their roles in pelagic marine food webs. *Zool. Stud.* **43**, 255–266.

Tyler, P.A. (1977). Seasonal-variation and ecology of gametogenesis in genus Ophiura (Ophiuroidea Echinodermata) from Bristol Channel. *J. Exp. Mar. Biol. Ecol.* **30**, 185–197.

Underwood, A.J. (1997). Experiments in Ecology: Their Logical Design and Interpretation using Analysis of Variance. Cambridge University Press.

Underwood, A.J., and Chapman, M.G. (2006). Early development of subtidal macrofaunal assemblages: Relationships to period and timing of colonization. *J. Exp. Mar. Biol. Ecol.* **330**, 221–233.

Vale, P., and Sampayo, M.A.D. (2002). Evaluation of marine biotoxin's accumulation by *Acanthocardia tuberculatum* from Algarve, Portugal. *Toxicon* **40**, 511–517.

Vander Zanden, M.J., Shuter, B.J., Lester, N., and Rasmussen, J.B. (1999). Patterns of food chain length in lakes: A stable isotope study. *Am. Nat.* **154**, 406–416.

Verling, E. PhD thesis, unpublished

Verling, E., Crook, A.C., Barnes, D.K.A., and Harrison, S.S.C. (2003). Structural dynamics of a sea-star (*Marthasterias glacialis*) population. *J. Mar. Biol. Assoc. UK* **83**, 583–592.

Verling, E., Barnes, D.K.A., and Crook, A.C. (2005). Smashing tests? Patterns and mechanisms of adult mortality in a declining echinoid population. *Mar. Biol.* **147**, 509–515.

Vevers, H.G. (1956). Observations on feeding mechanisms in some echinoderms. *Proc. Zool. Soc. Lond.* **126**, 484–485.

Vitousek, P.M., Mooney, H.A., Lubchenco, J., and Melillo, J.M. (1997). Human domination of Earth's ecosystems. *Science* **277**, 494–499.

Ward, J.E., MacDonald, B.A., and Thompson, R.J. (1993). Mechanisms of suspension feeding in bivalves: Resolution of current controversies by means of endoscopy. *Limnol. Oceanogr.* **38**, 265–272.

Ward, J.N., Pond, D.W., and Murray, J.W. (2003). Feeding of benthic foraminifera on diatoms and sewage-derived organic matter: An experimental application of lipid biomarker techniques. *Mar. Environ. Res.* **56**, 515–530.

Warén, A. (1983). A generic revision of the family Eulimidae (Gastropoda, Prosobranchia). *J. Molluscan Stud.* **13**, 1–96.

Warner, G.F., and Woodley, J.D. (1975). Suspension-feeding in brittle-star *Ophiothrix fragilis*. *J. Mar. Biol. Assoc. UK* **55**, 199–210.

Warren, P.H. (1989). Spatial and temporal variation in the structure of a fresh-water food web. *Oikos* **55**, 299–311.

Warren, P.H., and Lawton, J.H. (1987). Invertebrate predator–prey body size relationships—An explanation for upper-triangular food webs and patterns in food web structure. *Oecologia* **74**, 231–235.

Watson, D.I., and Barnes, D.K.A. (2004). Temporal and spatial components of variability in benthic recruitment, a 5-year temperate example. *Mar. Biol.* **145**, 201–214.

Werner, I. (1997). Grazing of Arctic under-ice amphipods on sea-ice algae. *Mar. Ecol. Prog. Ser.* **160**, 93–99.

West, G.B., Brown, J.H., and Enquist, B.J. (1997). A general model for the origin of allometric scaling laws in biology. *Science* **276**, 122–126.

White, E.P., Ernest, S.K.M., Kerkhoff, A.J., and Enquist, B.J. (2007). Relationships between body size and abundance in ecology. *Trends Ecol. Evol.* **22**, 323–330.
Whittaker, J.E. (1974). On Semicytherura nigrescens (Baird). *Stereo-Atlas Ostracod Shell.* **2**, 69–76.
Whittaker, J.E. (1978). On Loxoconcha aurantia (Baird). *Stereo-Atlas Ostracod Shell.* **5**, 27–34.
Wigham, G.D. (1976). Feeding and digestion in the marine prosobranch Rissoa parva (Da Costa). *J. Molluscan Stud.* **42**, 74–94.
Wildish, D.J., and Kristmanson, D.D. (1984). Importance to mussels of the benthic boundary-layer. *Can. J. Fish. Aquat. Sci.* **41**, 1618–1625.
Williams, R.J., and Martinez, N.D. (2000). Simple rules yield complex food webs. *Nature* **404**, 180–183.
Williams, S.I., and Walker, D.I. (1999). Mesoherbivore–macroalgal interactions: Feeding ecology of sacoglossan sea slugs (Mollusca, Opisthobranchia) and their effects on their food algae. *Oceanogr. Mar. Biol.* **37**(37), 87–128.
Wilson, K. (1984). A bibliography of Lough Hyne (Ine) 1687–1982. *J. Life Sci. R. Dublin Soc.* **5**, 1–11.
Woodward, G., and Hildrew, A.G. (2001). Invasion of a stream food web by a new top predator. *J. Anim. Ecol.* **70**, 273–288.
Woodward, G., and Hildrew, A.G. (2002). Body-size determinants of niche overlap and intraguild predation within a complex food web. *J. Anim. Ecol.* **71**, 1063–1074.
Woodward, G., Speirs, D.C., and Hildrew, A.G. (2005). Quantification and resolution of a complex, size-structured food web. *Adv. Ecol. Res.* **36**, 85–135.
Woodward, G., Papantoniou, G., Edwards, F., and Lauridsen, R.B. (2008). Trophic trickles and cascades in a complex food web: Impacts of a keystone predator on stream community structure and ecosystem processes. *Oikos* **117**, 683–692.
Woodward, G., Benstead, J.P., Beveridge, O.S., Blanchard, J., Brey, T., Brown, L., Cross, W.F., Friberg, N., Ings, T.C., Jacob, U., Jennings, S., Ledger, M.E., *et al.* (2010). Ecological networks in a changing climate. *Adv. Ecol. Res.* **42**, 71–138.
Worm, B., and Duffy, J.E. (2003). Biodiversity, productivity and stability in real food webs. *Trends Ecol. Evol.* **18**, 628–632.
Worm, B., Barbier, E.B., Beaumont, N., Duffy, J.E., Folke, C., Halpern, B.S., Jackson, J.B.C., Lotze, H.K., Micheli, F., Palumbi, S.R., Sala, E., Selkoe, K.A., *et al.* (2006). Impacts of biodiversity loss on ocean ecosystem services. *Science* **314**, 787–790.
Yodzis (1998). Local trophodynamics and the interaction of marine mammals and fisheries in the Benguela ecosystem. *J. Anim. Ecol.* **67**, 635–658.
Yonge, C.M. (1928). Feeding mechanisms of invertebrates. *Biol. Rev.* **3**, 21–76.
Yonge, C.M. (1954). Food of invertebrates. *Tabulae Biol.* **11**, 25–45.
Yvon-Durocher, G., Montoya, J.M., Emmerson, M.C., and Woodward, G. (2008). Macroecological patterns and niche structure in a new marine food web. *Cent. Eur. J. Biol.* **3**, 91–103.
Yvon-Durocher, G., Reiss, J., Blanchard, J., Ebenman, B., Perkins, D.M., Reuman, D.C., Thierry, A., Woodward, G., and Petchey, O.L. (2010). Across ecosystem comparisons of size structure: Methods, approaches, and prospects. *Oikos.*
Zavattari, E. (1920). Osservazioni etologiche sopra l'anfipodo tubicolo *Ericthonius brasiliensis* (Dana). *Mem. R. Comitato Talassografico Ital.* **77**, 1–25.
Zimmerman, R., Gibson, R., and Harrington, J. (1979). Herbivory and detritivory among gammaridean amphipods from a Florida USA seagrass community. *Mar. Biol. (Berlin)* **54**, 41–48.

Index

Advances in Ecological Research Volume 1–42

Cumulative List of Titles